# STATPHYS 19

**The 19th IUPAP International Conference on Statistical Physics**

# STATPHYS 19

## The 19th IUPAP International Conference on Statistical Physics

Xiamen, China July 31–August 4, 1995

Editor

**Hao Bailin**

*Institute of Theoretical Physics*
*Academia Sinica, China*

**World Scientific**
*Singapore • New Jersey • London • Hong Kong*

*Published by*

World Scientific Publishing Co. Pte. Ltd.
5 Toh Tuck Link, Singapore 596224
*USA office:* 27 Warren Street, Suite 401-402, Hackensack, NJ 07601
*UK office:* 57 Shelton Street, Covent Garden, London WC2H 9HE

**British Library Cataloguing-in-Publication Data**
A catalogue record for this book is available from the British Library.

**STATPHYS 19**
**Proceedings of the 19th IUPAP International Conference on Statistical Physics**



ISBN-13 978-981-02-2314-4
ISBN-10 981-02-2314-5
ISBN-13 978-981-02-2579-7 (pbk)
ISBN-10 981-02-2579-2 (pbk)

一生二，二生三，三生万物。　　　　—老子

One generates two, two generates three, three generates all things. —Laozi

## PREFACE

The 19th IUPAP International Conference on Statistical Physics was held in the city of Xiamen, on the campus of Xiamen University, from July 31 to August 4, 1995. This was the first time when an IUPAP STATPHYS Conference took place in China and the second in Asia since the Kyoto Conference in 1968. About 700 participants from 47 countries and regions came to Xiamen.

Statistical physics enjoys a very special position among subfields of physics. Its principle holds whenever there are many (more than three?) particles or subunits, interacting with each other to make a "complex" system. Its subjects vary from microscopic and mesoscopic scales to macroscopic and cosmic scales. Its working tool ranges from rigorous mathematics to numerical simulation. Its application extends to almost all branches of natural sciences and, perhaps, economy and sociology. This explains why STATPHYS Conferences have always been an interdisciplinary gathering of scientists.

The thriving status of statistical physics was further witnessed by the fact that, besides the main Conference in Xiamen, there were 7 satellite meetings. They were:

1. **Coherent Approach to Fluctuations**, organized by Prof. M. Suzuki in Kyoto from 16 to 20 July, 1995.
2. **Nonlinear and Random Processes**, organized by Dr. C.-K. Hu in Taipei from 18 to 24 July, 1995.
3. **Future of Fractals**, organized by Prof. S. Miyazima in Nagoya from 25 to 27 July, 1995.
4. **Complex Fluids and Monte Carlo Methods**, organized by Prof. Ping Sheng in Hong Kong from 25 tp 29 July, 1995.
5. **Nonlinear Dynamics and Chaos**, organized by Prof. H. T. Moon in Pohang from 26 to 28 July, 1995.
6. **Dynamics of Complex Systems**, organized by Prof. S. Puri in Calcutta from 6 to 11 August, 1995.
7. **Statistical Models, Yang-Baxter Equations and Related Topics**, organized by Profs. F. Y. Wu and Mo-lin Ge in Tianjin from 8 to 10 August, 1995.

The presentations on STATPHYS 19 were organized around 12 topics:

1. General equilibrium and non-equilibrium statistical physics
2. Exact and rigorous results, relation with field theory
3. Phase transitions and critical phenomena
4. Transport and relaxation
5. Fluctuation phenomena and random processes
6. Disordered systems and heterogeneous systems
7. Simple and complex fluids; liquid crystals and polymers
8. Classical chaos and turbulence
9. Quantum chaos and mesoscopic systems
10. Structures and patterns; fractals; neural networks; cellular automata; self-organized criticality
11. Computer simulation and visualization in statistical physics
12. Statistical physics in cross-disciplinary physics, living systems

There were 12 plenary invited talks and 36 invited talks in 3 parallel sessions. A total of 168 15-minute oral presentations and more than 500 posters were given during the 5-day conference. Most of the invited talks were included in this *Proceedings* in the order that they were scheduled on the Conference. Prof. K. Kawasaki carefully prepared his invited talk, but had to cancel the trip in the last minute. We are happy to include his contribution in this *Proceedings*. In order to help the readers to orientate an *Analytic Subject Index* is given at the end of the volume.

Prof. R. Kubo, a former Boltzmann Medalist and a member of the International Advisory Committee of STATPHYS 19, passed away on 31 March, 1995. A Kubo Memorial Symposium was organized by Prof. M. Suzuki in the evening of 31 July.

Although an uninvited guest, the 1995 Typhoon No. 4, tried to overtake our first invited speaker, the Conference went successfully in general. The Xiamen Meteorological Service even succeeded in predicting that we were going to have a rainless Wednesday afternoon for the Conference excursion in spite of the heavy rainfall brought about by the Typhoon.

The success of STATPHYS 19 would not be possible without the support and help from many organizations and scientists. On behalf of the Local Organizing Committee I would like to thank all the members of the IUPAP C3 Commission on Statistical Physics as well as our International and Regional Advisory Committees. Advices from Drs. Amnon Aharony, Éduoard Brézin, Michael Fisher, Siegfried Hess, Joel Lebowitz, Gene Stanley, Yu Lu , and many other colleagues were especially invaluable for making the program.

STATPHYS 19 was co-organized or/and sponsored by the following international and domestic institutions:

- Association of Asia Pacific Physics Societies
- Chinese Physics Society
- Chinese Academy of Sciences
- Climbing Plans, State Commission of Science and Technology of China
- Fujian Physical Society
- Fujian Provincial Government
- Institute of Theoretical Physics, Chinese Academy of Sciences
- International Centre for Theoretical Physics (ICTP)
- International Union of Pure and Applied Physics (IUPAP)
- Lee Foundation of Singapore
- National Natural Science Foundation of China
- State Commission of Education of China
- State Commission of Science and Technology of China
- Tan Kah Kee International Society (TKKIS)
- Theoretical Physics Association of South-Eastern Asia
- UNESCO Regional Office for Science and Technology for Southeast Asia
- World Scientific Publishing Company
- Xiamen Municipality
- Xiamen University

Owing to generous support from TKKIS, Lee Foundation, IUPAP, ICTP, and ROSTSEA/UNESCO, more than 160 participants from 41 countries and regions obtained some form of financial help, ranging from full support including international travel to a minimal local support of a free lodging on the university campus.

The Conference also obtained various support, donation, and help from the following enterprises and organizations in the Xiamen region:

- Anhai Township Chamber of Commerce, Jinjiang
- Center for International Exchange, Xiamen University
- Chendai Township Government, Jinjiang
- China Travel Service of Xiamen
- Fujian Petroleum Company of Xiamen
- Fujian Ocean Fishery Materials Supply Company
- International Trust and Investment Company of Xiamen
- Shenghui Advertisement and Arts Company
- Xiamen Branch of China Insurance Company
- Xiamen Company of Jinke Group

- Xiamen Fengsheng Trade Co., Ltd.
- Xiamen Grand World Laser Label Product Co., Ltd
- Xiamen Luxin Electron Co., Ltd
- Xiamen Meteorological Service
- Xiamen Municipal Bureau of Culture
- Xiamen Overseas Chinese Electron Joint-Stock Co., Ltd
- Xiamen PICO Company
- Xiamen Rongbin Group
- Xiamen Special Economic Zone Trade Co., Ltd
- Xiamen University America Alumni Association
- Xiamen University 1948 Classmates Association
- Xiamen Xingsha Group
- Xunxing Metal Die-Casting Co., Ltd, Jinjiang
- Zhonghuan Economic and Technical Development Company

The Organizing Committee thanks all these organizations.

Our special thanks go to more than one hundred young students-volunteers from the Departments of Physics and Foreign Languages, Xiamen University, and from the Institute of Theoretical Physics, Chinese Academy of Sciences. Their tireless work at various posts, visible or invisible to the participants, was crucial for the success of the Conference. I would like also to thank Drs. Qiu Zhao-ming, Cheng Xi-you, and Hu Zhan-ning for preparing the *List of Participants* and the *List of Oral and Poster Presentations* for this volume.

Last but not least, I am grateful to all colleagues in the Local Organizing Committee for the pleasant four-year long cooperation since 1991 when we prepared the Proposal to IUPAP C3 Commission to host STATPHYS 19 in Xiamen. In fact, I had prepared a Proposal to have STATPHYS 17 in China with the support of Joel Lebowitz, F. Y. Wu, Chia-Wei Woo, and many others. The Proposal was withdrawn in the last minute before going to Boston to attend STATPHYS 16 for purely technical reasons. It was a bitter and "lucky" anti-coincidence for me. Only my colleagues in LOC knew how much labour and trouble we must take in order to run a big international conference such as STATPHYS 19 in a developing country as China. However, the participation of more than 180 Chinese physicists, many of whom would not have a chance to take part in such a high-level scientific meeting elsewhere, is a great award for us.

Bai-lin Hao
Conference Chairman

Beijing, October 1995

# CONTENTS

# STATPHYS 19 — OPENING ADDRESS

Edouard Brézin
*Ecole Normale Supérieure, Paris*
*Chairman, IUPAP C3 Commission on Statistical Physics*

On behalf of the International Union of Pure and Applied Physics, I would like to welcome all the participants to STATPHYS 19 in Xiamen.

Among the various goals of IUPAP the development and promotion of international cooperation in physics stands very high. For that purpose IUPAP sponsors suitable meetings, in which the free circulation of scientists is a necessary condition. In particular to secure IUPAP sponsorship the organizers of a conference have to provide assurance that no scientists will be excluded on the grounds of national origin, nationality or political considerations unrelated to science. The IUPAP sponsorship means some material support for the Conference, but it imposes stringent rules in terms of freedom of circulation, limits on Conference fees, reduced rates for students and scientists from developing countries, supply of copies of the proceedings free of charge for distribution to libraries in developing countries, etc.

These principles are more than mere words. They imply that scientists, through an organization such as IUPAP, have found ways to know each other, to work in common, to appreciate each other, to express their mutual solidarity, which go far beyond the existing international treaties between states. Consequently, it implies that if a country enters IUPAP, it has to bear with its principles, and it is not the science community which complies with the policy of any particular state.

The C3 Commission on Statistical Physics sponsors a regular series of general meetings, now held every three years, as well as occasional topical meetings or regional conferences. The name Statistical Physics was substituted in 1990 to the old Thermodynamics and Statistical Mechanics which it had held since 1945. This change of names was the recognition of a very wide field of activities which range from the statistical and thermodynamics studies of states of matter to applications of various kinds, chemical kinetics, polymers, colloids, liquid crystals, heterogeneous media, disordered systems, glassy states, neural networks, physical and biological membranes, various aspects of PIB: physics inspired by biology, mathematical methods related to all these problems, etc. I am sure that the list is even much longer and it is the recognition of the wide scope of situations which could be modeled through a statistical approach which is the common feature of our community.

During the 1994 – 1996 period two meetings only have received the sponsorship of the C3 Commission, this general one, as well as the topical conference on nonlinear and random processes which took place a few days ago in Taipei. This was decided in 1992 when our Commission met in Berlin; it had been recommended in Rio in 89 that the 95 meeting would go to Asia, in an effort to balance the distribution of locations for these general meetings among the various continents. We then met the difficult task of choosing between the proposals presented by the Academia Sinica in Taipei and Dr. Hao Bai-lin from the Institute of Theoretical Physics in Beijing. We all had in our mind the sad events of the spring of 89, and we, as scientists, are particularly sensitive to attacks against the basic freedom of opinion and speech. Professor Hao brought to us the assurance, which he had confirmed with the Chinese authorities, that all IUPAP rules would be strictly followed in this meeting and we made it clear that the IUPAP sponsorship would not bear with any breach of its principles. I would like to thank Dr. Hao Bai-lin and the Organizing Committee that he has chaired, not only for doing a great job, you cannot imagine how much work and personal involvement and sacrifice goes into the organization of such a meeting, but also for his extremely fair, honest and open cooperation with the IUPAP Commission, and for fully complying with IUPAP regulations.

Our C3 Commission will meet during this Conference and it will have to examine the proposals for STATPHYS 20 in 1998. It is in fact time for you to start thinking about the first Statistical Physics Conference of the 21st century, which will be accidentally named STATPHYS 21. I encourage you to send proposals to our Commission, as well as proposals for sponsorship of local and topical Conferences. Let me add a few words about our general STATPHYS meetings. It is said occasionally by respectable scientists that such large meetings, open by definition to a wide range of subjects, are not very useful and that one should favor the smaller format of topical meetings. Indeed I remember myself having sometimes this reaction, but I would plead in favor of keeping the STATPHYS meetings for a number of reasons. It is true that in some very large scientific centres, the flood of informations, seminars, visitors, makes it look necessary to have anything but workshops. I believe that this view is not shared by those of us who work in smaller centres. Second we live under the growing influence of electronic bulletin boards or archives, which are extraordinarily convenient. Among their many advantages these new means of communication are now available for institutions which could not even afford to buy journals. However, in spite of efforts such as cross-listing, these electronic means are highly specialized, and I think it is worth being exposed from time to time to a wider range of topics. Furthermore, the main STATPHYS meeting is accompanied by more specialized satellite meetings, nine in the case of this one, which in most cases could not have taken place otherwise. Finally, I do believe that for young scientists, the need to explore the various possible subjects, to put their work in perspective and to make it noticed from the community, justifies general meetings.

The most pleasant task of the Commission, if not the easiest, is to select the recipient of the Boltzmann medal. In a few minutes we shall honor Professor Samuel Edwards, the tenth recipient of this award. The names of these ten recipients, since the institution of the medal in 1975, shows how prestigious are the scientists distinguished by this award. The medalist is chosen through a vote by a panel which consists of the C3 Commission and the previous Boltzmann medalists. It is during the period of this voting process that we learnt with great sorrow, that the 1977 medalist, the highly respected Professor R. Kubo, had passed away. He will be remembered as one of the great minds and teachers of our times.

For the next medal self-candicacies are not warmly welcome but I would like to encourage all of you to participate in the nomination of a scientist, whose work you regard as truely important. This is done simply by sending a letter to the Chairperson of the C3 Commission which mentions the reasons of your choice. For the next medal this letter should reach him (or her) before September 1997.

I have been a constant witness since 1992, of the efforts that the Conference Chairman, Dr. Hao Bai-lin, and the Local Organizing Committee, have made to organize this STATPHYS 19. This ranges, from a long and careful consultation of the community on the best possible topics and speakers to details of the material organization. I am quite grateful to all of them and, on behalf of IUPAP, I express to Dr. Hao and his colleagues our profound gratitude.

Before the Conference starts let me read an excerpt of an IUPAP recommendation of October 94 on Languages at International Conferences.

"It frequently happens that only one language is used at international conferences. For many of the audience and of the speakers, this is not the mother tongue, and this gives rise to difficulties in expression and comprehension ⋯. Thoughtfulness in the preparation and delivery of papers is of assistance ⋯. Attempts should be made to keep the words on slides and transparencies to a minimum. Text and figures should be clear and sufficient time allotted for digesting the material shown. In particular, the speaker should present his paper slowly ⋯. Thus, less text should be prepared for a given time at an international conference than for a monolingual group. Chairmen of sessions can be extremely helpful by keeping this difficulty in mind, and reminding the speakers to address the conference in a very deliberate fashion."

I wish you now a very interesting, stimulating and fruitful meeting which will help all of you to bring our beloved science further.

**Note added**: In its meeting on August 1, the C3 Commission has decided to locate STATPHYS 20 in Paris at the end of July 1998. The Conference will be organized by Drs. Yves Pomeau, Daniel Iagolnitzer and Michel Moreau.

# THE BOLTZMANN MEDAL

Presentation by Edouard Brézin
*Chairman, IUPAP C3 Commission on Statistical Physics*

The Boltzmann Medal is awarded every three years since 1975 by the IUPAP C3 Commission on Statistical Physics, in recognition of outstanding contributions to statistical physics.

The list of the previous medallists consists of

K.G. Wilson (1975), R. Kubo (1977), R.J. Baxter (1980), M.E. Fisher (1983), D. Ruelle and Y. Sinai (1986), L. Kadanoff (1989), J. Lebowitz and G. Parisi (1992).

On behalf of the IUPAP Commission, I have the great honor of presenting the Boltzmann medal for 1995 to Professor Samuel Edwards.

## Samuel Edwards: Boltzmann Medallist 1995

The work of Prof. Sir Sam Edwards provides a cornerstone to the modern theory of disordered systems. He was instrumental in bringing quantum field theory into statistical physics; he stimulated — and was at the forefront of — a new understanding of the statics and dynamics of polymeric systems, and he co-founded the subject of spin glasses.

Edwards' work has been marked by a series of deep physical insight into difficult problems, coupled with major technical innovations, many of which have become a standard tool for workers in polymer theory and condensed matter physics. He introduced quantum field theory, in its formulation in terms of functional integrals, to condensed matter physics. This led him in 1965 to model polymer chains as continuous flexible curves with point-like repulsions; this has now become standard but it has been clearly a very novel and important idea. The self-consistent field formalism to study screening effects in many chain systems was also a very important step. This led to the first quantitative theories of polymer adsorbtion and remains the method of choice for studies of concentrated solutions and melts; new predictions still flow from it, 25 years after its introduction.

A second major series of contributions has concerned systems with quenched disorder. In a seminal collaboration with P.W. Anderson, the replica method was discovered in the context of spin glasses, magnetic systems with quenched random interactions, which are clearly the prototype of new states of matter and of many problems in which disorder leads to "frustation" of the ordered state.

Edwards studied the effects of chain entanglements in rubbery materials and introduced the idea of a "tube" around a particular molecule, representing the topological constraints of its neighbours. This led to a reformulation of the theory of rubber elasticity based on the replica method. The impact of this work on subsequent theories of quenched disorder has been profound and far reaching: for instance much of neural network theory descends from it.

A third major area of impact has been in predicting the dynamics and viscoelastic properties of dense polymer systems. These phenomena are now systematically understood in terms of tubes models of polymer entanglement. Edwards' idea of this tube was essential for PG de Gennes' work on "reptation", the snake-like motion of a chain in the medium, explicitly visualized nowadays in the experimental settings of S. Chu with optical tweesers. It was Edwards who had the insight to apply reptation directly to polymer melts, in a brilliant series of papers with Doi, in which the concept of reptation is scrutinized, formalized and worked out in all its consequences. This leads to predictions of such scope and importance that has taken the polymer science community a decade or more to assimilate them. Let us quote also the work of Doi and Edwards on the dynamics of entangled rod-like molecules.

Over the past decade Prof. Edwards has made several major contributions to the theory of aggregated granular matter. In 1982 he gave the first theoretical treatment of interfacial structure and roughness in granular materials, which became a new field of study. He was able there to take up ideas which he had developed in turbulence around 64 and he gave the first accurate description in this context, of the strong coupling fixed point in the Kardar–Parisi–Zhang equation, which is in fact a generalization of the earlier Edwards–Wilkinson model for randomly moving interfaces.

Let me briefly describe the various steps which have marked Prof. Edwards' career. Educated in Cambridge (UK) and Harvard, where he made his PhD under the supervision of Julian Schwinger, Prof. Edwards started as a member of the Institute for Advanced Study in Princeton, then took a position at Birmingham University in 53, became a Professor of theoretical physics at Manchester University in 63, and at Cambridge in 72, the Cavendish Professor for theoretical physics in 84. Prof. Edwards has held various governmental positions: Chairman of the SRC 73-77, Chairman of the Defense Scientific

Council 77-80, Chief Scientific Adviser of the Department of Energy 83-88. Prof. Edwards has been the Vice-President of the Royal Society, of the Institute of Physics, the President of the Institute of Mathematics; he is a foreign member of the French Académie des Sciences, and he has received a number of international awards and honorary degrees. Let me also mention that Prof. Edwards has been actively militant in favour of human rights, in particular at the time of the former Soviet Union.

This Boltzman medal goes therefore to a scientist of exceptional vision which has allowed him to identify and fruitfully apply methods of seemingly unrelated fields to other areas. This has allowed him to make important contributions to some of the most outstanding and significant problems in statistical physics over more than 30 years.

S. F. Edwards

# CONFERENCE COMMITTEES

## IUPAP Commission on Statistical Physics (C3):

E. Brezin ............. France, Chairman
P. C. Hohenberg .... USA, Vice-Chairman
J. Frohlich ........ Switzerland, Secretary
G. Casati .......................... Italy
B. Chirikov ...................... Russia
D. Dhar .......................... India
E. Domany ....................... Israel
M. H. J. Ernst ............... Netherlands
J. Hertz ...................... Denmark
K. Kawasaki ..................... Japan
M. A. Moore ........................ UK
G. Ropke ...................... Germany
J. Solyom ...................... Hungary

## Associate Members of C3:

A. Aharony ....................... Israel
T. V. Ramakrishnan .............. India
Ya. G. Sinai ...................... Russia
J. T. M. Walraven .......... Netherlands

## The International Advisory Committee

R. J. Baxter .................... Australia
D. D. Betts ..................... Canada
K. Binder ...................... Germany
M. H. Cohen ...................... USA
P. G. de Gennes ................. France
C. di Castro ....................... Italy
J. Feder ........................ Norway
M. E. Fisher ...................... USA
P. C. Hemmer ................... Norway
S. Hess ........................ Germany
L. P. Kadanoff ..................... USA
R. Kubo .............. Japan (deceased)
J. L. Lebowitz ..................... USA
T. D. Lee .......................... USA
V. P. Mineev ...................... Russia
G. Parisi .......................... Italy
I. Prigogine .................... Belgium
D. Ruelle ........................ France
H. E. Stanley ...................... USA
R. B. Stinchcombe .................. UK
M. Suzuki ........................ Japan
C. Tsallis ......................... Brazil
J. M. J. van Leeuwen ........ Netherlands
M. G. Velarde ..................... Spain
B. Widom .......................... USA
F. Y. Wu .......................... USA
C. N. Yang ........................ USA

## Regional Advisory Committee

Chen Wu Po ................ Hong Kong
K. C. Chou ...................... Beijing
Feng Duan ...................... Nanjing
Ge Mo-lin ....................... Tianjin
Gong Chang-de .................. Nanjing
Hong Yong-shi .................. Xiamen
C.-K. Hu ........................ Taipei
Huang Zu-qia .................... Beijing
S. C. Lee ........................ Taipei
Peng Huan-wu ................... Beijing
K. K. Phua .................... Singapore
Pu Fu-cho ....................... Beijing
Tao Rui-bao .................... Shanghai
Wang Liang-tuan ................. Fuzhou
Chia-Wei Woo .............. Hong Kong
Kenneth Young ............. Hong Kong
Yu Lu ........................... Beijing
Zhao Kai-hua .................... Beijing

## Local Organizing Committee (LOC)

Cai Shou-guo, Chen Chuan-hong, Chen Li-xuan, Fang Jiang-ling, Hao Bai-lin, Hu Gang, Huang Mei-chun, Liu Ji-xing, Liu Rui-tang, Ou-Yang Zhong-can, Qiu Zhao-ming, Tian Guang-shan, Wang Rong, Xu Qiao-zhen, Yu Yang-zheng, Zheng De-juan, Zheng Wei-mou

# GLASS TRANSITIONS

S. F. EDWARDS
*Cavendish Laboratory*
*Cambridge CB3 0HE UK.*

## 1. Introduction

It is a great honour to give the Boltzmann lecture at this IUPAP meeting in Xiamen. I have chosen to speak on simple ideas on glass transitions, because they illustrate my own interests in statistical physics.

Very roughly speaking statistical physics falls into two groups: problems which have an absolutely clear definition e.g. an Ising model, where the investigator devotes great care to very precise results, which although presenting a great intellectual challenge, usually refer to very limited physical circumstances, and the other group based on taking difficult physical situations and modelling them, often crudely, but adequately for the purpose e.g. the flow of foam or of sand, or the modulus of a polycrystalline material.

My interest in statistical physics is in this second region; one could almost call it theoretical chemical engineering rather than theoretical physics. I have chosen glass transitions for here is a subject with clear laws unclearly understood, where modelling seems the right way forward provided one can find cases where simple models have validity.

I can put forward this philosophy in this plan:

THEORY OF GLASS

### Principles

1. All glasses are similar, so choose a simple one.

2. Small molecules form glasses with difficulty. Long molecules form glasses with ease. So choose polymeric glasses.
3. Polymeric glasses are easily visualised and are validly studied by self-consistent methods.
4. Look for clear laws well founded in experiment to test theories. Polymeric glasses provide these.

Our model with polymeric glass where there really is a theory of liquids (unlike small molecule liquid theory which is very complicated). The flow diagram is now:

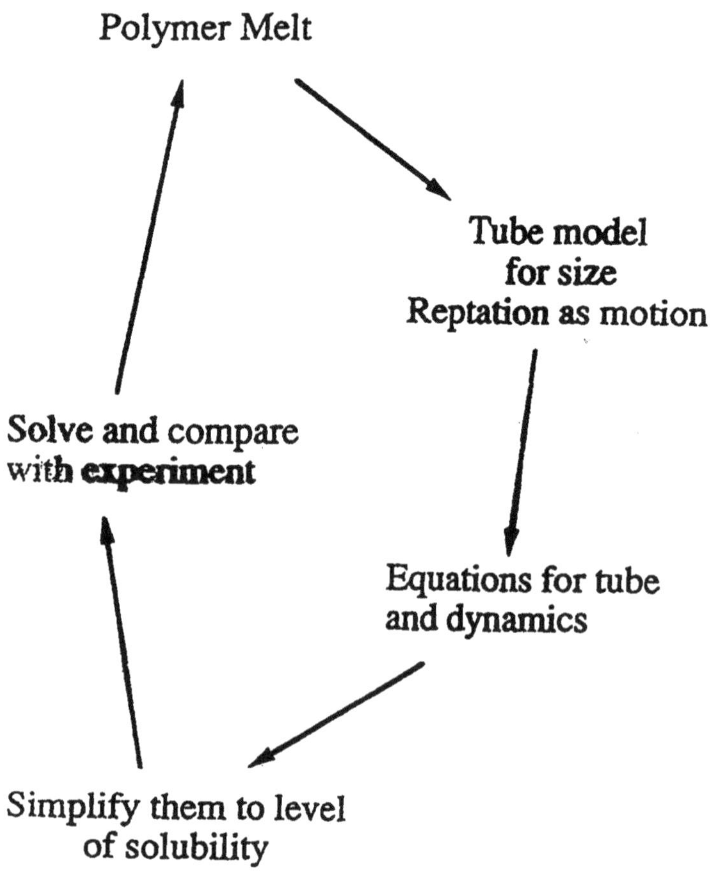

Finally if one uses the theory of the flow chart (given in detail in the book of Doi and Edwards) one can work through the final flow chart:

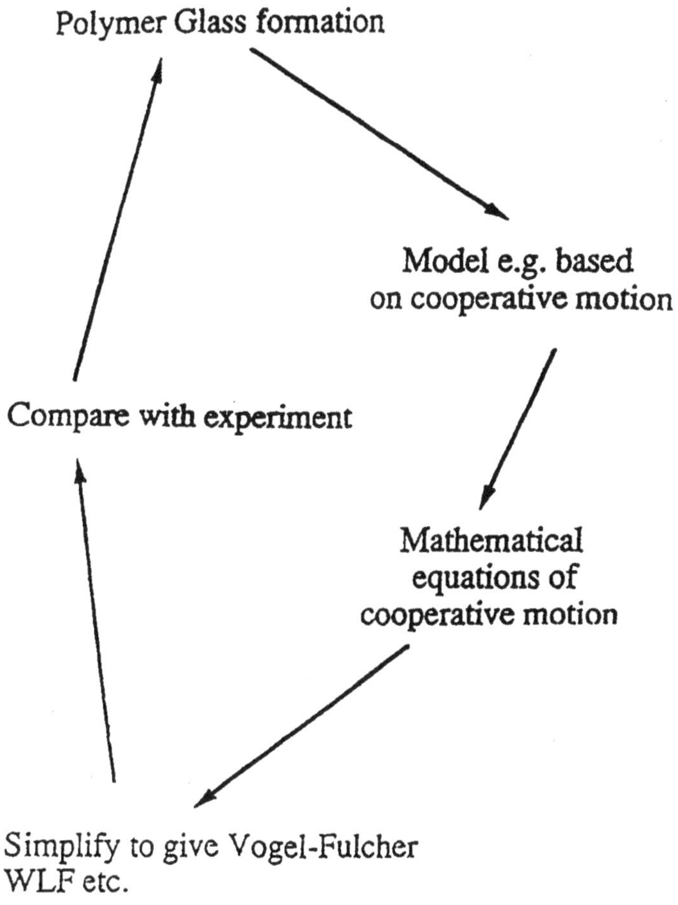


It is to be noted that by varying the molecular weight a new variable appears where properties like $T_g$ depend on Mand not on other features of the molecule, in contrast to small molecules where the chemical similarity of say alcohols has no simple consequence for $T_g$, likewise for blends. We can list problems:

## 2. Universal Results in Glasses

**Problem 1.** When a liquid cools its viscosity and its inverse diffusivity increase in a law discovered and rediscovered by many investigators: Vogel, Fulcher,

Williams, Landel, Ferry, Dolittle, and no doubt many others. The structure is known to mathematicians as an essential singularity

$$\eta = A \exp\left(+\frac{B}{T - T_g}\right) .$$

Thus when $(\log \eta/A)^{-1}$ is plotted a straight line results. Although sometimes the plot shows some curvature which would correspond to $(T - T_g)^\alpha$, this formula does seem to have a vast validity and in the form used by WLF

$$(\log \eta)^{-1} = \frac{T - T_g}{B + (T - T_g) \log A}$$

is used as the standard form for polymeric systems. A central position in glass theory must be given to this formula. A theory which cannot give a reasonably direct and simple derivation of the DVFWLF formula is a dubious basis for the study of more complex systems.

**Problem 2.** The glass transition depends on the short time scale on the rate of cooling, i.e. $T_g(\dot{T})$ is the simplest way to express this. One can argue that the degrees of freedom of the material can be divided into those of short and of long relaxation times and the long relaxation times relate to configurational rearrangements. There should then be the same kind of quasi-thermodynamics for systems that have two bands of relaxation times and, although obviously it can only be approximate, this should allow some quantification of how $T_g$ depends on $\dot{T}$.

Turning now to problems which are specific to polymeric systems there are two problems that are well known.

**Problem 3.** Polymers form glasses easily, and blends of compatible polymers do also. A new variable now characterises the glass transition: the relative densities of the two components. Suppose the proportions of polymers 1 and 2 are $\phi_1$ and $\phi_2$

$$\phi_1 + \phi_2 = 1 .$$

Then it is well established that

$$\frac{1}{T_g(\text{blend})} = \frac{\phi_1}{T_{g_1}} + \frac{\phi_2}{T_{g_2}}$$

is a good approximation to the dependence of $T_g$ on the blend proportions. Can we derive this?

**Problem 4.** Accurate distributions of molecular weights are available in many cases, to the extent that one can study the dependence of the glass transition on the molecular weight, i.e. the length of the polymer. The law given by Flory and Fox[3-8] is a reasonable approximation.

$$T_g(M) = T_g(\infty) - \frac{1}{M} \text{ constant} .$$

Can we derive this law, and evaluate the constant involved.

## 3. Model of Polymeric Systems

The model to be adopted is the tube model, introduced to give a simple description of concentrated polymeric systems. This model is capable of giving rise to useful constitutive equations for melts and offers a way of coping .with the topological constraints of polymers which whilst being intuitively obvious are virtually impossible on a rigorous mathematical basis.

The model is illustrated in the diagrams below. Imagining the other polymers represented by dots as they pass through the plane of the polymer of interest (the 'plane' being taken wlg for simplicity), the (long) polymer of interest behaves as if confined by a tube. The tube is roughly speaking a random walk of step length $a$ and of radius also of order $a$. The polymer between the kinks of the tube is said to have the entanglement molecular weight $Me$. These concepts are spelled out in detail in the book of Doi and Edwards[9] and the reader is referred to this for deeper justification of the utility of the model.

The complex movement of the polymer between its neighbours is reduced by Doi and Edwards to the movement along the axis of the tube, called the primitive path, by the polymer which is reduced to a line polymer called the primitive chain. The complex movement of reptation is thus reduced to the sliding of a smooth chain backwards and forwards in Brownian motion, but with one degree of freedom.

Thus we have the sequence

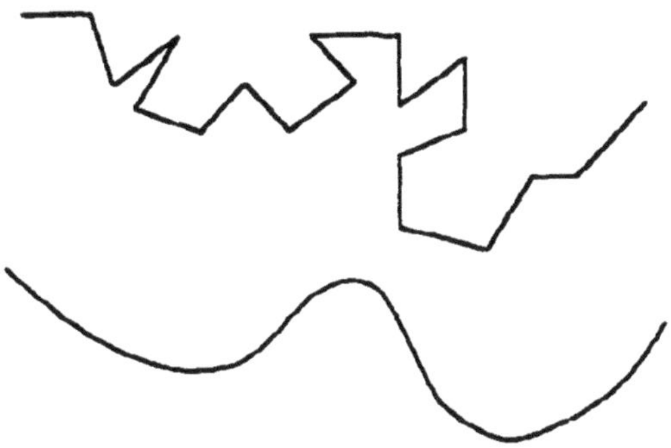

In order to approach problem 1, the sequence is taken one stage further by making the tube and primitive chain rectilinear. This will be inadequate for problem 4 but, perhaps surprisingly, it is good enough to solve problems 1 and 3. Thus the sequence now goes

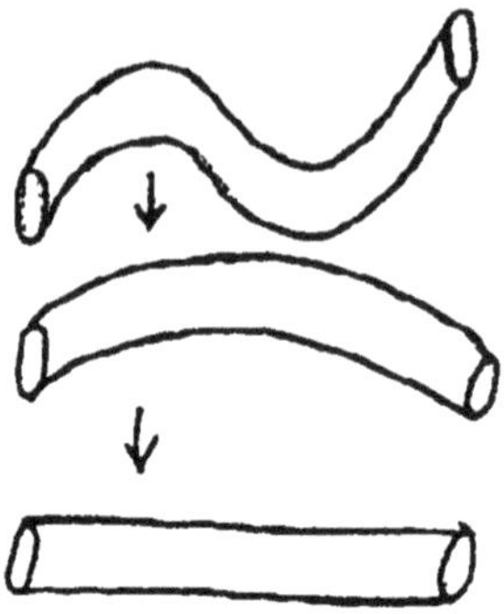

The refinement required to handle problem 4 is to note that in addition to obstruction experienced by the polymer in exiting from its tube (thereby creating new tube in the melt) if it meets an obstacle $\underline{--\rightarrow}$ there are also obstacles along the path, e.g.

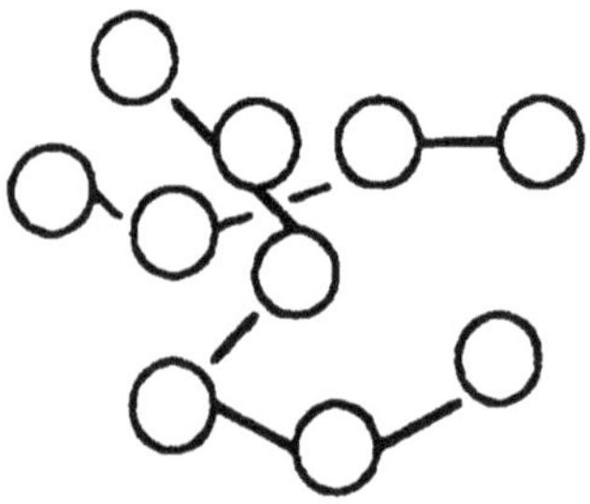

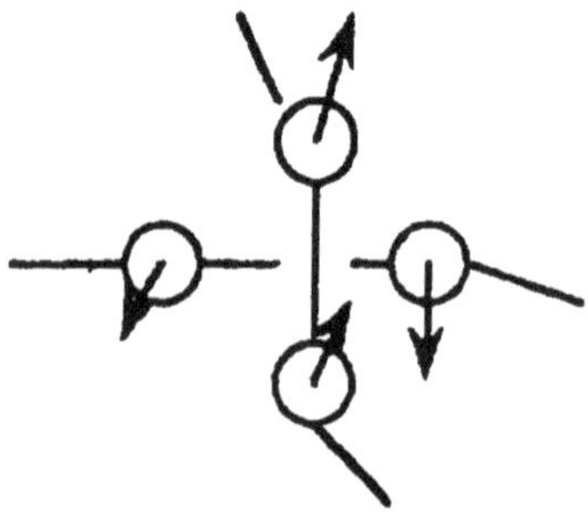

The monomers can only pass if, for example, arrows are followed. An assembly of perfectly smooth straight cylinders will be shown to give rise to the V–F law, but not to the Flory–Fox law.

## 4. Derivation of the Vogel–Fulcher Law

Before deriving the law, it is to be emphasised again that since so many diverse systems show the glass transition, it is worth thinking out the simplest. It is well known in phase transition problems that the closer one can get to mean field conditions, the easier it is to calculate. Thus a crystal in $d \geq 4$ dimensions, are a crystal with a large number of nearest neighbours (mathematically, even if not physically possible) is rigorously solved by mean field methods. For spheres the number of nearest neighbours is $N_n \leq 12$, but for rods $N_n$ can be made as large as one likes. This means that the rod molecule can be rigorously described by the tube of its neighbours, and slides backwards and forwards under Brownian motion, in some pervading background fluid.

It was shown by Edwards and Evans[8] that the mean diffusion constant for such a system of randomly orientated sliding rods was[5]

$$D = D_0 \left(1 - \frac{3}{8\pi}(cdL^2)^{3/2}\right)$$

where $D_0$ is the diffusion constant when dilute, $c$ the concentration of rods of diameter $d$ and length $L$. It will be appreciated that if $d$ is a consequence of the effect of hard and soft forces and $\kappa T$, and $D_0$ is to expected to the proportional to $T$, this form is equivalent to

$$D = D_\infty \left(\frac{T}{T_g} - 1\right)$$

or in terms of pressure

$$D = D_0 \left(\frac{P}{P_g} - 1\right) .$$

It was further shown by Edwards and Vilgis that if corrections to the above picture were envisaged such that for example the configuration

could by 'shuffling', i.e. slightly bending the rods, be got into the otherwise inaccessible configuration one could add up such effects in terms of the number of ways a 'shuffle' could be constructed.

The result is to modify the series expansion of $D^{-1}$

$$D = D_0(1 - A)$$

from i.e. $D^{-1} = D_0^{-1}(1 + A + A^2 + \cdots + A^N)$ into

$$D^{-1} = D_0^{-1}\left(1 + A + \ldots + \left(A + \frac{B}{\sqrt{N}}\right)^N + \ldots\right)$$

and by summing parametrically

$$\begin{aligned} D^{-1} &= D_0^{-1} \int_0^{\infty - x(1-A) + \sqrt{x}B} e dx \\ &= D_0^{-1} \exp\left(\frac{B^2}{4(1-A)}\right) \quad \text{by steepest descent.} \\ &\text{as } A \to 1 \,. \end{aligned}$$

This immediately allows an attack on problem 3, because when types 1 and 2 are present the '$A$' parameter becomes

$$A_1 = \left(\frac{3}{8\pi}\right)(cd_1 L_1)^{3/2} \text{ with weight } \phi_1$$

and the first order perturbation effect is

$$D^{-1} = D_0^{-1}(1 + \phi_1 A_1 + \phi_2 A_2 + \ldots)$$

careful counting shows a series which is indeed geometric in

$$\phi_1 A_1 + \phi_2 A_2$$

so that

$$D = D_0(1 - \phi_1 A_1 - \phi_2 A_2) \,.$$

To put these manipulations into temperature form one needs to write them in the form given above

$$D = D_\infty\left(\frac{T}{T_g} - 1\right) \text{ or } -D_\infty\left(1 - \frac{T}{T_g}\right) \,.$$

Where now the expansion will give rise to

$$\frac{\phi_1}{T_{g_1}} + \frac{\phi_2}{T_{g_2}} = \frac{1}{T_g}$$

as expected.

This is a rather crude argument and a fuller analysis with a direct derivation relying on the interactions of the molecules appearing in the effective diameter $d$, hence $d$ is $d(T)$, gives rise to more complicated forms. Nevertheless there is here an outline derivation of the form valid over a modest range of $\phi$ values. The arguments given here extend to the V–F region near $T_g$ also.

## 5. Molecular Weight Dependence of $T_g$

A physical model which has considerable success is to regard the entanglements which give rise to the tube as represented by hoops at entanglements points, i.e. a picture like

Crude as this is it is capable of giving interesting constitutive equations for rubber for example.

The glass transition can now be thought of as being caused by blockages at the ends of the tube and by blockages at the entanglement points. The total number of problem points is now $\frac{L}{a}+1$ of which two are ends and $\frac{L}{a}-2$ interior points. These will make different contributions to the value of the glass transition temperature, so that, with $T_g(\infty)$ the value of $T_g$ as $L \to \infty$,

$$\begin{aligned} T_g(L) &= T_g(\infty)\left(1-\gamma\left(\frac{a}{L}\right)\right) \\ &= T_g(\infty)\left(1-\gamma\left(\frac{Me}{M}\right)\right) \end{aligned}$$

where $\gamma$ is a numerical factor.

The essence of this argument is that whatever causes the glass transition does it in a way which recurs down the chain varying only at the ends. To get $\gamma$ one will need a molecular picture. The problem is similar to that used in the relaxation of stress. The original Doi–Edwards picture was that stress only relaxed at the ends in the registration process. This has been challenged by des Cloizeaux who suggests that when the end of a polymer disengages from neighbours, not only does that polymer relax, but the neighbours also. This would be like putting $\gamma\left(\frac{L}{a}\right)$ in the above formula, whereas the D–E picture would be $\gamma = 0$. In this problem neither picture is adequate.

## 6. The Problem of the Cooling Rate (Problem 2)

Suppose one can make the enormous simplification of dividing the degrees of freedom of a liquid into fast modes and slow modes. Suppose one now argues that as the liquid cools the fast modes, by far the most numerous, give rise to the temperature, but the slow modes essentially the rearrangement of molecular configurations are not in equilibrium and relax slowly. Suppose the liquid is glass forming, and if cooled rapidly forms a glass but one which is not yet in thermal equilibrium which is only achieved after a further time.[10]

A crude description of the way the system relaxes is to find an appropriate single variable which characterises the departure of the configurational entropy from that of equilibrium. If one thinks of the number of ways of packing the molecules into a volume $V$, $S(v)$ the appropriate intensive variable is $\frac{\partial V}{\partial S}$ in analogy to temperature $\frac{\partial E}{\partial S}$.[8,9]

In thermal equilibrium

$$\frac{\partial V}{\partial S} = \frac{\partial V}{\partial E}\frac{\partial E}{\partial S} = \frac{T}{P},$$

so if we denote $\frac{\partial V}{\partial S}$ by $X$, calculated from the actual number of configurations available in $V$, $Z = X - \frac{T}{P}$ is the measure suggested. Thus if the cooling rate is $\dot{T}$ and the initial temperature, retained by the configuration, is $T_0$

$$\begin{aligned}
&\left(\frac{\partial V}{\partial S}\right) \text{ at } T_0 - \left(\frac{T}{P}\right) \text{ at } T \\
&= \left(\frac{\partial V}{\partial S}\right)_{T+\tau\dot{T}} - \left(\frac{T}{\dot{P}}\right) \\
&= \beta\dot{T},
\end{aligned}$$

where the time of cooling is $\tau$ and $\beta$ is a constant characteristic of the material. Thus

$$\beta\dot{T} = \tau\dot{T}\left(\frac{\partial^2 V}{\partial S \partial T}\right)$$

and

$$\frac{\partial Z}{\partial t} + \frac{1}{\tau}Z = \beta\dot{T}$$

is the equation governing relaxation from the initial glass temperature to the final.

## Conclusion

Crude as the models of this lecture may seem they bring out the key point that progress can be made by choosing particular systems to model universal situations.

## References

1. J. D. Ferry, Viscoelastic Properties of Polymers. Wiley, New York (1980).
2. I. M. Ward, Mechanical Properties of Solid Polymers. Second Ed., (1983), Wiley (Chichester).
3. T. G. Fox and P. J. Flory, *J. App. Phy.* **21**, 581 (1950).
4. L. Mandelkern and G. M. Martin, *J. Pol. Sci.* **14**, 315 (1954).
5. F. A. Quinn, *J. Res. Natl. Bur Stand* **58**, 137 (1959).
6. T. G. Fox and S. Loshaek, *J. Pol. Sci.* **15**, 371 (1955).
7. M. Doi and S. F. Edwards, Theory of Polymer Dynamics, OUP (1986).
8. S. F. Edwards and K. E. Evans, *JRSC Farad. Trans. II*, **78**, 113 (1982).
9. S. F. Edwards and T. A. Vilgis, *Physica Scripts* T13, (1986).
10. S. F. Edwards, *Proc. Annals of NY Acad. Sci.* **371**, 210–217 (1981).
11. A. Mehta, Granular Matter: Interdisciplinary Approach. Springer–Verlag (in press) 1993.

# Remarks About Some Developments in Statistical Mechanics

(Talk given at the Xiamen Conference, July 31-August 4, 1995)

Chen Ning Yang

The Chinese University of Hong Kong, Hong Kong

## 1. Gibbs and Equilibrium Statistical Mechanics

In 1901, at the age of 62, Gibbs (1839-1903) published a book (Fig. 1) called *Elementary Principles in Statistical Mechanics.* It was remarkable in several ways. First, it had as a subtitle *The Rational Foundation of Thermodynamics.* Gibbs chose this subtitle because he knew his theory did not agree with experiments, as he emphasized in the preface to the book. Yet he believed there were no other possible rational basis to thermodynamics. Second, the style of writing of the book has an elegance akin to a long poem, which is quite unique in the history of physics. Third, it was to become the fundamental basis of twentieth century equilibrium statistical mechanics.

A year before Einstein's death in 1955, he was asked who were the most powerful thinkers he had known. He[1] replied, "Lorentz," and added, "I never met Willard Gibbs; perhaps, had I done so, I might have placed him beside Lorentz."

## 2. Origin of Bose-Einstein and Fermi-Dirac Statistics

In the last chapter of Gibbs' book mentioned above, he found it necessary to insert a factor $N!$ in the formula for the free energy:

$$F = -kT\ell n\frac{1}{N!}\int\cdots\int e^{-E/kT}dq_1\cdots dq_N\, dp_1\cdots dp_N.$$

This factor made $F$ an extensive quantity. Furthermore it was convenient for treating mixtures of atoms. Later on the factor $h^N$ was inserted in the denominator to make the quantity under the logarithm dimensionless. But the factor $N!$ remained mysterious and was called the problem of the "absolute entropy constant". It was believed to be related to the fact that for identical particles, the phase space should not be multiply counted, but the precise meaning of this factor, especially for small $N$, was not clarified.

In June 1924 Bose's paper was published. He derived Planck's radiation law by counting states of photons in a novel way. As soon as Einstein saw this paper, he generalized it to the counting of states of atoms, thereby predicting the phenomena of Bose-Einstein condensation, a most daring and insightful extrapolation which has only now been brilliantly experimentally confirmed.

A year and a half after Bose's and Einstein's papers, upon reading Pauli's article on the exclusion principle, Fermi realized in 1926 that he had now the concepts in hand to discuss the thermodynamics of a collection of electrons. The results were such fundamental concepts like the Fermi sea, the Fermi energy, etc.. According to Rasetti[2], Fermi was not influenced by the earlier work of Bose and of Einstein.

Also in 1926, Heisenberg pointed out that the difference between the singlet and triplet energy levels of the (1s)(2s) states of He was due to the difference of the symmetry of the space wave function, which in turn was caused by the requirement of antisymmetrization of the total wave function of the two spinning electrons.

Finally in August 1926 Dirac developed the general theory of the symmetry of wave functions of Bose-Einstein and Fermi-Dirac particles.

Incidentally this chain of papers Einstein-Fermi-Heisenberg-Dirac reveals in a dramatic way the differences of styles of these four great physicists: (1) Einstein's prediction of Bose-Einstein condensation of *free* particles was against all intuitive concepts of phase transitions at that time. To make such a prediction, without full mathematical rigor, based on a novel counting method extrapolated from photons to atoms, required a perception and a boldness that was uniquely Einstein's. (2) Fermi's paper formed the basis of all subsequent theories about condensed matter physics. It has the hallmark of Fermi's physics: the ability to capture the fundamentals of the problem at hand and extract from it the essence that will affect all future developments. Solidity and imagination marched hand-in-hand in Fermi's work. (3) Heisenberg's work produced the key idea that on the one hand linked the symmetry of wave functions to Pauli's exclusion principles and on the other hand resolved the great puzzle of how the spin alignment of two electrons can affect the Coulomb energy of the He atom. Furthermore the idea of the "exchange integral" which originated in this paper later produced another great achievement of Heisenberg's: the mechanism of ferromagnetism. However, Heisenberg's paper was long on originality but short on elegance and precision, a characteristic of all of Heisenberg's papers. (4) In contrast, Dirac's papers were always elegant and precise. They also tend to be the final word in the problems that they address. In the case of his 1926 paper, very little can be added later to his masterly analysis.

## 3. Early Discussions of Phase Transitions in Statistical Mechanics

The development of quantum mechanics in 1925-1927 removed the difficulties that Gibbs had to face in checking his *rational foundation* against experimental facts.

Thus was born quantum statistical mechanics. In the late 1920's and early 1930's, physicists and chemists applied the new quantum statistical mechanics to many problems in dilute gases and dilute solutions with great success.

Then in the mid 1930's, because of the discovery of peaks in the $C_p$ versus $T$ curve in alloys, the theory of order-disorder transformations became quite fashionable. An "order parameter" was introduced for the discussion of phase transitions. Such theories are now called mean field theories, which actually was first used by Weiss[3] in 1907 for describing ferromagnetic transitions.

In 1937 J. Mayer attempted to formulate a theory of liquid-gas transition without introducing mean fields by examining the convergence properties of the virial series. There was a "vigorous discussion" of his ideas at the Van der Waals Centenary Congress[4] on November 26, 1937 in Amsterdam, followed in the next months by elaborations/alternations of Mayer's theory. Incidentally, my own entry into statistical mechanics was related in a way to Mayer's theory. I was an undergraduate student in Kunming in 1941-1942 when Professor J.S. Wang gave several lectures on this theory of Mayer's and on subsequent developments. I did not then quite understand the complexities of the theories, but became fascinated with the subject[5]. That led to my working with Wang for my Master degree thesis on statistical mechanics. As to Mayer's theory, some ten years later in two papers[6] Lee and I cleared up the confusion in this field.

## 4. Onsager and the "Ising Disease"

In 1944, in an amazing paper, Onsager solved the two dimensional Ising model rigorously. It was the first in a field which undoubtedly will be covered in many presentations at this Conference. His paper was very difficult to read because he did not describe his strategy. He seemed to have a predilection for calculating the commutators of every other expression in sight without telling what he was aiming at. I still remember vividly today how I was frustrated in trying to understand that paper, first when I was a graduate student in China and then a graduate student in Chicago. It was many years later, in March 1965, that I finally learned[7] how it had come about that Onsager was so fond of calculating those commutators.

Young physicists today may find it surprising, even unbelievable, that in the 1950's the Ising model and similar problems were not deemed important by most physicists. They were considered arcane exercises, narrowly interesting, mathematically seducing, but of little real consequence. There was the phrase[8], for example, of "contracting the Ising disease." In a recent article by Dyson in my Festschrift (edited by S.T. Yau, published by the International Press) he recalled how, in 1952, when he read my article about the magnetization of the Ising model, he was impressed by the beautiful complexity of the calculation and the beautiful simplicity of the result, but felt I was wasting my time.

The situation dramatically changed around 1960 because of several developments: (1) the experimental discoveries[9] of divergences of specific heats near various phase transition points; (2) theoretical work on the critical exponents led gradually to the concept of universality and to some very useful inequalities among the critical exponents; and (3) the proposal of a scaling law[10].

With all these developments equilibrium statistical mechanics finally became respectable and was considered physically important in the 1960's.

## References

1. A. Pais, *Subtle is the Lord, the Scientific Biography of Albert Einstein*, (1982).

2. R. Rasetti in *Collected Papers of Enrico Fermi*, vol. 1, p.178 (University of Chicago Press, 1965).

3. P. Weiss, J. of Physics **6** (1907) 667.

4. M. Born and K. Fuchs, Proc. Roy. Soc. **A166** (1938) 391.

5. Chen Ning Yang, Int. J. Mod. Phys. **B2** (1988) 1325.

6. C. N. Yang and T.D. Lee, Phys. Rev. **87** (1952) 404; T.D. Lee and C. N. Yang, Phys. Rev. **87** (1952) 410.

7. Chen Ning Yang, *Selected Papers 1945-1980 with Commentary*, p. 11-13 (Freeman, 1983).

8. A. Pais, Nuclear Physics **5** (1958) 297.

9. W.M. Fairbank, M.J. Buckingham and C.F. Kellers, Proceedings of the Fifth International Conference on Low Temperature Physics (Madison, Wisconsin) p.50. See also W.M. Fairbank and C.F. Kellers in "Critical Phenomena", National Bureau of Standards Miscellaneous Publication 273 (1965); W.K. Robinson and S.A. Friedberg, Physical Review, **117** (1960) 402; M.I. Bagatskii, A.V. Voronel and V.G. Gusak, Journal of Experimental and Theoretical Physics (in Russian), **43** (1962) 728. [English translation: Soviet Physics JETP, vol. 16, p. 517.]

10. B. Widom, J. Chem. Phys. **12C** (1965) 3898.

ELEMENTARY PRINCIPLES

IN

STATISTICAL MECHANICS

DEVELOPED WITH ESPECIAL REFERENCE TO

THE RATIONAL FOUNDATION OF
THERMODYNAMICS

BY

J. WILLARD GIBBS

*Late Professor of Mathematical Physics in Yale University*

DOVER PUBLICATIONS, INC.
NEW YORK NEW YORK

**Figure 1**

# DIRECTED COMPACT LATTICE ANIMALS, RESTRICTED PARTITIONS IN NUMBER THEORY, AND THE INFINITE-STATE POTTS MODEL

F. Y. Wu
*Department of Physics, Northeastern University, Boston, MA 02115, USA*

We consider a directed compact site lattice animal problem on the $d$-dimensional hypercubic lattice, and establish its equivalence with i) the infinite-state Potts model and ii) the enumeration of $(d-1)$-dimensional restricted partitions of an integer. The directed compact lattice animal problem is then solved exactly in $d = 2, 3$, and the maximum number of animals of size $n$ is found to grow as $\exp\left(n^{(d-1)/d}\right)$. The solution of the infinite-state Potts model suggests a conjectured limiting form for the generating function of restricted partitions for $d > 3$.

## 1 Directed Compact Site Lattice Animals

We first define a directed compact site lattice animal problem.

Consider a $d$-dimensional hypercubic lattice of size $L_1 \times L_2 \times \cdots \times L_d$ whose sites are labelled by coordinates $\{n_1, n_2, ..., n_d\}$. Starting from the origin $\{1, 1, ..., 1\}$ a site animal grows in the directions of increasing $n_i$ subject to the rule that a site $\{n_1, n_2, ..., n_d\}$ is occupied only when the $d$ sites $\{n_1, ..., n_i - 1, ..., n_d\}, i = 1, 2..., d$ are all occupied. (In applying this rule sites with coordinates $n_i = 0$ are regarded as being occupied.) Compared to the usual lattice animals,[1] the present model generates animals whose configurations are compact, since it excludes unoccupied interior sites. In addition, by keeping $L_i$ finite, there exists a maximum animal size of $L_1 L_2 \cdots L_d$.

Let $A_n(L_1, L_2, \cdots, L_d)$ be the number of distinct $n$-site animals that can grow on the lattice. One is interested in finding the asymptotic behavior of $A_n$ for large $n$. Define the generating function

$$G(L_1, L_2, \cdots, L_d; t) = 1 + \sum_{n=1}^{L_1 L_2 \cdots L_d} A_n(L_1, L_2, \cdots, L_d) t^n. \tag{1}$$

For example, the generating function for the $3 \times 3$ lattice is

$$G(3, 3; t) = 1 + t + 2t^2 + 3t^3 + 3t^4 + 3t^5 + 3t^6 + 2t^7 + t^8 + t^9. \tag{2}$$

It is observed that $A_n(L_1, L_2)$ reaches a maximum at $n \sim L_1 L_2/2$.

## 2 Restricted Partitions of an Integer

In this section we show that the directed compact lattice animal problem formulated in the preceeding section is equivalent to the problem of restricted partitions of an integer.

For simplicity, consider first the case of $d = 2$ for which we have a simple quartic lattice of size $L_1 \times L_2$. Let $h_i$, $i = 1, 2, ..., L_1$, be the number of occupied sites in the $i$th column of the lattice. One observes that the animal growth rule implies the restriction

$$L_2 \geq h_1 \geq h_2 \geq \cdots \geq h_{L_1} \geq 0. \tag{3}$$

In addition, one has the sum rule

$$\sum_{i=1}^{L_1} h_i = n, \tag{4}$$

where $n$ is the animal size. It is convenient to regard (4) as specifying the partitions of a positive integer $n$ into sums of integers $h_i$, and the condition (3) ensures that all partitions are distinct. Then $A_n(L_1, L_2)$ is precisely the number of distinct ways that an integer $n$ is partitioned into at most $L_1$ parts with each part less or equal to $L_2$. This leads to the classic restricted *partitio numerorum* problem dating back to Gauss [2].

In a similar fashion and considering the rule of generating animals, one can regard $A_n(L_1, L_2, \cdots, L_d)$ as the number of distinct partitions of a positive integer $n$ into sums of nonnegative integers $m(n_1, n_2, ..., n_{d-1})$ associated with vertices $\{n_1, n_2, ..., n_{d-1}\}$ of a $(d-1)$-dimensional hypercubic lattice, or, explicitly,

$$n = \sum_{n_1=1}^{L_1} \sum_{n_2=1}^{L_2} \cdots \sum_{n_{d-1}=1}^{L_{d-1}} m(n_1, ..., n_{d-1}), \tag{5}$$

such that

$$0 \leq m(n'_1, n'_2, ..., n'_{d-1}) \leq m(n_1, n_2, ..., n_{d-1}) \leq L_d, \tag{6}$$

whenever $n_1 \leq n'_1$, $n_2 \leq n'_2, \cdots$, $n_{d-1} \leq n'_{d-1}$. This defines a $(d-1)$-dimensional restricted partition. The generating function (1) now generates $(d-1)$-dimensional restricted partitions of $n$. This is a classic problem in number theory whose exact solution is known only for $d = 2$ and 3. Particularly, for $d = 2$, one finds [3,4]

$$G(L_1, L_2; t) = (t)_{L_1+L_2}/(t)_{L_1}(t)_{L_2}, \tag{7}$$

for which the generating function (7) is known as the Gaussian polynomial, and for $d = 3$, one finds [3,4]

$$G(L_1, L_2, L_3; t) = \frac{[t]_{L_1+L_2+L_3-1}[t]_{L_1-1}[t]_{L_2-1}[t]_{L_3-1}}{[t]_{L_1+L_2-1}[t]_{L_2+L_3-1}[t]_{L_1+L_3-1}}, \tag{8}$$

where

$$[t]_L \equiv \prod_{p=1}^{L} (t)_p, \qquad (t)_L \equiv \prod_{p=1}^{L} (1 - t^p). \tag{9}$$

Having known the explicit expressions of $G_n$, one can in principle invert the generating function using the Cauchy integral

$$A_n(L_1, \cdots, L_d) = \frac{1}{2\pi i} \oint \frac{1}{z^{n+1}} G(L_1, \cdots, L_d; z) dz \tag{10}$$

where the integration is taken over a contour inside $|z| = 1$ enclosing the origin. The asymptotic behavior of $A_n(L_1, L_2, \cdots, L_d)$ for large $n$ can then be deduced by applying saddle point analyses to (10). In the case of $d = 2$ when the rows and columns of $\mathcal{L}$ are not fully filled, namely, $n < \min\{L_1, L_2\}$, the partition of $n$ is actually without restrictions. Then, a careful analysis[5] of (10) yields the celebrated Hardy-Ramanujan[6] result

$$A_n(L_1, L_2) \sim \frac{1}{4n\sqrt{3}} \exp\left(\pi \sqrt{\frac{2n}{3}}\right), \qquad n < \min\{L_1, L_2\}. \tag{11}$$

However, the asymptotic behavior is different at $n \sim L_1 L_2 / 2$ where $A_n$ is the maximum.

To find the leading behavior of the maximum $A_n$ which we assume to occur at $n \sim L_1 L_2 \cdots L_d / 2$ (a fact very difficult to prove even for $d = 2$), one can use the approximation

$$A_n(L_1, L_2, \cdots, L_d) \propto G(L_1, L_2, \cdots, L_d; 1). \tag{12}$$

It can be shown[3] from (7) that we have

$$G(L_1, L_2; 1) = \binom{L_1 + L_2}{L_1}. \tag{13}$$

The closed-form expression of $G(L_1, L_2, L_3; 1)$ is a bit harder to compute, but its asymptotic expression is known.[7] Using (12), (13), and the result of Ref. 7, one obtains

$$A_n(L_1, L_2) \propto \exp\big(c(\alpha)\sqrt{n}\big), \qquad n \sim L_1 L_2 / 2, \tag{14}$$

$$A_n(L_1, L_2, L_3) \propto \exp\big[c(\alpha_1, \alpha_2, \alpha_3)\, n^{2/3}\big], \qquad n \sim L_1 L_2 L_3 / 2, \tag{15}$$

where

$$c(\alpha) = \sqrt{2}\left[\sqrt{\frac{1}{\alpha}}\ln(1+\alpha)+\sqrt{\alpha}\ln\left(1+\frac{1}{\alpha}\right)\right], \qquad \alpha = L_1/L_2,$$

$$c(\alpha_1,\alpha_2,\alpha_3) = 2^{-1/3}\left[\left(\frac{\alpha_1}{\alpha_2}\right)^{1/3}+\left(\frac{\alpha_2}{\alpha_3}\right)^{1/3}+\left(\frac{\alpha_3}{\alpha_1}\right)^{1/3}\right]t(\alpha_1,\alpha_2,\alpha_3),$$

$$t(\alpha_1,\alpha_2,\alpha_3) = \left(x_1x_2+x_2x_3+x_3x_1\right)^{-1}\sum_{i=1}^{3}\left[x_i^2\ln x_i-(1-x_i)^2\ln(1-x_i)\right]$$

with $x_i = (1+\alpha_j+1/\alpha_k)^{-1}$, $\alpha_i = L_j/L_k$, $i,j,k$ in cyclic order of $1,2,3$. Particularly, for $L_1 = L_2 = L_3 = L$, one has

$$c(1,1) = 2\sqrt{2}\ln 2 = 1.9605...$$
$$c(1,1,1) = 2^{2/3}\,(9\ln\sqrt{3}-3\ln 4) = 1.2459... \tag{16}$$

Expressions (14) and (15) suggest the asymptotic behavior

$$A_n(L_1,L_2,\cdots,L_d) \propto \exp(c\,n^{(d-1)/d}), \qquad n \sim L_1L_2\cdots L_d/2 \tag{17}$$

for general $d$, where $c$ is a constant depending on the geometry of the lattice. These results are to be compared with the known asymptotic behavior $3^n n^{-3/2}$ of the usual directed animals on an infinite lattice in $d = 2$.[1]

## 3 The Infinite-State Potts Model

We now show that both the directed compact lattice animal and the restricted partition problems, in any $d$, are equivalent to the infinite-state Potts model. The known solution of the latter problem[8] then suggests a solution to the unsolved problem of restricted partitions for $d > 3$ in the thermodynamic limit.

Introduce an extra site to the hypercubic lattice, which is connected by edges to every site in the $d$ $(d-1)$-dimensional hyperplanes $n_i = L_i$. Particularly, the extra site is connected to the site $\{L_1, L_2, ..., L_d\}$ by $d$ edges.[9] We regard the resulting lattice as a graph $\mathcal{L}$, and there are altogether $N+1$ vertices and $Nd$ edges in $\mathcal{L}$, where $N = L_1L_2\cdots L_d$. Consider the $q$-state Potts model on $\mathcal{L}$, with each vertex of $\mathcal{L}$ occupied by a spin which can be in $q$ different states. Let two neighboring spins in states $\sigma, \sigma' = 1, 2, ..., q$ interact with a (reduced) energy $K\delta(\sigma,\sigma')$. The partition function is

$$Z = \sum_{\sigma_i=1}^{q}\prod_{\text{edges}} e^{K\delta(\sigma,\sigma')}, \tag{18}$$

where the summations are taken over the spin states at all vertices $i$, and the product taken over all edges of $\mathcal{L}$.

Using the identity

$$e^{K\delta(\sigma,\sigma')} = 1 + (e^K - 1)\delta(\sigma,\sigma'), \tag{19}$$

one can rewrite the partition function (18) as a graphical expansion.[10] This leads to the expression

$$Z = \sum_{G\subseteq\mathcal{L}} (e^K - 1)^{e(G)} q^{n(G)}, \tag{20}$$

where the summation is taken over all subgraphs $G \subseteq \mathcal{L}$, $e(G)$ is the number of edges in $G$, and $n(G)$ the number of connected clusters in $G$.

Introducing

$$x = (e^K - 1)q^{-1/d}, \tag{21}$$

we can rewrite the partition function (20) as

$$Z_N(q,x) = \sum_{G\subseteq\mathcal{L}} q^{n(G)+e(G)/d} x^{e(G)}. \tag{22}$$

We consider $Z_N(q,x)$ in the limit of $q \to \infty$ with $x$ fixed.

For $q$ large the leading term in $Z_N(q,x)$ is of the order of $q^{N+1}$ with a coefficient in the form of a polynomial of degree $N$ in $x^d$. This permits us to introduce the reduced partition function

$$Y_d(x^d) \equiv \lim_{q\to\infty} q^{-(N+1)} Z_N(q,x). \tag{23}$$

We now establish the identity

$$Y_d(t) = G(L_1, L_2, \cdots, L_d; t) \tag{24}$$

where $G(L_1, L_2, \cdots, L_d; t)$ is the generating function (1).

To establish (24), we consider the generation of the reduced partition function by identifying terms in (22) having the factor $q^{N+1}$. These terms come from the subgraph having no edges with $\{n, e\} = \{N+1, 0\}$, the proper subgraph $G = \mathcal{L}$ with $\{n, e\} = \{1, Nd\}$, and other subgraphs obtained by perturbing from either of these two.

Starting from the subgraph $G = \mathcal{L}$, for example, which has the weight $x^{Nd}$, one can generate the reduced partition function term by term by removing edges systematically. Generally, to hold the number $n + e/d$ constant, the minimum one can do is to increase $n$ by 1 and thus necessarily decreasing $e$ by $d$. Therefore, one looks for vertices connected to exactly $d$ neighboring vertices.

For $d > 1$, there is only one such vertex in $\mathcal{L}$, namely, the vertex $\{1,1,...,1\}$ connected to the $d$ vertices $\{2,1,...,1\}$, $\{1,2,...,1\}$, and $\{1,1,...,2\}$. Removing the $d$ edges connecting to $\{1,1,...,1\}$, one creates a subgraph $\mathcal{L}_1$ consisting of one isolated vertex with $\{n,e\} = \{2,(N-1)d\}$. This generates a correction factor $x^{-d}$ to the fully covered weight $x^{Nd}$, and hence a term $x^{(N-1)d}$ in the reduced partition function (23).

Repeating this process, one next looks for in $\mathcal{L}_1$, in which the vertex $\{1,1,...,1\}$ is isolated, vertices which are connected to exactly $d$ neighboring vertices. There are now exactly $d$ such vertices, namely, the $d$ vertices original connected to $\{1,1,...,1\}$. By removing the $d$ edges connected to any of these vertices, one finds the next term in the reduced partition function having $\{n,e\} = \{3,(N-2)d\}$. The resulting subgraphs $\mathcal{L}_2$ now have two isolated vertices contributing to a term $dx^{(N-2)d}$ in (23).

Continuing in this fashion, the process of generating terms in (23) is seen to be the same as generating subgraphs containing isolating vertices by removing $d$ edges at a time. Denote subgraphs containing $n$ isolated vertices constructed in this manner by $\mathcal{L}_n$, and let there be $c_n(L_1,L_2,...,L_d)$ such subgraphs. For example, we have $c_1 = 1, c_2 = d$, and $c_3 = d(d+1)/2$ if $L_i \geq 3$. Thus, we have

$$\begin{aligned} Y_d(x^d) &= x^{Nd}\left[1+\sum_{n=1}^{N} c_n(L_1,L_2,...,L_d)x^{-nd}\right] \\ &= x^{Nd}Y_d(x^{-d}). \end{aligned} \tag{25}$$

Here, the second equality in (25) is established by considering generating $Y_d(x^d)$ by perturbing from the subgraph with no edges.

The identity (24) now follows from the second line of (25) and the identity

$$c_n(L_1,L_2,...,L_d) = A_n(L_1,L_2,...,L_d). \tag{26}$$

To establish (26), one observes that, by considering the set of the $d$ integers

$$\left\{n_1,n_2,\cdots,n_{d-1},m(n_1,n_2,\cdots,n_{d-1})\right\} \tag{27}$$

as specifying the coordinates of isolated vertices, and vice versa, one has a bijection between subgraphs $\mathcal{L}_n$ (in which there are $n$ isolated vertices) and the set of nonnegative integers $m(n_1,n_2,\cdots,n_{d-1})$ satisfying (5). Furthermore, the rule of removing the $d$ edges connected to a vertex $\{n_1,n_2,...,n_d\}$ is that the $d$ vertices $\{n_1,n_2,...,n_i-1,...,n_d\}$, $i=1,2,...,d$, are already isolated. But this is precisely a statement of the condition (6). Thus we have established the identity (26) and hence the identity (24).

## 4 A Limiting Solution of Restricted Partitions for General $d$

The problem of $(d-1)$-dimensional restricted partitions of a positive integer for $d > 3$ is an outstanding unsolved problem in number theory. The known solution of the infinite-state Potts model [8] now suggests a solution in the thermodynamic limit.

Define the "per-site" generating function in a thermodynamic limit

$$f_d(t) \equiv \lim_{L_1\to\infty} \lim_{L_2\to\infty} \cdots \lim_{L_d\to\infty} \left[\frac{1}{N} \ln Y_d(t)\right]. \tag{28}$$

Using the explicit expressions of $Y_2$ and $Y_3$, one can evaluate $f_2(t)$ and $f_3(t)$ by noting that the zeroes of the polynomials $Y_2(t)$ and $Y_3(t)$ in the complex $t$ plane are simply superpositions of zeroes of $(t)_L$ which are all on the unit circle $|t| = 1$. Let the zeroes of $Y_d(t)$ be located at $e^{i\theta_j}$. Then we can write

$$Y_d(t) = \prod_j (e^{i\theta_j} - t) \qquad d = 2, 3. \tag{29}$$

Now the zeroes of $(t)_L$ are distributed uniformly on the unit circle in the limit of $L \to \infty$.[11] It follows that zeroes of $Y_2(t)$ and $Y_3(t)$ are also distributed uniformly on the unit circle. This leads to, for $d = 2, 3$,

$$f_d(t) = \frac{1}{2\pi} \int_0^{2\pi} \ln(e^{i\theta} - t) d\theta. \tag{30}$$

Thus, for the $d = 2, 3$ Potts model, we have after carrying out the integration

$$\begin{aligned} f_d(x^d) &= \frac{1}{2\pi} \int_0^{2\pi} \ln(e^{i\theta} - x^d) \\ &= \ln|x^d|, \qquad |x| > 1 \\ &= 0, \qquad |x| < 1. \end{aligned} \tag{31}$$

This leads to the occurrence of a first-order transition at $x = 1$, a transition previously known to exist for $q > 4$ in two dimensions[12] and for $q \to \infty$ in any $d$.[8]

However, as the corresponding explicit expression for $Y_d$ is not known for finite lattices, this procedure cannot be carried through for $d > 3$, On the other, the infinite-state Potts model on a $d$-dimensional hypercubic lattice is known[8] to exhibit a first order transition in the thermodynamic limit at $x = 1$ with the free energy (31). Thus, this suggests that the generating function (1) can be evaluated for any $d$ in the thermodynamic limit $L_i \to \infty$, and the result is given by (31).

Finally, we remark that in our consideration of the Potts model we have taken the $q \to \infty$ limit before taking the thermodynamic limit. But it can be verified[9] by explicitly considering the $d = 1$ solution that the two limits commute under the free boundary condition of $\mathcal{L}$ introduced in our analysis. The validity of (31) for $d > 3$, then, rests on the assumption that the two limits commute for all $d$.

## Acknowledgments

This work is supported in part by NSF grant DMR-9313648.

## References


1. D. Dhar, Phys. Rev. Lett. **49**, 959 (1982).
2. C. F. Gauss, Werke, Vol. 2, Königliche Gesellschaft der Wissenschaften, Göttingen (1870).
3. For properties of restricted partitions, see, for example, G. E. Andrews, Theory of Partitions, in Encyclopedia of Mathematics and Its Applications, Ed. G.-C. Rota (Addison-Wesley, Reading, Massachusetts, 1976) Chaps. 3 and 11.
4. P. A. MacMahon, Combinatory Analysis, Vol. 2 (Cambridge, 1916).
5. H. Rademacher, Proc. London Math. Soc. **43**, 241 (1937).
6. G. H. Hardy and S. Ramanujan, Proc. London Math. Soc. (2) **17**, 75 (1918).
7. V. Elser, J. Phys. A 17, 1509 (1984).
8. P. A. Pearce and R. B. Griffiths, *Potts model in the many-component limit*, J. Phys. A **13**, 2143-2148 (1980).
9. For details, see F. Y. Wu, appear in J. Math. and Comp. Modelling.
10. For a review on the Potts model and its physical relevance, see F. Y. Wu, Rev. Mod. Phys. **54**, 235 (1982).
11. F. Y. Wu, G. Rollet, H. Y. Huang, J. M. Maillard, C. K. Hu and C. N. Chen, preprint.
12. R. J. Baxter, J. Phys. C 6, L445 (1973).

# INTEGRABILITY AND STRUCTURE IN THE $1/r^2$ MODELS

B. SRIRAM SHASTRY
*Department of Physics, Indian Institute of Science*
*Bangalore 560012, India*
bssphysics.iisc.ernet.in

## ABSTRACT

In this talk, we will discuss a wide variety of $1/r^2$ class of models of interacting particles in 1 dimension. We summarize some of the ideas behind their recent exact solutions and discuss the reasons for their solvability. We examine the notion of Quantum Integrability in this context. In contrast to the well studied class of problems satisfying Yang Baxter equations, this class does not have a local algebraic structure, and several variants of Quantum Lax representations have been proposed. We summarize some of these and explore interrelations as well as some open problems.

## • Introduction

In this talk I would like to introduce these varieties of $1/r^2$ models, that have been of great recent interest. Several recent works can be traced from the proceedings of the Taniguchi symposium[1] and I will not aim at a comprehensive review or referencing, but rather at communicating a selection of developments that might be of interest to an audience having a broad interest in Exactly Integrable Systems. The talk is in the following sections.

- **The $1/r^2$ Models** Here I will discuss the incidence and origin of these models. The family of $1/r^2$ models was quite small from 1962 [2, 3, 4, 5] until 1987, when it underwent a boom following the work in Refs.[6, 7]. I will list some of these models and indicate what made this possible.

- **Exact Solutions of the Models** Here I will discuss the ideas behind the solution of a very general model in this class. The solution is exact in the thermodynamic limit, and obtained by the Asymptotic Bethe Ansatz (ABA) invented many years ago by Sutherland [4]. The solutions is particularly straightforward for the $1/r^2$ model, and one finds that the model maps on to an almost free particle system.

- **Correlation Functions** The dynamical correlations of this class of models has been the subject of great activity in the immediate past, and I feel quite incompetent to talk about this, so I will content myself by directing attention to the literature.

- **Algebraic Structures** Here I will talk about the various alternative schemes that have sprung up to describe the Integrability of these models. Unlike the case of the well known and powerful Yang Baxter algebra[9] that applies to a wide variety of models, here one lacks a good local algebraic structure, and hence there is no proof of Integrability that is as efficient. We list a few topics here.

  **Classical Lax Equations:** These were set up by Moser[5] to demonstrate the Integrability in the classical case.

  **Quantum Lax Equations:** Here one has a few alternative schemes for extending Moser's results to the quantum case.

  **Annihilators:** This class of operators is particularly natural for the $1/r^2$ models, and although the annihilators are not simultaneously Hermitean and mutually commutative, they do seem to be promising.

- **The $1/r^2$ Models**

The origin of these models, I believe, is due to F.J.Dyson[2], who studied the Brownian motion of random matrices, in order to endow the Wigner-Dyson "Log Coulomb Gas" with dynamics. For example the Gaussian ensembles are characterized by a probability distribution function (pdf)

$$P(H) = \prod_{i<j} \exp(-2H_{i,j}^2/g_{i,j}) \ = \exp(-TrH^2) \tag{1}$$

where $g_{i,j} = 1 + \delta_{i,j}$, leading to the familiar eigenvalue distribution

$$p(\{\lambda_1, \ldots, \lambda_n\}) = \prod_{i<j}(\lambda_j - \lambda_i)^\beta \ \exp(-\sum \lambda_i^2). \tag{2}$$

The "three fold way" [2] implies that the value of $\beta$ is either $1, 2$ or $4$, depending on whether the matrices are real symmetric, complex hermitean or symplectic. Brownian motion is induced by making the matrices functions of a new variable "t", such that we have a stochastic equation with $\eta$ a $\delta$ correlated noise:

$$\frac{\partial}{\partial t} H_{i,j}(t) = -H_{i,j} + \eta_{i,j} \tag{3}$$

leading to a Fokker Planck Equation for the probability

$$\frac{\partial}{\partial t} p(H_{i,j}) = \frac{\partial}{\partial H_{i,j}} \Big[ g_{i,j} \frac{\partial}{\partial H_{i,j}} + H_{i,j} \Big] p(H_{i,j}). \tag{4}$$

The crucial result of Dyson is that this equation has invariance under a certain group whereby the equation for the pdf of the eigenvalues can be written down immediately. This equation, after a slight massaging $P(t) = \sqrt{P(\infty)}\ \psi(t)$, is a Euclidean

Schroedinger Equation $H\psi = -\frac{\partial}{\partial t}\psi$, with

$$H = -\sum_n \frac{\partial^2}{\partial \lambda_n^2} + c\sum \lambda_n^2 + \beta(\beta/2-1)\sum_{i<j} \frac{1}{(\lambda_j - \lambda_i)^2}. \tag{5}$$

The inverse second power of the separation, is essentially due to level repulsion in Quantum Mechanics, and this is one of the main reasons for the ubiquity of the model. A similar situation holds for the circular ensembles of Dyson, where the random matrices are unitary, such as $\exp(iM)$, with Hermitean $M$, so that the unimodular eigenvalues $\lambda_n$ are $\exp(i2\pi x_n/L)$. We again end up with a Hermitean Hamiltonian and a Schroedinger equation (with $\phi = \pi/L$)

$$H = -\sum_n \frac{\partial^2}{\partial x_n^2} + \beta(\beta/2-1)\sum_{i<j} \frac{\phi^2}{\sin^2(\phi(x_j - x_i))}. \tag{6}$$

These problem have a ready interpretation in terms of $N$ interacting particles either in a quadratic confining well or on a ring of length $L$. These Hamiltonians were explicitly written down by Calogero[3] and Sutherland[4] for *all* $\beta$, not just 1, 2 and 4, and solved exactly for various thermodynamic quantities. These models are known as the Calogero Sutherland Moser models (CSM), since Moser[5] provided a proof of integrability in the classical limit. Very recently these models have been revived in the context of parametric correlations in perturbed Quantum Chaotic systems by Altshuler and coworkers[8], a point of view that seems very close to that in Dyson's original work.

- **Discretizing the $1/r^2$ Models**

The next step in generalizing these models was taken in 1987 in independent works by Shastry[6] and Haldane[7] who proposed and solved a new spin 1/2 quantum antiferromagnetic problem on a *lattice* (with $x_i$ being integers on a lattice of L sites),

$$H = \sum_{i<j} J(x_i - x_j)\vec{S}_i.\vec{S}_j \ ; \ J(x) = \frac{\phi^2}{\sin^2(\phi x)} \sim \frac{1}{x^2}. \tag{7}$$

This model has come to be called the Haldane Shastry Model (HSM), and has several points of interest. We can view it as a model of interacting hard core bosons, for which the ground state wave functions is

$$|\psi(x_1,\ldots,x_m)>= \sum_{x_1<x_2..<x_m} \exp\left(i\pi\sum x_j\right)\prod_{i<j}\sin^2\phi(x_j - x_i) b_{x_1}^\dagger..b_{x_m}^\dagger|0>. \tag{8}$$

Since $2\vec{S}_i.\vec{S}_j + \frac{1}{2}$ is equal to the permutation operator $P_{i,j}$, the exchange Hamiltonian may be extended to multicomponent systems. This is where the richness of the HSM comes about, the statistics of the constituent particles is irrelevant in the CSM models, the interaction is too repulsive at the origin and hence if we start with an ordered configuration of particles, then these are stuck there. In the HSM model, the $P_{i,j}$ operator picks up and switches particles (since the difference operator consists of *all* powers of the differential operator) and hence the statistics is inherently important. Further, we can recover the continuum models by going to the low density limit. The correlation functions in this model, known from the work of Gebhard and Vollhardt [10], have the same asymptotic decay rate as those of the Heisenberg model with nearest nbrs, the Bethe chain. The model is closely connected to the $\beta = 4$ case in the continuum, i.e. the symplectic case. The original motivations of the authors lay in analogies with the Quantum Hall wavefunctions, and with strongly correlated fermi systems, where the above wave function is the Gutzwiller wave function in 1-d. A very general form of the model is due to Polychronakos[11] and reads ( with $\lambda = \beta/2$)

$$H = -\sum_n \frac{\partial^2}{\partial x_n^2} + 2\lambda \sum_{i<j} (\lambda - P_{i,j}) V(x_j - x_i), \tag{9}$$

where, specializing to $V(x) = 1/\sin^2 x$, we have the usual model whereas setting $V(x) = g^2/\sinh^2(gx)$, with a variable $g$ gives us an interpolation between the $1/r^2$ and the nearest neighbour model. The point here is that allowing $\lambda$ to grow large, we crystallize the particles and recover the lattice HSM model as an effective low energy Hamiltonian. We indicate below the recent exact solution of this class of problems[12, 13].

- **Exact Solution by Asymptotic Bethe's Ansatz**

Here we will indicate the solution of Eq(9) by the ABA[12, 13]. The ABA requires that the system should have the following (a) Integrability (b) Positive pressure (i.e. $\partial E/\partial L < 0$) (c) a convergent virial expansion. We also need the explicit form of the 2-body phase shift. This model is integrable (see later) and the phase shift requires us to study a simple problem:

$$\left[ -\frac{d^2}{dx^2} + \frac{\lambda(\lambda - P)}{x^2} \right] \psi = E\psi \tag{10}$$

with the mirror operator $Pf(x) = f(-x)$. A small calculation shows that this problem is *reflectionless* at *all energies*, two particles, say a green particle falling on a blue particle goes through completely, with only a phase shift $\theta(k) = \lambda\pi sign(k)$. This is moreover democratic in the sense that the identity of the target is irrelevant. In the general case of Eq(9) $g \neq 0$, the reflection amplitude is non zero and the problem is much harder, it has recently been solved exactly by an adaptation of the Yang Bethe Ansatz[13], but I will only discuss the simpler case further. The Bethe wave function $\psi = \sum_{Q,P} A[Q|P] \exp i \sum x_{Q_i} k_{P_i}$ with $Q$ standing for one of the $N!$ sectors $x_{Q_1} < ... < x_{Q_N}$ has parameters $k_i$ that are determined by taking a particle around the loop, picking up a phase factor $\theta(k_i - k_j)$ for every other particle encountered. There is no consistency equation for the A's to be solved here in view of the reflectionless scattering. Peridic b.c's give

$$k_n = k_n^0 + \frac{\pi\lambda}{L}\sum_m{}' sign(k_n^0 - k_m^0), \tag{11}$$

with $k_n^0 = \frac{2\pi}{L} I_n$ the free particle quantum numbers. Note that Eq(11) is democratic in the species identity, and has huge degeneracies associated with free particle q.nos. We can solve these immediately and find $k_n = k_n^0 + 2\pi\lambda/L(n-(N+1)/2)$. A macroscopic thermodynamic state is specified by giving the occupation numbers $\nu_\alpha(k^0)$ of the various free particle states by particles of species $\alpha$. These numbers are $\nu_\alpha(k^0) = 0, 1$ for fermions, and $\nu_\alpha(k^0) = 0, 1, 2, \ldots$ for bosons. The total energy can be written as

$$E_{tot}/L = const + \frac{1}{4\pi}\sum_\alpha \int_{-\infty}^{\infty} dk\ k^2\ \nu_\alpha(k) + \frac{\lambda}{8\pi}\sum_{\alpha,\beta}\int_{-\infty}^{\infty}\int_{-\infty}^{\infty} dk\ dk'\ |k-k'|\nu_\alpha(k)\nu_\alpha(k'). \tag{12}$$

The entropy is given by a standard argument

$$S/L = \frac{1}{2\pi}\sum_\alpha \int_{-\infty}^{\infty} dk\ \{(\nu_\alpha(k) \pm 1)\log(1 \pm \nu_\alpha(k)) - \nu_\alpha(k)\log(\nu_\alpha(k))\} \tag{13}$$

and maximizing the pressure

$$P\,L = TS - E - \sum_\alpha \mu_\alpha N_\alpha \quad ; \frac{\delta P}{\delta \nu_\alpha(k)} = 0 \tag{14}$$

we find a sufficient number of equations to determine the thermodynamic densities $\nu_\alpha(k)$. The results are amusing for the case of large $\lambda \to \infty$ and low density $d \equiv N/L \to 0$ such that $\lambda d^2 \to J$ where we find

$$H_{continuum} = \lambda J N \pi^2/8 + J H_{lattice} + o(1/\lambda), \tag{15}$$

with the lattice model

$$H_{lattice} = \sum_{n<m} \frac{\pi^2/N^2}{\sin^2((n-m)\pi/N)}[(a_n^\dagger - a_m^\dagger)(a_n - a_m) + 1 - P_{n,m}]. \tag{16}$$

In $H_{lattice}$, the $a$'s are the phonon operators of the small oscillations around the crystalline state, and $P_{n,m}$ gives us the HSM. This total lattice model $H_{lattice}$ has a simple spectrum,

$$E_{lattice} = J \sum_{j=1,N} p_j q_j \tag{17}$$

where $p_j = 2\pi(j - 1/2)/N$ are the free fermi ground state q.nos and $q$'s are free particle numbers $2\pi I/N$. Strictly speaking, the thermodynamic limit of the spectrum corresponds to that of the above formula, but it is well known that for the $1/\sin^2$ system, the ABA gives the exact results even for a finite system, unlike say the case of the Toda Lattice. The ferromagnet corresponds to two species of bosons and the antiferromagnet to two species of fermions. The partition function

$$Z_{HSM} = Z_{lattice}/Z_{phonons}, \tag{18}$$

and since $Z_{lattice}$ is hardly more difficult than $Z_{phonons}$ to calculate, we have expressed the difficult problem of $Z_{HSM}$ in a "happy formula".

- **Correlation Functions**

This is a field of considerable recent activity, and I will merely mention a few salient references. The first surprising result in this field came in 1993, when Altshuler and co workers[8] found an exact result for the *dynamical* correlation function of the CSM model at $\beta = 1, 2, 4$, by using supersymmetry techniques of Efetov. As a sample of their results, the $T = 0$ dynamical correlation function for the symplectic ensemble $\beta = 4$, $d = N/L = k_f/\pi$

$$\begin{aligned} < 0|\rho(r,t)\rho(0)|0 > \;&=\; d^2 + d/2 \int_1^\infty d\lambda \int_{-1}^1 d\lambda_1 \int_{-1}^1 d\lambda_2 \cos[2k_f r(\lambda - \lambda_1\lambda_2)] \exp itE \\ &\qquad \frac{(\lambda^2-1)^2(\lambda-\lambda_1\lambda_2)^2}{(\lambda^2+\lambda_1^2+\lambda_2^2-1-2\lambda\lambda_1\lambda_2)^2} \end{aligned} \tag{19}$$

with $E = 4k_f^2(\lambda^2 + \lambda_1^2 + \lambda_2^2 - 1 - 2\lambda_1^2\lambda_2^2)$. These expressions have a very natural interpretation in terms of quasiparticle content of bare particles. Using similar techniques, the correlation function for the HSM was found[14]

$$G(q,t) \equiv < 0|S^z(-q)S^z(q,t)|0 > = \frac{1}{4}\int_0^\alpha d\psi \, \exp it\frac{\pi^2}{2}a(\cosh\psi - a), \tag{20}$$

where $a = 1 - |q|/\pi$ and $\alpha = -\log a$. More recently, the exact dynamical correlations for the case of integer $\beta$ or rational $\beta$ have been found by Forrester[15] and Ha[16], using rather novel techniques involving Jack's polynomials and Selberg integrals. These methods are very original and involved, giving no clue, at least to me, of why they work.

- **Algebraic Structures**

- **Classical Integrability:** We now discuss the Integrability of this class of models. Part of the interest here is that these are emerging alternatives to the current paradigm of Quantum Integrable Systems, namely the Yang Baxter equations[9]. All the methods have been inspired by Moser's treatment of the classical problem, which we first describe. Moser[5] defines, following P. D. Lax, the Lax matrix

$$L_{i,j} = \delta_{i,j}p_i + (1 - \delta_{i,j})i\chi_{i,j} \tag{21}$$

and ask the famous question, is there a $M_{i,j}$ matrix such that

$$\{L_{i,j}, H\}_{P.B.} = i[L, M]_{i,j}. \tag{22}$$

The Hamiltonian $H$ is taken as the kinetic energy plus the usual sum over two body interaction model. This leads to a functional equation for the various functions, which has a large class of solutions, of which the most general one is $V(x) = 1/sn^2(x)$, the Elliptic case. In these cases, $M_{i,j} = \delta_{i,j} \sum_k a_{i,k} - b_{i,j}$ is a function of coordinates alone, as is $\chi_{i,j}$. If $M$ exists then we see that there are infinite conservation laws, since

$$\dot{\overline{Tr(L^n)}} = Tr\{L...L.(L.M - M.L).L...L\} = Tr\{L^n.M - M.L^n\}. \tag{23}$$

We may use the cyclicity of trace to assert the vanishing of above.

• **Quantum Integrability:** Quantum mechanically, we could borrow Moser's matrices and follow through most of above, but we see that since $[L_{i,j}, M_{k,l}] \neq 0$, for general $i, j, k, l$ the cyclicity of trace does not hold, and we must rethink the programme. Calogero and coworkers[17] gave a prescription of quantizing Eq(22), replacing the Poisson brackets by quantum commutators and the right hand side by symmetrized products, i.e. $ab \to 1/2(ab + ba)$. They claim to prove that the determinant of the Lax matrix $d(\lambda) = Det[L_{i,j} - \lambda 1_{i,j}]$, (which has no ordering ambiguities) is conserved $[d(\lambda), H] = 0$, as well as in involution with $d(\lambda')$, i.e. $[d(\lambda), d(\lambda')] = 0$.

Leaving the elliptic case aside, for all the other cases, we have now a variety of new and relatively satisfactory demonstrations of integrability. One method, due to Polychronakos[11] defines commuting Hermitean operators

$$q_i = p_i - i \sum_j \chi_{i,j} \Pi_{i,j} \quad ; \quad H = \sum q_i^2, \tag{24}$$

with $[q_i, q_j] = 0$, and the Majorana exchange operator $\Pi_{i,j}$ acting on spatial coordinates alone.

Another method is due to Shastry and Sutherland[18] is inspired by the Moser form, seeks to express the very asymmetric Lax equation as a simple commutator. This is achieved by enlarging the Hilbert space, by adding fermi degrees of freedom for each original particle. Thus the Lax operator Eq(21), with $p's$ non commuting now, is promoted to a super Lax operator

$$\mathcal{L} = \sum_{i,j} L_{i,j} C_i^\dagger C_j \tag{25}$$

and a super Hamiltonian

$$\mathcal{H} = H_b + \sum_{i,j} M_{i,j} C_i^\dagger C_j, \tag{26}$$

where $H_b$ is the original problem. It is found that the condition for commutation $[\mathcal{L}, \mathcal{H}] = 0$ implies

$$\sum_{i,j} C_i^\dagger C_j \{[L_{i,j}, H_b] - [L, M]_{i,j}\} = 0, \tag{27}$$

with the commutator being the regular ordered object. Thus the Lax equation is just a simple commutator. However this does not quite prove integrability, we need one more operator that commutes with $\mathcal{H}$, say a fermi operator

$$\mu = \sum_i C_i. \tag{28}$$

The condition $[\mathcal{H}, \mu] = 0$ implies $[f(\mathcal{L}, \mu), \mathcal{H}] = 0$, i.e. any operator valued function of $\mathcal{L}$ and $\mu$ commutes with $\mathcal{H}$. The "bosonic projection" $\tilde{f} =< 0|\mu f[\mathcal{L}, \mu]\mu^\dagger|0 >$, with $|0>$ the fermi vaccuum, then commutes with our $H_b$. Thus enlarging the space and projecting down makes two constants of motion (two non abelian symmetries) proliferate in the bosonic problem. The same ideas work for the HSM with an appropriate choice of $L = \chi_{i,j} P_{i,j}$. Two examples of these functions are given, the first $f = \mathcal{L}^n$ giving $\tilde{f} = L_{1,2}L_{2,3}L_{3,4}..L_{n-1,n}$ i.e. an almost trace form. This class of integrals was also found independently by Ujino et. al. [19] for the CSM model. We have another very natural class $f = \{\exp(i\mathcal{L}t)\ \mu\ \exp(-i\mathcal{L}t), \mu^\dagger\}$, which on expanding in 't' gives a series of operators, including the $H_b$. We repeat that although these proofs are very automated, they do not cover the elliptic case, which does remain a slight problem.

• **Annihilators:** Jastrow functions of the type $\prod_{i<j} \phi(x_j - x_i)$ are natural candidates for satisfying a first order differential equation $q_i\psi = 0$, with $q_i = p_i + i\sum_j \partial \log \phi(x_i - x_j)\partial x_i$. We could form a Hermitean and permutationally invariant operator $H = \sum_i q_i^\dagger q_i$, which is typically of the form[20] $H = \sum p_i^2 + \sum_{i,j} V_{i,j} + \sum_{i,j,k} W_{i,j,k}$. The condition for the vanishing of the inconvenient three body term $W$ leads to the same functional equations that arise in the integrability game. These operators have remained a curiosity to a large extent, but only now are we begining to find interesting applications of these.

One application[21] is for calculating *current* correlation functions for the operator $J(r) = \sum_i (p_i\delta(r - x_i) + \delta(x - x_i)p_i)/2m$, where the annihilators give a beautiful and useful alternative representation $J(r) = \sum_i (q_i^\dagger \delta(r - x_i) + \delta(x - x_i)q_i)/2m$, from which very interesting identities follow, such as

$$< 0|J(r)J(0)|0 >= -\frac{\lambda}{r^2} < 0|\rho(r)\rho(0)|0 > . \tag{29}$$

Another class of applications is for the HSM, where it was shown[20] that the wavefunction Eq(8) is annihilated by a large class of operators, amongst which are the local "spin current" operators

$$(\vec{L}_n - i\vec{\psi}_n)|\psi >= 0 \ \text{(any n)} \tag{30}$$

with $\vec{L}_n = \sum_m \chi_{n,m} \vec{S}_n \times \vec{S}_m$ and $\vec{\psi}_n = \sum_m \chi_{n,m} \vec{S}_m$. Further we have a decomposition of the s=1/2 HSM Hamiltonian

$$H = E_0 + \frac{1}{9}\sum_n (\vec{L}_n + i\vec{\psi}_n).(\vec{L}_n - i\vec{\psi}_n) + 3\phi^2(L+1)/2S_{tot}^2. \tag{31}$$

It is in fact easy to see that

$$-i[\vec{L}_n, H] = 1/2 \sum{}' \chi_{n,m} V_{n,m} (\vec{S}_n - \vec{S}_m) + \vec{S}_n \sum{}' \chi_{n,m} V_{n,k} (\vec{S}_k . \vec{S}_m) - \sum{}' \chi_{k,m} V_{n,k} \vec{S}_k (\vec{S}_n . \vec{S}_m), \tag{32}$$

from which it follows that all three components of the "Yangian" $\vec{\Omega} \equiv \sum_n \vec{L}_n$ commute with $H_{HSM}$, without commuting with each other[22], this non abelian extra symmetry exists for finite systems as well, unlike the case of the Bethe chain, wherein such a symmetry only manifests in the thermodynamic limit. The local spin current operators $\vec{L}_n$ may thus also be viewed as the densities of the Yangian $\vec{\Omega}$. These operators are very closely connected to mutually commuting operators $y_n$ that occur in the Yang Baxter theory: $y_n = \lambda_n \sum_m \vec{S}_n . \vec{S}_m / (\lambda_n - \lambda_m)$ via the relation

$$\vec{L}_n = y_n \vec{S}_n + \text{trivial terms.} \tag{33}$$

The annihilator Eq(30) expresses a "step down" relation, rather like a Ward Identity, and is expected to be useful in calculating C.F.'s. It becomes useful when we write it in momentum space as $\vec{L}_q = 2i \frac{\partial}{\partial q} \dot{\vec{S}}_q + \vec{\xi}_q$, with a well characterized $\vec{\xi}_q$. Eq(30) becomes

$$\vec{L}_q |\psi> = i(\pi sign(q) - q) \vec{S}_q |\psi> \tag{34}$$

from which the Greens function of Eq(20) satisfies the equation

$$\frac{\partial^2}{i \partial t \partial q} G(q,t) - (\pi sign(q) - q) G = R(q,t) \tag{35}$$

with $R(q, 0) = 0$. This relation inspires us to look at the rhs from Eq(20), and we find a linear non-autonomous PDE satisfied by G, namely

$$\left[ \frac{\partial^2}{i \partial t \partial q} - (\pi sign(q) - q) \right] G(q,t) = -\frac{i \pi^3 t}{2} \left( 1 - \frac{|q|}{\pi} \right) \left[ \frac{1}{2} \left( 1 - \left( 1 - \frac{|q|}{\pi} \right)^2 \right) + \frac{i3}{\pi^2} \frac{\partial}{\partial t} \right] G(q,t). \tag{36}$$

If we suppliment Eq(36) with the value[10] of $G(q, 0) = -\frac{1}{4} \log(1 - |q|/\pi)$ and $G(0, t) = 0$, then the function is in fact determined. It seems to us that such PDE's are likely in most cases studied and may give added clues to the solvability of the CF's.

• **Summary:** It is becoming increasingly clear that this class of models is the best understood non trivial exactly solvable many body problem. The many interesting features may inspire interesting higher dimensional solutions/approximations. I do believe that some problems remain, and perhaps, at this stage, we just need to understand better what has been done by the large community of $1/r^2$ players. To this community and to my collaborators, I would like to express my thanks for making it such an exciting season.

## References


[1] Proceedings of the 16th Taniguchi symposium, ed. A. Okiji and N. Kawakami (Springer Verlag, Berlin 1994).

[2] F. J. Dyson, J. Math Phys. **3** 1191, 140, 157, 166 (1962).

[3] F. Calogero, J. Math. Phys. **10**, 2191, 2197 (1969).

[4] B. Sutherland, Phys. Rev. A**4** 2019 (1971), J. Math. Phys. **12** 246, 251 (1971).

[5] J. Moser, Adv. Math. **16** 197 (1975).

[6] B. S. Shastry, Phys. Rev. Letts. **60** 639 (1988).

[7] F. D. M. Haldane, Phys. Rev. Letts. **60** 635 (1988).

[8] B. Simons, B. L. Altshuler and P. A. Lee, Phys. Rev. Letts. **70** 4063 (1993).

[9] *Yang Baxter Equations in Integrable Systems*, ed. M Jimbo, World Scientific, Singapore (1989).

[10] F. Gebhard and D. Vollhardt, Phys. Rev. Letts. **59** 1472 (1987).

[11] A. Polychronakos, Phys. Rev. Lets. **69** 703 (1992).

[12] B. Sutherland and B. S. Shastry, Phys. Rev. Letts. **71** 5 (1993).

[13] B. Sutherland, R. Roemer and B. S. Shastry, Phys. Rev. Letts. **71** 2154 (1994).

[14] F.D. M. Haldane and M. Zirnbauer, Phys. Rev. Letts. **71** 4055 (1993).

[15] P. J. Forrester, Nucl. Phys. **B** (1994).

[16] Z.N.C. Ha, Phys. Rev. Letts. **73** 1574 (1994).

[17] F. Calogero, M. Ragnisco and C. Marchioro, Lett. Al. Nuovo Cimento **13** 383 (1975).

[18] B. S. Shastry and B. Sutherland, Phys. Rev. Letts. **70** 4031 (1993).

[19] H. Ujino, K. Hikami and M. Wadati, J. Phys. Soc. Japan **61** 3425 (1992).

[20] B. S. Shastry, Phys. Rev. Letts. **69** 164 (1992).

[21] N. Taniguchi, B. S. Shastry and B. Altshuler, preprint (1995).

[22] F. D. M. Haldane, *et al.*, Phys. Rev. Letts. **69** 2021 (1992).

# THE DILUTE LATTICE MODELS AND WHAT THEY TELL US ABOUT CRITICAL BEHAVIOUR

Paul A. PEARCE
*Mathematics Department, University of Melbourne*
*Parkville, Victoria 3052, Australia*
E-mail: pap@mundoe.maths.mu.oz.au

ABSTRACT

The dilute lattice models are families of exactly solvable two-dimensional lattice models which exhibit some new properties not found in previous exactly solvable lattice models. A general review is given of these models and their significance in understanding critical behaviour. The models are important for at least two reasons. First, the off-critical models yield solvable models in symmetry-breaking fields including a model in the universality class of the Ising model in a magnetic field. Second, these models provide lattice realizations of all the unitary minimal theories of conformal field theory with central charge $c < 1$. In other words, these models give exactly solvable representatives of all possible universality classes for $c < 1$.

## 1. Dilute $A$–$D$–$E$ Lattice Models

The dilute $A$–$D$–$E$ lattice models are families of exactly solvable two-dimensional lattice models. They were introduced independently by Warnaar, Nienhuis and Seaton[1] and by Roche[2] in 1992 before the StatPhys18 meeting in Berlin. At criticality, the models are parametrized by trigonometric functions. The dilute $A$ models also admit an elliptic parameterization[1] and can be solved[3] off-criticality. These models are significant for a number of reasons. We summarize these reasons here and refer to the original literature for more in depth discussions.

Like the $A$–$D$–$E$ models of Pasquier[4], the states $a, b, c, d = 1, 2, \ldots, L$ of the dilute $A$–$D$–$E$ models take values on the $A$–$D$–$E$ Dynkin diagrams of the classical $A$–$D$–$E$ Lie algebras shown in Figure 1. Associated with each $A$–$D$–$E$ graph is an adjacency matrix

$$A_{a,b} = \begin{cases} 1, & a,\, b \text{ connected} \\ 0, & \text{otherwise.} \end{cases} \tag{1.1}$$

Denote the Coxeter number of an $A$–$D$–$E$ graph by $h$. The nonnegative elements $S_a$ of the Perron-Frobenius eigenvector of the adjacency matrix are then given by

$$\sum_b A_{a,b} S_b = 2\cos\left(\frac{\pi}{h}\right) S_a, \qquad h = \begin{cases} L+1, & A_L \\ 2L-2, & D_L \\ 12, 18, 30, & E_{6,7,8}. \end{cases} \tag{1.2}$$

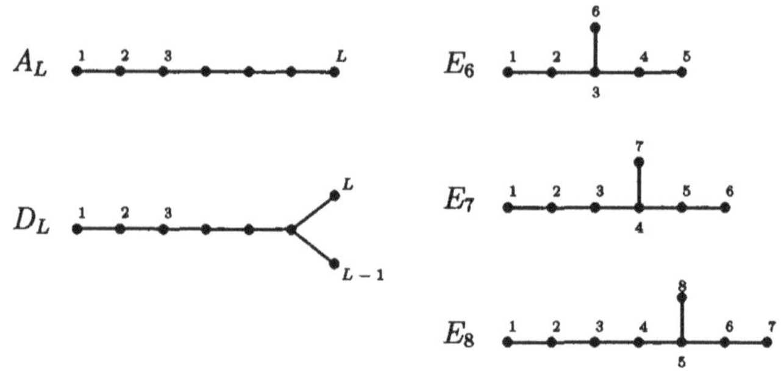


Figure 1. The adjacency graphs of the classical $A$–$D$–$E$ models.

The face weights of the dilute $A$–$D$–$E$ models are defined, in terms of the data of the $A$–$D$–$E$ graph, by

$$\begin{aligned} W\left(\begin{matrix} d & c \\ a & b \end{matrix}\middle| u\right) &= \rho_1(u)\delta_{a,b,c,d} + \rho_2(u)\delta_{a,b,c}A_{a,d} + \rho_3(u)\delta_{a,c,d}A_{a,b} \\ &+ \sqrt{\frac{S_a}{S_b}}\rho_4(u)\delta_{b,c,d}A_{a,b} + \sqrt{\frac{S_c}{S_a}}\rho_5(u)\delta_{a,b,d}A_{a,c} + \rho_6(u)\delta_{a,b}\delta_{c,d}A_{a,c} \\ &+ \rho_7(u)\delta_{a,d}\delta_{c,b}A_{a,b} + \rho_8(u)\delta_{a,c}A_{a,b}A_{a,d} + \sqrt{\frac{S_aS_c}{S_bS_d}}\rho_9(u)\delta_{b,d}A_{a,b}A_{b,c}. \end{aligned} \tag{1.3}$$

The trigonometric functions are

$$\begin{aligned} \rho_1(u) &= 1 + \frac{\sin u \sin(3\lambda - u)}{\sin(2\lambda)\sin(3\lambda)}, & \rho_2(u) &= \rho_3(u) = \frac{\sin(3\lambda - u)}{\sin(3\lambda)} \\ \rho_4(u) &= \rho_5(u) = \frac{\sin u}{\sin(3\lambda)}, & \rho_6(u) &= \rho_7(u) = \frac{\sin u \sin(3\lambda - u)}{\sin(2\lambda)\sin(3\lambda)} \\ \rho_8(u) &= \frac{\sin(2\lambda - u)\sin(3\lambda - u)}{\sin(2\lambda)\sin(3\lambda)}, & \rho_9(u) &= -\frac{\sin u \sin(\lambda - u)}{\sin(2\lambda)\sin(3\lambda)} \end{aligned} \tag{1.4}$$

and the generalized Kronecker delta is

$$\delta_{a,b,c,\dots} = \begin{cases} 1, & a = b = c = \dots \\ 0, & \text{otherwise.} \end{cases} \tag{1.5}$$

Here the spectral parameter $u$ lies in the interval $0 < u < 3\lambda$ where there are two choices for the parameter $\lambda$

$$\lambda = \frac{(h \pm 1)\pi}{4h}. \tag{1.6}$$

Notice that the effective adjacency condition for the dilute models is given by $I + A$ which is analogous to adding a loop to each node of the $A$–$D$–$E$ graph describing the admissible states.

## 2. Dilute Algebra

Let us consider the Yang-Baxter algebra generated by the face operators $X_j(u)$ of the dilute $A$–$D$–$E$ models. If $a = \{a_1, a_2, \ldots\}$ and $a' = \{a'_1, a'_2, \ldots\}$ are consecutive diagonal rows of spins, the local face operators are defined by

$$\langle a|X_j(u)|a'\rangle = W\begin{pmatrix} a_{j-1} & a'_j \\ a_j & a_{j+1} \end{pmatrix}\bigg| u \bigg) \prod_{k\neq j} \delta(a_k, a'_k)\,. \tag{2.1}$$

These face operators satisfy the Yang-Baxter equation

$$X_{j+1}(u)X_j(v)X_{j+1}(v-u) = X_j(v-u)X_{j+1}(v)X_j(u) \tag{2.2}$$

and the locality condition

$$X_j(u)X_k(v) = X_k(v)X_j(u), \qquad |j-k| \geq 2. \tag{2.3}$$

These two relations are the defining relations of a Yang-Baxter algebra.

The Yang-Baxter or face operators of the dilute $A$–$D$–$E$ models can be decomposed as

$$X_j(u) = \sum_{n=1}^{9} \rho_n(u) X_j^{(n)}. \tag{2.4}$$

Using a convenient graphical notation[5] we can write the $u$-independent operators $X_j^{(n)}$ as

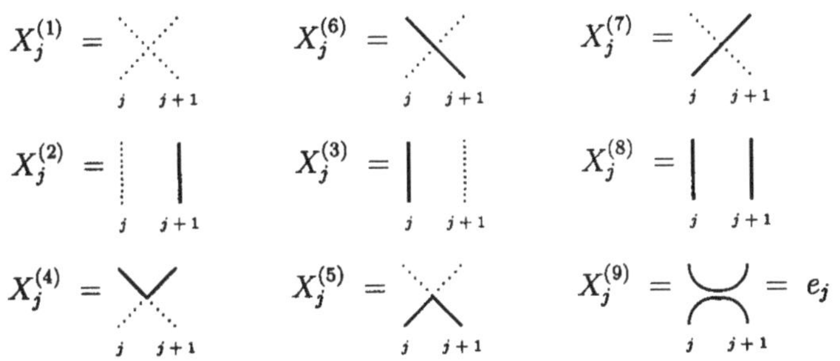

(2.5)

(2.6)

(2.7)

It is easily seen that the operators $X_j^{(9)} = e_j$ satisfy the defining relations of a Temperley-Lieb algebra[6]:

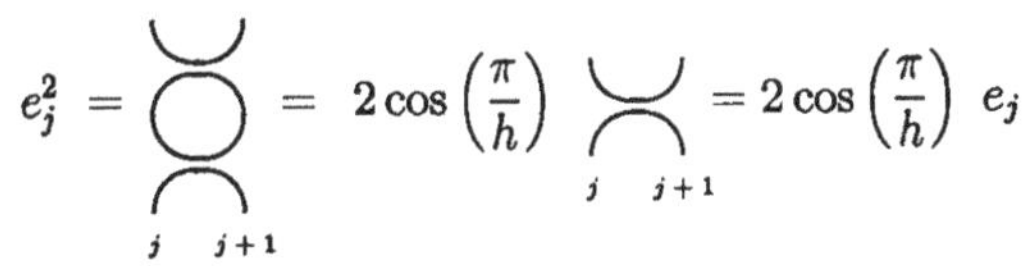

(2.8)

$$e_je_{j+1}e_j =$$ 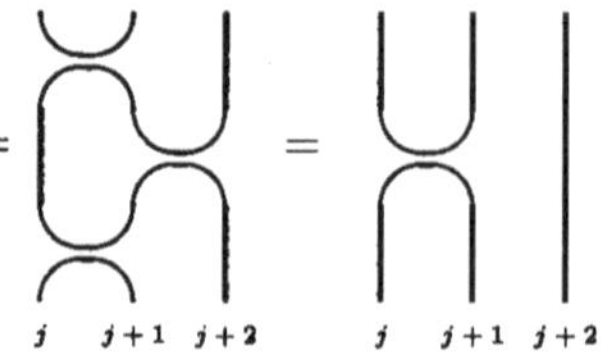 $$= e_j \tag{2.9}$$

The Temperley-Lieb algebra is therefore a subalgebra of the algebra generated by the operators $X_j^{(n)}$. The operators $X_j^{(n)}$ in fact generate a dilute Temperley-Lieb algebra, which is a particular example of more general multicolour braid-monoid algebras[7], and satisfy many other relations as suggested by their graphical representations such as

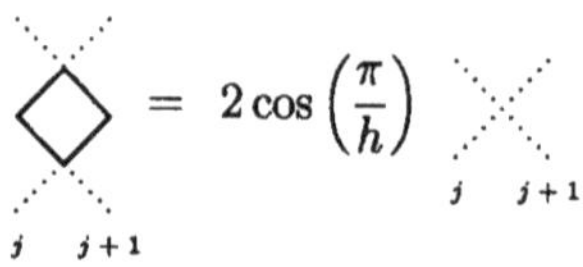 (2.10)

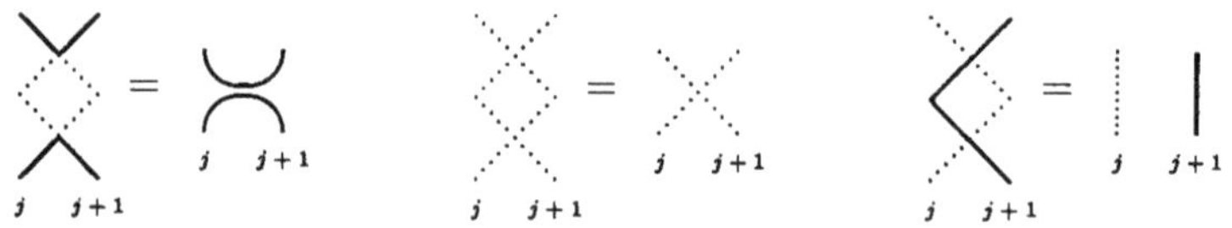 (2.11)

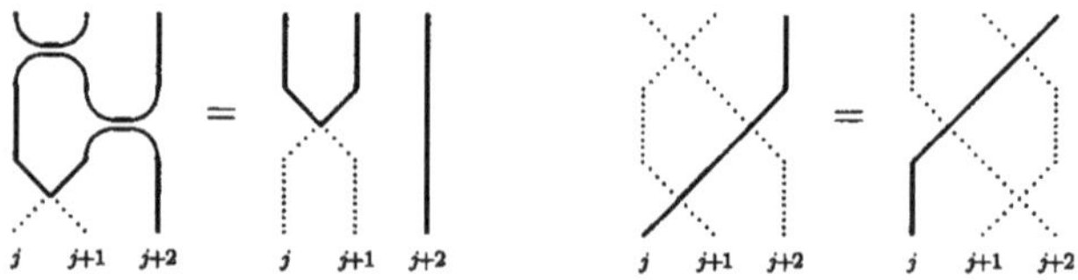 (2.12)

These relations have a natural interpretation in terms of isotopy of strings in knots and links. The defining relations of the dilute Temperley-Lieb algebra suffice to ensure that the dilute $A$–$D$–$E$ models satisfy the Yang-Baxter equations.

## 3. Unitary Minimal Series

In appropriate regimes, the critical behaviours of the $A$–$D$–$E$ models are described by the unitary minimal series of conformal field theories[8,9] with central charges $c < 1$. More specifically, the modular invariant partition functions (MIPFs) of these models on a torus belong to either of two series classified[10] by a pair $(G, G')$ of $A$–$D$–$E$ Lie algebras:

| Critical | Tricritical |
|---|---|
| $\lambda = \frac{(h+1)\pi}{4h}, \quad c = 1 - \frac{6}{h(h-1)}$ | $\lambda = \frac{(h-1)\pi}{4h}, \quad c = 1 - \frac{6}{h(h+1)}$ |

$$(A,G') = \begin{cases} (A_{h-2}, A_{h-1}) \\ (A_{h-2}, D_{(h+2)/2}) \\ (A_{10}, E_6) \\ (A_{16}, E_7) \\ (A_{28}, E_8) \end{cases} \qquad (G,A') = \begin{cases} (A_{h-1}, A_h) \\ (D_{(h+2)/2}, A_h) \\ (E_6, A_{12}) \\ (E_7, A_{18}) \\ (E_8, A_{30}) \end{cases} \tag{3.1}$$

Here the Coxeter number $h$ belongs to the Lie algebra of $A$–$D$–$E$ type.

It has been known for some time that the MIPFs of Pasquier's $A$–$D$–$E$ models all belong to the critical series. Recently, however, it has been shown numerically[11] that the MIPFs of the dilute $A$–$D$–$E$ models exhaust both the critical and the tricritical series. This result can be paraphrased as follows:

> *The dilute $A$–$D$–$E$ lattice models provide an exactly solvable representative for each universality class of the theories characterized by $c < 1$.*

In particular, the dilute $D_4$ model with $\lambda = 7\pi/24$ lies in the universality class of the critical 3-state Potts model, has $c = 4/5$ and the MIPF is given by $(A_4, D_4)$. By comparison, the dilute $D_4$ model with $\lambda = 5\pi/24$ lies in the universality class of the tricritical 3-state Potts model, has $c = 6/7$ and the MIPF is given by $(D_4, A_6)$.

## 4. Off-Critical Solution in Symmetry-Breaking Field

The dilute $A_L$ models admit an elliptic extension[1] and can be solved[3] off-criticality for the free energies, local height probabilities and order parameters. It is therefore possible to obtain the associated critical exponents directly. Unlike the $A_L$ models of Andrews, Baxter and Forrester[12] (ABF) where the elliptic nome plays the role of a temperature-like variable $t$, the elliptic nome of the dilute $A_L$ models (at least when $L$ is odd) plays the role of a symmetry breaking field $h$. Specifically, results[12,13] for the order parameters of the ABF models give

$$R^{(k)} \sim t^{\beta_k}, \qquad \beta_k = \frac{(k+1)^2 - 1}{8L}, \qquad k = 1, 2, \ldots, L-2 \tag{4.1}$$

which for $L = 3$ and $k = 1$ yields the familiar result for the magnetization of the Ising model

$$m \sim t^{\beta}, \qquad \beta = 1/8. \tag{4.2}$$

In remarkable contrast, when $L$ is odd, results[3] for the order parameters of the dilute $A_L$ models give

$$\overline{R}^{(k)} \sim h^{1/\delta_k}, \qquad \delta_k = \frac{3L(L+2)}{(k+1)^2-1}, \qquad k = 1, 2, \ldots, L-2. \tag{4.3}$$

The dilute $A_3$ model in fact[14] lies in the universality class of the Ising model in a magnetic field and is therefore of particular interest. In this case the elliptic nome is identified as the leading magnetic field and, accordingly, the direct calculation[3] of the magnetization as a function of field leads for $L = 3$ and $k = 1$ to the celebrated critical exponent

$$m \sim h^{1/\delta}, \qquad \delta = 15. \tag{4.4}$$

Previously, this critical exponent had only been obtained by invoking scaling laws.

*Dilute $A_3$ lattice model* = 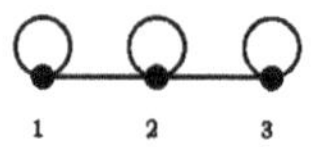 = *Universality class of Ising model in a magnetic field*



## 5. Magnetic Ising Model, Dilute $A_3$ and $E_8$

The connections between the dilute $A_3$ model, the magnetic Ising model and the exceptional Lie algebra $E_8$ are particularly interesting. In 1989 Zamolodchikov[15] showed that the critical Ising model in a magnetic field is an integrable quantum field theory containing 8 particles with masses related to $E_8$. Subsequently, in 1994 Bazhanov, Nienhuis and Warnaar[16] showed that, subject to a certain string hypothesis, a Bethe ansatz study of the dilute $A_3$ model in the scaling limit leads to thermodynamic Bethe ansatz (TBA) equations[17] with an $E_8$ structure.

In a seemingly unrelated development in 1993, Kedem, Klassen, McCoy and Melzer[18] conjectured fermionic representations for the $c = \frac{1}{2}$ Virasoro characters $\chi^{(3,4)}_{r,s}$ by considering the coset conformal field theory based on $E_8$. In particular, they conjectured the $E_8$ Rogers-Ramanujan identity

$$q^{1/48}\chi^{(3,4)}_{1,1}(q) = \sum_{n_1,\ldots,n_8=0}^{\infty} \frac{q^{\boldsymbol{n}^T C^{-1}_{E_8} \boldsymbol{n}}}{(q)_{n_1}\cdots(q)_{n_8}} = \frac{1}{(q)_\infty}\sum_{j=-\infty}^{\infty}\left\{q^{12j^2+j} - q^{12j^2+7j+1}\right\} \tag{5.1}$$

where $(q)_m = \prod_{k=1}^{m}(1-q^k)$ for $m \geq 1$, $(q)_0 = 1$ and $C_{E_8}$ is the Cartan matrix

$$C_{E_8} = 2I - A_{E_8}, \qquad C^{-1}_{E_8} = \begin{pmatrix} 2 & 3 & 4 & 5 & 6 & 4 & 2 & 3 \\ 3 & 6 & 8 & 10 & 12 & 8 & 4 & 6 \\ 4 & 8 & 12 & 15 & 18 & 12 & 6 & 9 \\ 5 & 10 & 15 & 20 & 24 & 16 & 8 & 12 \\ 6 & 12 & 18 & 24 & 30 & 20 & 10 & 15 \\ 4 & 8 & 12 & 16 & 20 & 14 & 7 & 10 \\ 2 & 4 & 6 & 8 & 10 & 7 & 4 & 5 \\ 3 & 6 & 9 & 12 & 15 & 10 & 5 & 8 \end{pmatrix}. \tag{5.2}$$

This conjecture has been proved[19] by taking the limit $L \to \infty$ starting with the finite polynomial identity

$$\sum_{(\boldsymbol{n},\boldsymbol{m})_L} q^{\boldsymbol{n}^T C_{\mathrm{E}_8}^{-1} \boldsymbol{n}} \prod_{i=1}^{8} \begin{bmatrix} n_i+m_i \\ n_i \end{bmatrix}_q$$

$$= \sum_{j,k=-\infty}^{\infty} \left\{ q^{12j^2+j+k(k+8j)} \begin{bmatrix} L \\ k, k+8j \end{bmatrix}_q - q^{12j^2+7j+1+k(k+8j+2)} \begin{bmatrix} L \\ k, k+8j+2 \end{bmatrix}_q \right\} \quad (5.3)$$

and using the results

$$\lim_{N\to\infty} \begin{bmatrix} N \\ m \end{bmatrix}_q = \frac{1}{(q)_m}, \qquad \lim_{L\to\infty} \sum_{k=0}^{\infty} q^{k(k+a)} \begin{bmatrix} L \\ k, k+a \end{bmatrix}_q = \frac{1}{(q)_\infty}. \quad (5.4)$$

Here the Gaussian polynomials are given by

$$\begin{bmatrix} N \\ m \end{bmatrix}_q = \frac{(q)_N}{(q)_m (q)_{N-m}}, \qquad \begin{bmatrix} N \\ m_1, m_2 \end{bmatrix}_q = \frac{(q)_N}{(q)_{m_1} (q)_{m_2} (q)_{N-m_1-m_2}} \quad (5.5)$$

and the constrained sum is over all solutions to the TBA system $(\boldsymbol{n}, \boldsymbol{m})_L$:

$$\boldsymbol{n} + \boldsymbol{m} = \frac{1}{2}(A_{\mathrm{E}_8}\boldsymbol{m} + L\boldsymbol{e}_1) \quad (5.6)$$

$$\boldsymbol{n} = (n_1, n_2, \ldots, n_8),\ \boldsymbol{m} = (m_1, m_2, \ldots, m_8),\ \boldsymbol{e}_1 = (1, 0, \ldots, 0) \in \mathbb{Z}^8.$$

The polynomials on the right-side of (5.3) are precisely the (bosonic) polynomials that occur in the corner transfer matrix calculation[3] of the magnetization of the dilute $A_3$ models. The (fermionic) representation[19] of these same polynomials on the left-side of (5.3) makes explicit the hidden $E_8$ structure in the dilute $A_3$ lattice model and the critical Ising model in a magnetic field. Similarly, the dilute $A_4$ and $A_6$ reveal $E_7$ and $E_6$ structures respectively but the situation for the general case of dilute $A_L$ is completely mysterious.

## 6. Conclusion

Perhaps surprisingly, there are still many new solvable lattice models being discovered. Among these, the dilute lattice models exhibit some very interesting properties. We have summarized some of these in this article. The dilute lattice models in fact consist of many families of solvable lattice models. Indeed, Warnaar and Grimm[21,22] have recently constructed higher rank dilute lattice models which generalize the higher rank models of Jimbo, Miwa and Okado[23]. No doubt when these models are studied further they will also exhibit some interesting properties.

## Acknowledgements

It is a pleasure to acknowledge fruitful collaborations with U. Grimm, B. Nienhuis, D.L. O'Brien, K.A. Seaton, S.O. Warnaar and Yu-kui Zhou. I thank Ole Warnaar for reading the manuscript. This work is supported by the Australian Research Council.

## References

1. S. O. Warnaar, B. Nienhuis and K. A. Seaton, Phys. Rev. Lett. **69** (1992) 710.
2. Ph. Roche, Phys. Lett. **B4** (1992) 929.
3. S. O. Warnaar, P. A. Pearce, K. A. Seaton and B. Nienhuis, J. Stat. Phys. **74** (1994) 469.
4. V. Pasquier, Nucl. Phys. **B28** (1987) 162; J. Phys. A **20** (1987) L1229, 5707.
5. L. Kauffman, Topology **26** (1987) 395.
6. H. N. V. Temperley and E. H. Lieb, Proc. Roy. Soc. (London) **A322** (1971) 251.
7. U. Grimm and P. A. Pearce, J. Phys. A **26** (1993) 7435.
8. A. A. Belavin, A. M. Polyakov and A. B. Zamolodchikov, Nucl. Phys. **B241** (1984) 333.
9. D. Friedan, Z. Qiu and S. Shenker, Phys. Rev. Lett. **52** (1984) 1575; in "Vertex Operators in Mathematics and Physics", eds. J. Lepowsky, S. Mandelstam and I.M. Singer, Springer, 1984.
10. A. Capelli, C. Itzykson and J.-B. Zuber, Nucl. Phys. **B280** (1987) 445; Comm. Math. Phys. **113** (1987) 1.
11. D. L. O'Brien and P. A. Pearce, Lattice Realizations of Unitary Minimal Modular Invariant Partition Functions, submitted to J.Phys. A (1995).
12. G. E. Andrews, R. J. Baxter and P. J. Forrester, J. Stat. Phys. **35** (1984) 193.
13. D. A. Huse, Phys. Rev. B **30** (1984) 3908.
14. S. O. Warnaar, B. Nienhuis and K. A. Seaton, Int. J. Mod. Phys. **B7** (1993) 3727.
15. A. B. Zamolodchikov, Adv. Stud. in Pure Math. **19** (1989) 1; Int. J. Mod. Phys. A 4 (1989) 4235.
16. V. V. Bazhanov, B. Nienhuis and S. O. Warnaar, Phys. Lett. **B322** (1994) 198.
17. V. V. Bazhanov and N. Yu. Reshetikhin, Prog. Theor. Phys. Suppl. **102** (1990) 301.
18. R. Kedem, T. R. Klassen, B. M. McCoy and E. Melzer, Phys. Lett. **B304** (1993) 263; **B307** (1993) 68.
19. S. O. Warnaar and P. A. Pearce, J. Phys. A **27** (1994) L891.
20. U. Grimm, J. Phys. A **27** (1994) 5897; Lett. Math. Phys **32** (1994) 183;
21. S.O. Warnaar, Nucl. Phys. B. **B435** 463 (1995).
22. S. O. Warnaar and U. Grimm, Nucl. Phys. **B435** (1995) 482; Yang-Baxter algebras based on the two-colour BWM algebra, hep-th/9506119.
23. M. Jimbo, T. Miwa and M. Okado, Commun. Math. Phys. **116** (1988) 507–25.

# STATISTICAL PROBLEMS IN THEORY OF KNOTS AND NONCOMMUTATIVE RANDOM WALKS

Sergei NECHAEV

*L.D. Landau Institute for Theoretical Physics,*
*117940 Moscow, Russia*
E-mail: nechaev@landau.ac.ru

ABSTRACT

The probabilistic problems in topology have been considered on the following examples: 1. Limit distribution of Markov chains on local noncommutative discrete groups; 2. Statistics of random walks on multiconnected manifolds; 3. Distribution of randomly generated lattice knots over different topological states characterized by Kauffman algebraic invariants. Spin-glass-like features of statistical systems with "topological disorder" have been discussed.

## 1. Entropy of knots

The great progress during last decade in construction of topological invariants of knots and links (Jones, HOMFLY, Vassiliev) and their deep relation to the statistical physics of integrable systems made the subject of invention of new series of knot and link invariants very popular.

There is, however, completely different aspect of the problem, which is hardly ever touched in the mathematical literature, but started recently to attract attention of physicists. We call this aspect "the problem of knot entropy". In other words, we are aimed in calculation of the probabilitiy distribution associated with different homotopy classes of randomly generated knots.

One of possible approaches to this huge task suggests to deal with slightly different but more well-defined problem. Namely, one can represent knots by braids and consider the distribution of corresponding topological invariants of knots generated by "random braids" (i.e. by braids with uniform random choice of generators). Thus we can study the connection between limit distribution for products of noncommutative random matrices – generators of the braid group – and limit distribution of powers of algebraic knot invariants. On this way we can take a look on simplest problems of knot statistics from the point of view of the random walks on hyperbolic groups.

One of our goals is as follows: we show that a lot of non-trivial properties of statistics of randomly generated knots can be explained in context of the random walks over the elements of some local noncommutative group [1].

Another reason which forces us to consider the limit distributions (and conditional limit distributions) of Markov chains on local noncommutative discrete groups is due to the fact that this class of problems could be regarded as a first step of a consistent harmonic analysis on multiconnected manifolds (like Teichmüller space). For that class of problems some results concerning statistics of random walks entangled with regular array of topological obstacles on the plane are obtained and corresponding nonabelian topological invariants are constructed using the conformal methods.

It is noteworthy that topological problems mentioned above could be reformulated in spirit of statistical physics of ordinary spin systems with quenched disorder. In order to show that we compute the probability for a lattice knot to be in specific topological state under a random path closure. For identification of topological state of randomly generated paths we use Kauffman representation of Jones algebraic invariants.

The connections between problems of knot entropy and statistics of entangled random walks can be schematicaly shown as follows:

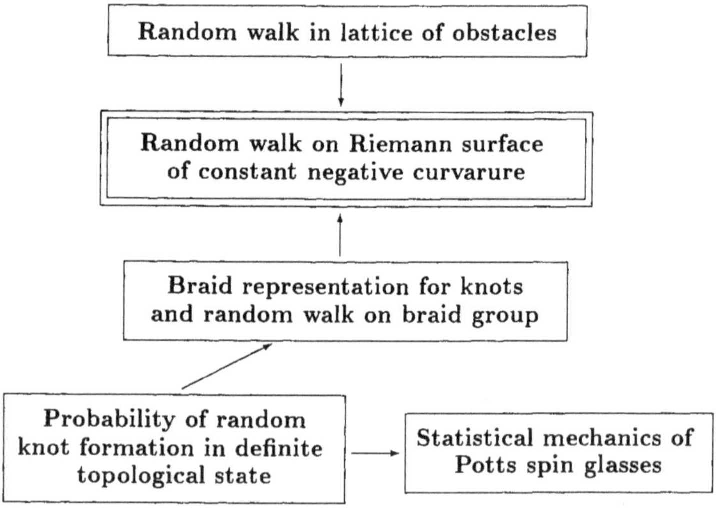


## 2. Formulation of the problem and statistical models

Let us characterize the knot by the topological invariant, $G_0$, which depends only on the topological state of the knot and does not depend on its shape. Entropy, $S$, of some knot, characterized by $G_0$, equals to

$$S(G_0) = \ln Z(G_0)$$

where $Z(G_0)$ is the partition function, i.e. available volume in phase space for closed random trajectories in 3D with fixed topological state; $Z(G_0)$ can be formally written as follows

$$Z(G_0) = \int \delta\left(G\{\Gamma\} - G_0\right) D\{\Gamma\} \tag{1}$$

The path integral Eq.(1) means that the summation is spreaded over all conformations $\Gamma$ of trajectories in 3D and $\delta$-function should be understood as an indicator:

$$\delta\left(G\{\Gamma\} - G_0\right) = \begin{cases} 1 & \text{if } G\{\Gamma\} = G_0 \\ 0 & \text{if } G\{\Gamma\} \neq G_0 \end{cases}$$

i.e., it cuts off the conformations with prescribed topology (or, more rigorous, with prescribed topological invariant). Hence we come to the problem of calculation of the set of trajectories with specific topological state.

We consider the posed general problem on example of three different topological models which allow rather complete analytic examination.

### *2.1. Random Walks on Local Groups*

The investigation of probabilistic properties of algebraic knot invariants we begin with the consideration of statistics of the random loops ("brownian bridges") on the simplest noncommutative groups. In most general way the problem can be formulated as follows.

Take the discrete group $\mathcal{G}_n$ with a fixed finite number of generators $\{g_1, \ldots, g_{n-1}\}$. Let $\nu$ be the uniform distribution on the set $\{g_1, \ldots, g_{n-1}, g_1^{-1}, \ldots, g_{n-1}^{-1}\}$. For convenience we suppose $h_j = g_i$ for $j = i$ and $h_j = g_i^{-1}$ for $j = i + n - 1$; $\nu(h_j) = \frac{1}{2n-2}$ for any $j$. We construct the (right-hand) random walk (the random word) on $\mathcal{G}_n$ with a transition measure, $\nu$, i.e., the Markov chain $\{\xi_n\}$, $\xi_0 = e \in \mathcal{G}_n$ and $\text{Prob}(\xi_j = u|\xi_{j-1} = v) = \nu(v^{-1}u) = \frac{1}{2n-2}$. It means that with the probability $\frac{1}{2n-2}$ we add the element $h_{\alpha_N}$ to the given word $h_{N-1} = h_{\alpha_1}h_{\alpha_2}\ldots h_{\alpha_{N-1}}$ from the right-hand side [a].

**Definition 1** *The random word $W$ formed by $N$ letters taken independently with the uniform probability distribution $\nu = \frac{1}{2n-2}$ from the set $\{g_1, \ldots, g_{n-1}, g_1^{-1}, \ldots, g_{n-1}^{-1}\}$ is called* **the brownian bridge** (BB) *of length $N$ on the group $\mathcal{G}_n$ if the length of the shortest noncontractible* (**primitive**) *word* [b] *of $W$ is identical to unity.*

The most attention is paid to the following two questions:

1. What is the probability $P(N)$ of the brownian bridge on the group $\mathcal{G}_n$.
2. What is the conditional probability distribution $P(\kappa, m|N)$ of the fact that the subword $W'$ consisting of first $m$ letters of the $N$-letter word $W$ has the primitive length $\kappa$ under the condition that the whole word $W$ is the brownian bridge on the group $\mathcal{G}_n$. (Below the conditional distribution on BB is denoted as $P(\kappa, m|N)$).

[a] Analogously we can construct the left-hand side Markov chain.

[b] The *length* of the word is the total number of used letters while the *primitive word* is the shortest word remaining after applying of all group relations.

**Braids**. The braid group $B_n$ of $n$ strings has $n-1$ generators $\{\sigma_1, \sigma_2, \ldots, \sigma_{n-1}\}$ with the following relations:

$$\begin{aligned} &\sigma_i\sigma_{i+1}\sigma_i = \sigma_{i+1}\sigma_i\sigma_{i+1} && (1 \le i < n-1) \\ &\sigma_i\sigma_j = \sigma_j\sigma_i && (|i-j| \ge 2) \\ &\sigma_i\sigma_i^{-1} = \sigma_i^{-1}\sigma_i = e && \end{aligned} \tag{2}$$

Let us mention that:

– The word written in terms of "letters" — generators from the set $\{\sigma_1, \ldots, \sigma_{n-1}, \sigma_1^{-1}, \ldots, \sigma_{n-1}^{-1}\}$ gives a particular *braid.*

– The *length* of the braid is the total number of used letters, while the *minimal irreducible length* called below as the "primitive word" is the shortest noncontractible length of a particular braid which remains after applying of all possible group relations Eq.(2). Diagramatically braid can be represented as a set of crossing strings going from top to bottom appeared after subsequent "gluing" of braid generators.

– The closed braid appears after gluing the "upper" and the "lower" free ends of the braid on the cylinder.

– Any braid corresponds to some knot or link, but this correspondence is not mutually single valued and each knot or link can be represented by infinite series of different braids.

It is well known that for the free group the corresponding problem can be mapped to the investigation of the random walks on simply connected (Cayley) tree. In case of braid group $B_n$ the more complicated group structure does not allow us to use the same simple geometrical image directly. Nevertheless the problem of the limit distribution for the random walks on $B_n$ can be reduced to the consideration of the random walk on some graph $\Gamma$. In case of the group $B_3$ we are able to construct this graph evidently, while for the group $B_n$ $(n \ge 4)$ we give upper estimations for the limit distribution of the random walks considering the walks on so-called *locally free groups.*

**Definition 2** *The group $\mathcal{F}_{n+1}(d)$ we call* **the locally free** *if generators, $\{f_1, \ldots, f_n\}$ obey the following commutation relations:*

(a) *Each pair $(f_j, f_k)$ generates the free subgroup of the group $\mathcal{F}_n$ if $|j-k| < d$;*

(b) *$f_j f_k = f_k f_j$ for $|j-k| \ge d$*

(Below we restrict ourselves with the case $d = 2$ only).

**Theorem 1** *The limit probability distribution for the $N$-step random walk on the group $\mathcal{F}_{n+1}$ to have the minimal irreducible length $\mu$ is*

$$\begin{aligned} \mathcal{P}(\mu, N) &= \frac{9}{2\pi\sqrt{6\pi N^3}} e^{-N/6} \mu \sinh\mu \exp\left(-\frac{3\mu^2}{2N}\right) && (n=3) \\ \mathcal{P}(\mu, N) &= \frac{\text{const}}{N^{3/2}} \left(h(pq)^{1/2}\right)^N \left(\frac{q}{p}\right)^{\mu/2} \exp\left(-\frac{\mu^2 h}{4N}\right) && (n \gg 1) \end{aligned} \tag{3}$$

where $h = 2 + \dfrac{r}{(pq)^{1/2}}$ and the values of $p$, $q$, $r$ are as follows:

$$r = \frac{2n-8}{2n-1}; \qquad p = \frac{7}{8(2n-1)}; \qquad q = \frac{49}{8(2n-1)}$$

**Corollary 1** *The* Eq.(3) *gives the upper estimation for the limit distribution of the primitive word on the group $B_n$ (see the definition below) for $n \gg 1$.*

ALGEBRAIC INVARIANTS OF KNOTS. Take a knot diagram $K$ in general position on the plane. Let $f[K]$ be the topological invariant of the knot $K$. One of possible ways of the Alexander knot invariant construction using the braid group representation can be realized in the following steps.

Write the generators of the braid group in the so-called Magnus representation

$$\sigma_j \equiv \hat{\sigma}_j = \begin{pmatrix} 1 & 0 & \cdots & & \\ 0 & \ddots & & & \\ \vdots & & \boxed{\text{A}} & & \vdots \\ & & & \ddots & 0 \\ & & \cdots & 0 & 1 \end{pmatrix} \leftarrow j\text{th row}; \quad A = \begin{pmatrix} 1 & 0 & 0 \\ t & -t & 1 \\ 0 & 0 & 1 \end{pmatrix} \tag{4}$$

Now the Alexander polynomial of the knot represented by the closed braid $W = \prod_{j=1}^{N} \sigma_{\alpha_j}$ of the length $N$ one can get as follows

$$(1+t+t^2+\ldots+t^{n-1})\,\nabla(t)\{A\} = \det\left[\prod_{j=1}^{N} \sigma_{\alpha_j} - e\right], \tag{5}$$

where the index $j$ runs "along the braid", i.e. labels the number of used generators, while the index $\alpha = \{1,\ldots,n-1,n,\ldots,2n-2\}$ marks the set of braid generators ("letters") ordered as follows $\{\sigma_1,\ldots,\sigma_{n-1},\sigma_1^{-1},\ldots,\sigma_{n-1}^{-1}\}$. In our further investigations we repeatedly address to that representation.

Let us stress that in general the length of the minimal irreducible length of the braid, introduced above, is not related directly to any topological knot invariants but we show below that newertheless the "primitive word" can be served as a well defined characteristic of the "knot complexity". The "primitive word" has the simple topological sense which can be expressed in the following necessary condition. If the "primitive word" of some closed braid of $n$ srtings has the unit length then this braid belongs to the "trivial" class and the corresponding knot is represented by the set of $n$ disjoint unentangled trivial loops for sure.

We are interested in the limit behavior of knot or link invariants when the length of the corresponding braid tends to infinity, i.e., when the braid "grows". In that case we can rigorously define some more simple topological characteristics than the algebraic invariant which we call the "knot complexity".

**Definition 3** *Call the* **knot complexity**, $\eta$, *the power of some algebraic invariant,* $f_K(t)$ (Alexander, Jones, HOMFLY) (see also [2])

$$\eta = \lim_{|t| \to \infty} \frac{\ln f_K(t)}{\ln |t|} \tag{6}$$

**Remark.** By definition, the "knot complexity" takes one and the same value for rather broad class of topologically different knots corresponding to algebraic invariants of one and the same power, being from that point of view more weak topological characteristics than complete algebraic polynomial.

The polynomial invariant can give rather complete information about the knot topology, but dealing with statistics of randomly generated knots very often we do not require such detailed data and look for more rough characteristics of "topologically similar" knots.

Let us summarize the advantages of the quantity introduced in Eq.(6) with respect to the corresponding topological invariant $f_K(t)$:

i) One and the same value of $\eta$ characterizes a narrow class of "topologically similar" knots which is however much broader than the class represented by the polynomial invariant $X(t)$. This allows one to introduce the smoothed measures and distribution functions for $\eta$.

ii) The knot complexity $\eta$ describes correctly, (at least from the physical point of view) the limit cases: $\eta = 0$ corresponds to "weakly entangled" trajectories while $\eta \sim N$ matches the system of "strongly entangled" paths. The later case has been discussed in details in [1-2].

iii) The knot complexity keeps all nonabelian properties of the polynomial invariants.

Our goal here concerns the estimation of the limit probability distribution of "knot complexity" $\eta$ for knots obtained by randomly generated closed braids of length $N$. Let us stress that we essentially simplify the general problem "of knot entropy": namely, we insert an additional requirement that the knot should be represented by a braid from the group $B_n$ without fail.

**Statement 1** *The complexity* $\eta$ *of any known algebraic invariants* (Alexander, Jones, HOMFLY) *for knots represented by* $B_n$*-braids of length* $N \gg 1$ *with uniform distribution over generators has the following limit behavior:*

$$P(\eta, N) \sim \frac{\text{const}}{N^{3/2}} \exp\left(-\frac{(\eta - \alpha(n)N)^2}{\beta(n)N}\right) \tag{7}$$

*where* $\alpha(n)$ *and* $\beta(n)$ *are some numerical constants depending on* $n$ *only.*

### *2.2. Random Walks in Regular Lattices of Obstacles*

Let us apply conformal methods for calculating of the distribution function for the random walk in the regular lattice of topological obstacles on the complex plane

$w = u + iv$; the elementary cell of the lattice is the equal-sided triangle with the side length $c$ (see [3] for details).

Introduce the distribution function $P(w_0, w, L|\text{hom})$ which defines the probability of the fact that the trajectory of random walk starting at the point $w_0$ comes after "time" $L = Na$ to the point $w$ and *all paths going from $w_0$ to $w$ belong to the same homotopy class with respect to the lattice of obstacles.* Formally we can write the diffusion equation

$$\frac{a}{4}\Delta_w P(w, L|\text{hom}) = \frac{\partial}{\partial L} P(w, L|\text{hom}) \tag{8}$$

with the initial and the normalization conditions:

$$P(w, L = 0|\text{hom}) = \delta(z_0);$$

$$\sum_{\text{hom}} P(w_0, w, L|\text{hom}) = \frac{1}{\pi a L} \exp\left(-\frac{|w - w_0|^2}{aL}\right)$$

but it is very hard even to formulate the problem analytically because of the absence of well-defined representation of the topological invariant. However one can use the conformal methods to find the asymptotic solution of Eq.(8) when $L \gg a$. In particular, we are interested in the probability to find the closed path of length $L$ to be unentangled in the lattice of obstacles.

We construct for the multiconnected complex plane $w$ the *universal covering space* $\zeta = \zeta(w)$; $\zeta = \xi + i\eta$, which does not contain the branching point in any finite domain. Due to conformal invariance of Brownian motion, the new random process in the covering space is random again but in metric-dependent "new time". This covering space coincides with the so-called free group $\Gamma_2$ and can be realized in the upper half-plane $\text{Im}(\zeta) > 0$ of the plane $w$ as a discrete subgroup of group of motions of Lobachevskii plane [c]. The halfplane $\text{Im}(\zeta) > 0$ does not contain any branching points and consists of infinite set of Riemann sheets, each of them having the form of zero-angled triangle. These Riemann sheets correspond to the fibre bundle of $w$.

The conformal approach gives us a well defined nonabelian topological invariant — the noneuclidean distance between initial and final points of the trajectory in the covering space. Thus, the diffusion equation for the distribution function $P(\zeta, L)$ in the covering space $\zeta$ with given initial point $\zeta_0$ reads

$$\frac{a}{4}\frac{\partial^2}{\partial\zeta\partial\bar{\zeta}} P(\zeta, \zeta_0, L) = |w'(\zeta)|^2 \frac{\partial}{\partial L} P(\zeta, \zeta_0, L) \tag{9}$$

where we took into account that under the conformal transform the Laplace operators is transformed as follows

$$\Delta_w = \left|\frac{d\zeta}{dw}\right|^2 \Delta_\zeta \qquad \text{and} \qquad \left|\frac{d\zeta}{dw}\right|^2 = \frac{1}{|w'(\zeta)|^2}$$

[c]The group $\Gamma_2$ is the free subgroup of the modular group $\mathsf{PSL}(2, \mathbb{Z})$ and has topological structure of Cayley tree.

In particular, the value $P(\zeta = \zeta_0, \zeta_0, L)$ gives the probability for the path of length $L$ to be unentangled (i.e. contractible to the point) in the lattice of obstacles.

The expression for the Jacobian $|w'(\zeta)|^2$ one can find using the properties of Jacobi Theta-functions. Finally we get

$$|w'(\zeta)|^2 = c^2 h^2 \left|\theta_1'\left(0, e^{i\pi\zeta}\right)\right|^{8/3}; \qquad h = \frac{\pi}{B\left(\frac{1}{3}, \frac{1}{3}\right)} \tag{10}$$

where

$$\theta_1(\chi, e^{i\pi\zeta}) = 2e^{i\frac{\pi}{4}\zeta} \sum_{n=0}^{\infty} (-1)^n e^{i\pi n(n+1)\zeta} \sin(2n+1)\chi \tag{11}$$

Now perform the conformal transform of the upper halfplane $\mathrm{Im}\zeta > 0$ to the interior of the unit circle on the complex plane $\tilde{\zeta}$ in order to use the symmetry properties of the system:

$$\zeta(\tilde{\zeta}) = e^{-i\frac{\pi}{3}} \frac{\tilde{\zeta} - e^{i\frac{2\pi}{3}}}{\tilde{\zeta} - 1} - 1 \tag{12}$$

The Jacobian $|w'\left(\tilde{\zeta}\right)|^2$ takes now the form

$$g(r, \psi) = |w'(\tilde{\zeta})|^2 = \frac{3c^2h^2}{|1 - \tilde{\zeta}|^4} \left|\theta_1'\left(0, e^{i\pi\zeta(\tilde{\zeta})}\right)^{8/3}\right|, \tag{13}$$

where $\tilde{\zeta} = r\, e^{i\psi}$.

The gain of such representation becomes clear if we make averaging of the function $g(r, \psi)$ with respect to $\psi$. The numerical integration gives us:

$$\lim_{r \to 1} \frac{1}{2\pi} \int_0^{2\pi} g(r, \psi) d\psi \simeq \frac{\mathrm{const}}{(1 - r^2)^2} \tag{14}$$

Thus we can see that for rather large $r$ the diffusion is governed by the Laplacian on the surface with the constant negative curvature — the Lobachevskii plane. The representation of the Lobachevskii plane in the unit circle is called the Poincare model. Changing the variables $\mu = \ln \frac{1+r}{1-r}$ we get the unrestricted random walk on so-called 3-pseudosphere. The corresponding distribution function $P(\mu, L)$ reads

$$P(\mu, L) = \left(\frac{c^2h^2}{2\pi La}\right)^{3/2} \exp\left(-\frac{La}{4c^2h^2}\right) \int_{\mu}^{\infty} \frac{\xi \exp\left(-\frac{\xi^2 c^2 h^2}{4La}\right)}{\sqrt{\cosh\xi - \cosh\mu}} d\xi \tag{15}$$

The physical meaning of length of geodesics on 3-pseudosphere, $\mu$, is rather straightforward: $\mu$ coincides with the length of so-called "primitive path" in the lattice of obstacles, i.e., of the path which remains after all topologically allowed contractions of the random trajectory in the lattice of obstacles. Hence, $\mu$ can be considered as a nonabelian topological invariant, much more powerfull than the Gauss linking number. This invariant is not complete except one point $\mu = 0$ where it precisely classifies the trajectories belonging to the trivial homotopy class.

The connection of the problem under consideration with statistics of Markov chains on locally free groups (see Section 2.1) can be understood as follows. Compare the distribution function of the primitive path $\mu$ to the distribution function of previously defined "knot complexity", $\eta$, (see Eq.(6)). We may conclude that both of these invatiants have one and the same meaning: actually, the random walk in the covering space constructed for the lattice of obstacles can be reduced to the random walk in hyperbolic space. At the same time, we showed in Section 2.1 that the knot complexity is proportional to the length of the shortest noncontractible word written in terms of generators of the braid group $B_n$, i.e., is proportional to the length of geodesics in some space of constant negative curvature.

### *2.3. Algebraic Invariants and Entropy of Lattice Knots*

We consider 2D square lattice and suppose that it represents the plane projection of the knot embedded in 3D-space. The crossings on projection are lattice vertices, their total number is $N$. We distinguish two kinds of crossings, $b_i$, corresponding to different configurations of knot: a) $b_i = +1$ -|- and b) $b_i = -1$ ┼ .

There are $2^N$ different microrealizations of the lattice. We suppose that the variable $b_i$ can take values $\pm 1$ independently in all verices. Let us estimate the part $P_0(N)$ of unknotted (i.e. topologically isomorphic to trivial ring) paths on the lattice among $2^N$ possible ones.

**Statement 2** *Fixed topological state of knot can be regarded as a quenched disorder for ordinary vertex (in particular, Potts) model. Fixation of the knot topology means freezing of all spins $b_i$, i.e. quenching of some disorder pattern of these spins on the lattice.*

For the identification of the topological state of the knot we use the Kauffman algebraic invariant, $K(A)$, which is the Laurent polynomial in $A$ variable [d]. We have shown the Kauffman invariant to be equal to the partition function of some special *disordered Potts model* [2,4]. The number of equivalent states, $q$, and the nearest neighbor interaction constant, $J_{kl}$, are defined as follows:

$$q = \left(A^2 + A^{-2}\right)^2, \qquad J_{kl} = b_{kl} \ln\left[-A^{-4b_{kl}}\right]$$

The variables $b_{kl}$ play a role of disorder on the edges of the dual lattice connected to the variables $b_i$

$$b_{kl} = \begin{cases} -b_i & \text{if } (kl)\text{-edge is vertical} \\ +b_i & \text{if } (kl)\text{-edge is horizontal} \end{cases}$$

[d] We consider for simplicity the invariants of regular isotopic knots only.

METHODS OF SOLUTION AND RESULTS. We consider the variables $b_i$ to be independent and randomly distributed with the the density

$$Q\{b_i\} = \frac{1}{2}\left[\delta(b_i + 1) + \delta(b_i - 1)\right] \tag{16}$$

Thus the path integration in Eq.(1) is replaced by simple averaging of $\delta(\ldots)$ over a set of independent variables $b_i$ with the distribution Eq.(16).

We have constructed the upper estimation for the probability $P_0$ to find the trivial knot. Since Kauffman invariant is not complete some nontrivial knots with $K(A) = 1$ may exist ($K(A) \equiv 1$ for the trivial knot). Then $P_0$ is less than the probability to find a knot with $K(A) = 1$ *for all* $A$-values; which, in turn, is less than the probability to find a knot with invariant $K(A)$ being equal to 1 *for some fixed* value $A = A^*$:

$$P_0(N) < \int \delta\left(\ln K^2_{\{b\}}(A^*)\right) Q\{b\} D\{b\} \tag{17}$$

Using the integral representation of usual scalar $\delta$-function we can rewrite Eq.(17) as follows

$$P_0(N) < \frac{1}{\pi} \int_{-\infty}^{+\infty} \left\langle \left[K_{\{b\}}(A^*)\right]^{2iy} \right\rangle dy \tag{18}$$

where $< \ldots >$ means averaging of type $\int \ldots Q\{b\}D\{b\}$. Thus our problem is reduced to averaging of moments of the partition function over realizations of lattice disorder, what is well known problem in the theory of spin glasses and other systems with *quenched* disorder. Such averaging can be performed in frameworks of *replica trick* for an integer exponent like $\langle Z^n \rangle$ and then continued to non-integer complex values $n = iy$.

The Potts variables $q$ and $J$ are the functions of $A$. Thus, changing the $A$-value we are moving along a certain curve on $(q, J)$-plane of Potts phase space. The special value $A = A^*$ at this curve is choosen from the condition of minimization of the free energy of the Potts system.

The calculation of averaged moments of partition function (Eq.(18)) in the thermodynamic limit near the spin-glass transition point gives the upper estimation for the part of unknotted paths [4]

$$P_0(N) < \exp(-N/N_0); \qquad N_0 \approx 2.6$$

where $N$ is the number of lattice vertices, i.e. the number of crossings on the knot 2D projection.

## 3. Discussion

The problems considered in Sections 2.1 – 2.2 admit reformulation in spirit of the spin-glass-like problems discussed in Section 2.3. Suppose one has the closed path of length $L$ which we randomly drop on the plane with the regular array of removed points and ask for the probability to find the path in the definite topological state. Looking on that problem from the point of view of Section 2.3, we can regard the topological state of the path as a sort of quenched disorder. To find the corresponding distribution function we have to average the moments of the topological invariants over the Gaussian distribution (the measure for paths on the plane). On this way we could apply to the function $-|w'(\zeta)|^2$ (inverse of the Jacobian Eq.(10)) the sense of an ultrametric "phase space" where each valley corresponds to some given topological state—see fig.1). As closer Im$\zeta$ is to the real axis, as higher are the barriers between

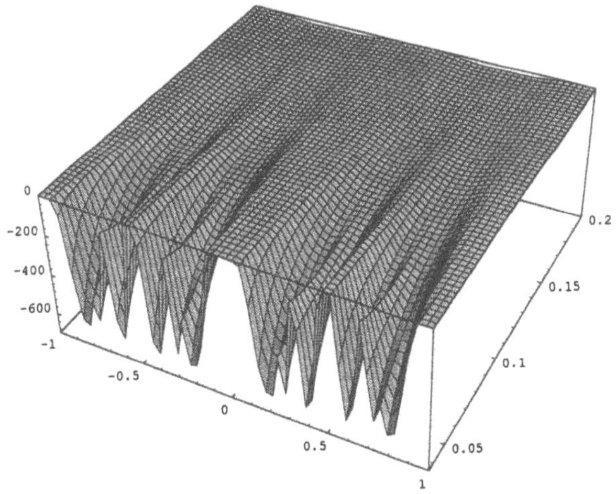


Figure 1: Ultrametric structure of conformal mapping $-|w'(\zeta)|^2$

neighboring valleys. Thus, all rather long ($La \gg c^2$) random trajectories in such potential becomes "localized" in some heavily entangled state, what means that the probability of spontaneous disentanglement of trajectory of length $La$ is of order of $\exp\left(-\text{const}\frac{La}{c^2}\right)$. Probably this analogy could be useful in the ordinary theory of spin-glasses because of explicit expression of Parisi ultrametric potential constructed via analytic functions.

## 4. References

1. S. Nechaev, A. Grosberg, A. Vershik, Preprint—IHES/P/95/49

2. A. Grosberg, S. Nechaev, *Europhys. Lett.* **20** (1992) 613

3. S. Nechaev, *Int. J. Mod. Phys.* **B4** (1990) 1809

4. A. Grosberg, S. Nechaev, *J. Phys. A: Math. Gen.* **25** (1992) 4659

**Inverse Problems in Statistical Physics and Möbius inversion**

Chen Nan-xain
*CCAST(World Lab),P.O.Box 8730,100080,Beijing China and*
*Beijing University of Science and Technology,100083,China* [1]

**Abstract**

The Möbius inversion in the number theory has been applied to solving some important inverse problems for both the Boson system and Fermion system such as the inverse blackbody radiation problem, inversion of specific heat for phonon density of states, inversion of carrier density for electronic density of states. In addition, the Möbius method is extended to the Gaussian and Eisenstein's rings of integers and the inequally weighted series with examples for solving the two- and three- dimensional inverse lattice problems, which builds a bridge from *ab initio* electronic structure calculation to interatomic potentials. This is important for calculation of the mechanical properties of materials. Of more importance, all the above show the potential applications of the number theory and the ring theory to statistical physics.

# 1 Introduction

The original one-dimensional Möbius inversion is already an old and venerable subject in classical number theory[1]. The Möbius inversion formula for continuous functions states that[1] if

$$F(x) = \sum_{n=1}^{\infty} f(nx), \tag{1}$$

then

$$f(x) = \sum_{n=1}^{\infty} \mu(n) F(nx), \tag{2}$$

vice versa, where $\mu(n)$ is the Möbius function defined as

$$\mu(n) = \begin{cases} 1, & \text{if } n = 1, \\ (-1)^s, & \text{if n is a product of s distinct primes,} \\ 0, & \text{otherwise.} \end{cases} \tag{3}$$

[1] Mail address.

The Möbius inversion has many different versions with important applications in statistical physics. This short review will concentrate on the version with continuous functions.

The inverse cohesive energy problem is to determine the interatomic potential based on experimentally measurable or *ab initio* calculated cohesive energy curve. Assume that the cohesive energy E(x) for each atom in an infinite atomic chain can be expressed as the sum of interatomic pair potentials $\Phi(x)$ as

$$E(x) = \frac{1}{2}\sum_{n\neq 0}\Phi(nx) = \sum_{n=1}^{\infty}\Phi(nx)$$

where x represents the interval between nearest neighbour atoms. Then a simple and neat enough physical result is given as

$$\Phi(x) = \sum_{n=1}^{\infty}\mu(n)E(nx).$$

# 2 Some applications of the Möbius transform

## 2.1 Inverse specific heat problem[3,4]

The specific heat of lattice vibrations can be expressed as

$$C_v(T) = rk\int_0^{\infty}\frac{(h\nu/kT)^2 e^{h\nu/kT}}{(e^{h\nu/kT}-1)^2}g(\nu)d\nu. \tag{4}$$

where h and k represent the Planck and Boltzmann constants respectively, r is the number of atoms per unit cell, and $g(\nu)$, the phonon density of states, is normalized to 3N:

$$\int_0^{\infty} g(\nu)d\nu = 3N. \tag{5}$$

The problem is to recover the $g(\nu)$ based on the experimentally measurable $C_v(T)$. Introducing the "coldness" as u=h/kT, thus (4) becomes

$$C_v(\frac{h}{ku}) = rk\int_0^{\infty}\frac{(u\nu)^2 e^{-nu\nu}}{(e^{u\nu}-1)^2}g(\nu)d\nu. \tag{6}$$

By using the Taylor's expansion, one can find that

$$C_v(h/ku) = rk\sum_{n=1}^{\infty}\int_0^{\infty}n(u\nu)^2 e^{-nu\nu}g(\nu)d\nu$$

$$= rku^2 \sum_{n=1}^{\infty} n \int_0^{\infty} \nu^2 g(\nu) e^{-n\nu u} d\nu$$

$$= rku \sum_{n=1}^{\infty} (nu) L[\nu^2 g(\nu); \nu \to nu],$$

where L[.] represents the Laplace transform. Therefore,

$$uL[\nu^2 g(\nu); \nu \to u] = \sum_{n=1}^{\infty} \mu(n) \frac{C_v(h/nku)}{rknu}$$

or

$$g(\nu) = \frac{1}{rk\nu^2} \sum_{n=1}^{\infty} \mu(n) L^{-1}[\frac{C_v(h/nku)}{nu^2}; u \to \nu]. \tag{7}$$

Different from Einstein and Debye's work, this solution is an exact closed-form one.

## 2.2 Inverse blackbody radiation problem[3]

The inverse blackbody radiation problem is important for remote sensing. The equation for the area-temperature a(T) is

$$W(\nu) = \frac{2h\nu^3}{c^2} \int_0^{\infty} \frac{a(T)dT}{e^{h\nu/kT} - 1}, \tag{8}$$

in which the total radiation power spectrum W($\nu$) is measurable, $\nu$ is the frequece, c is the velocity of light. Introducing two variables "coldness" u and "area-coldness" A(u) so that

$$u = h/kT \; and \; A(u)du = -a(T)dT, \tag{9}$$

then the solution of eq.(10) can be written as

$$A(u) = \frac{c^2}{2h} \sum_{n=1}^{\infty} \mu(n) L^{-1}[\frac{W(n\nu)}{(n\nu)^3}; \nu \to u]. \tag{10}$$

The equations (7) and (10) have been applied to temperature distributions for both the interstitial dust and the shell of black hole[5]. The same method has also been applied to solving the inverse problem of a Fermi system.

# 3 2D inverse lattice problems

The inverse problem of simple square lattice can be expressed as

$$E(x) = \frac{1}{2} \sum_{(m,n)\neq(0,0)} \Phi(\sqrt{m^2+n^2 x}), \tag{11}$$

where, say, E(x) is the measurable or calculatable cohesive energy, $\Phi(x)$ is the unknown pair potential between two atoms at a distance of x, and the vertices of an infinite lattice are all occupied by interacting atoms.

## 3.1 Square lattice problem and ring of Gaussian integers

The ring of Gaussian integers is the set $G = \{z = m + ni | m, n \in \mathbf{Z}\}$ on a complex plane, where m,n are the integers[1,6]. For our purpose, the multiplicative semigroup is taken as $G^* = G - \{0\}$. Instead to only one unit element 1 in the set of natural numbers, there exists a set of unit elements $U = \{1, i, -1, and - i\}$. Therefore, there are four associated Gaussian integers in $G^*$ for any $z \in G^*$, namely, z, iz, -z and -iz. Originally the Möbius function $\mu(n)$ was given based on the unique factorization of any natural number, namely,

$$n = p_1^{r_1(n)} p_2^{r_2(n)} ... p_k^{r_k(n)} \tag{12}$$

and

$$p_i \neq p_j \ when \ i \neq j, \tag{13}$$

where $p_i$ is the prime, and $r_i \geq 1$. Now for any $z \in G^*$ the uniqueness of the factorization is still held

$$z = \epsilon P_1^{k_1} P_2^{k_2} ... P_i^{k_i} ... P_t^{k_t}, \tag{14}$$

where $P_i$ and $P_j$ are not associated when $i \neq j, \epsilon \in U$, and $k_i \geq 1$. Therefore , one can define the Möbius function on $G^*$ similarly, namely,

$$\mu(z) = \begin{cases} 1, & \text{if z} \in \text{U}, \\ (-1)^t, & \text{if the factorization(14) of z includes all the } \text{k}_\text{i} = 1, \\ 0, & \text{otherwise.} \end{cases} \tag{15}$$

Obviously, the Möbius functions of associated elements are equal to each other, and a sum rule of $\mu(z)$ is given as

$$\sum_{d|z} \mu(d) = \begin{cases} 0, & \text{if z} \notin \text{U}, \\ 4, & \text{if z} \in \text{U}, \end{cases} \tag{16}$$

which is similar to the sum rule of the Möbius function $\mu(n)$ on the nutural numbers, namely

$$\sum_{d|n} \mu(d) = \delta_{n1}. \tag{17}$$

By applying the Möbius inversion formula on the algebraic ring of integers, the solution of eq.(11) can be obtained as

$$\Phi(x) = \frac{1}{8} \sum_{(m,n)\neq(0,0)} \mu(m+ni)E(\sqrt{m^2+n^2}x)$$

$$= \frac{1}{2} \sum_{m=1}^{\infty} \sum_{n=0}^{\infty} \mu(m+ni)E(\sqrt{m^2+n^2}x). \tag{18}$$

**Table 1. Möbius functions on Gaussian ring**

| m | n→ | 0 | 1 | 2 | 3 | 4 | 5 | 6 | 7 | 8 |
|---|---|---|---|---|---|---|---|---|---|---|
| 1 | | 1 | -1 | -1 | 1 | -1 | 1 | -1 | 0 | 1 |
| 2 | | 0 | -1 | 0 | -1 | 0 | -1 | 0 | -1 | 0 |
| 3 | | -1 | 1 | -1 | 1 | 0 | 1 | 1 | 1 | -1 |
| 4 | | 0 | -1 | 0 | 0 | 0 | -1 | 0 | 1 | 0 |
| 5 | | 1 | 1 | -1 | 1 | -1 | -1 | -1 | 1 | -1 |
| 6 | | 0 | -1 | 0 | 1 | 0 | -1 | 0 | 1 | 0 |
| 7 | | -1 | 0 | -1 | 1 | 1 | 1 | 1 | 1 | -1 |
| 8 | | 0 | 1 | 0 | -1 | 0 | -1 | 0 | -1 | 0 |

Obviusly, the exact solution (18) is much simpler for analysis and easier for calculation than all the previous results.

## 3.2 Applications

This has also been used for 2D AFT in parallel processing. The inverse problem of hexagonal graphite sheet corresponds to the Möbius inversion on the Eisenstein's ring[6].

# 4 3D inverse lattice problems[7,8]

The inverse cohesion problem for three dimensional lattice can be solved based on the Möbius inversion for an inequally weighted sum. Assume that the total potential E(x) for each atom in a crystal structure can be expressed as the sum of pair potentials $\Phi(x)$ between atoms as

$$E(\gamma x) = \frac{1}{2}\sum_{\vec{\alpha}\in L}\Phi(|\vec{\alpha}|x) = \frac{1}{2}\sum_{\alpha} r(\alpha^2)\Phi(\alpha x)$$

$$= \frac{1}{2}\sum_{n} r(n)\Phi(\sqrt{n}x), \tag{19}$$

where $\gamma$x is the lattice constant, $\vec{\alpha}$ represents the displcement vector of each atom from the origin. $\{n\}$ is the set of natural numbers which is an extension of $\{\alpha^2\}$. $r(n)$ or $r(\alpha^2)$ represents the number for lattice points with the same square distance n or $\alpha^2$. L denotes the set of lattice points. Obviously, $\alpha$(L) and r(n) are structure dependent.

## 4.1 fcc structure

Now (19) can be rewritten as

$$E(\gamma x) = \frac{1}{2}\sum_{n=1}^{\infty} r(\beta n)\Phi(\sqrt{\beta n}x), \tag{20}$$

the result is

$$\Phi(x) = 2\sum_{m=1}^{\infty}\mu(m)E(\gamma\sqrt{\frac{m}{\beta}}x), \tag{21}$$

where $\mu(m)$ can be obtained by the recursion relation

$$\sum_{n|k} r(\beta n)\mu(k/n) = \delta_{k1}. \tag{22}$$

In fact, the right side of eq.(21) can be expressed as

$$R(21) = 2\sum_{m=1}^{\infty}\mu(m)E(\gamma\sqrt{\frac{m}{\beta}}x) = \sum_{m=1}^{\infty}\mu(m)\sum_{n=1}^{\infty} r(\beta n)\Phi(\sqrt{mn}x)$$

$$= \sum_{k=1}^{\infty}[\sum_{m=1}^{\infty} r(\beta n)\mu(k/m)]\Phi(\sqrt{k}x) = \sum_{k=1}^{\infty}\delta_{k1}\Phi(\sqrt{k}x) = \Phi(x)$$

**Table 2. The modified Möbius function for fcc structure**

| n | 1 | 2 | 3 | 4 | 5 | 6 | 7 | 8 | 9 | ... |
|---|---|---|---|---|---|---|---|---|---|---|
| $\mu(n)$ | 1/12 | -1/24 | -1/6 | -1/16 | -1/6 | 1/ 9 | -1/ 3 | 1/ 32 | 1/ 12 | ... |

## 4.2 $L1_2$ structure

Now the total potential can be expressed as

$$E(\gamma x) = \frac{1}{2}\sum_{n} r(\beta(2n-1))\Phi(\sqrt{\beta(2n-1)}x), \tag{23}$$

and the corresponding result is

$$\Phi(x) = 2\sum_{m=1}^{\infty} \mu(m)E(\gamma\sqrt{\frac{2m-1}{\beta}}x), \tag{24}$$

with

$$\sum_{2m-1|\beta(2k-1)} r(\frac{\beta(2k-1)}{2m-1})\mu(m) = \delta_{k1}. \tag{25}$$

**Table 3. The Möbius function for $L1_2$ structure**

| n | 1 | 2 | 3 | 4 | 5 | 6 | 7 | 8 | 9 | ... |
|---|---|---|---|---|---|---|---|---|---|---|
| $\mu(n)$ | 1/12 | -1/6 | -1/6 | -1/3 | 1/12 | -1/6 | -1/2 | 1/3 | -1/3 | ... |

## 4.3 Calculation of physical properties of $Ni_3Al$[7]

Based on the above, we have calculated pair potentials $\Phi_{Ni}(x)$, $\Phi_{Al}(x)$ and $\Phi_{Ni-Al}(x)$. Thus elastic modules and phonon dispersion of $Ni_3Al$ are evaluated, and the result is in good agreement with experiment[7].

Noted that for solving the inverse problems related to bcc and hcp structures the different iterative methods with the Möbius inversion formulas on the classical number theory[8]. Now, the challenge for physists proposed by Maddox has been addressed[9]. Recently, this procedure has been used to modify the embeded atom method to recover the n-body interatomic potentials[10]. In conclusion, it shows the potential application of the number theory and the ring theory to physical sciences. In addition, this short review did not go details of discrete Möbius inversion to parallel processing[1,11] and Witten index[12], and the applications of the Möbius inversion on partially ordered set[13], and so on.

## 5 Acknowledgements

The author would like to thank C.B.Pan, Z.D.Chen, Y.N.Shen, M.Li, S.J.Liu and Q.Xie for their very helpful suggestions and encouragements. This work was supported in part by the National Foundation of Natural Science in China, in part by the National Committee of Advanced Materials in China.

## 6 References

1. M.R.Schroeder,*Number Theory in Science and Communication*, second English edition, corrected printing, Springer-Verlag, New York (1990)
2. N.X.Chen, Phys. Rev. Lett., **64**(1990)1193.
3. A.E.Carlsson, C.D.Gelatt and H.Ehrenreich, Phil. Mag., **A41** (1980)241.
4. N.X.Chen and Y.Chen and G.Y.Li, Phys. Lett. **A149**(1990)357.
5. T.L.Xie, P.F.Goldsmith and W.Zhou, AstrophysJ., **371** (1991) L81;
T.L.Xie, P.F.Goldsmith, R.L.Shell and W.Zhou, AstrophysJ., 402 (1993) 216;
H.Rosu, IL Nuovo Cimento, 108(1993) 1333.
6. W.Narliewicz, Elementary and Analytic Theory of Algebraic Numbers, 2nd ed., Springer- Verlag, 1989, New York
7. N.X.Chen, M.Li, S.J.Liu and Z.D.Chen, Phys.Lett., **A195** (1994)135.
8. N.X.Chen and G.B.Ren, Phys. Rev. **B45**(1992)8177.
9. G.C.Rota, Z.Wahrsch. 2(1964) 340; T.Morita, J.Stat.Phys. 59(1990) 819.
10. J.Maddox, NatureLondon**344**(1990)**377**.
11. Q.Xie and N.X.Chen, Phys.Rev.B51( 1995) 15 85 6
12. I.S.Reed, D.W.Tufts, X.Yu, T.K.Truong, M.T.Shih, and X.Yin, IEEE Trans. Accoust. Speech Signal Process. 38(1990) 458
13. D.Spector, Commun. Math. Phys. 127(1990) 239

# DRIVEN LATTICE GASES: TYPICAL FLUCTUATIONS AND LARGE DEVIATIONS

HERBERT SPOHN

*Theoretische Physik, Ludwig-Maximilians-Universität*
*Theresienstr. 37, D - 80333 München, Germany*
E-mail: spohn@stat.physik.uni-muenchen.de

ABSTRACT

Driven lattice gases are studied with the goal to understand whether at all and if so how far the very general principles of local thermodynamic equilibrium (LTE) extend to non-LTE systems characterized by broken detailed balance.

## 1. Introduction

My contribution is based on joint work together with G.L. Eyink and J.L. Lebowitz [1].

Traditionally macroscopic continuum equations are based on local thermodynamic equilibrium (LTE), which means that on a scale where the macroscopic fields are approximately constant the statistics of the particles are governed by the Gibbs-Boltzmann distribution. The transport coefficients are then defined through the equilibrium linear response in the currents due to driving gradients, which leads to the famous Green-Kubo formulae and Onsager reciprocity relations. There are many physical situations where the assumption of LTE is overoptimistic. E.g. radiative transport in hot, low density stellar atmospheres is markedly non-LTE [2] and a considerable effort has been invested to go beyond an LTE approximation [3]. In semiconductor devices electrons and holes can be driven by such strong fields that LTE fails (hot electrons) [4]. Surface growth is a further point in case: the deposition of atoms on the surface produces non-LTE conditions and leads to an enhanced roughness [5].

Given these examples one may ask how the "sacred" principles, if applicable at all, are to be modified under non-LTE conditions. To clarify such a basic point, it seems useful to study a sufficiently simple and well-defined model system, not so much troubled by the complexities of real life. As model we propose here driven lattice gases at sufficiently high temperatures. Such a restriction is necessary because at low temperatures driven lattice gases have a complicated phase diagram and associated with it a wealth of dynamical phenomena hardly explored yet [6]. For lattice gases the number of particles is the only locally conserved field. Thus under LTE-conditions the macroscopic equation for the density is just a nonlinear diffusion equation,

$$\frac{\partial}{\partial t}\rho(\mathbf{r},t) = \nabla \cdot L(\rho(\mathbf{r},t))\nabla\lambda(\rho(\mathbf{r},t)) \tag{1.1}$$

written in Onsager force-flux form. $\lambda(\rho)$ is the chemical potential and $L(\rho)$ the symmetric Onsager matrix which in our context coincides with the (Ohmic) conductivity. The bulk diffusion matrix is then

$$D = L\chi^{-1} \tag{1.2}$$

with $\chi = d\rho/d\lambda$ the static compressibility.

For driven lattice gases one would expect that (1.1) goes over to the drift-diffusion equation

$$\frac{\partial}{\partial t}\rho(\mathbf{r},t) = -\nabla \cdot [\mathbf{j}(\rho(\mathbf{r},t)) - L(\rho(\mathbf{r},t))\nabla\lambda(\rho(\mathbf{r},t))]. \tag{1.3}$$

Under what conditions is (1.3) valid? What is the meaning of $\lambda$? How are $L$ and $\mathbf{j}$ determined microscopically from the lattice gas dynamics? Which principles govern small fluctuations around and large deviations away from (1.3)? We will indicate some answers. For a complete account the reader is referred to [1].

## 2. Driven Lattice Gases, Stationary Measures

We consider particles with hard core exclusion on the $d$-dimensional lattice $\mathbb{Z}^d$. The occupation variables are $\eta(x), x \in \mathbb{Z}^d$, with $\eta(x) = 0$ for site $x$ empty and $\eta(x) = 1$ for site $x$ occupied. A whole particle configuration is denoted by $\eta$. For simplicity we choose the standard energy

$$H(\eta) = -J \sum_{\langle x,y\rangle} \eta(x)\eta(y) \tag{2.1}$$

with $\langle x,y\rangle$ a pair of nearest neighbor sites. $J > 0$ is an attractive and $J < 0$ a repulsive lattice gas. Particles jump to nearest neighbor sites. Thus the dynamics are completely determined by specifying the *exchange rates* $c(x,y,\eta)$ depending on the configuration $\eta$ through the sites next to $x,y$. The rates are assumed to be translation invariant and nondegenerate in the sense that $c(x,y,\eta) > 0$ whenever $(\eta(x) - \eta(y))^2 = 1, \mid x-y \mid = 1$. The generator for the dynamics is then

$$\mathfrak{L}f(\eta) = \sum_{\langle x,y\rangle} c(x,y,\eta)[f(\eta^{xy}) - f(\eta)] \tag{2.2}$$

acting on local functions $f$. Here $\eta^{xy}$ is the configuration $\eta$ with occupancies at sites $x,y$ exchanged. $(e^{\mathfrak{L}t})(\eta,\eta')$ is the probability to have the configuration $\eta'$ at time $t$ given it was initially $\eta$, $\sum_{\eta'}(e^{\mathfrak{L}t})(\eta,\eta') = 1$.

LTE is *defined* through the condition of detailed balance,

$$c(x,y,\eta) = c(x,y,\eta^{xy})\exp[-\beta(H(\eta^{xy}) - H(\eta))]. \tag{2.3}$$

Physically, this condition fixes the temperature $\beta^{-1}$, whereas the density may still vary. In particular the Gibbs state $\mu_\lambda = Z^{-1}\exp[-\beta H(\eta) + \lambda\sum_x \eta(x)]$ is invariant under $\mathfrak{L}$, i.e. $\mathfrak{L}^*\mu_\lambda = 0$, for an arbitrary chemical potential $\lambda$. As a further consequence of (2.3), in equilibrium, a particular history and its time reversal have the same probability. In this sense a LTE lattice gas is time reversal invariant.

To drive the lattice gas we simply replace the rates $c(x, y, \eta)$ in (2.3) by

$$c_E(x, y, \eta) = c(x, y, \eta) \exp[-\frac{1}{2}\beta q \mathbf{E} \cdot (x - y)(\eta(x) - \eta(y))]. \tag{2.4}$$

This corresponds to adding in (2.1) locally a linear external potential which biases the jumps along the driving field $\mathbf{E}$ [7]. The strength of the driving is not restricted, but in this note $|\beta J|$ is assumed to be sufficiently small. Although (2.4) is physically fairly natural, there are of course many other choices to break detailed balance.

The macroscopic equation (1.1) describes a situation where the density is slowly varying on the scale of the lattice. Because of the conservation law the density varies then also slowly in time. Physically we would expect then that in the driven case the role of local equilibrium is taken over by states which are locally stationary for $\mathfrak{L}$. Stationary states are measures $\mu$ on $\{0, 1\}^{\mathbf{Z}^d}$ which satisfy

$$\mu(\mathfrak{L} f) = 0 \tag{2.5}$$

for every local function $f$. Unfortunately (2.5) has no explicit solution, the main exception being the asymmetric simple exclusion process (ASEP). Its exchange rates are defined by $c(x, y, \eta) = \eta(x)(1 - \eta(y))p(y - x) + (1 - \eta(x))\eta(y)p(x - y)$, $p(x) \geq 0, \sum_x p(x) < \infty$. The stationary measures for the ASEP are Bernoulli, i.e. the $\{\eta(x)\}$ are independent and $< \eta(x) >_\rho = \rho$. Despite our difficulties with (2.5) there are two general results, which are quite remarkable.

Let us assume that for every (average) density $\rho \in [0, 1]$ there is a unique steady state $\langle \cdot \rangle_\rho$ with some clustering. Then for dimension $d \geq 2$ the steady state two point function decays as

$$| < \eta(x)\eta(y) >_\rho - \rho^2 | \cong |x - y|^{-d} , \tag{2.6}$$

which is supported by theoretical considerations [8] and by MC simulations [9]. The two point function has no definite sign and looks like an electrostatic quadrupole potential. As the ASEP shows, (2.6) must have some exceptions, a phenomenon further studied by using a RG flow at high temperatures [10].

The second result [11] is an extension of [12] to lattice gases and proves that the stationary measures can be written in a form familiar from equilibrium, namely

$$\langle f(\eta) \rangle_{\rho(\lambda)} = Z^{-1} \sum_\eta \exp[-\tilde{H}(\eta) + \lambda \sum_x \eta(x)] f(\eta) \tag{2.7}$$

$\tilde{H}$ has nothing to do with $H$ in (2.1). The crucial point is that $\tilde{H}$ *does not depend* on the one-body potential $\lambda$ (neither on $\rho$). Of course (2.7) is understood for the infinite volume. At finite volume $< \cdot >_\rho$ can always be written in the form (2.7). But then the many-body potentials building up $\tilde{H}$ still depend on the volume. The precise

statement is as follows:
*Let $\mu$ be stationary for $\mathfrak{L}$, translation invariant, and Gibbs for some summable potential $\{\phi_A\}$ , $\sum_{A\ni 0}|\phi_A| < \infty$. Then any other stationary, translation invariant state is Gibbs with the same potential except for $\phi_{\{0\}} = -\lambda$.*
As a consequence slow density variations can be imposed through variations in the "chemical potential". Furthermore, if $n_\Lambda$ denotes the number of particles in the large box $\Lambda$, then we can define an "entropy" through the large deviations

$$\mu_{\tilde{\rho}}(\{n_\Lambda = N\}) \cong \exp[|\Lambda|s(N/|\Lambda|)]. \tag{2.8}$$

Here $\tilde{\rho}$ is some reference density. The entropy is concave with $s(\rho) \leq 0, s(\tilde{\rho}) = 0$. If the reference density is changed to another one, then $s$ is shifted by a linear term. In particular, we have $\lambda = -\partial s/\partial\rho$ and $\chi = \partial\rho/\partial\lambda = \sum_x(< \eta(x)\eta(0) >_\rho -\rho^2)$. Note that the steady states might not be Gibbsian at all in which case (2.7) is no longer meaningful. If they are Gibbs, it would be of interest to determine $\tilde{H}$ in a perturbative expansion around $\beta J = 0$.

## 3. Drift-Diffusion Equation, Typical Fluctuations

We return to the drift-diffusion equation (1.3) which describes the density of the lattice gas on a scale large compared to the lattice constant. The first term on the r.h.s. of (1.3) should then be identified with the steady state current

$$j_\alpha(\rho) = < c(0, e_\alpha, \eta) >_\rho, \tag{3.1}$$

$\alpha = 1, ..., d, e_\alpha = (0, ..., 1, ..., 0)$ at the $\alpha$-th entry. As explained the chemical potential is still a meaningful quantity. A linear response computation yields indeed an Onsager matrix, $L$, microscopically given by a Green-Kubo type formula. In fact, for a one component system, as (1.3), only the symmetric part of $L$ appears in the macroscopic equation. Linear response provides us however with the full Onsager matrix. It satisfies

$$L_{\alpha\beta} = L^{(r)}_{\beta\alpha}\ , \tag{3.2}$$

where $L^{(r)}$ is the same as $L$ but evaluated for the time reversed dynamics. (3.1) looks like Onsager reciprocity but it is hardly as useful, since the time-reversed process does not correspond to merely replacing $\mathbf{E}$ by $-\mathbf{E}$. It can be constructed only from the knowledge of the stationary measure(s), which is not available in general.

The drift-diffusion equation (1.3) can be further supported through a formal derivation by means of the Zubarev nonequilibrium distribution method. However such formal considerations, although providing some insight, give little indication on the range of validity of (1.3). A more stringent test is to consider small fluctuations around a uniform steady state. If we write the density as $\rho + \xi(\mathbf{r}, t)$, then to linear order $\xi$ satisfies

$$\frac{\partial}{\partial t}\xi(\mathbf{r}, t) = -\nabla \cdot [\mathbf{j}'(\rho)\xi(\mathbf{r}, t) - L(\rho)\lambda'(\rho)\nabla\xi(\mathbf{r}, t) - \mathbf{J}(\mathbf{r}, t)], \tag{3.3}$$

$' = \partial/\partial\rho$. We added the fluctuating current $\mathbf{J}$ with zero average and covariance $< J_\alpha(\mathbf{r},t)J_\beta(\mathbf{r}',t') >= R_{\alpha\beta}\,\delta(\mathbf{r}-\mathbf{r}')\delta(t-t')$, where $R_{\alpha\beta}$ is the space-time integral over the microscopic current-current correlation properly truncated. On this basis the structure function is predicted as

$$S(\mathbf{k},t) = S(\mathbf{k})\exp[i\mathbf{k}\cdot\mathbf{j}'(\rho)t - \mathbf{k}\cdot D\mathbf{k}|t|], \tag{3.4}$$

$D = L\chi^{-1}$, and is to be compared with the true structure function

$$S(\mathbf{k},t) = \sum_x e^{i\mathbf{k}\cdot x}(< \eta_t(x)\eta_0(0) >_\rho - \rho^2). \tag{3.5}$$

In one dimension driven lattice gases are equivalent to the KPZ equation [13,14,15], for which very detailed information is available. The static structure function $S(k)$ is smooth indicating exponentially decaying correlations. If $j''(\rho) \neq 0$, then the dissipative term in (3.4) is of order $|k|^{3/2}|t|$. If $j''(\rho) = 0, j'''(\rho) = 0$, it is $k^2|t|$. However, if $j''(\rho) = 0, j'''(\rho) \neq 0$, there are logarithmic corrections [16,17]. Similarly for $d = 2$ one obtains logarithmic corrections of the form $k^2|t(\log t)^{2/3}|$ in (3.4) [14] and only for $d \geq 3$ the diffusive $k^2|t|$ is expected. Comparing with the prediction (3.4) we conclude that in general the drift-diffusion equation (1.3) can hold only for dimension $d \geq 3$.

Our picture is supported by beautiful rigorous results. Varadhan and Xu [18] study the ASEP with $p(x) \neq p(-x)$, thus the breaking detailed balance condition $p(x) = p(-x)$, but such that $\sum_x p(x)x = 0$ which implies $\mathbf{j}(\rho) = 0$. They prove the validity of (1.3) in *any* dimension. Esposito, Marra, and Yau [19] investigate the ASEP with $\mathbf{j}(\rho) = \rho(1-\rho)\sum_x p(x)x \neq 0$. In their case $d \geq 3$ is needed to prove the drift-diffusion equation (1.3). Further extensions are given in [20,21,22]. In fact, as for LTE there is a variational principle characterizing the diffusion matrix which is proved to be enhanced by the driving.

Even on the level of a formal derivation there is a difficulty not completely satisfactorily resolved yet. As hinted at in (2.6) the static structure function for small $\mathbf{k}$ has the form

$$S(\mathbf{k}) = \frac{n_\parallel(k_\parallel)^2 + n_\perp(k_\perp)^2}{\tau_\parallel(k_\parallel)^2 + \tau_\perp(k_\perp)^2} \tag{3.6}$$

with suitable coefficients $n_\parallel, n_\perp, \tau_\parallel, \tau_\perp$, where $k_\parallel$ points parallel and $k_\perp$ transverse to the driving field $\mathbf{E}$. In general $(n_\parallel, n_\perp) \neq c(\tau_\parallel, \tau_\perp)$ and $S(\mathbf{k})$ is discontinuous at $\mathbf{k} = 0$. The limit

$$\chi(\rho, \hat{\mathbf{k}}) = \lim_{\mathbf{k}\to 0}\sum_x e^{i\mathbf{k}\cdot x}(< \eta(x)\eta(0) >_\rho - \rho^2) \tag{3.7}$$

will exist provided the direction vector $\hat{\mathbf{k}} = \mathbf{k}/k$ is specified. In the Kadanoff-Martin-Zwanzig type derivation of (3.4) it is always assumed that $S(\mathbf{k})$ is continuous at $\mathbf{k} = 0$. In our context we have to carefully keep track of the $\hat{\mathbf{k}}$ dependence and argue [1] that $D$ in (3.4) is nevertheless independent of $\hat{\mathbf{k}}$. However, in the large deviation (2.8) and in the formal derivation of (1.3) the $\hat{\mathbf{k}}$ dependence is yet to be displayed properly. In fact, if $(n_\parallel, n_\perp) \neq c(\tau_\parallel, \tau_\perp)$ one could worry whether the driven lattice gas responds to a small density gradient according to standard hydrodynamics as e.g. assumed in (1.3). The rigorous results mentioned above are not much of a comfort because in these cases $S(\mathbf{k}) = \rho(1-\rho)$ independent of $\mathbf{k}$. At present this issue remains largely unexplored.

## 4. Large Deviations

Large deviations refer to events which have an exponentially small probability and one is tempted to conclude that they are of no physical interest. An example of a large deviation is Eq. (2.8). The reference measure $\mu_{\tilde{\rho}}$ has the typical density $\tilde{\rho}$. Configurations with density $|\Lambda|^{-1}\sum_{x\in\Lambda}\eta(x) \neq \tilde{\rho}$ have a probability which is exponentially small in the size of the box $\Lambda$ and the corresponding rate is the entropy function $s(\rho)$ normalized such that $s(\tilde{\rho}) = 0$. Thus events of small probability are linked to the thermodynamics of the lattice gas. Put differently: there is a physical control parameter, namely the chemical potential, which allows to tune in (in approximation to condition on) events of small probability. We argue here that a similar structure is present in time-dependent situations. The role of the chemical potential is now taken over by slowly varying time-dependent forces. These are not as readily available, but the theoretical structure as such is of interest.

To keep track of the order of magnitude of the various terms we take now a lattice spacing $\epsilon$ and speed up time by a factor $\epsilon^{-1}$. $\rho(\mathbf{r},t)$ is then slowly varying on the scale of the lattice. The entropy of a given density profile, $\rho(\mathbf{r})$, is

$$S([\rho(\mathbf{r}]) = \int d^d r \; s(\rho(\mathbf{r})) \tag{4.1}$$

with $s(\rho)$ from (2.8). Following [23] we first write down a non-linear Langevin equation, valid on a mesoscopic scale, with the condition that $\exp[-\epsilon^{-d}S([\rho(\mathbf{r})])]$ is stationary. Such a structure has been explored by Graham [24] which we use here to arrive at

$$\frac{\partial}{\partial t}\rho = -\nabla\cdot[\mathbf{j}(\rho) - \epsilon L\nabla\lambda(\rho) + \epsilon^{(d+1)/2}\sum_{\alpha=1}^{d}\mathbf{g}^\alpha(\rho)\zeta_\alpha], \tag{4.2}$$

where $\mathbf{g}^\alpha$ are the eigenvectors of $(L+L^*)/2$, the symmetric part of $L$, and $\zeta_\alpha$ is standard white noise, $<\zeta_\alpha(\mathbf{r},t)\zeta_{\alpha'}(\mathbf{r}',t')>= \delta_{\alpha\alpha'}\delta(\mathbf{r}-\mathbf{r}')\delta(t-t')$. On the basis of (4.2) we can now determine the probability of some *prescribed* history $t\mapsto\rho^*(\mathbf{r},t)$: For this purpose we first have to find a time dependent-potential $U(\mathbf{r},t)$ and the associated force $\mathbf{F}(\mathbf{r},t) = -\nabla U(\mathbf{r},t)$ such that the given $\rho^*$ solves

$$\frac{\partial}{\partial t}\rho^*(\mathbf{r},t) = -\nabla\cdot[\mathbf{j}(\rho^*(\mathbf{r},t)) - \epsilon L(\rho^*(\mathbf{r},t))\nabla\{-U(\mathbf{r},t)+\lambda(\rho^*(\mathbf{r},t))\}]. \tag{4.3}$$

Then the probability for the history $t\mapsto\rho^*(\mathbf{r},t)$ is given by

$$\begin{aligned}&\mathrm{Prob}([\rho^*(\mathbf{r},t)])\\ &= \exp[-\epsilon^{-d}S([\rho^*(\mathbf{r},0)])]\exp[-\epsilon^{-(d-1)}\frac{1}{4}\int_0^\infty dt\int d^d r\mathbf{F}(\mathbf{r},t)\cdot L(\rho^*(\mathbf{r},t))\mathbf{F}(\mathbf{r},t)].\end{aligned} \tag{4.4}$$

The first piece is the probability to have the required $\rho^*(\mathbf{r}, 0)$ at $t = 0$. The second term is one-fourth of the Ohmic dissipation due to the driving forces.

The simplicity of (4.4) hides a physical problem. In LTE any weak driving force enters the macroscopic equation automatically as in (4.3) because of the Einstein relation. However for non-LTE the microscopic force has to be chosen carefully just so to make (4.3) valid. There seems to be no immediate physical procedure how to achieve this and merely to adopt (2.4) will not work.

## 5. Conclusion

A fairly substantial part of the structure so familiar from reversible (LTE) lattice gases extends to driven (nonreversible) lattice gases. Perhaps the most remarkable feature is that the dynamically correct procedure to induce small density variations is still through introducing a slowly varying chemical potential. On the other hand Onsager symmetry, at least in its directly applicable form, and the Einstein relation, as a link between mechanical and thermodynamic driving, are lost. It would be of interest to have a better understanding whether the long range static correlations show up on the macroscopic scale.

**Acknowledgement** A travel grant from the Deutsche Forschungsgemeinschaft is gratefully acknowledged.

## References

[1] G.L. Eyink, J.L. Lebowitz, H. Spohn, Hydrodynamics and fluctuations outside local equilibrium: driven diffusive systems, J. Stat. Phys., to appear

[2] D. Mikulas, Stellar Atmospheres. Freeman and Co., San Fransisco 1978

[3] A.W.A. Pauldrach et. al., Astron. Astrophys. J. **283**, 525 (1994)

[4] P.J. Price, in: Fluctuation Phenomena in Solids, ed. R.E. Burgess, p. 355. Academic Press, New York 1965

[5] A.L. Barabási, H.E. Stanley, Fractal Concepts in Surface Growth. Cambridge University Press 1995

[6] B. Schmittmann, R.K.P. Zia, Statistical mechanics of driven diffusive systems. In: Phase Transitions and Critical Phenomena, eds. C. Domb and J.L. Lebowitz, Vol. 17. Academic Press, London, 1995

[7] S. Katz, J.L. Lebowitz, H. Spohn, J. Stat. Phys. **34**, 497 (1984)

[8] P.L. Garrido, J.L. Lebowitz, C. Maes, H. Spohn, Phys. Rev. **A42**, 1954 (1990)

[9] M.Q. Zhang, J.S. Wang, J.L. Lebowitz, J.L. Vallés, J. Stat. Phys. **52**, 1461 (1988)

[10] R.K.P. Zia, B. Schmittmann, Z. Physik B **97**, 327 (1995)

[11] A. Asselah, in preparation

[12] H. Künsch, Z. Wahrscheinlichkeitstheorie verw. Geb. **66**, 407 (1984)

[13] M. Kardar, G. Parisi, Y.C. Zhang, Phys. Rev. Lett. **56**, 889 (1986)

[14] H. van Beijeren, R. Kutner, H. Spohn, Phys. Rev. Lett. **53**, 806 (1985)

[15] J. Krug, H. Spohn, in: Solids Far From Equilibrium, ed. C. Godrèche, p. 479. Cambridge University Press 1991

[16] P. Devillard, H. Spohn, J. Stat. Phys. **66**, 1089 (1992)

[17] M. Paczuski, preprint 1994

[18] L. Xu, Diffusion limit for lattice gases with short range interactions. Ph.D. Thesis, New York University (1993)

[19] R. Esposito, R. Marra, H.T. Yau, Rev. Math. Phys. **6**, 1233 (1994)

[20] C. Landim, S. Olla, H.T. Yau, First order correction for the hydrodynamic limit of the asymmetric simple exclusion process in dimension $d \geq 3$, preprint 1995

[21] R. Esposito, R. Marra, H.T. Yau, Navier-Stokes equations for a stochastic particle system on the lattice, preprint 1995

[22] C. Landim, S. Olla, H.T. Yau, Some properties of the diffusion coefficient for asymmetric simple exclusion processes, preprint 1995

[23] D.N. Zubarev, V.G. Morosov, Physica **A120**, 411 (1983)

[24] R. Graham, Z. Physik **B26**, 397 (1977)

# DEPOSITION AND EVAPORATION OF $k$-MERS: DYNAMICS OF A MANY-SECTOR DECOMPOSABLE SYSTEM

MUSTANSIR BARMA and DEEPAK DHAR
*Tata Institute of Fundamental Research*
*Homi Bhabha Road, Bombay 400 005, India*

**Abstract**

We study the stochastic dynamics of deposition and evaporation of $k$ adjacent particles at a time on a linear substrate of length $L$. For $k = 2$, a single conservation law operates and the phase space breaks into $L+1$ disjoint sectors, in each of which the density autocorrelation function decays diffusively. For $k \geq 3$, the system is many-sector decomposable, that is, the number of sectors grows exponentially with $L$, and the autocorrelation function shows very different long-time decays in different sectors. This is due to an infinity of conservation laws, which can be encoded compactly in a nonlocal construct, the Irreducible String (IS). The IS helps in counting numbers and sizes of sectors, and provides a way to explain the different decay laws found in different sectors in terms of a simpler model of hard-core random walkers with conserved spin.

# 1 Introduction

The presence of a conservation law influences the dynamical properties of a many-particle system in two related but distinct ways. Firstly, as the system evolves, it is confined to a part of phase space (called a sector) in which the conserved quantity takes the same value. In lattice models, where the phase space is discrete, the number of distinct sectors typically grows as a power of the size of the system. The second consequence pertains to the behaviour of time-dependent correlation functions for evolution within a sector. Implemented on a local scale, the conservation law leads to a slow, power-law decay, whose form is generically the same in all sectors. Perhaps the most familiar example is the conservation of particle number $N$. Configurations with different $N$ are dynamically disjoint, and the conservation law leads to the diffusion equation and thus to nonexponential decays of density-density correlation functions.

In this paper, we are interested in systems which, in the thermodynamic limit, possess an *infinite* number of conservation laws. In such a case, the number of disjoint sectors grows exponentially with the system size, i.e. ergodicity is broken strongly. We shall refer to the general class of such systems as Many-Sector Decomposable (MSD) systems. Like a system with a single conservation law, an MSD system also exhibits slow decays of correlation functions, but unlike it, the form of the long-time decay (e.g. the power characterizing the decay) can vary from one sector to another. The study of such systems can help in understanding the effect of broken ergodicity in a simple setting, without the additional complication of quenched disorder as in spin glasses [1].

The example of MSD systems we discuss in this paper is one of deposition and evaporation of $k$ particles at a time on a linear substrate of length $L$ [2, 3]. The elementary kinetic step involves either the deposition of $k$ atoms at a time onto $k$ adjacent unoccupied sites, or the evaporation of $k$ atoms from adjacent sites of the linear chain. These $k$-mers can reconstitute freely, i.e. the $k$ atoms that leave need not be the same $k$ atoms that fell in together. Reconstitution is a crucial feature of deposition-evaporation (DE) dynamics. As we shall see in subsequent sections, for the case $k = 2$ (dimers) there is a single conservation law which operates, and concomitant diffusive behaviour holds in all sectors. For $k \geq 3$ (trimers and above), there is an infinity of conservation laws, and the DE system is many-sector decomposable. Correspondingly, different decay laws hold in different sectors.

In Section 2, we define DE dynamics, and discuss the partitioning of the phase space into disjoint sectors. The key step is the definition of the Irreducible String (IS), a nonlocal construct obtained by applying a deletion algorithm to a configuration. Its usefulness stems from the fact that it provides a compact way of encoding an infinite number of conservation laws. These conservation laws are maximal in the sense that two configurations having the same IS necessarily belong to the same sector. Results for sector numbers and sizes are summarized in Section 2. We turn to dynamical properties in Section 3. The IS influences also the time evolution, and we explain the source of the different time dependences in different sectors in terms of the diffusive movement of its conserved but mobile elements. In the concluding

Section 4, we briefly discuss related MSD systems.

# 2 Phase Space Decomposition

The case $k = 2$ corresponds to the deposition and evaporation of dimers. Let us associate a pseudospin $S_i = \pm 1$ with site $i$, with $+1$ (denoted by $\uparrow$) corresponding to an occupied site and $-1$ (denoted by $\downarrow$ ) corresponding to an empty site. In a small time $dt$, an elementary deposition event $(|\downarrow\downarrow\rangle \rightarrow |\uparrow\uparrow\rangle)$ occurs with probability $\epsilon dt$ while evaporation $(|\uparrow\uparrow\rangle \rightarrow |\downarrow\downarrow\rangle)$ occurs with probability $\epsilon' dt$. Dividing the linear chain into odd and even sublattices $A$ and $B$ respectively, it is evident that under DE dynamics $M_A - M_B$ is conserved, where $M_A = \sum_{i\epsilon A} S_i$ and $M_B = \sum_{i\epsilon B} S_i$. The consequence of this single conservation law is the partitioning of phase space into $L+1$ sectors, and the values of $M_A - M_B$ provide a distinct label for each of these sectors [2, 3].

In strong contrast to the case $k = 2$, the $k$-mer DE system for $k \geq 3$ belongs to the MSD class. The case $k = 3$ (trimers) is typical of all larger $k$, and in the following we shall restrict ourselves to discussing this problem.

The existence of the large number of dynamically disconnected sectors can be understood in terms of a nonlocal construct, the *irreducible string* (IS). The IS is defined as follows [4]. Consider a spin configuration $C \equiv \{S_1, S_2, ...S_L\}$ on the linear chain. Starting from one end (say the left), delete triplets of consecutive parallel spins, reducing the size of the spin string by three by each such reduction. After completing a pass through the full string, repeat the procedure till no further reduction is possible. The resulting IS evidently has no parallel triplets of spins. The process of reduction and the IS are illustrated in Fig. 1.

The elementary step of deposition or evaporation changes three successive $\uparrow$ spins to three successive $\downarrow$ spins, or vice versa; evidently, the IS is invariant under this. Also, it is possible to prove that if two configurations have the same IS, they are necessarily reachable from each other by deposition-evaporation dynamics [5]. It follows that the IS is a constant of the motion and provides a distinct label for each sector of phase space.

Given this labelling scheme for sectors in phase space, the number of sectors and their sizes can be described in a fairly complete way.

We briefly summarize some of the results below.

There are quite a few completely jammed configurations (sectors consisting of a single configuration with no transition to other states), for which the length of the IS is the length $L$ of the lattice. The number of such sectors is $2F_L$, where $F_L$ is the $L$'th Fibonacci number, which increases as $\mu^L$ as $L \to \infty$ [2, 3]. The total number of sectors $N_L$ is the number of IS's of length $L$ or less, and clearly obeys the recursion relation $N_L = N_{L-3} + 2F_L$. Consequently, $N_L \approx (1+\mu)\mu^L$ for large $L$ [4].

The size $D(L,I)$ of a sector with irreducible string $I$ is the number of configurations in that sector. It turns out that $D$ depends on $I$ only through its length $\ell(I)$. An explicit formula [4] can be obtained for the generating function $g(I) \equiv \sum_{L=\ell}^{\infty} x^L D(L,I)$. It is then possible to find the asymptotic growth law for $D(L,I)$ for given $I$. For instance, in the sector labelled by the null irreducible string $I \equiv \phi$ (which contains the set of all fully reducible configurations $\{\mathcal{C}\}_\phi$), we find $D(L,\phi) \sim (27/4)^{L/3}$ for large $L$. For a sector with an IS of length $\ell = \rho L$, the size $D(L,I)$ increases as $\kappa(\rho)^L$ where

$$\kappa(\rho)^{-3} = (1-\rho)^{1-\rho}(2+\rho)^{2+\rho}/27. \tag{1}$$

For a configuration $\mathcal{C}_{ran}$ chosen totally at random from amongst the $2^L$ possible configurations, $\rho$ is equal to $1 - 3\mu^{-2}/2$ with probability 1 in the limit of large $L$.

The Irreducible String encapsulates an infinite number of conservation laws. A procedure for finding these conservation laws is now sketched. For an arbitrary configuration of spins, say $\{\uparrow\downarrow\uparrow\downarrow\uparrow\uparrow\uparrow\downarrow\downarrow\uparrow\downarrow$ ...$\}$ form an operator product of the form $\Pi = PQPQPPPQQPQ\ldots$, where we write down linear operators $P$ or $Q$ in the same sequence as the up or down spins in the configuration ($P$ for $\uparrow$ and $Q$ for $\downarrow$). If the operators satisfy the condition $P^3 = Q^3 = 1$, and do not commute with each other, the product $\Pi$ for the full configuration is evidently the same as that associated with the corresponding product for the IS. Concretely, $P$ and $Q$ may be represented as matrices; examples of $2 \times 2$ representations are given in Ref. 4. If $P$ and $Q$ are functions of a continuous parameter $x$, the operator product is then a matrix $\Pi(x)$ whose elements are functions of $x$. We may expand a typical element $f(x)$ in a power series $\sum f_m x^m$. The coefficients $f_m$ constitute an infinite, commuting set of conserved quantities, each a function of the spins $\{S_i\}$. For instance with the parameterization used in Ref.

4, $f_1$ is linear in the $S_i$, and its real and imaginary parts imply the conservation of $M_A - M_B$ and $M_B - M_C$, where $M_A, M_B$ and $M_C$ are the sublattice magnetizations when the line is broken into three sublattices $A, B$ and $C$. Higher order $f_m$'s involve strings of spins. Since it is the IS which determines the order of the uncollapsed $P$ and $Q$ operators in the product, it carries information about $f(x)$, and therefore all the $f_m$ as well. Thus, the IS encodes an infinite number of conservation laws in a compact form.

# 3 Time Dependence: Dynamical Diversity

Under time evolution, an initial configuration evolves into the steady state of the corresponding sector. It is interesting to ask about the behaviour of time-dependent correlations in the steady state such as the autocorrelation function

$$C(t) = \langle S_i(0)S_i(t)\rangle - \langle S_i\rangle^2, \tag{2}$$

and its dependence on $k$.

For the case of dimers ($k = 2$) on a linear chain (with sublattices labelled $A$ and $B$), it is useful to make a particle-hole transformation on the $A$ sublattice [3]. Under this mapping, a configuration $\mathcal{C}$ goes into $\mathcal{C}'$, where $S_i(\mathcal{C}') = (-1)^i S_i(\mathcal{C})$. The conservation of $M_A - M_B$ then becomes the conservation of particle number in the transformed problem. The elementary steps of DE dynamics become particle-hole interchanges at neighboring sites. If $\epsilon \neq \epsilon'$, there is also a staggered external potential. On coarse graining, the effects of this potential are irrelevant, and therefore $C(t)$ decays diffusively, $C(t) \sim C(0)/\sqrt{Dt}$ at large $t$ where the diffusion constant $D$ can be evaluated explicitly [3]. If $\epsilon = \epsilon'$, the stochastic evolution operator can be exactly mapped to the Hamiltonian of the spin-$\frac{1}{2}$ Heisenberg model [2, 3]. This allows $C(t)$ to be evaluated in closed form for all $t$, and provides confirmation of the diffusive behaviour discussed above.

For the case of trimers, representative of all $k \geq 3$, Monte Carlo simulations have shown that the asymptotic behaviour of $C(t)$ can be quite different in different sectors, a characteristic of MSD systems [6]. Figure 2 shows the behaviour of $C(t)$ obtained from simulations in the

following representative sectors.
(1) a random initial configuration
(2) an IS formed by repeating [↑↓↓] $L/6$ times
(3) an IS formed by repeating [↑↓] $L/4$ times
(4) an IS formed by repeating [↑↑↓↑↓↓] $L/12$ times
(5) the null sector $\phi$ with vanishing IS.

The following general points are evident from the Monte Carlo results.

*Sublattice Dependence*: If the lattice is divided into 3 sublattices $A, B, C$, the autocorrelation function $C_i(t)$ can depend on the sublattice to which site $i$ belongs. In some sectors, the functional form of the long-time decay can be different on different sublattices. In Fig. 2, such behaviour is displayed in Sectors 2 and 4. Thus the long-time dynamics on each sublattice appears to proceed independently of the others, even though the elementary step of trimer deposition or evaporation necessarily involves all three sublattices together.

*Dynamical Diversity*: In almost all sectors, $C_i(t)$ decays as $t^{-1/4}$. However, one can find many sectors where $C_i(t)$ decays as a power law $\sim t^{-\theta}$, with a sector-dependent power $\theta$. The number of such sectors is an exponentially increasing function of $L$, even though their fractional number goes to zero for large $L$. Specifically, in Sector 1, $\theta \simeq 0.25$; in Sector 2 and on two sublattices of Sector 4, $\theta \simeq 0.5$; in Sector 5, $\theta \simeq 0.6$. In some sectors (*e.g.* Sector 3 and one sublattice in Sector 4), the decay follows a stretched exponential form.

How one can understand these unusual features of the dynamics? An answer is provided by the Irreducible String, for besides being a label on sectors, it has a certain dynamical significance. Under DE dynamics, configurations evolve in time, inducing a motion of the elements of the IS (see Fig. 1).

Consider time evolution in a sector labelled by the IS $\{\alpha_1, \alpha_2, \cdots, \alpha_\ell\} \equiv \{\alpha_j\}$. Let $C(t)$ be the full string corresponding to the configuration at time $t$, and consider applying the deletion algorithm to it. If $X_j(t)$ is the location of the $j^{\text{th}}$ site (from the left) that remains undeleted, we may look upon the set $\{X_j(t)\}$ as the positions of $\ell$ interacting random walkers on a line. Each walker carries a conserved spin label $S_{X_j(t)} = \alpha_j$ for all $t$. Since the order of spins in $\{\alpha_j\}$ is invariant, effectively the walkers cannot cross each other, i.e. $X_{j+1}(t) > X_j(t)$, for all $j$, all $t$. Under time evolution $X_j(t)$ changes in jumps, by a multiple of 3 spacings at a time.

Since the sequence of spins in the IS is an invariant of the motion, it is plausible that their movement constitutes the slow modes of the system, and would suffice to explain the long-time behaviour. We conjecture that at long times, the behaviour of $C_i(t)$ in a particular sector of the DE model is in the same universality class as the spin-spin autocorrelation function $C_i^{RW}(t)$ in a simpler problem with the same spin sequence. This problem is defined as follows: There are $\ell$ random walkers with each walker carrying a spin $\alpha_j$ which is unchanged under dynamics. Each walker can jump left or right by 3 steps only if no other walkers intervene. The motion of this system of hard core random walkers with conserved spin (HCRWCS) is simpler than the motion of the IS elements. This is because in the original DE model, there are effective long-ranged interactions between walkers, induced by integration over the remaining degrees of freedom [4]. These are ignored in the HCRWCS. The conjectured similarity of $C_i(t)$ and $C_i^{RW}(t)$ is plausible as the HCRWCS model has the exactly the same conservation laws as the original deposition evaporation model.

It is then easy to understand the sublattice dependence discussed above. As each element $\alpha_n$ of the IS moves by multiples of 3 lattice spacings, it always stays on the same sublattice, say $A$. This is reflected in the jump rules of the HCRWCS model. The long-time behaviour of $C_i^{RW}(t)$ and thus $C_i(t)$ for $i \in A$ is governed only by the subset $\{\alpha_{j'}\}_A$ on sublattice $A$, independent of $\{\alpha_{j'}\}_B$ and $\{\alpha_{j'}\}_C$.

Since the spin of the walkers does not affect their motion at all in the HCRWCS model, fluctuations in the number of walkers are like those in the symmetric exclusion process [7]. Thus in the steady state of the HCRWCS model, the mean squared difference between the number of particles between two distant points, at times $t_0$ and $t_0+t$, grows as $t^{1/2}$. To test for a similar behaviour in the DE system, we monitored $\sigma^2(t) \equiv \langle [N(r,t_0+t) - N(r,t_0)]^2 \rangle$ where $N(r,t)$ is the length of irreducible string between points 0 and $r$ at time $t$. This quantity is roughly equal to the number of walkers to the left of $r$, but remains well defined even in the sector where there are no walkers, $\rho = 0$. For all sectors with $\rho$ finite, $\sigma^2(t)$ is found to grow as $t^{1/2}$, lending strong support to the conjecture of equivalence of the DE and HCRWCS models [6]. In the sector where the IS vanishes ($\rho = 0$), on the other hand, $\sigma^2$ is found to grow as $t^{2\beta}$ with $\beta \simeq 0.19$.

The large $t$ behaviour of $C_i^{RW}(t)$ may be found as follows. In terms of the spins $\{\alpha_j\}$ and coordinates $\{X_j\}$ of the walkers, we have

$S_i(t) = \sum_j \alpha_j \delta(X_j(t) - i)$, where $\delta$ denotes a Kronecker delta function. Recalling that only every third particle visits a particular sublattice $n$, the autocorrelation function $C_i^{RW}$ then becomes

$$C_i^{RW}(t) = \sum_{j,m} \alpha_j \alpha_{j+3m} \Big\langle \delta(X_j(t_0) - i)\delta(X_{j+3m}(t_0 + t) - i) \Big\rangle \tag{3}$$

where the angular brackets denote average over evolution. Thus we may write, for site $i$ on the $n$'th sublattice,

$$C_i^{RW}(t) = \sum_{m \epsilon n} \gamma_n(m)\, G(m|t) \tag{4}$$

where $G(m|t)$ is the spin-independent HCRWCS correlation function within angular brackets in Eq. (3). $\gamma_n$ is an average over the spin sequence, which is time-independent but sector-dependent

$$\gamma_n(m) = \frac{3}{\ell} \sum_{j \epsilon n} \alpha_j \alpha_{j+3m} \qquad n = 1, 2, 3. \tag{5}$$

The long-time behaviour of $G(m|t)$ is the same as for the symmetric exclusion process. Using the large-$t$ equivalence of the latter to a stochastic harmonic theory [8, 9], $G(m|t)$ can be evaluated. The result is

$$G(m|t) \approx \frac{1}{\sqrt{2\pi\Delta^2}}\, e^{-m^2/2\Delta^2}. \tag{6}$$

The width $\Delta$ is given by [9]

$$\Delta^2 = |m|\; erf\left(\frac{|m|}{\sqrt{t}}\right) + \sqrt{\frac{t}{\pi}}\, e^{-m^2/t}. \tag{7}$$

The upshot is that $G(m|t) \sim t^{-1/4}$ for $m \ll t^{1/2}$ while $G(m|t) \sim \exp(-|m|)$ for $m \gg t^{1/2}$.

Putting Eqs. (4), (5) and (6) together, we may deduce the behaviour of $C_i^{RW}(t)$ in various sectors of HCRWCS model. The convolution over $m$ in Eq. (4) produces very different forms for $C_i^{RW}(t)$, as $\gamma_n(m)$ varies substantially from sector to sector.

Sector 1 (random initial condition): In this case $\gamma_n(m)$ is significant only for very small values of $m$. Thus $C_i^{RW}(t)$ varies as $t^{-1/4}$. This result is also true for other sectors in which $\gamma_n(m) \to 0$ as $|m| \to \infty$, and $\sum_m \gamma_n(m)$ is nonzero and finite.

Sector 2 [↑↓↓]: In this case $\alpha_k = \alpha_{k+3m}$ for all $m$. Hence $C_i^{RW}(t)$ reduces to a density-density correlation which decreases as $t^{-1/2}$ for large $t$. The same behavior is expected in all periodic IS in which each sublattice magnetization is nonzero.

Sector 3 [↑↓]: In this case $\gamma_n(m) = (-1)^m$, and this implies that $C_i^{RW}(t)$ has the stretched exponential form $\exp[-(t/\tau)^{1/2}]$ at large time at all sites. The same behaviour would occur whenever the IS is periodic with zero net magnetization on each sublattice. For instance this is also the case when the IS is [↑↑↓↓] repeated $L/8$ times. We have verified by Monte Carlo simulation that stretched exponential behaviour is obtained again.

Sector 4 [↑↑↓↑↓↓]: On sublattices $A$ and $C$ we have $\alpha_j = \alpha_{j+3m}$, and thus as in sector 2, we have $C_i^{RW}(t) \sim t^{-1/2}$. On sublattice $B$, $\gamma_n(m)$ is $(-1)^m$, and as in sector 3 we get stretched exponential decay.

A rich variety of other power laws is also possible. If the IS consists of a spin sequence with slowly decaying correlations, this decay influences the power characterizing the behaviour of the autocorrelation function. If the correlations in the $IS$ imply that $\tilde{\gamma}_n(q)$, the Fourier transform of $\gamma_n(m)$, varies as $|q|^\phi$ for small $q$, then $C_i^{RW}(t)$ is predicted by Eq. 3 to vary as $t^{-(1+\phi)/4}$.

The HCRWCS predictions for $C_i^{RW}(t)$ accord very well with the Monte Carlo results for $C_i(t)$ shown in Fig. 1. We conclude that in all sectors with a nonzero density $\rho$ of the IS, the long-time behaviour of $C_i(t)$ is determined primarily by the diffusive motion of the elements of the IS.

However, in the null-IS sector consisting of the set of all fully reducible

configurations$\{C\}_\phi$, the dynamics is different, as it is the fluctuations in the separations between the deleted spins which matter (these were the fast modes in sectors with $\rho \neq 0$, and could be ignored there as a result). Numerical data from Monte Carlo simulations [6], and exact diagonalization of the transfer matrix for finite size systems [10] suggests that in this case fluctuations are slowly decaying and sub-diffusive in nature. The estimated value of the dynamical exponent $z \simeq 2.5$ is quite different from the value for the Kardar-Parisi-Zhang type of stochastic growth models [11] which in one dimension are known to have $z = 3/2$, even though the trimer model can be viewed as a matrix generalization of the KPZ model [5].

# 4 Conclusion

An interesting and important point about $k$-mer deposition-evaporation systems is that for $k \geq 3$, they belong to the many-sector decomposable (MSD) class, characterized by strong breaking of ergodicity and a rich variety of dynamical behaviours. Central to the understanding of both the phase space decomposition into exponentially many sectors, and also the diversity of asymptotic dynamical behaviour, is the Irreducible String. The IS provides a compact and convenient means of labelling disjoint sectors. Besides, the diffusive movement of IS elements through the lattice (with an implied hard core constraint between different elements) constitute the slow modes of the system; this allows us to characterize the asymptotic decay laws of the autocorrelation function, and understand their different behaviour in different sectors.

There are several other examples of systems in the MSD class.
(i) The dynamics of the 3-state Potts antiferromagnet at $T = 0$ with exchange (Kawasaki-like) moves can be mapped onto that of the trimer deposition-evaporation model [12].
(ii) A deposition-evaporation model for dimers with $q$ colours also shows the MSD property [13]. For this model, the reduction to HCRWCS-like systems by integrating over the remaining degrees of freedom can be done exactly, and the dynamics of random walkers is independent of their spin configurations.
(iii) A one-dimensional model of hopping dimers with reconstitution also belongs to this class [14]. It can be analysed by mapping it to the exclusion process with two types of particles.

So far, most examples of MSD systems which have been studied are one dimensional, and suitably defined Irreducible Strings play a central role in their analysis. It would be very interesting to demonstrate and study MSD behaviour in higher-dimensional systems where our techniques using the IS fail. A 2-dimensional example is the single-spin-flip Ising model with energy conserved locally (the so-called Q2R automaton) [15] in which the nonergodicity effects appear to be important below a critical temperature but not above. There is also a DE system in two dimensions, where the number of jammed states can be shown to grow exponentially in system size. Monte Carlo studies of dynamics in a particular unjammed sector suggest a diffusive behaviour [16]. The full phase space structure and associated dynamical properties remain to be explored.

# References


[1] R.G. Palmer, *Adv. in Phys.* **31**, 669 (1982).

[2] M. Barma, M.D. Grynberg and R.B. Stinchcombe, *Phys. Rev. Lett.* **70**, 1033 (1993).

[3] R.B. Stinchcombe, M.D. Grynberg and M. Barma, *Phys. Rev. E* **47**, 4018 (1993).

[4] D. Dhar and M. Barma, *Pramana - J. Phys.* **41**, L193 (1993).

[5] M. Barma and D. Dhar, in *Proceedings of the International Colloquium on Modern Quantum Field Theory II* edited by S. R. Das, G. Mandal, S. Mukhi and S.R. Wadia (World Scientific, Singapore, to appear).

[6] M. Barma and D. Dhar, *Phys. Rev. Lett.* **73**, 2135 (1994).

[7] T.M. Liggett, *Interacting Particle Systems* (Springer, New York, 1985).

[8] S. Alexander and P. Pincus, *Phys. Rev. B* **18**, 2011 (1978).

[9] S.N. Majumdar and M. Barma, *Physica A* **177**, 366 (1991).

[10] P.B. Thomas, M.K. Hari and D. Dhar, *J. Phys. A* **27**, L831 (1994).

[11] M. Kardar, G. Parisi and Y. Zhang, Phys. Rev. Lett. **56**, 889 (1986); J. Krug and H. Spohn in *Solids far from Equilibrium* ed. C. Godrèche (Cambridge Univ. Press, Cambridge, 1991).

[12] D. Dhar, *J. Indian Inst. Science* **75**, 297 (1995).

[13] M.K. Hari and D. Dhar, *J. Phys. A*, to appear.

[14] G.I. Menon, M. Barma and D. Dhar, preprint TIFR/TH/95-58.

[15] C. Moukarzel and N. Parga, *J. Phys. A* **22**, 943 (1989), and references cited therein.

[16] N. N. Chen, M. D. Grynberg and R. B. Stinchcombe. *J. Stat. Phys.***78**, 971 (1995).

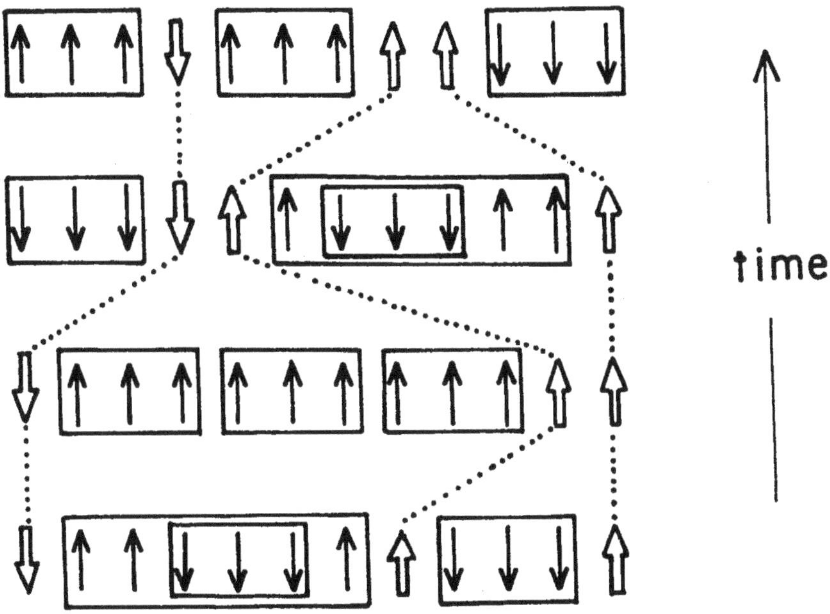


Fig. 1 Illustration of the process of reduction and the identification of the irreducible string (IS). Rectangular boxes are drawn around reducible triplets, with reduction done from left to right. The remaining spins (shown as thick arrows) constitute the IS, which is a constant of the motion. The world lines (shown dotted) of these spins do not cross, and resemble those of hard-core random walkers with spins.

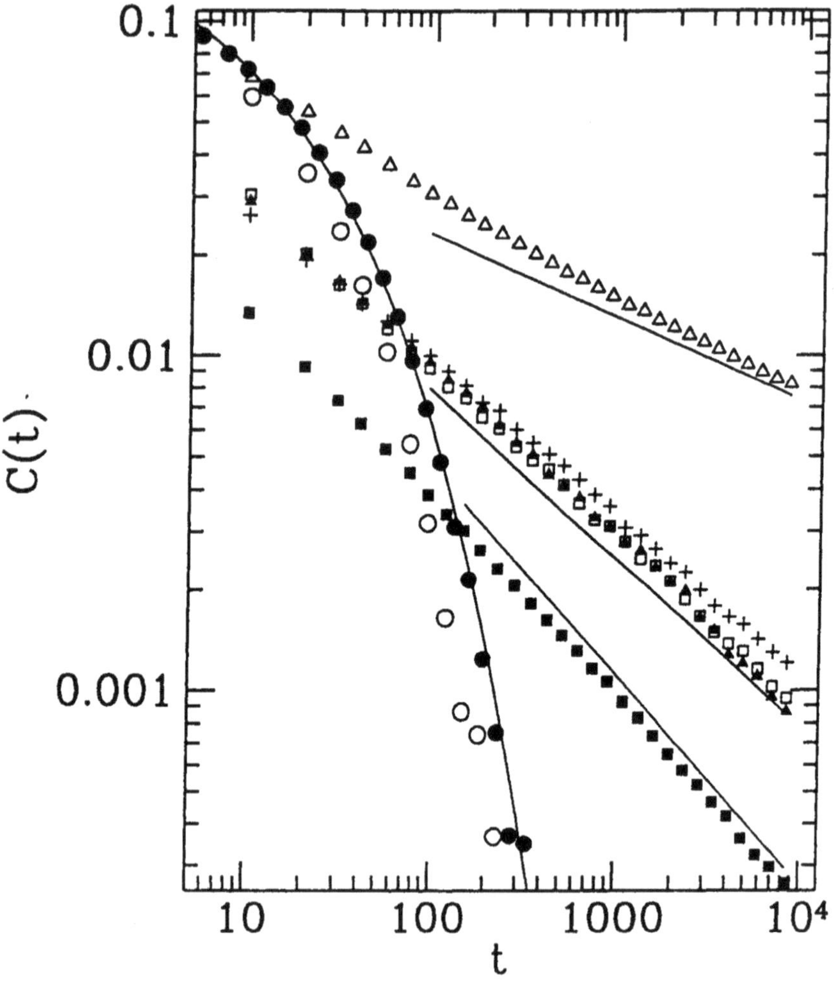


Fig. 2 Variation of the decay of the autocorrelation function $C_i(t)$ in Sectors 1-5 (defined in text). Sector 1 (open triangles); Sector 2, sublattice A (+), sublattices B and C (filled triangles); Sector 3 (filled circles); Sector 4, sublattices A and C (open squares), sublattice B (open circles); Sector 5 (filled squares). The straight lines show power law decays $t^{-\theta}$ with $\theta = 0.25$ (top), 0.50 (middle), 0.62 (bottom); the curve shows a stretched exponential ($\sim \exp(-\sqrt{t/\tau})$) decay.

# RANDOMLY CHARGED POLYMERS: EXCESS CHARGE DEPENDENCE OF SPATIAL CONFIGURATIONS

YACOV KANTOR
*School of Physics & Astronomy, Tel Aviv University*
*69978 Tel Aviv, Israel*
E-mail: kantor@orion.tau.ac.il

and

MEHRAN KARDAR
*Department of Physics, Massachusetts Institute of Technology*
*Cambridge, MA 02139, U.S.A.*
E-mail: kardar@mit.edu

ABSTRACT

Spatial configurations of randomly charged polymers, known as polyampholytes (PAs), are very sensitive to the overall excess charge $Q$. Analytical arguments, supported by Monte Carlo simulations[5,7] and exact enumeration studies,[8] lead to the following picture: For $Q < Q_c \approx q_0\sqrt{N}$ ($q_0$ is the elementary charge, $N$ is the number of monomers in the polymer), the radius of gyration $R_g$ of the polymer decreases with decreasing temperature $T$ and the polymer becomes compact, while for $Q > Q_c$ the polymer stretches with decreasing $T$. At low $T$, the dense states are described by Debye–Hückel theory, while the expanded states resemble a *necklace* of globules connected by strings. At such temperatures, the transition between the dense and the expanded states with increasing $Q$, is reminiscent of the breakup of a charged drop.

## 1. Introduction

Statistical mechanics of randomly charged polymers, called *polyampholytes* (PAs), is a challenging subject because it embodies an interesting combination of long range interactions and randomness. While the physics of homogeneous polymers has a reasonably firm basis,[1] considerably less is known about heteropolymers, although the latter present an extremely rich problem of biological significance.[2] In this presentation we review some properties of PAs.

The spatial extent of a polymer is characterized by the critical exponent $\nu$, which relates its radius of gyration (r.m.s. size) $R_g$ to the number of monomers $N$, by the power law $R_g \propto N^\nu$. The polymer will be called "compact" if $\nu = 1/d$, where $d$ is the dimension of the embedding space, and "stretched" if $\nu = 1$. The simplest model of PAs is as a flexible chain of $N$ monomers, each of which has a charge $\pm q_0$ selected from a well defined ensemble of quenches. The polymer has a characteristic microscopic length scale $a$, such as the range of excluded volume interactions, or the nearest neighbor distance along the chain. The monomers of the PA interact both via (short range) excluded volume interactions and long range (unscreened) Coulomb interactions. In the simplest ensemble of quenches, each monomer takes a charge $q_i = \pm q_0$ independently of all the others; i.e. $\overline{q_i q_j} = \delta_{ij} q_0^2$, where the overline indicates averaging over quenches. While the average excess charge $Q \equiv \sum_i q_i$ of such PAs is zero, a "typical" sequence has Q of about $\pm Q_c$, with $Q_c \equiv q_0 N^{1/2}$. This statement, as well as the definition of $Q_c$, is unrelated to the embedding dimension $d$. However, the importance of charge fluctuations (both for the overall polymer, or for large segments of it) does depend on the space dimension. The electrostatic energy of the excess charge, spread over the characteristic size of an ideal polymer ($R_g \propto N^{1/2}$), grows as $Q^2/R_g^{d-2} \sim N^{(4-d)/2}$. This simple dimensional argument shows that for $d > 4$ weak electrostatic interactions are irrelevant. (The excluded volume effects are also irrelevant in $d > 4$.) Thus, in $d > 4$ and at high temperatures, the PA behaves as an ideal polymer with an entropy–dominated free energy of the order of $-NT$. However, on lowering temperature it collapses into a dense state, taking advantage of a condensation energy of the order of $-Nq_0^2/a^{d-2}$.

We are primarily interested in the behavior of PAs in $d = 3$. We present both numerical and analytical evidence that the behavior of PA in $d < 4$ is extremely sensitive to the value of $Q$. In particular, we show that for $Q < Q_c$ the PA assumes compact configurations, while for $Q > Q_c$ it is in an expanded state. We suggest a "necklace" model which qualitatively describes the latter state.

## 2. Properties of Polyampholytes

### *2.1. Neutral Heteropolymers: Short–Range* vs. *Long–Range Interactions*

It is interesting to compare and contrast the behaviors of heteropolymers with short and long range interactions. Consider a polymer represented by a self–avoiding chain carrying a quenched sequence of random charges which interact only at *short* distances, with similar charges repelling each other and the opposite charges attracting.[3] In $d = 3$, at high temperatures $T$, the presence of such short range interactions does not influence the behavior of the polymer and it behaves as a self–avoiding walk (SAW). If $Q$ is small, i.e. the positive and negative charges are almost balanced, upon decrease of temperature, charges of opposite sign are more likely to be in the vicinity of each other. This introduces an effective attraction which increases in strength with decreasing temperature, and at a certain temperature the polymer collapses into a globule, as in a regular $\theta$–transition. When the imbalance $Q$ between the two types of charges increases, the transition temperature decreases, as demonstrated in Fig. 1a, and for sufficiently large imbalance the transition disappears altogether.[3] Since the behavior of the system varies very gradually with increasing $Q/N$, it does not really matter whether the ensemble of chains consists only of quenches with exactly vanishing $Q$, or is obtained by independently choosing each $q_i = \pm q_0$ with equal

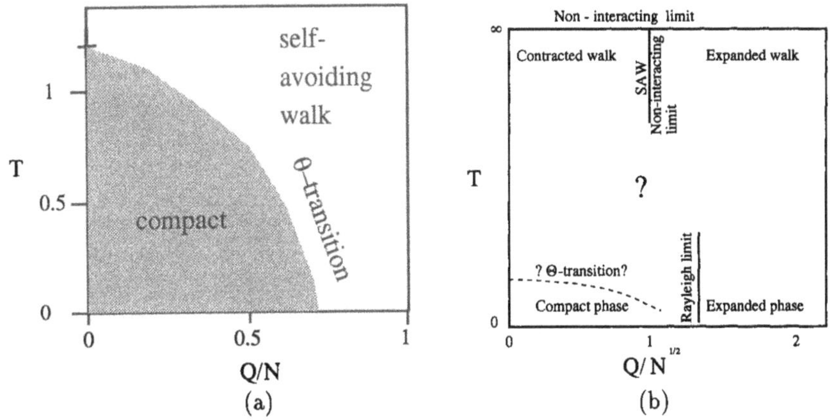


Fig. 1. Phase diagrams for random polymers with (a) short–range, and (b) Coulomb interactions, in $d = 3$. Note the different scaling of the excess charge $Q$ in the two figures.

probability. In the latter case $Q$ may deviate from zero by an amount of order $q_0 N^{1/2}$; such fluctuations have a negligible contribution to $Q/N$ for large $N$.

The long range Coulomb interactions in random PAs are always relevant at $d < 4$. An approximate treatment using a Debye–Hückel type theory[4] leads to the conclusion that a PA minimizes its free energy by assuming a configuration in which positive charges are predominantly surrounded by negative charges, causing screening as in a regular electrolyte. A detailed calculation suggests[4,5] the following picture: At a finite $T$, the PA can be divided into segments containing $n$ monomers each, where $n$ defines the length scale at which electrostatic interactions become of order $T$, i.e. $Q^2(n)/R_g(n) \sim q_0^2 n/(an^{\nu_0}) = T$ in $d = 3$, where $\nu_0 \approx 0.59$. The overall configuration is then composed of $n$–mer blobs: Inside the blob the PA behaves as an uncharged polymer, while the blobs form a compact structure resembling an electrolyte. The resulting PA is compact (i.e., has $R_g \sim N^{1/3}$) at all temperatures, with a $T$–dependent density. On the other hand, scaling arguments[6] assume that the $N$–dependence of the $R_g$ of a random PA is determined from the relation $Q^2/R_g \sim q_0^2 N/R_g \approx T$, leading to the conclusion that $R_g \sim N$ in $d = 3$. This apparent contradiction is resolved by noting[5] that the spatial configurations of PAs are extremely sensitive to the overall charge $Q$. In particular, it can be shown[5,7] that there is a critical charge $Q_c = q_0 N$, such that for $Q < Q_c$ the PAs are compact, while for $Q > Q_c$ they are stretched. The diagram in Fig. 1b indicates such separation between contracted and expanded regimes of the PA. In the following subsections we shall present different arguments

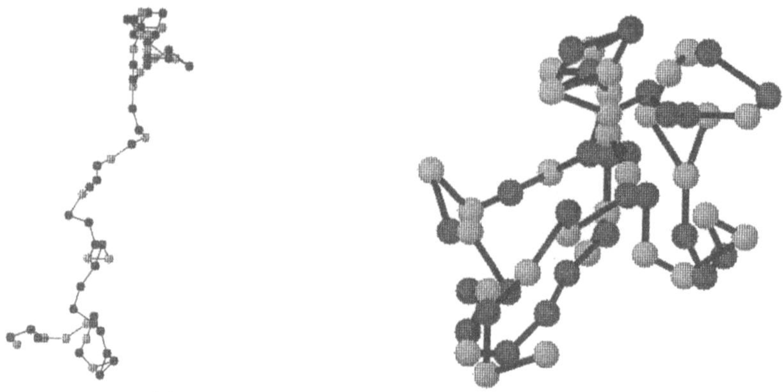

(a) (b)

Fig. 2. Low $T$ configuration of 64–monomer PAs with (a) $Q = 12$, and (b) $Q = 0$. Dark and bright shades indicate opposite charges.

and numerical results supporting this conclusion.

### 2.2. Polyampholytes with Excess Charge: High Temperature Arguments

For $d < 4$, electrostatic interactions are relevant and the high temperature phase is no longer a regular self-avoiding walk. At high $T$, the behavior of the polymer can be explored perturbatively. For the ensemble of uncorrelated charges ($\overline{q_i q_j} = 0$), the lowest order ($1/T$) correction to the quench-averaged $R_g^2$ vanishes.[5] However, restricting the ensemble to yield fixed $Q$, slightly modifies the quench-averaged charge–charge correlations. In particular, the two-point correlation function becomes $\overline{q_i q_j} = (Q^2 - Q_c^2)/N^2$ for $i \neq j$. This small (order of $1/N$) correction to the correlation function causes a significant change in $R_g^2$ due to the long range nature of the Coulomb interaction. A $1/T$–expansion[5] has a first order term proportional to $Q - Q_c$. Thus the size of a PA tends to decrease upon lowering temperature if $Q$ is less than $Q_c$, and increases otherwise.

Confirming the $1/T$–expansion, a strong $Q$–dependence of $R_g$ has been seen in Monte Carlo simulations:[7] It has been shown[5] that for $Q < Q_c$ the PA contracts with decreasing $T$, while for $Q > Q_c$ it expands. Fig. 2 depicts low temperature equilibrium configurations of two PAs, one of them has $Q$ slightly larger than $Q_c$ and is strongly stretched, while the other has $Q = 0$ and forms a globule. The two behaviors are separated by the vertical line at the top of the phase diagram in Fig. 1b. It should be noted that the arguments leading to location of this line are independent of $d$, as long as $d < 4$.

*2.3. Polyampholytes with Excess Charge: Low Temperature Arguments*

Monte Carlo results indicate that PAs with $Q = 0$ form dense globules at low temperatures. We can use this observation as a starting point for the investigation of the dependence of low–$T$ shapes on $Q$ in $d = 3$. The energy (or rather the quench–averaged free energy) of the PA is phenomenologically related to its shape by

$$E = -\epsilon_c N + \gamma S + \frac{bQ^2}{R}. \tag{1}$$

The first term is a condensation energy proportional to the volume (assumed compact) and $\epsilon_c \sim q_0^2/a$, the second term is proportional to the surface area $S$ (with a surface tension $\gamma \sim q_0^2/a^3$), while the third term represents the long–range part of the electrostatic energy due to an excess charge $Q$ ($b$ is a dimensionless constant of order unity). The correctness of separating the energy into the 3 terms appearing in Eq. 1 is not self–evident in *random* systems. A justification for the assumptions implicit in this equation is provided by an exact enumeration study of the ground states of PAs defined on a lattice:[8] Fig. 3 depicts the ground state energies of *all* 2080 possible quenches for 12–step (13–atom) chains. (The horizontal axis represents an arbitrary numbering of the quenches.) The energies are clearly separated into 7 bands, corresponding to excess charges of $Q = 1, 3, 5, \cdots, 13$. (There is only one quench with $Q = 13$.) While each band has a finite width, we see that the energy of a PA can be determined rather accurately by only specifying its net charge $Q$, while the additional details of the quench appear only in the width of the band.

The optimal shape is obtained by minimizing the overall energy in Eq. 1. The first term is the same for all compact shapes, while the competition between the surface and electrostatic energies is controlled by the dimensionless parameter

$$\alpha \equiv \frac{Q^2}{16\pi R^3 \gamma} = \frac{Q^2}{12V\gamma} \equiv \frac{Q^2}{Q_R^2}\,. \tag{2}$$

Here, $R$ and $V$ are the radius and volume of a spherical drop of $N$ particles, and we have defined the *Rayleigh charge* $Q_R$. We note that in the case of PAs[7,8] $Q_R \approx q_0 N^{1/2} = Q_c$. The behavior of a system described by Eq. 1 has been analyzed in the past in the contexts of charged conducting drops, as well as liquid drop models of atomic nuclei. If one restricts the possible shapes of a globule to spheroids, then one finds that for sufficiently large values of $\alpha$ (of order one or more), the energy of the system can be decreased by distorting the drop into a prolate spheroidal shape. Both MC simulations[7] and experimental results[9] indicate that for $Q < Q_R$, the $R_g$ of PA at low temperature is almost independent of $Q$, while it increases extremely fast as a function of $Q$ for larger charges.

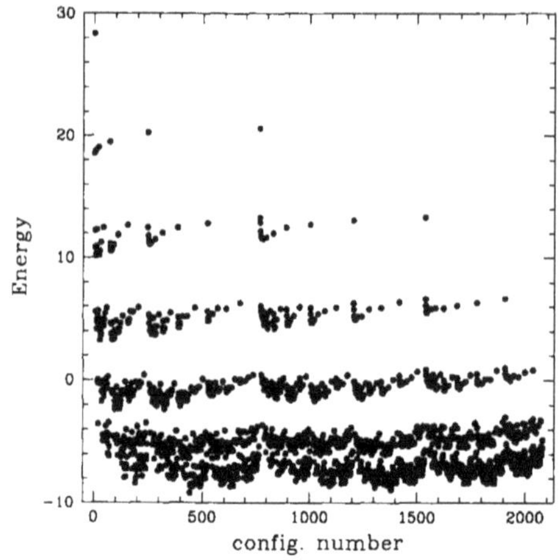


Fig. 3. Ground state energies of (arbitrarily numbered) quenches.

The transition between compact and expanded states with increasing $Q$ is represented by the vertical line at the bottom of the diagram in Fig. 1b. The presence of this line signifies the instability of the spherical shape, but provides no indication of what the stable state of the PA looks like. This question is taken up in the following section. It should be stressed, that although both high and low temperature arguments seem to provide a consistent picture of transition between compact and extended states, the equality between $Q_c$ and $Q_R$ depends on the space dimension. For $d < 3$ ($d > 3$) $Q_R$ increases with $N$ slower (faster) than $Q_c$, and therefore the high–$T$ criterion will not coincide with the low–$T$ limit.

## 3. The Necklace Model

### *3.1. Uniform Charges*

Although the ellipsoidal globule may have a lower energy than the spherical one, the former is *not* the minimum energy configuration of a charged drop: A uniformly charged drop minimizes its energy by *splitting* into two equal, infinitely removed, droplets for $\alpha > 0.3$, and the number of such droplets increases with $\alpha$. Obviously, the PA must maintain its connectivity. We can constrain the overall shape to remain singly connected by linking the droplets via narrow tubes of total length $L$ and diameter $a$. As long as $La^2 \ll R^3$, most of the charge remains in the spheres. The total electrostatic energy is proportional to $Q^2/L$, while the surface energy cost grows as

$\gamma aL$; equating the two gives $L \propto Q$. For large $Q$, the PA will look like a *necklace* of globules connected by narrow strands. The configuration depicted in Fig. 2a indeed bears some resemblance to such a necklace.

It should be noted that $R_g$ in this simplified discussion is proportional to the typical charge $Q$, and therefore proportional to $N^{1/2}$. This serves as an indication that the quench–averaged configurations are not compact. The picture has one important shortcoming for PAs: it assumes that the excess charge is uniformly distributed along the chain and disregards the strong charge fluctuations. It thus applies to such problems as an alternating sequence of charges[10] to which some additional charge has been added, or to a weakly charged homogeneous polymer in which the attraction (and, therefore, the condensation energy) is provided by a different mechanism, such as short range attractions.

### *3.2. Effects of Randomness*

We now illustrate the difficulties caused by randomness for the case of an unrestricted PA. Since $\overline{Q^2} = q_0^2 N$, where the overline denotes an average over the ensemble of all quenches, we have $\overline{\alpha} = 1$. A uniformly charged drop is unstable to splitting already for $\alpha \approx 0.3$, and thus a typical random PA is expected to form several globules connected by narrow tubes. Now consider splitting sequences of $N$ monomers with total charge constrained by a particular $\alpha$ into two equal subchains of charges $Q_1$ and $Q_2$. It is easy to show that each segment has $\overline{\alpha_{\text{subchain}}} = (1+\alpha)/2$, while the mean product of the charges is $\overline{Q_1 Q_2} = q_0^2 N(\alpha - 1)/4$. The subchains have, on average, values of $\alpha$ close to unity. Also, for $\alpha = 1$, the average value of the product of charges vanishes. We thus have the paradoxical situation in which most spherical shapes are unstable, while there is on average no energetic gain in splitting the sphere into two equal parts.

Therefore, charge inhomogeneities drastically modify the necklace picture. The resulting PA is probably still composed of rather compact globules connected by a (not necessarily linear) network of tubes. The globules are selected preferentially from segments of the chain that are approximately neutral (or at least below the instability threshold), while the tubes are from subsequences with larger than average excess charge. It can be shown[11] that the probability of finding a very large neutral segment in a random sequence of charges is large, i.e. it is possible to build a configuration consisting of one very large neutral globule with highly charged "tails" sticking out of it.

While the necklace model provides a convenient starting picture for the behavior of PAs, it does not encompass the entire complexity of the problem. The probability distribution of the $R_g$ of the ground states of PAs obtained from exact enumeration study[8] has a large peak for small $R_g$, indicating that most of the configurations are compact, as well as a slowly decaying tail at large $R_g$. This tail represents the expanded configurations, and determines the behavior of the quench-averaged $R_g$. Indeed, for $N \leq 13$ we found[8] that 80% of all quenches can be classified as compact, i.e. their $R_g$ increases as $N^{1/3}$, while the remaining 20% have quench-averaged $R_g$s increasing with $N$ at least as fast as that of a SAW.

We can summarize our current knowledge of the behavior of random PAs as follows: The exact enumeration and Monte Carlo studies confirm that random PAs have (on the average) expanded spatial conformations, although further work is needed to verify whether $\nu = 1$. The necklace model provides a useful qualitative view of the ground state configurations. Further studies are needed to put the model on a more quantitative basis.

## 4. Acknowledgements

This work was supported by the US–Israel BSF grant No. 92–00026, and by the NSF through grant No. DMR–94–00334 at MIT's CMSE.

## 5. References


1. P.G. de Gennes, *Scaling Concepts in Polymer Physics*, Cornell Univ. Press, Ithaca (1979).
2. D.L. Stein, Proc. Natl. Acad. Sci. USA **82**, 3670 (1985); J.D. Bryngelson and P.G. Wolynes, Proc. Natl. Acad. Sci. USA **84**, 7524 (1987); H.S. Chan and K.A. Dill, Physics Today **46**(2), 24 (1993); T. Garel and H. Orland, Europhys. Lett. **6**, 307 (1988); E.I. Shakhnovich and A.M. Gutin, Europhys. Lett. **8**, 327 (1989); M. Karplus and E.I. Shakhnovich, in *Protein Folding*, ed. by T.E. Creighton, ch.4, p. 127, (Freeman & Co., New York, 1992).
3. Y. Kantor and M. Kardar, Europhys. Lett. **28**, 169 (1994).
4. P.G. Higgs and J.-F. Joanny, J. Chem. Phys. **94**, 1543 (1991); J. Wittmer, A. Johner and J.F. Joanny, Europhys. Lett. **24**, 263 (1993).
5. Y. Kantor, H. Li, and M. Kardar, Phys. Rev. Lett. **69**, 61 (1992); Y. Kantor, M. Kardar, and H. Li, Phys. Rev. **E49**, 1383 (1994).
6. Y. Kantor and M. Kardar, Europhys. Lett. **14**, 421 (1991).
7. Y. Kantor and M. Kardar, Europhys. Lett. **27**, 643 (1994); Y. Kantor and M. Kardar, Phys. Rev. **E51**, 1299 (1995).
8. Y. Kantor and M. Kardar, Phys. Rev. **E52**, in press (1995).
9. X.-H. Yu, A. Tanaka, K. Tanaka, and T. Tanaka, J. Chem. Phys. **97**, 7805 (1992); Yu X.-H., Ph. D. thesis, MIT (1993).
10. J.M. Victor and J.B. Imbert, Europhys. Lett. **24**, 189 (1993).
11. Y. Kantor and D. Ertaş, J. Phys. **A27**, L907 (1994); D. Ertaş and Y. Kantor, Phys. Rev. **E**, submitted (1995).

# SELF-AVOIDING POLYMERIZED MEMBRANES

François DAVID
*Service de Physique Théorique, CE de Saclay*
*91191, Gif-sur-Yvette CEDEX, FRANCE*

Recent progresses in the understanding of the scaling behavior of self-avoiding flexible polymerized membranes (tethered manifolds) are reviewed (joint works with B. Duplantier and E. Guitter, and with K. Wiese). They rely on a new general renormalization group approach for a class of models with non-local singular interactions. This approach allows to prove the existence of a $\epsilon$-expansion for the scaling exponents, and validates the one loop results obtained by direct renormalization methods. Applications of the method to polymerized membranes at the tricritical $\Theta$-point are presented.

## 1 What are polymerized membranes ?

Polymerized membranes, also called tethered membranes, are an interesting model for two-dimensional flexible polymerized films [1]. In particular, they generalize the well-known problem of the statistical behavior of long chains, i.e. of polymers. However, even simple properties, such as the statistical properties at thermal equilibrium of a single membrane, are quite difficult to study theoretically, and require the development of new techniques. I shall review some recent progresses made with B. Duplantier and E. Guitter [2], and some applications of these techniques with K. Wiese [3].

Let me consider an ideal flexible two dimensional network fluctuating in three dimensional space, characterized by: *(i)* a fixed and regular internal structure with a fixed connectivity (for instance a triangular lattice), and *(ii)* negligible bending rigidity. The only forces between elements of the network are the elastic forces (described for instance by a harmonic potential between nearest neighbors). If self-avoiding interactions are not taken into account, one deals with a "phantom" membrane, which is known to be, for vanishing or small enough bending rigidity, in a crumpled phase. The large scale fluctuations are described by a Gaussian model. Such a model is physically different from the random surfaces used in models of fluid two-dimensional films (for instance for vesicles or micro-emulsions). For these models, the bending rigidity is important and the fluid character of the film is reproduced by a fluctuating connectivity of the lattice.

Taking into account the self-avoiding interactions (which are short ranged in the three dimensional embedding space, but involve elements of the lattice which can be arbitrarily far apart in term of the internal lattice distance) one expects that the long range properties of the membrane will be changed. For instance, the average size $\langle R \rangle$ of the membrane in physical space is expected to scale with the internal size $L$ of the lattice as

$$\langle R \rangle \propto L^{\nu} \qquad 0 < \nu < 1 \tag{1}$$

if the membrane is still crumpled, but swollen by self-avoidance. The membrane may even become flat (this is observed in numerical simulations of self-avoiding tethered membranes in three dimensions); the exponent $\nu$ is then equal to 1.

## 2 An Edwards Model for $D$-dimensional polymerized membranes:

In the continuous model introduced in [4,5], the field $\vec{r}(x)$ describes the embedding of a $D$-dimensional space, whose points are labeled by the coordinate $x$, into the external $d$-dimensional space, whose points are labeled by the vector $\vec{r}$. The Hamitonian $H$ (the free energy for a configuration $r$) is the sum of a Gaussian elastic energy and of a 2-body repulsive interaction, proportional to the coupling constant $b$:

$$H[\vec{r}] = \int d^D x \, \frac{1}{2}(\nabla_x \vec{r})^2 + b \int d^D x \int d^D y \, \delta^d(\vec{r}(x) - \vec{r}(y)) \, . \tag{2}$$

The internal dimension $D$ may be taken as a continuous parameter, interpolating between polymers ($D = 1$) and membranes ($D = 2$). The issue is to compute with this model critical exponents describing the scaling behavior of large membranes, for instance the exponent $\nu$ (related to the fractal dimension $d_f$ of the membrane by $\nu = D/d_f$), and the configuration exponent $\gamma$, related to the scaling of the partition function $Z$ of a finite membrane with internal extent $L$ by

$$Z \propto L^{\gamma-1} \, \mathrm{constant}^{L^D} \, . \tag{3}$$

The mean field exponents are obtained by setting $b = 0$ and one obtains $\nu_0 = (2-D)/2$ and $\gamma_0 = 1 - d(2-D)/2$.

Dimensional analysis shows that mean field theory is invalid if $\epsilon$, the engineering dimension of $b$, is positive

$$[b] = \epsilon = 2D - d(2-D)/2 > 0 \, . \tag{4}$$

The general picture of the expected behavior as a function of the internal dimension $D$ and of the external dimension $d$ is presented in Fig. 1

A natural idea is to compute the corrections to mean field by a $\epsilon$-expansion à la Wilson-Fisher. Using the so-called direct renormalization, inspired from

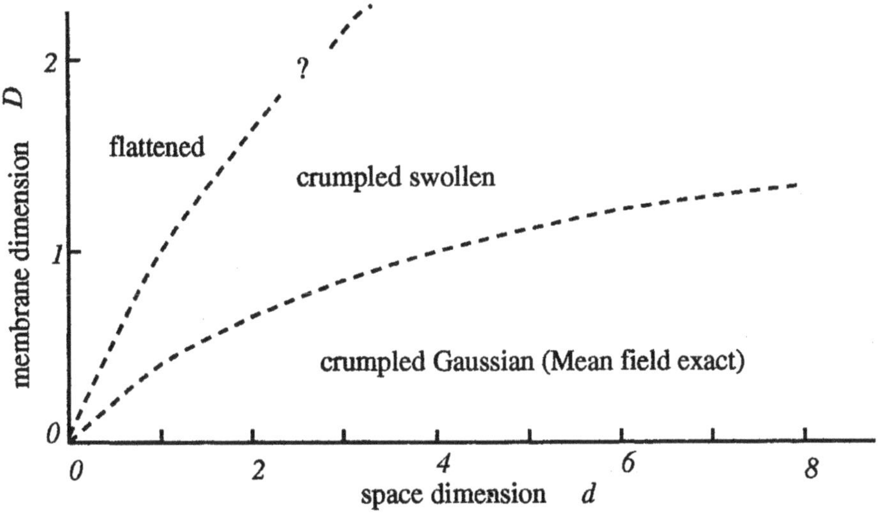


Figure 1: $d$-$D$ plane

polymer theory[6], the first explicit calculations have been performed to first order in $\epsilon$ by Aronowitz & Lubensky[4] and by Kardar & Nelson[5]. The basic idea of this method is to perform explicit perturbative calculations for a finite membrane and for $\epsilon > 0$. Perturbation theory is then UV and IR finite, but has UV divergences when $\epsilon \to 0$. These poles in $1/\epsilon$ can be removed by reexpressing the observables in terms of adequate dimensionless renormalized quantities, such as the 2nd virial coefficient. The internal size $L$ of the membrane plays the role of a renormalization $mass^{-1}$ scale, and the renormalization group equations can thus be obtained, by writing the $L$ dependence of the renormalized theory. The consistency of this procedure at first order has been checked by explicit calculations by Duplantier, Hwa & Kardar[8], but its consistency at all orders cannot be proved by the de Gennes trick valid for polymers (the scaling limit of the self-avoiding walk can be mapped into a local field theory with $O(n)$ symmetry in external $d$-dimensional space, in the limit $n \to 0$, and standard renormalization group theory is then appliquable).

## 3 Renormalization Theory and Renormalization Group for non-local Theories:

Instead of using direct renormalization, we succeeded in proving the renormalizability of the model (2), by considering it as a non-local field theory in $D$

dimensions[2]. Perturbation theory is obtained by expanding the observables as power series in $b$. The bi-local "interaction vertex" (in field theoretic language I denote it a bilocal operator) is written in Fourier transform as

$$= \delta^d(\vec{r}(x) - \vec{r}(y)) = \int d^dk \; e^{i\vec{k}(\vec{r}(x)-\vec{r}(y))} \,. \tag{5}$$

It can be viewed in a Coulomb gas representation as the integral over the "charge" $\vec{k}$ of a neutral "dipole" with charge $+\vec{k}$ at $x$ and charge $-\vec{k}$ at $y$. The term of order $b^K$ in the perturbative expansion of the partition function (as well as of other observables) involves $K$ dipoles $(x_1, y_1), \ldots (x_K, y_K)$. The integration over the charges $\vec{k}_1, \ldots \vec{k}_K$ gives an integral over the positions of the dipoles of the determinant of the "dipole energy" quadratic form $Q$

$$\int \cdots \int \prod_{i=1}^{K} d^Dx_i \, d^Dy_i \; \det\left[Q[x_i, y_i]\right]^{-d/2} \,. \tag{6}$$

$Q$ is a $K \times K$ matrix such that $\sum_{i,j=1}^{K} \vec{k}_i Q_{ij} \vec{k}_j$ is the Coulomb energy (in $D$-dimensions) of the $K$ dipoles. Each $Q_{ij}$ is a linear combination

$$Q_{ij} = G_0(x_i, x_j) + G_0(y_i, y_j) - G_0(x_i, y_j) - G_0(x_j, y_i) \,. \tag{7}$$

of the Coulomb potentials $G_0$ between endpoints of the dipoles $i$ and $j$

$$G_0(x, x') = \langle r(x) r(x') \rangle_0 = \frac{\Gamma((D-2)/2)}{4\pi^{D/2}} \, |x - x'|^{2-D} \,. \tag{8}$$

The Coulomb potential is properly defined for $0 < D < 2$ by analytic continuation. It is then negative, it vanishes for $x = x'$ (for $D > 2$ it diverges), but as for $D > 2$ is decreases at large distances. The integration over the $2K$ points in non-integer $D$-dimensional space can also be defined properly by analytic continuation in $D$ and the use of distance geometry. This amounts to replace the integration over the $2K \times D$ coordinates by an integration over the $K \times (2K - 1)$ scalar distances between these points.

One can show that short distance UV singularities occur in the integrals, when the determinant $\det[Q]$ vanishes. This occurs if and only if some of the end-points of (not necessarily the same) dipoles coincide, forming "atoms", and so that the dipoles form "molecules, and moreover if one can assign non-zero charges $\vec{k}_i$ to the dipoles while the atoms stay neutral. This condition is more easily depicted graphically on Figure 2.

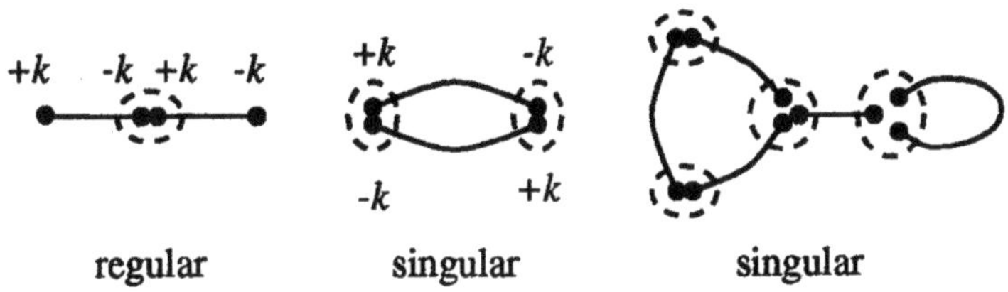


**Figure 2: A non-neutral regular configuration and two neutral UV singular configurations**

The associated singularities of these integrals are related to the behavior at short distance of the expectation value (with respects to the free Gaussian model) of products of bilocal operators as given by Equ. 5. One can show that this short distance behavior is encoded in a multilocal product expansion (MOPE), which generalizes Wilson's operator product expansion. Let me give two examples:

When the two points $x$ and $y$ of the bi-local interaction operator tend towards a single point, this operator can be expanded in terms of local operators involving derivatives of the field $\vec{r}$. The first terms of the expansion are explicitly (omitting unimportant coefficients)

$$
\begin{aligned}
= \quad & |x-y|^{\epsilon-2D}\ 1 \\
& + |x-y|^{\epsilon-D-2}(x^\alpha - y^\alpha)(x^\beta - y^\beta)\ :\nabla_\alpha \vec{r}\, \nabla_\beta \vec{r}: \\
& + \cdots
\end{aligned}
\tag{9}
$$

1 is the identity operator (its expectation value is 1), the : : in the operator $:\nabla_\alpha \vec{r}\nabla_\beta \vec{r}:$ denotes the normal ordering subtraction prescription required to deal properly with the UV singularities contained in $\nabla_\alpha \vec{r}\nabla_\beta \vec{r}$.

The second example is less simple, and shows that when the end-points of two bilocal operators tend pairwise towards two different points, this generates again bilocal operators

$$
= \quad \left[|x_1 - x_2|^{2-D} + |y_1 - y_2|^{2-D}\right]^{-d/2}
$$

$$
+ \cdots \tag{10}
$$

This structure is generic, and products of local and bilocal operators gen-

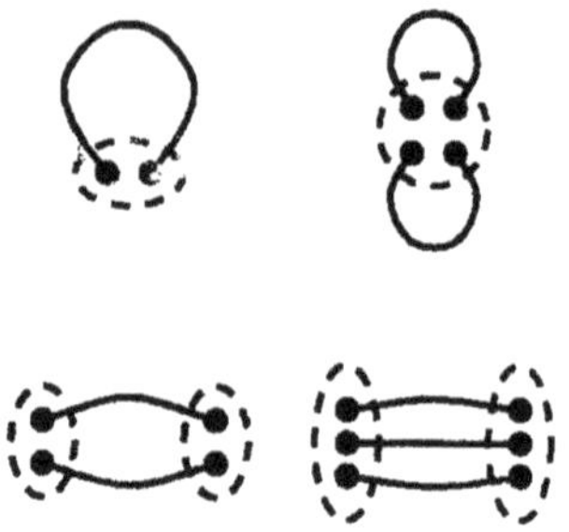

$1,\ (\nabla\vec{r})^2$

$\delta^d(\vec{r}(x)-\vec{r}(y))$

Figure 3: UV divergent configurations and the associated relevant operators

erate multilocal operators of the general form

$$\Phi\{x_1,\cdots,x_P\} = \int d^d\vec{r}_0 \prod_{i=1}^{P} \left[(\nabla_{\vec{r}_0})^{m_i}\ \delta^d(\vec{r}_0-\vec{r}(x_i))\ A_i(x_i)\right] . \tag{11}$$

where the $A_i(x_i)$ are local operators, which can be decomposed into products of multiple $x$-derivatives of $\vec{r}$. The $m_i$ are integers. For $P=1$ and $m=0$ one recovers local operators $A(x)$ ($m>0$ gives 0). For $P=2$, $m_1=m_2=0$ and $A_1=A_2=1$ one recovers the bilocal interaction operator, etc... These operators have a very special form: they can be viewed as a local convolution in the external $d$-dimensional $\vec{r}$ space of a non-local product (in the internal $D$-dimensional space) of the $P$ local operators $A_i$.

The MOPE implies that the formalism of renormalization theory and of renormalization group equations, which has been developed for local quantum field theories, can be adapted for this model. One is in fact interested in the IR scaling behavior of the lattice model, when some length scale $L$ goes to $\infty$. This lattice model is described by the Hamiltonian (2), with a short distance lattice cut-uff $a$. To study this IR limit it is equivalent to look at the UV continuum limit of the model when the physical length scale $L$ is kept fixed, while the UV cut-off $a$ goes to 0. In this limit one can construct, via renormalization, a finite renormalized theory with $a=0$, which obeys renormalization group equations. From these equations, one recovers the large distance behavior of the lattice model I started from. The procedure works well in perturbation theory when one is close to the upper critical dimension, i.e. for $\epsilon$ small, and it leads to the $\epsilon$-expansion.

In our case, the MOPE can be used to determine, by power counting, which multilocal operators are relevant and give UV singularities (poles in $1/\epsilon$). Then one can also show that these poles can be subtracted by adding

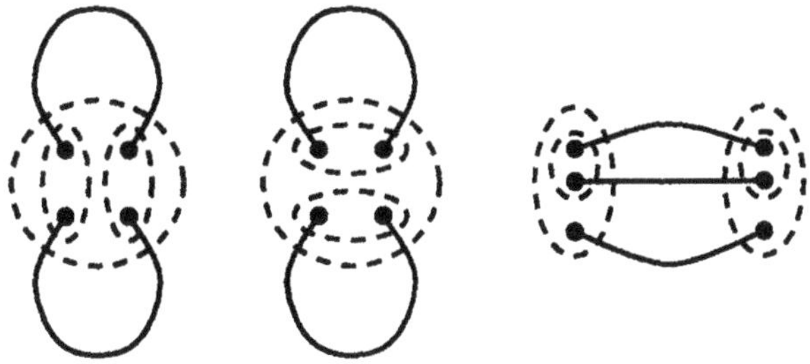

Figure 4: Examples of nested singular configurations

to the Hamiltonian (2) counterterms proportional to the marginally relevant multilocal operators, leading to the UV finite renormalized theory. For the model of self-avoiding surfaces, this analysis shows that the UV divergences are associated only with local and bilocal operators, as depicted on Fig. 3, and that only three operators are relevant: the identity operator 1, the elastic energy operator $(\nabla\vec{r})^2$ and the bilocal operator $\delta^d(\vec{r}(x) - \vec{r}(y))$. 1 is strongly relevant, and gives power-like UV divergences proportional to $a^{-D}$ ($a$ being a short-distance cut-off). The two other operators are superficially relevant, they give logarithmic UV divergences or equivalently poles in $1/\epsilon$ at $\epsilon = 0$.

The fact that the so-called superficial divergences, associated to a global contraction of points towards a singular configuration, can be subtracted by counterterms is a consequence of the MOPE. A complete proof of the renormalizability of the theory is possible, but much more delicate. It requires a control of the subdivergences coming from successive contractions associated to nested singular configurations, such as those depicted of Fig. 3 .

## 4 Scaling for infinite self-avoiding membranes

A first application of this formalism is the derivation of scaling laws. Since the model is renormalizable (at least perturbatively), it can be made UV finite (for $\epsilon \simeq 0$) by introducing two counterterms in the Hamiltonian. The new renormalized Hamiltonian is of the form

$$H[\vec{r}] = \frac{Z}{2} \int d^Dx\, (\nabla\vec{r})^2 + b_R\, \mu^\epsilon\, Z_b \int\!\!\int d^Dx\, d^Dy\, \delta(\vec{r}(x) - \vec{r}(y)) \ . \quad (12)$$

$b_R$ is the dimensionless renormalized coupling constant (the perturbative expansion in $b_R$ is UV finite order by order). $Z$ is a "wave-function" renormalization factor and $Z_b$ a coupling constant renormalization factor, both are perturbative series in $b_R$, with poles up to degree $1/\epsilon^{K-1}$ at order $K$. $\mu$ is the

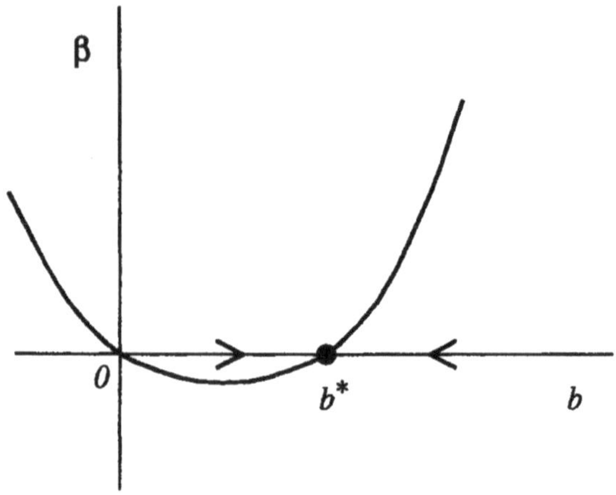


Figure 5: The $\beta$-function and the RG flow for $\epsilon > 0$.

renormalization momentum scale. As for ordinary local theories, such as the Landau-Ginzburg-Wilson $\Phi^4$ Hamiltonian, one can change $b_R$ and $\vec{r}$ in Equ. 12 into bare quantities in order to rewrite the renormalized Hamitonian as a bare Hamiltonian given by Equ. 2. The renormalization group $\beta$-function and the anomalous dimension $\gamma$ of the field $\vec{r}$ are now defined in the standard way

$$\beta(b_R) = \mu \frac{\partial}{\partial \mu} b_R \Big|_{\text{bare}} \quad ; \quad \gamma(b_R) = -\frac{1}{2}\, \mu \frac{\partial}{\partial \mu} \ln Z \Big|_{\text{bare}} . \tag{13}$$

The $\beta$-function is found to be of the form

$$\beta(b_R) = -\epsilon\, b_R + c\, b_R + \mathcal{O}(b_R^2) \quad ; \quad \text{c positive constant} , \tag{14}$$

and therefore there is, at least for small $\epsilon > 0$, an IR attractive fixed point $b_r^* = \mathcal{O}(\epsilon)$, which governs the scaling behavior of self-avoiding polymerized surfaces at large distance. The existence of this fixed point ensures the universality of this non-trivial scaling for $\epsilon >$ 0i, and that no new interactions, possibly non-local in external space, are generated by the RG transformations.

The explicit calculation leads to the same results for the scaling exponents $\nu$ and $\gamma$ than the direct renormalization method at first order in $\epsilon$. With this method higher order calculations are in principle feasible, but technically quite difficult. In particular, already at second order the RG functions cannot be expressed analytically, and numerical integration methods have to be developed. Work is in progress to compute the scaling exponents at order $\epsilon^2$.

## 5 Finite Size Scaling and Direct Renormalization

The model given by Equ. 2 describes an infinite membrane with flat internal metric, corresponding to an infinite and regular flexible lattice. Finite membranes are described by a similar model, but the $D$-dimensional membrane $M$ is now embodied with a fixed non-trivial Riemannian metric $g_{\alpha\beta}(x)$, with curvature $R$ (examples are closed membranes with the topology of the sphere $\mathcal{S}_D$ or the torus $\mathcal{T}^D$), and may have a boundary $\partial M$ (open membrane with the topology of the disk for instance). A similar analysis can be performed for such models, and the MOPE structure of short distance singularities is still valid, but new local operators $A(x)$, which depend on the internal metric on $M$ and on the boundary $\partial M$, such as the curvature $R$, appear in the MOPE and in Equ. (11). The renormalized Hamiltonian now contains at least five operators and five independent renormalization factors $Z$

$$H[\vec{r}] = \int_M Z\,1 + \int_M Z\,(\nabla\vec{r})^2 + \int\int_M Z\,b\,\delta^d(\vec{r}-\vec{r}) + \int_M Z\,R + \int_{\partial M} Z\,1\ . \tag{15}$$

The curvature operator $\int_M R$ is superficially relevant only for $D = 2$ and the boundary operator $\int_{\partial M} \mathbf{1}$ only for $D = 1$. When these additional terms are not relevant, the first three renormalization factors $Z$ are the same for finite curved membranes than for the infinite flat membrane. This property is analogous to the renormalization property of local field theories in finite geometries, which justifies the finite scaling laws for critical systems in finite geometries, and it has two very important consequences: *(i)* The scaling hypothesis at the basis of the direct renormalization approach, which relies explicitly on calculations with finite membranes, is shown to be valid to all orders in perturbation theory; *(ii)* for "abstract" membranes with dimension $D < 2$, with the only exception of open polymers ($D = 1$ open surface), the following hyperscaling relation [9] relating the configuration and the $\nu$ exponents holds:

$$\gamma = 1 - \nu d\ . \tag{16}$$

## 6 Self-avoiding Membrane at the tricritical Θ-point:

Finally, let me briefly discuss recent results obtained with K. Wiese on the scaling behavior of polymerized membranes at the Θ-point [3]. This point separates the swollen phase, where the self-avoidance repulsive forces that I considered previously dominate, from the dense collapsed phase, where short ranged attractive forces dominates. At the Θ-point the effective two body repulsive

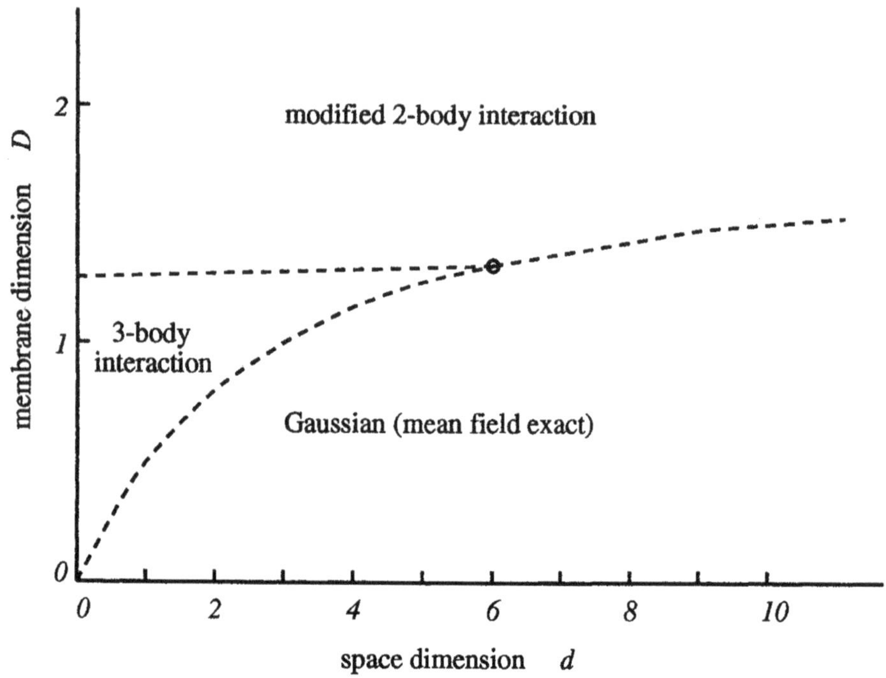


Figure 6: Relevant interactions for the Θ-point in the $d$-$D$ plane.

coupling $b$ vanishes, and two different interactions may become relevant. The first one is the 3-body contact repulsion, which is usually considered for polymers

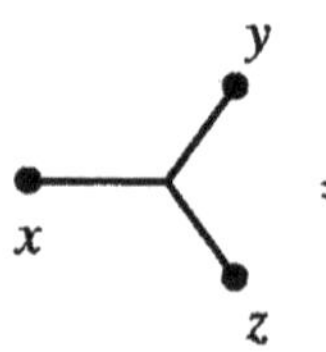


$$= \int\int\int d^D x\, d^D y\, d^D z\, \delta^d(\vec{r}(x)-\vec{r}(y))\,\delta^d(\vec{r}(x)-\vec{r}(z)) \ . \quad (17)$$

The second one is a modified 2-body interaction, repulsive at short range but attractive at larger range ($\Delta_{\vec{r}}$ is the $d$-dimensional Laplacian)

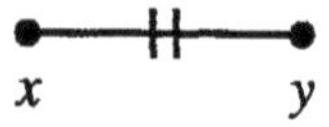


$$= -\int\int d^D x\, d^D y\, \Delta_{\vec{r}}\,\delta^d(\vec{r}(x)-\vec{r}(y)) \ . \quad (18)$$

Calculations at first order are not feasible analytically, and already require numerical evaluations of complicated integrals. The results of such one loop calculations are schematically depicted on Fig. 6, where the domains where the 3-body and modified 2-body terms are respectively relevant are shown. This indicates that the last modified 2-body term is the relevant one for 2-dimensional membranes in any external dimension $d$. There is also a quite interesting and non-trivial crossover between the two terms around $D = 4/3$ $d = 6$, which must be studied by a double $\epsilon$-expansion.

## 7 Conclusion:

The theoretical study of the scaling behavior of polymerized flexible membranes leads to the development of new multilocal continuum field theories, and to new applications of renormalization group methods. I hope that these methods will lead to a quantitative progress in the understanding of the behavior of real 2-dimensional polymerized membranes. This requires results beyond first order in the $\epsilon$-expansion (recall that $D = 2$ correspond to $\epsilon = 4$), and a better understanding of the relation between this RG approach and more heuristic or approximate methods, such as variational methods or approximate recursion relations. The sophisticated renormalization theory for multilocal models should hopefully also find applications in other problems of statistical physics, or in other areas of theoretical physics.

## References:


1. *Statistical Mechanics of Membranes and Surfaces*, Proceedings of the Fifth Jerusalem Winter School for Theoretical Physics (1987), D. R. Nelson, T. Piran and S. Weinberg Eds., World Scientific, Singapore (1989).
2. F. David, B. Duplantier and E. Guitter, Phys. Rev. Lett. **72** (1994) 311.
3. K. Wiese and F. David, *Self-avoiding tethered membranes at the tricritical point*, Saclay preprint cond-mat/9503126, to appear in Nucl. Phys, B.
4. J. A. Aronowitz and T. C. Lubensky, Europhys. Lett. **4** (1987) 395.
5. M. Kardar and D. R. Nelson, Phys. Rev. Lett. **58** (1987) 1289.
6. J. des Cloizeaux, J. de Physique **42** (1981) 635.
7. M. Kardar and D. R. Nelson, Phys. Rev. Lett. **58** (1987) 2280 (E); Phys. Rev. **A 38** (1988) 966.
8. B. Duplantier, T. Hwa and M. Kardar, Phys. Rev. Lett. **64** (1990) 2022.
9. B. Duplantier, Phys. Rev. Lett. **58** (1987) 2733; and in [1].

# THE GEOMETRY OF CRITICAL PERCOLATION AND CONFORMAL INVARIANCE

M. AIZENMAN

*Departments of Physics and Mathematics,*
*Princeton University, Princeton, NJ 08544, USA*

**The connectivity properties of typical configurations of a critical, or nearly – critical, percolation model are expressed, in the scaling limit, through the *critical percolation web*. The *web* is a random geometric object formulated over the continuum. It exhibits interesting behavior under conformal maps. The article presents the *web's* construction and describes some of its remarkable properties.**

## 1. Introduction

### 1.1. *The scaling limit*

In percolation models close to their percolation thresholds, connected clusters are seen on scales which are much larger than those of the individual constituents of the system. This presentation will focus on the *Percolation Web* which emerges in the corresponding scaling limit.

It is a common experience that phenomena which are suitable subjects for physics are found on different scales. A structure which seems rather elementary on its scale is often a manifestation of the collective modes of a finer one. Viewed from the finer perspective, the appearance of such modes in the scaling limit — the term used here to indicate the change of scale in the observation of a system, and also for the limiting object — could be due to varying combinations of interactions and/or fluctuations. Remarkably, often the physics of the coarser scale is well described by autonomous laws, which may include symmetries not found in the finer description.

Examples where scaling limits play obvious roles are seen in hydrodynamic equations (large scale behavior of molecular matter), Brownian motion, field theory, and in the critical behavior of systems described by statistical mechanics. The analysis of the scaling limit is a challenge for our intuitive and mathematical understanding of the relevant system. In only few cases has it been accomplished to a high degree of satisfaction.

The critical percolation web (or just the *web*) is a random geometric object, formulated over the continuum, expressing the connectivity properties of typical configurations of a critical, or nearly – critical, percolation model. In this presentation I discuss its construction, as a well–defined mathematical object, its basic connectivity properties and its conjectured conformal invariance. The new results reported here are derived in Ref. 1. They are related to recent developments which were reviewed in R. Langlands et. al. [2].

Much has been written on percolation, and the fractal and graph theoretic aspects of its large scale features have been studied quite extensively (H.E. Stanley [3], B. Mandelbrot [4]). The subject has, however, been typically approached from the perspective of the lattice. The formulation of a continuum object is particularly desirable because of the existing indications that the limit is conformally invariant. The 2D *web* demonstrates interesting behavior under conformal maps, not all of which seems to be expressed through local field theories [5].

The interest in the subject of critical percolation has recently been stimulated by the numerical work and the questions raised by Langlands et.al. [6]. After a proposal by the author which explained some aspect of the numerical results by conjecturing conformal invariance, J. Cardy proposed an explicit equation which provided a superb fit with the data [7]. Cardy's proposal was motivated by arguments which make better sense for $q = 2, 3$ Potts models. Considerations of the *web's* structure shed light on Cardy's equation rooting it within the percolation model. Although that has not yet resulted in a rigorous derivation, the new perspective suggest extensions beyond the original argument.

While the geometric nature of the *web* is quite distinct, it shares a number of broad features with other scaling limits:

- *continuum* emerges as the appropriate medium
- *fractal* nature
- *enhanced symmetry* — invariance under *conformal maps*
- *differential equations* (of finite order) — for some of the connectivity functions.
- some of the *characteristic exponents* can be determined through appropriate differential equation

Although we focus here on independent percolation, the construction can be extended to the correlated Fortuin–Kasteleyn random cluster models, which are related to a number of quite different systems. That range includes the $\phi^4$ field theory, which describes critical Ising systems, and the quantum Heisenberg antiferromagnetic spin (1/2) chain, where the collective spin modes in the ground state can be described as the patterns of intersections of a line ($t = 0$) with the boundaries of the connected clusters of a critical q=4 FK-Potts model [8].

For the discussion which follows we adapt the perspective of the continuum. One may regard the unit length to be that of our desk-top, on which we have a system for which the elementary connections are determined on a microscopic scale, characterized by length $a \to 0$.

### 1.2. *An analogy: the Brownian paths*

The simple random walk, whose scaling limit is the Brownian motion, offers an instructive analogy. It is conveniently formulated on a $d$ dimensional square lattice (spacing $a$) with equal transition probabilities from each site to any of its $2d$ nearest–neighbors. Large scale features are seen by scaling $a \to 0$ and $b(t) = x([t/\sqrt{a}])$ , with $x(n)$ being the position of the walk after $n$ steps. The limiting process is non-degenerate, and is related to the Brownian motion studied by A. Einstein and J. Perrin. As a mathematical object, it is given by a probability measure on the space of continuous functions, constructed by N. Wiener. Of particular interest to us is the trail marked by the random walk, which we start at some point $x$ in a domain $D$ and continue until it reaches the boundary $\partial D$. The domain $D$ is regarded as independent of $a$.

The random trail exhibits the features listed above —

- various associated hitting probabilities are given by harmonic functions, satisfying $\Delta f = 0$,
- the lattice orientation does not show in the limiting probability distribution, which is invariant under dilations, rotations, and indeed conformal maps (of which there is quite a rich class in two dimensions).

An explicit expression of the trail's remarkable conformal invariance will naturally start with words like "let $\eta$ represent the random trail, and let $\mu_{D,x}(d\eta)$ be its probability distribution". Such declarations are meaningful only after identification of a space which supports these structures. In practical terms, that requires some "hard" information on the nature of "typical" paths. In the case of Brownian motion, a suitable probability space can be formulated, as indicated below, and one finds that

$$\mu_{TD,Tx}(d\eta) = \mu_{T,x}(d\, T^{-1}\eta), \tag{1}$$

for each conformal map $T : R^d \to R^d$ which is univalent on $D$. (There is an analogous equation for the time–dependent process b(t). That, however, involves an adjustment of the motion's rate.)

Brownian motion demonstrates also two other relevant features:

- the *Markov property*.
- the importance of a *regularity condition* (e.g., Hölder continuity).

Let me expand on the last point, which echoes an important element in the construction of the percolation *web* .

The mathematical construction of the Brownian motion can be approached in two different ways:

i. The *stochastic process approach* — $b(t)$ is described by an uncountable collection of random variables, one for each $t \geq 0$, with values in $R^d$. The probability measure is characterized by the joint probability distributions for all finite collections $\{b(t_1), ..., b(t_k)\}$, $k < \infty$. This setup permits a relatively quick construction of the scaling limit, but the initial result is rather "leaky". E.g., these joint distributions do not determine whether the paths of the process are *continuous* or not. (Path continuity is not, initially, a "measurable event"; common marginal distributions are shared by different versions of the probability measure, some with erratic glitches at a random Cantor-type set of times. )

Or -

ii. *Random function* — $b(t)$ is given by a single random variable which takes values in the space of continuous functions $C([0,\infty), R^d)$ (with simple interpolation for $a > 0$). In this setting more information is needed up-front in order to know the scaling limit exists, since that implies the persistence of the path's continuity. The conclusions are, however, considerably stronger. E.g., one can deduce that Brownian motion yields a good approximation even for the random walk's trail.

Brownian motion is known to be Hölder continuous to any order $0 < \lambda < 1/2$, in the sense that for, say, all $0 \leq t < t' \leq 1$:

$$|b(t) - b(t')| \leq K_\lambda(b)\ |t - t'|^\lambda, \tag{2}$$

with a finite, realization–dependent, $K_\lambda(b)$. Such information for random walks, with a uniform bound of the form

$$Prob_a(K_\lambda(b) > r) \leq g_\lambda(r) \tag{3}$$

with $g_\lambda(r) \to 0$ for $r \to \infty$ (uniformly in $0 < a < 1$), permits the extraction of the stronger convergence, in a sense corresponding to (ii), from the weaker one (i).

An analogous Hölder regularity condition is now known for the 2D critical percolation, and it plays an important role in the analysis discussed below.

## 2. Critical Percolation Models

The scaling limit of critical percolation is expected not to depend on many of the details of the model. A convenient example is provided by the bond percolation model. In it, the bonds linking neighboring sites of the $d$-dimensional lattice are occupied at random, independently, with probability $0 \leq p \leq 1$ (which is our control parameter). The object of study are the realization–dependent maximal clusters of lattice sites connected by the occupied bonds. In line with our main theme, we set the unit length to be of the scale on which the clusters are studied and consider the case where the lattice spacing is $a << 1$. Thus, what is seen on our 'desk–top' are the large scale features of clusters for which the notion of connectivity is set on a minute scale, $a \to 0$.

As the model's control parameter(s) vary, there is an abrupt transition in the nature of the connected clusters. The sharpness of this transition has now been proven at considerable generality, by theorems which apply to any (non-degenerate) translation–invariant independent percolation model in $d > 1$ dimensions (Aizenman–Barsky [9], Menshikov [10]). The proof was first achieved for the 2D model (Kesten [11]), for which also more is known (including the scaling relations, established by Kesten [12]).

On our desk–top scale, one sees the following:

$p < p_c$: In the low–density phase there are no visible connected clusters. Very seldom would a given region (our "desk–top" of linear size $R$) include a cluster with diameter exceeding (Const.(p) $a \log(R/a)$). More explicitly: for $B$ large enough

$$Prob_p(\{ \max_{x \in [0,R]^d} diam\ C(x)\ > B\ a\ log(R/a)\}) \ \leq \ A(a/R)^{\mu B}, \qquad (4)$$

with some $A, \mu > 0$.

$p > p_c$: In the high–density phase there is, typically, one connected (sponge like) cluster of macroscopic size, which is dense everywhere. The size of the maximal spherical void in that cluster satisfies a bound similar to Eq. 4 [13]. (Though in slabs, and in the higher dimensions $d \geq 3$ the vacancies also percolate, as long as $p$ is not too large. Such bi-percolating phases are of interest, as we heard in the talk of Prof. D. Chandler.)

$p \approx p_c$: Large scale fluctuations, i.e., realization–dependent features, are observed in critical or nearly critical models, for which either $p = p_c$ or $p \to p_c$ (fast enough), as $a \to 0$.

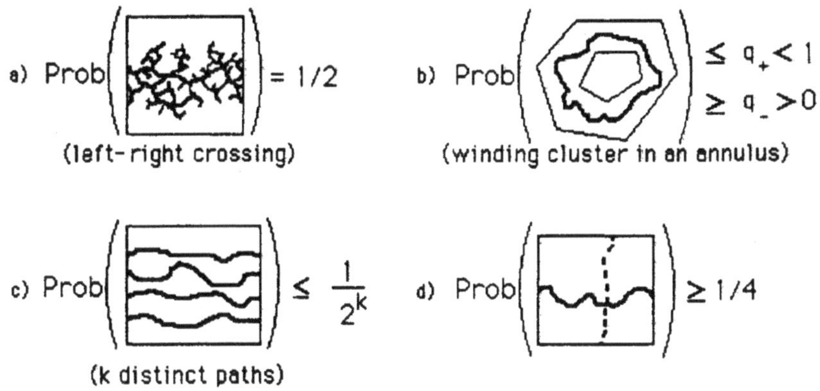


Fig. 1. Crossing probabilities for the 2D critical percolation model ($p = p_c$). The solid lines depict connected paths along occupied bonds. The dotted lines are paths of the dual, disrupting, bonds. The bounds are uniform in the lattice spacing, as $a \to 0$. In (d) the existence of a spanning cluster is sensitive to the state of a single bond.

We shall now focus on the two–dimensional case, for which a number of known bounds on crossing probabilities are depicted in Fig. 1. These bounds yield considerable insight:

a), b) Spanning clusters occur, but are not to be taken for granted. The probability that opposite faces of a macroscopic shape are connected by a path of occupied bonds is bounded away from both 0 and 1 (for the square it is 1/2 by duality). Same applies to the probabilities that prescribed annuli contain winding connected paths (d).

c) The probability that there are $k$ disjoint connecting lines decays exponentially. Thus, the number of such paths — or the order of the connection, is seldom large.

d) The large connected clusters can be broken to large disconnected pieces by overturning few bonds, often just a single "pivotal bond". Such choke elements are also referred to as "red bonds" — since they would heat up first were a current to be driven across the sample [3].

Expanding on the last point, the clusters exhibit also loops, and plenty of "dangling ends" [3]. One of the challenges in the construction of the scaling limit is to find an efficient method for managing the information needed for the description of the clusters' complex structure.

Further information on the critical model is provided in Fig. 2, where the symbol $\approx$ means that with the insertion of a suitable constant both an upper bound and a lower bound exist — uniformly in $a$. The bounds are expressed there in a form which stays meaningful in the continuum limit. The lattice version of cases (a) and (b) have been known and used before ((b) was shown in Ref. 12 to imply the bound $\rho \leq 1/3$, for the exponent which appears in Eq. 12). Case (c) appears in Ref. 1, where it provides an effective tool for the derivation of the regularity properties discussed next.

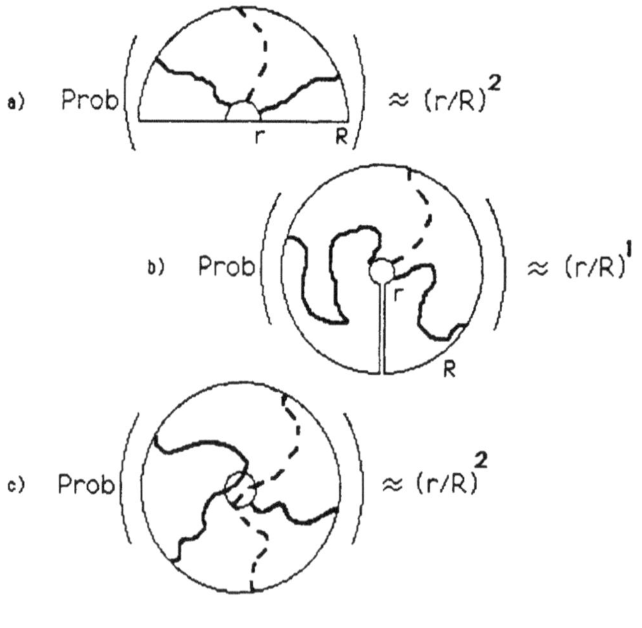


Fig. 2. Some scaling laws for the crossing probabilities, with $r/R$ the ratio of the inner radius to the outer one, which are valid at $p = p_c$ uniformly in the lattice cutoff $a \leq 1$. These particular characteristic exponents may be derived by elementary arguments. The relation of a) with b) is consistent with conformal invariance, through the map $z \mapsto z^2$.

## 3. Construction of the Percolation Web

In the construction of the scaling limit we focus here on the geometric extend of the connected clusters. Two examples of related connectivity functions are depicted in Fig. 3. We shall not discuss here all the percolation's relevant quantifiers, e.g., the "density" — with the appropriate scaling — of the sites connected to the boundary of the domain $D$, or the local resistance to current. Such quantities would be described by suitable random fields, supported on the *web* — but not fully determined by it.

Since the connected clusters have a rather complex morphology [3,4] (indicated in Fig. 1(a) ), it is not immediately clear what mathematical constructs will present a good descriptive tool. Often fractals are conveniently viewed as compact subsets of $R^d$. The space formed by the collection of such sets, equipped with the Hausdorff metric [14], has very convenient mathematical properties (it is complete, separable and compact). Such a description would not, however, do for us — since clusters which as elements of this space are infinitesimal distance apart (in the Hausdorff metric) may have very different connectivity properties: typically the connected clusters have points at which very small local changes (down to the lattice scale !) produce very drastic effects on the possible routing of connecting paths — see Fig. 1(d).

At this point we have some good news to report: *the connectivity properties of the connected clusters can be expressed through the collection of simple (Jordan) curves, describing the set of realized paths!* It even suffices to restrict the attention to the curves described by elements of the space $R_{\lambda,D}$ of functions $g(t) : [0,1] \to D$ satisfying the Hölder continuity condition:

$$|g(t) - g(t')| \leq \mathrm{K}_\lambda(\mathrm{g}) |\mathrm{t} - \mathrm{t}'|^\lambda \tag{5}$$

The order can be fixed at any value $\lambda < 1/2$.

On a lattice the above statement is obvious. Its validity at the scaling limit may, however, come as a surprise for two reasons:

(i) Typical configuration could, conceivably, exhibit connecting self–avoiding paths which are so rough that Hölder continuous parametrization, in the sense of Eq. 5, is possible only with an exceedingly large continuity constant. In fact, for $p > p_c$ the typical range for $\mathrm{K}_\lambda$ diverges at least as $|a \log a|^{-\lambda}$, for $a \to 0$.

(ii) There typically exist connecting paths which make contact on the scale of the continuum, but fail to be linked down at the lattice scale (as in the situation depicted in Fig. 1(d) ). Thus, a description of the connected clusters based on paths needs to be supplemented with information specifying which of the intersections are actually realized as connection events.

These concerns are, however, answered by the regularity properties of typical configurations of the percolation model, expressed in the following two Propositions. The first result is that the constant $\mathrm{K}_\lambda(\omega)$ seldom needs to be large. The second implies that paths which seem to intersect, at other than end–points, do intersect. The regularity statements refer to percolation models on a lattice of positive spacing $a > 0$, but are formulated in a way which makes sense even at the limit $a = 0$.

**Proposition 1 (Hölder continuity of connecting paths).** For each configuration $(\omega)$ of the bond percolation model, let $\mathrm{K}_\lambda(\omega)$ be the minimal value permitting each self-avoiding path along occupied bonds (of $\omega$), in the region $[0,1]^2$, a parametrization satisfying Eq. 5. Then, for any $\lambda < 1/2$:

$$Prob(\mathrm{K}_\lambda \geq \mathrm{t}) \leq \psi_\lambda(\mathrm{t}) \tag{6}$$

with $\psi_\lambda(t) \to 0$ for $t \to \infty$, uniformly in $a(\leq 1)$.

**Definition 1.** (i) Two connecting paths, $g_1(t)$ and $g_2(t')$, are said to have a **close encounter** at distance $r \leq 1$ and power $\alpha > 1$, if for some $t_1 < t < t_2$ and $t'_1 < t' < t'_2$

$$|g_1(t) - g_2(t')| \leq r \tag{7}$$

while

$$|g_1(t) - g_1(t_j)|, \ |g_2(t') - g_2(t'_j)| \geq r^{1/\alpha} \ , \ for \ \ j = 1,2 \ . \tag{8}$$

(ii) The close encounter is **of the first kind** if there is no connected path linking the two trajectories within the ball of radius $r^{1/\alpha}$ centered at $[g_1(t) + g_2(t')]/2$. If there is such a linking path, the encounter is **of the second kind.**

**Proposition 2 (Intersecting paths connect).** There is a value $\alpha > 1$ such that the probability of there being a close encounter of the first kind at distance $r \leq 1$, in the fixed region $[0,1]^2$, is less than $\hat{\psi}_\alpha(r)$, with

$$\hat{\psi}_\alpha(r) \to 0 \ \ for \ \ r \to 0 \ , \tag{9}$$

uniformly in the lattice spacing $a(\leq 1)$.

Due to this regularity we may describe the *web* by means of the collection of Hölder continuous curves, elements of the space $R_{\lambda,D}$, realized as connected paths. (We implicitly fix $\lambda$ at some value $< 1/2$.) To incorporate the natural consistency condition, let us denote by $S_{\lambda,D}$ the collection of subsets of $R_{\lambda,D}$ which are closed under: 1) uniform convergence, 2) reparametrizations which preserve the Hölder continuity, and 3) splicing of segments of paths which intersect at non-terminal times — i.e., times $0 < t_1, t_2 < 1$ beyond which $g(t)$ continues to change.

**Definition 2 (The percolation** *Web* **). Web configurations,** $W$, are described by the elements of $S_{\lambda,D}$, i.e., by subsets of the space $(R_{\lambda,D})$ of Hölder continuous curves satisfying the consistency conditions listed above.

For a given web configuration, $W \in S_{\lambda,D}$, paths with non-terminal intersections are regarded as **linked.**

The finest partition of $W$ which does not separate linked paths defines its decomposition into **connected clusters**. The **support** of the cluster is the union of sites visited by the cluster's paths. **Trivial clusters** are those consisting of only a constant function (alt.: those whose support is a one point set).

Finally, a **random web model** is described by a probability measure on $S$.

We can now report the following two existence results [1].

**Proposition 3 (Existence — rotation invariant measure).** There exists a random web model in $R^2$ whose probability distribution is:

1) independent in disjoint regions,

2) regular — in the sense that the crossing probabilities, of the events depicted in Fig. 1(a,b), are neither zero nor one,

3) invariant under translations, and

4) invariant under orthogonal rotations and reflections.

The measure is constructed as a scaling limit of a sequence of critical percolation models, using both regularity properties and compactness arguments (i.e., passage to sub-sequences). The scaling limit of the bond percolation is not proven yet to be rotation invariant. (Forcing the invariance by an average over rotations may spoil the independence property i) .) The construction of the random web of Proposition 3 is based, therefore, on another variant — that of the lily-pad percolation. However, basing the construction on the bond model has the advantage of manifest **self-duality**. We find that notion to be very useful in the argument for the differential equations discussed in Section 6.

**Proposition 4 (Existence — self dual measure).** There exists a random web model in $R^2$ whose probability distribution has the properties

1) - 3) independence, regularity, and translation invariance, as in Prop. 3, and is

4') invariant under the finite group of rotations and reflections which preserve the square lattice,

5) **self-dual** [in a natural sense, spelled out in Ref. 1].

We are still missing a uniqueness proof, though the widely expected universality of the scaling limit (Conjecture 1, below) would imply that these two measures are the same.

## 4. Typical *web* Configurations

The following is proven in ref. [1].

**Proposition 5 (*Web* 's properties).** The following holds with probability one (i.e., for almost every realization) for any random web satisfying i) - iii) of Prop. 3)

1) Lebesgue – almost every point $x \in R^2$ belongs to the support of only a trivial cluster.

2) There are only countably many non-trivial clusters.

3) The union of the supports of the non-trivial clusters — to be denoted as $U$ — is everywhere dense (in $R^2$), and its Hausdorff dimension is a non-random number, strictly smaller than 2:

$$1 < d_w = dim_H U < 2 \tag{10}$$

4) The web is of finite *articulation*. I.e., the maximal number of lines reaching disjointly a common point is given by a finite number (which, furthermore, is non–random).

This is a good point to address the natural question about the relation of the *Web* with the **incipient infinite cluster** (*Incip.Inf.Cl.*). The two are related but the concepts differ in significant ways.

The *Incip.Inf.Cl.*was introduced and has been studied within the context of a fixed lattice, looking at increasingly large cubes, $B_n$.

Since the percolation density vanishes at $p_c$, the *Incip.Inf.Cl.*could not be defined as the infinite cluster at the critical point. In lieu of that, the following three alternatives definitions have been pursued — all within the context of a fixed lattice, looking at increasingly large cubes $B_n$.

i. The largest cluster in a given large cube.

ii. The spanning cluster, connecting the cube's two opposite faces. (It has sometimes been implicitly, and even explicitly, assumed that this condition defines a single connected cluster. It does not.)

iii. The cluster of the origin — conditioned on the rare event that it reaches the boundary, $\partial B_n$.

Thus, *Incip.Inf.Cl.*has been viewed as a single connected cluster with large voids. In comparison, the *Web* has infinitely many clusters, and covers the continuum densely (in the topological sense — though it is sparse in measure theoretic sense).

Comparing the two concepts, one sees:

1. With the appropriate care, *Incip.Inf.Cl.*can be regarded as one of the clusters in the *Web* . The value of $d_w$ coincides with the dimension of the *Incip.Inf.Cl.*(widely believed to be $d_w = 91/48$, as proposed by Nienhuis et. al. [15,16]).

2. The *Web* resides in the continuum.

The construction of the continuum object opens new possibilities, as seen below.

## 5. Conjectured Behavior under Conformal Maps

We next turn to a number of pertinent conjectures.

**Conjecture 1.** 1) The properties 1) - 4), of Proposition 3, define a unique random web measure.

2) That measure is also invariant under dilations.

3) More generally, the measure is conformally invariant, in the sense that the random web associated with a domain $D \subset R^2$ maps as in Eq. 1 under conformal maps which are univalent on $D$.

4) The uniqueness persists even if the rotation invariance is relaxed to invariance under at least three independent reflections.

One may notice that the last statement would imply that the measure described above yields the scaling limit of the two dimensional critical bond, and site, percolation models. From a broader perspective, it seems that this common measure would be the fixed point of the relevant renormalization group transformation — had it only been given a more satisfactory definition. Various renomalization maps for percolation have been discussed in the past. However, the recent proposals and results of Langlands have led to revised views of the renormalization group analysis of percolation, with more appreciation of the implications of the existence of a meaningful fixed point.[17,18,19]

Since the work of Belavin Polyakov and Zamolodchikov [5] it has been generally appreciated that conformal symmetry is a very powerful condition. Let us discuss its possible consequences in our context — focusing on the two functions, F and G, described in Fig. 3.

The transformation law for the function $F$ is — under the assumption of conformal invariance, simply:

$$F(D, \alpha) = F(TD, \alpha) \ , \tag{11}$$

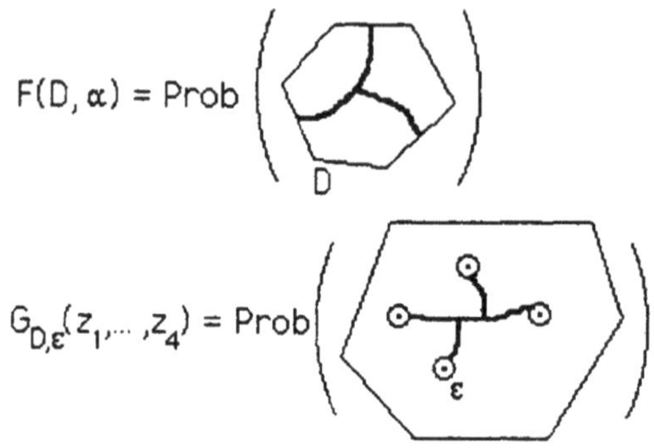


Fig. 3. Two function which quantify the connectivity properties of the critical percolation model (on a lattice of vanishing spacing). The connections are restricted to paths within the domain $D$, and the index $\alpha$ (seen in $F$) represents the connection pattern. For $G$, the discs are of radius $\epsilon$, and are centered at $z_i$.

for $T$ which is conformal and univalent on $D$. Simply connected domains $D$, can be mapped in this fashion onto the upper half plane. ($T$ being determined up to real Möbius transformations.) Such a map permit to express $F(D, \alpha)$ through a function $f_\alpha(x_1, ..., x_k)$ of a number of real variables (the images of the end–points of the boundary segments). More is said on this function below.

The function $G$ vanishes as $\epsilon \to 0$. One can prove power–law bounds, but along with the above conjecture it would be natural to expect the stronger statement, that with a suitable choice for $\rho$ [$= 1/(2 - d_w) =$ (?)48/5] the following limit exists (pointwise):

$$g_D(z_1, ..., z_k) = \lim_{\epsilon \to 0} \epsilon^{-k/\rho} \, G_{D,\epsilon}(z_1, ..., z_k) \quad . \tag{12}$$

Under the conformal invariance hypothesis, $G$ obeys the analog of Eq. 11. For $g_D$ that corresponds to:

$$g_D(z_1, ..., z_k) = \prod_{1 \le j \le k} |T'(z_j)|^{1/\rho} \; g_{TD}(Tz_1, ..., Tz_k). \tag{13}$$

Such transformation law has previously been seen for the correlators— the vacuum expectation values of products of field operators, within the context of conformal field theories. The appearance of this law here suggests that conformal covariance considerations are applicable in a broader arena. (Though the *Web* is not totally unrelated to conformal fields.)

## 6. Cardy's Equation and its Extension

At present, the most effective demonstration of the power of conformal invariance can be seen in the analysis of the functions $F(D, \alpha)$, and the corresponding functions of real variables $f_\alpha(x_1, ..., x_k)$ —which give the probability that certain intervals on the real line are connected through the upper half plane.

The recently renewed interest in percolation owes much to R. Langlands. One of the questions he posed and studied numerically is: for parallelograms with varying angles, in what way should the aspect ratio be adjusted in order to keep the crossing probability fixed? The author of this report has proposed that the answer should be determined by the condition of conformal equivalence (for which the maps are required to be analytic in the interior and continuous at the boundary). That suggestion was indeed supported by the data. J. Cardy produced a far more advanced proposal: suggesting a differential equation for the function $f_\alpha(x_1, ..., x_4)$ (to which the crossing probability is reduced through the conjectured conformal invariance). The (unique) solution of Cardy's equation, with the two natural boundary conditions, provided a superb fit for the numerical data of Langlands et. al [6].

### 6.1. *Cardy's proposal*

It is convenient to introduce the differential operators:

$$L_n = - \sum_{j=2,\ldots,k} (x_j - x_1)^{(n+1)} \frac{\partial}{\partial x_j} \tag{14}$$

whose commutation relations form the Virasoro algebra:

$$[L_n, L_m] = (n - m)L_{n+m} \ . \tag{15}$$

The equation which Cardy proposed for the case of $k = 4$ (by an argument which was not extended to higher values of $k$) is:

$$(L_{-1}^2 - 2/3 \ L_{-2})f_\alpha(x_1, ..., x_k) = 0 \ . \tag{16}$$

Any conformally invariant function depends on $\{x_1, ..., x_4\}$ only through the cross ratio:

$$u(x_1, ..., x_4) = \frac{(x_1 - x_2)(x_3 - x_4)}{(x_1 - x_3)(x_2 - x_4)} \ . \tag{17}$$

For $f_\alpha(x_1, ..., x_4) = \phi(u(x_1, ..., x_4))$ Eq. 16 reduces to [2]

$$u(1 - u)\phi''(u) + \frac{2}{3}(1 - 2u)\phi'(u) = 0 \ , \tag{18}$$

which is supplemented with the boundary conditions: $\phi(0) = 0$, and $\phi(1) = 1$. The equation can be integrated, and its solution (a hypergeometric function) can be given as:

$$\phi(u) = \int_0^u \frac{dx}{[x(1-x)]^{2/3}} \Big/ Norm. \tag{19}$$

### 6.2. *Cardy's equation from the web's perspective*

Cardy's proposal was based on the extrapolation to $q = 1$ of two features which originate within the context of the $q$ state Potts models: i) a relation between the crossing probability and the ratio of partition functions at different boundary conditions, which in turn is expressed as expectation value of a product of boundary field operators, and ii) null vectors, imposed by unitarity considerations. The two are well motivated for $q = 2, 3$, but are somewhat more speculative for $q = 1$. The derivation of Eq. 16 from within the model is a challenge which could also open the way to further relations, e.g., for $k > 4$. This task has not yet been accomplished. However, the *web* yields another way of looking at it which offers a plausible argument, and suggests generalizations.

First, one may observe that just conformal invariance implies that

$$L_1(L_{-1}^2 - 2/3\ L_{-2})f_\alpha(x_1, ..., x_k) = 0\ . \tag{20}$$

The derivation is elementary and purely algebraic — based on Eq. 15 and the invariance under $L_0$ and $L_1$. The conclusion is still rather weak, since the equation is obeyed by any conformally invariant smooth function. Nevertheless, this argument points to 2/3 as the only possible value for the constant with which an equation like Eq. 16 could hold (in our case).

The second step concerns the removal of the differential operator $L_1$ from the left side of Eq. 20. That is, of course, a major step since the resulting equation (with the obvious boundary conditions) completely determines the answer. The argument is more delicate. It involves the *web* 's geometry, considerations of the effects of the separation of scales, and self–duality. It is presented in ref. [1].

### 6.3. *Proposed extensions*

The analysis outlined above suggests that equation Eq. 16 is valid also for higher partitions of the boundary, e.g., the case depicted in Fig. 3, with $k = 6$. In explicit form, the proposed equations are: for $j = 1, ..., k$ —

$$[\ 1/2 \frac{\partial^2}{\partial x_j^2} + 1/3 \sum_{k \neq j} (x_k - x_j)^{-1} \frac{\partial}{\partial x_k}\ ] f_\alpha(x_1, ..., x_k) = 0\ . \tag{21}$$

In particular:

$$[\ 1/2\ \Delta + \underline{\nabla} V(x_1, ..., x_k) \cdot \underline{\nabla}\ ] f_\alpha(x_1, ..., x_k) = 0\ . \tag{22}$$

with $V(x_1, ..., x_k) = \log \prod_{i<j} |x_i - x_j|^{1/3}$.

Before concluding this discussion let us add that recently H. Pinson has proposed an expression for the probability of observing various winding numbers in percolation problems on compact Riemann surfaces, which were formulated by Langlands et. al. [2].

## 7. Summary and Further Directions

It was my intention in this article to explain some of the technical aspects of the construction of the *web* as a random geometric object, and to convey some of its remarkable properties.

It should be of interest to consider also the scaling limits of nearly critical models, which come with a length scale. Furthermore, it might be instructive to understand the equations (or differential inequalities [9,21]) governing the flow of such a theory under dilations.

The *web* construction can be carried also for interacting percolation. In particular it is of relevance for the scaling limit of the q–state Potts models, where the correlations are expressed through the connected clusters of the Fortuin – Kasteleyn random cluster models. Of special interest are: 1) The *web* corresponding to the case of $q = 2$, for which we expect a very simple relation with the $\phi^4$ field theory, in any dimension. 2) The *Web* corresponding to $q = 4$ (in $d = 2$ dimensions). The latter is related to the scaling limit of the ground states of the Heisenberg spin–1/2 antiferromagnetic spin chain [8].

## Acknowledgments

It is a pleasure to thank R. Langlands for stimulating discussions of the critical point, and S. Goldstein and J. Wehr for asking pointed questions about the web's construction. This work was supported by a grant from the US National Science Foundation.

## References

1. M. Aizenman. "The critical percolation web: construction and conjectured conformal invariance properties", in preparation.
2. R.P. Langlands, P. Pouiliot, and Y. Saint-Aubin. "Conformal invariance in two-dimensional percolation", *Bull. AMS* **30**, 1 (1994).

3. H.E. Stanley. "Fractal and multifractal approaches to percolation: some exact and not-so-exact results", in *Percolation Theory and Ergodic Theory of Infinite Particle Systems*, H. Kesten ed., Springer-Verlag (1987).
4. B. Mandelbrot. "Fractals in physics: Squig clusters, diffusions, fractal measures, and unicity of fractal dimensionality", *J. Stat. Phys.* **34**, 895 (1984).
5. A.A. Belavin, A.M. Polyakov, and A.B. Zamolodchikov. "Infinite conformal symmetry in two-dimensional quantum field theory", *Nucl. Phys.* B **241**, 333 (1984).
6. R.P. Langlands, C. Pichet, P. Pouiliot, and Y. Saint-Aubin. "On the universality of crossing probabilities in two-dimensional percolation", *J. Stat. Phys.* **67**, 553 (1992).
7. J. Cardy. "Critical percolation in finite geometries", *J. Phys.* A: *Math. Gen.* **25**, L201 (1992).
8. M. Aizenman and B. Nachtergaele. "Geometric aspects of quantum spin states", *Commun. Math. Phys.* **164**, 17 (1994).
9. M. Aizenman and D.J. Barsky. "Sharpness of the phase transition in percolation models", *Commun. Math. Phys.* **108**, 489 (1987).
10. M.V. Menshikov and A.F. Sidorenko. "Coincidence of critical points for Poisson percolation models", Th.Prob.Appl. **32**, 599 (in Russian; 544 in transl.) (1987).
11. H. Kesten. "The critical probability of bond percolation on the square lattice equals 1/2", *Commun. Math. Phys.* **74**, 41 (1980).
12. H. Kesten. "Scaling relations for 2D–percolation", *Commun. Math. Phys.* **109**, 109 (1987).
13. A. Pisztora. "Surface order large deviations for Ising Potts and percolation models", To appear in Prob.Th.Rel.Fields.
14. K. Falconer. *Fractal Geometry*, J. Wiley (1990).
15. B. Nienhuis, E.K. Riedel, and M. Schick. "Magnetic exponents of the two dimensional q–state Potts model", *J. Phys.* A: *Math. Gen.* **13**, L. 189 (1980).
16. M. den Nijs. "Extended scaling relations for the magnetic critical exponents of the Potts model",*Phys. Rev.* B **27**, 1674 (1983).
17. R. Ziff. "On the spanning probability in 2D percolation", *Phys. Rev. Lett.* **69**, 2670 (1992).
18. A. Aharoni and J-P Hovi. "Comment on 'Spanning probabilities in 2D percolation' ", *Phys. Rev. Lett.* **72**, 1941 (1994).
19. D. Stauffer, J. Adler, and A. Aharoni. "Universality at the three-dimensional percolation threshold". *J. Phys.* A: *Math. Gen.* **27**, L 475 (1994).
20. H. Pinson: "Critical percolation on the Torus", *J. Stat. Phys.* **75**, 1167 (1994).
21. G. Grimmett. *Percolation*, Springer-Verlag (1989).

# PHYLLOTAXIS : THE AVOIDANCE OF RATIONALS AND A RELATION WITH QUASI-CRYSTALS

Y. Couder and S. Douady

*Laboratoire de Physique Statistique, 24 rue Lhomond, 75231 Paris Cedex 05, France.*

ABSTRACT

The characteristic patterns observed in the organisation of vascular plants can be reproduced in a simple iterative system. The relation of the spiral modes with the Fibonacci series and the golden mean is shown to result from the system's trend to avoid rational organisations. The patterns thus spontaneously generated are shown to be related to the successive approximants of a one-dimensional quasi-crystal.

## 1. Introduction.

The investigation of the dispositions of leaves, bracts, sepals, petals... around the stem of vascular plants forms a domain of botany called phyllotaxis. It has a long history and was undertaken from several viewpoints. The observed organisations form regular patterns which are classified in only two categories.

In the first, the leaves form a spiral along the stem, each growing at a constant angle $\varphi$ from the previous one. This divergence angle $\varphi$ in the most common organisation has a constant value close to $\Phi=2\pi(2-\tau)=137.508°$ where $\tau$ is the golden mean $\tau = (-1+\sqrt{5})/2$.

In the second type of organisation (the whorled modes) a constant number n of leaves appear simultaneously at the same height on the stem. The leaves of a node are regularly disposed at an angle $\varphi=2\pi/n$ from each other. The leaves of a given node are shifted by $\varphi=\pi/n$ relatively to those of the previous and following ones.

### *1.1 Geometrical approach*

One of the pioneering works on phyllotaxis is due to A. and L. Bravais[1]. It was aimed at a characterisation of the geometry of the spiral and whorled organisations as observed on mature stems. By unrolling the cylindrical surface of the stem they reduced it to a two-dimensional strip and investigated the possible arrangement of lattices of points in that strip. In A. Bravais' career this investigation preceded his famous classification of the possible periodic lattices in three dimensional space[2]. It is interesting, from the point of view of the history of science, to notice that in this latter work A. Bravais used the same type of geometrical arguments about lattices of point-like elements that he had established in his botanical work.

The Bravais brothers demonstrated the equivalence of two different descriptions of the botanical patterns. In the first one, already used above, the position of each element is defined in relation with that of the elements formed previously in time. In the second the position of each element is related to its neighbours in space. In this case, when the elements are tightly packed (as in pine cones for instance), two sets of intersecting helixes (or spirals) are defined when each element is linked to its nearest (or contact) neighbours. They showed that if $\varphi=137.5°$, then the numbers i and j of these helixes (called "parastichies") necessary to cover the whole pattern are two successive terms of the main Fibonacci series : 1, 1, 2, 3, 5, 8, 13, 21, 34, ..., in which each term is the sum of the two previous ones (the first two terms being 1 and 1). These numbers of parastichies can easily be observed in pine cones or sunflower heads.

In the case of the whorled patterns, the same operation leads to the observation of two sets of parastichies of order i and of 2i of vertical rows often called the "orthostichies".

Van Iterson[3] extended the work of Bravais to the possible pilings of finite size disks in a strip and investigated the transitions from one organisation to the next. More recent works on the geometry of these patterns are due to Erickson[4], Rivier[5] and Rothen and Koch[6].

### *1-2 Dynamical approach*

A different approach is to investigate the dynamics of formation of these patterns. in the region of the tip of the stem. A smooth region of undifferentiated tissues, the apex, is situated at the extremity of the shoot. Around it, a specific mitotic activity creates small protrusions, the "primordia", which already form parastichies and will evolve into grown-up botanical elements: leaves, bracts, sepals, petals, stamens, florets.... In the reference frame of the apex, because of the growth of the shoot, the existing primordia drift away from the tip with a velocity V so that they leave room for the formation of new elements. W. Hofmeister[7] and M. and R. Snow[8] emitted two sets of hypotheses for the conditions of formation of a new primordium.

We showed in previous works (Douady and Couder[9,10]) that these two sets of hypotheses form the base of discrete iterative systems giving rise to the observed structures. The most general system can be summarised as follow:

- The process is 2D and occurs on the surface of a cone or of a plane.
- The central region of the apex is represented by a circle C of radius $R_0$.
- The primordia are circular, have a diameter $d_0$ and form at the periphery of this circle
- Once created they move away radially with a velocity V(r).
- A new primordium appears when and where an empty space of diameter $d_0$ has formed at the periphery of the apex

The organisation of the system derives from the choice of the position of formation of a new primordium on the circle which depends on the space left available by the previous ones. To implement this rule we assume in a simulation that each primordium is represented by a particle creating a repulsive energy E(d) (with d the distance to the particle). We then compute the potential resulting from the existing particles in all points of the circle C. When and where it becomes smaller than a fixed threshold $E_S$, a new particle is created. The value of $E_S$ fixes the size $d_0$ of the necessary space (i.e. the size of the primordium)

## 2 Results

The control parameter of this system is the ratio $\Gamma = d_o/R_o$. The first result of our simulations is that, for all values of Γ, steady regimes of the iterative system can be obtained. Both spiral and whorled modes of growth are obtained. We will consider them successively

### *2-1 The spiral modes*

The spiral patterns are characterised by their divergence angle φ. Figure 1 shows φ(Γ) for all the steady regimes obtained in our simulations. Half of the diagram is shown here as φ can be, with equal probability, positive or negative. The main feature of this diagram is that the number of different possible solutions increases with decreasing Γ. These solutions form disconnected curves in the diagram. Only one of these curves is continuous for all values of Γ. Along this curve, when Γ tends to zero, φ converges oscillatingly towards $\Phi = 2\pi(1-\tau)$.

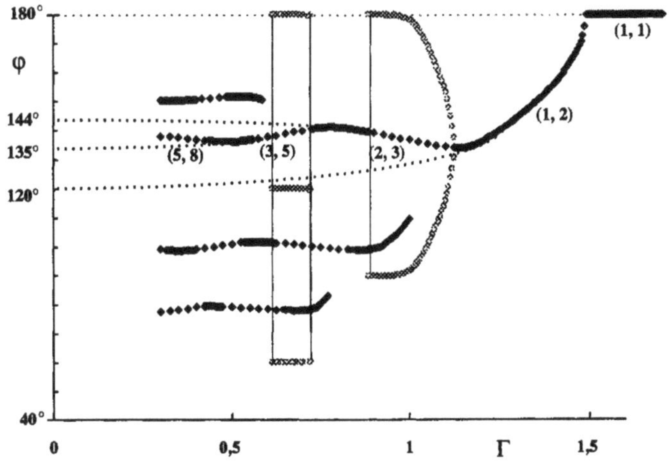


Fig. 1 Diagram of the values of the divergence φ obtained in the simulation as a function of Γ. Black diamonds : spiral modes, open circles : whorled modes. Each dotted lines shows the extrapolation at Γ=0 of a segment of the main curve to a rational angle.

At a given value of Γ, the observed spiral pattern has two parastichy numbers i and j (with j>i) which are consecutive numbers of the Fibonacci series. Figure 2 shows a disposition where i=5 and j=8. Inspection of the patterns obtained for decreasing values of Γ shows that near each extremum of the oscillation of the curve φ(Γ) (Fig.1) the pattern becomes locally hexagonal so that three sets of parastichies can be observed : (i, j, i+j). This corresponds to the value $\Gamma_{ij}$ for which there is transition from the parastichy pair (i, j) to the next (j, i+j). The convergence of φ towards Φ=137.508° results from these successive transitions. The predominance of the Fibonacci series in botany can be derived from this result, and from the fact that during the growth of a plant Γ is initially large and then undergoes a a continuous decrease. The other solutions located on secondary curves have a similar structure, each of them corresponding to a Fibonacci type of series built with different initial terms and converging towards one of the different irrational angles (99.502°, 77.955°, 151.135°, etc...) listed by A and L. Bravais[1].

In reference 10 Figure 1 is compared with two other diagrams. The first one was obtained by van Iterson[3] in a geometrical analysis of tilings by finite disks. The second one by Levitov[11] investigating a minimum energy condition on compressed helical lattices. The relation of this latter result with ours is similar to that which exists in classical crystallography between the energy of a periodic structure and its growth mechanism.

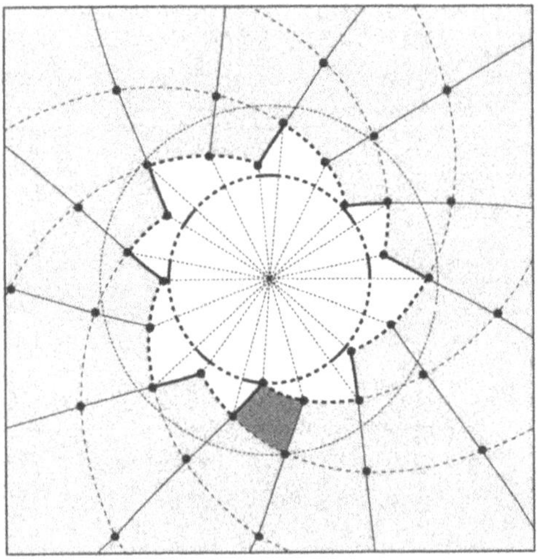

Fig. 2 A spiral pattern (8,13) obtained in the simulation. The points are the centres of the elements. The parastichies and the tiling of the central circle resulting from the projection are shown.

## *2-2 The avoidance of rationals*

It is possible to get an intuitive understanding of the self organisation in our system by a simple argument[10]. Let us consider the system at a value of Γ corresponding to the organisation (i, j). It can be shown[10] that this disposition is obtained when the position of a new particle n is determined by the repulsion of the j previous elements. Let us suppose for a moment that all the elements older than n-j have had their repulsive potential artificially switched off. Then if Γ is reduced there will be no bifurcation and the (i, j) branch would be followed all the way down to Γ=0. The angle which would be reached at Γ=0 would then be rational. For instance, as shown on Figure 1, an organisation (i=1,j=1) would lead to

$\varphi=(1/2)(2\pi)=180°$; (i=1,j=2) to $\varphi=(1/3)(2\pi)=120°$; (i=2, j=3) to $\varphi=(2/5)(2\pi)=144°$; (i=3, j=5) to $\varphi=(3/8)(2\pi)=135°$ etc... More generally following a branch (i, j) all the elements would tend when $\Gamma=0$ to be aligned on i+j radial lines and $\varphi$ would tend towards $(p/(i+j))(2\pi)$ where p is an integer called the encyclic number associated to i+j (see Bravais[1]). In reality this is not possible because the potentials are not switched off and the repulsion forbids the proximity of the elements that would result from such alignments. The bifurcation leading from (i, j) to (i, i+j) is the reaction of the system which prevents the elements n from coming too near to the elements n-i-j .

*The successive bifurcations can thus be seen as a sequence of "repulsions" by rational organisations. The convergence towards an irrational angle naturally results from the avoidance of successive rationals.*

It is worth noting that this situation is the opposite of a situation usually met by physicists when a non linear system is submitted to two frequencies[12, 13]. Generally the effect of non-linearity is to lock the two frequencies onto each other so that their ratio becomes a simple rational number. The width of the basin of attraction of the different rational numbers increases with the strength of the non-linear terms giving rise to a structure known as Arnold's tongues (Arnold[12]). These tongues have widths which correspond to the hierarchy of rationals : the simpler the rational the wider the tongue. In the diagram shown on Figure 1 a hierarchy of rationals is also exhibited : the simpler the rational the larger the value of $\Gamma$ at which it starts being avoided.

*2-3 Relation with quasi-crystals*

Let us recall that quasiperiodic patterns can be obtained by projections of periodic lattices into a space of smaller dimensionality. The simplest construction[14] is that of a 1-D tiling by a projection of a fixed 2-D square lattice of point-like elements onto a line. In the usual model two parallel straight lines $\Delta$ and $\Delta'$ are drawn passing respectively by two diagonally opposed vertices of a square of the lattice. All the elements located in the strip located between $\Delta$ and $\Delta'$ are then projected onto $\Delta$. If $\theta=45°$ a periodic tiling of periodicity $1/\sqrt{2}$ is obtained on the line. For all other angles a tiling of $\Delta$ by two types of segments - long (L) and short (S) - is obtained. If $\tan(\theta)$ is a rational number, say 5/8, a sequence (L, S, L, L, S, L, L, S, L, S, L, L, S) repeats itself along $\Delta$. If $\tan(\theta)$ is irrational $\Delta$ passes only by one element of the lattice and the sequence becomes infinite. A non periodic tiling of the line is obtained forming a 1D quasi-crystal. It is well known that the rational approximations of the golden mean are given by the ratio of two successive numbers of the Fibonacci series. The 1D tilings obtained for $\tan(\theta)$= 2/3, 3/5, 5/8, 8/13... are called quasi-crystals approximants.

We can apply to our radial phyllotaxic patterns a similar projection technique. On Fig. 2 the pattern is represented at the time of formation of the particle of order n. It is a classical result[1] that the neighbours along the parastichies of order i have orders of formation differing by i. The nearest neighbours of element n are thus of order n-i and n-j. We trace a circle of centre O and passing by the element placed diagonally relative to n which is of order n-i-j. An annulus is thus defined. All the elements located in this annulus are then projected radially onto the central circle. A tiling of the circle with elements of two lengths is obtained. For the mode (i=8, j=13) shown on Figure 2 the sequence of segments is, counterclockwise, (L, L, S, L, L, S, L, S, L, L, S, L, L, S, L, S, L, L, S, L, S).

We have seen above that the spiral lattice evolves continuously when $\Gamma$ is decreased. The evolution is characterised by a series of transitions in which a "square" lattice (i, j) transforms into another square lattice (j, i+j). At the transition the mode has three set of parastichies (i, j, i+j) and is therefore locally hexagonal. The relative positions of the elements above and below the transition are sketched on Figure 3. It is the corresponding evolution of the projection which interests us here. At the transition the neighbours of element n change so that the annulus has to be reduced and will now be limited by the two circles passing by element n and element n-(2j+i). A new sequence is obtained. It can be deduced from the previous one by noting (see Fig. 3) that the higher order parastichies of the initial pattern (of order j) have become the lower order parastichies of the new one. Therefore the segments S of the original sequence become segments L in the new one. As for the segments L they are split into (S, L) for a given transition and into (L, S) for the next. The successive sequences can be obtained by applying repeatedly these rules. The first spiral pattern (i=1, j=2) gives the sequence (L, L, S), the second (2, 3) gives (S, L, S, L, L) the third (3, 5) gives (L, L, S, L, L, S, L, S) etc.... In each of these tilings there are i short segments and j long ones. As $\Gamma$ decreases the spontaneous evolution of the 2D spiral lattice generates sequences which become longer and longer and the tilings evolve "naturally" as the successive approximants of a one dimensional 1D quasi-crystal.

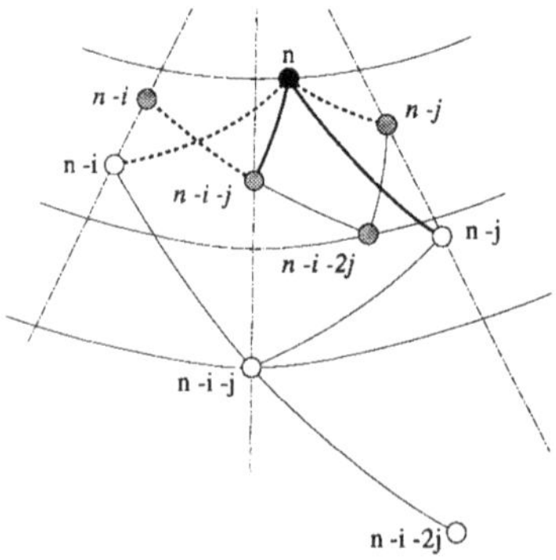


Fig. 3 Sketch of the positions of the centers of the neighbours of the incipient element n for two different values of $\Gamma$ respectively above (open circles) and below (grey circles) a transition at $\Gamma_{ij}$.

*2-4 Whorled modes*

Coexisting with the spiral modes our simulation also generates other types of patterns. There are ranges of values of $\Gamma$ (Fig.1) for which $\varphi$ can take several values: $\varphi=2\pi/n$ and $\varphi=\pi/n$. A number n of elements are formed simultaneously and a period $T_n$ separates their formation from that of the next group. The corresponding structures have the spatial dispositions observed in the botanical whorled modes in which n leaves form simultaneously at successive nodes along the stem. It is important to note that each of these organisations only exists in a finite range of values of Γ. The same projection technique used above gives on the circle a periodic lattice of 2n identical elements, thus a 1D crystalline structure.

## 3 - Conclusion

The iterative system simulating the dynamics of growth of the phyllotactic patterns reproduces the two types of organisations observed in botany. It can be shown[(10)] that for a given Γ the stablest pattern is that which provides the densest packing. The possible evolutions of the two types of patterns are however different. The spiral patterns undergo smooth transitions from one mode to the next due to the avoidance of successive rationals. The change in the pattern is then such that the projection on the central circle evolves *spontaneously* from one approximant of a 1-D quasi-crystal to the next. In contrast each whorled mode can only be observed in the narrow range of values of Γ where it provides a better packing than the spiral modes. Its projection corresponds to 1-D periodic tilings of the circle but there is no possible continuous transition from one of these modes to the next.

## 4 - References


(1) L. Bravais and A. Bravais, *Ann. Sci. Nat. second series* **7**, 42-110, 193-221, and 291-348, (1837), *Ann. Sci. Nat. second series* **8**, 11-42 (1837) ; *Ann. Sci. Nat. second series* **12**, 5-14, and 65-77 (1839).

(2) A. Bravais, A *Cahier du Journal de l'Ecole Polytechnique,* **33** (1850)

(3) G. Iterson, G. *Mathematische und microscopisch-anatomische Studien über Blattstellungen, nebst Betraschungen ûber den Schalenbau der Miliolinen,* Gustav-Fischer-Verlag, Iena (1907).

(4) R.O. Erickson, *Science*, **181**, 705-716, (1973).

(5) N. Rivier, *Mod. Phys. Lett. B*, **2**, 953-960, (1988)

(6) F. Rothen, and A.J. Koch,. *J. Physique* **50**, 633-657, (1989) and *J. Physique* **50**, 1603-1621, (1989)

(7) W. Hofmeister, *Allgemeine Morphologie der Gewachse, Handbuch der Physiologishen Botanik,* Engelman, Leipzig, 405-664 (1868)

(8) M. Snow, and R. Snow, *Proc. Roy. Soc. London. B*, **139**, 545-566 (1952) and *Phil. Trans. Roy. Soc. London Ser. B*, **244**, 483 (1962)

(9) S. Douady, and Y. Couder, *Phys. Rev. Lett.* **68**, 2098-2101(1992).

(10) S. Douady, and Y. Couder, Phyllotaxis as a self organized process Part I, II and III, To appear in *Journal of Theoretical Biology* (1995)

(11) L.S.Levitov, *Europhys. Lett.* **14**, 533-539 (1991)

(12) V. I. Arnold, Trans. *Am. Math. Soc.* **46**, 213, (1965)

(13) J. Stavans, F. Heslot and A. Libchaber, *Phys. Rev. Lett.* **55**, 596-599, (1985).

(14) M. Duneau and A. Katz, *J. Physique.* **47**, 181-196, (1986)

# FINITE-SIZE SCALINGS IN THE ASYMMETRIC SIX-VERTEX MODEL

DOOCHUL KIM and JAE DONG NOH
*Department of Physics, Seoul National University*
*Seoul, 151-742 South Korea*
E-mails: dkim@phya.snu.ac.kr, jdno@phyb.snu.ac.kr

ABSTRACT

Finite-size corrections for the energies of the general six-vertex model transfer matrix are considered using the Bethe ansatz solution. The natural parameter is $\tilde{\Delta}=(w_1w_2+w_3w_4-w_5w_6)/(w_1w_3+w_2w_4)$ where $w_i$ are the six vertex weights. The model is critical in the four-dimensional parameter space bounded by the Pokrovsky–Talapov transition surfaces and the stochastic surface $\tilde{\Delta}=1$. The scaling properties throughout the critical region are described by the Gaussian model. The universal Gaussian coupling constant $g$ is expressed in terms of the solution of an integral equation and also in terms of the Hessian of the free energy. The latter result relates the curvature of equilibrium crystal surfaces to $g$. Other non-universal parameters entering the finite-size scaling are also related to the thermodynamic quantities. At $\tilde{\Delta}=1$, the transfer matrix describes the deposition-evaporation model belonging to the Kardar–Parisi–Zhang (KPZ) universality class. For $\tilde{\Delta}$ close to but smaller than 1, a perturbative method is developed for a systematic expansion of the energies in powers of $1/N^{1/2}$, $N$ being the system size. The crossover scaling functions of the energy gaps are derived as a function of the scaling variable $s^{-1}(1-y^2)^{1/2}(1-\tilde{\Delta})N^{3/2}$ where $s$ is the growth rate and $y$ is the substrate slope. For $y$=0 and $\tilde{\Delta}=1$, the crossover scaling function from the KPZ to the Edwards–Wilkinson universality class is determined as a function of the scaling variable $u=s\sqrt{N}$ for large $u$.

## 1. Introduction

The six-vertex model was first introduced as a model for the residual entropy of ice and for related ferroelectric transitions[1] but more recently, several other physical applications are being found. In the body-centered solid-on-solid models, the six-vertex configuration is mapped to the surface configuration of the fcc and bcc faces and the free energy of the six-vertex model as a function of the horizontal and vertical fields depicts the equilibrium crystal shapes (ECS).[2,3,4] In this context, the critical region of the six-vertex model corresponds to the rounded region of the crystal surfaces. Also, on the stochastic surface where the vertex weights satisfy a certain relation, the transfer matrix of the six-vertex model can be regarded as the transition matrix of probabilistic cellular automata describing the dynamics of driven lattice gas systems and (1+1) dimensional surface growth models.[5,6] The growth models belong to the Kardar–Parisi–Zhang (KPZ) universality class with the dynamic exponent 3/2. Accordingly the energy gaps or the mass gaps of low lying energy levels scale with the system size $N$ as $1/N^{3/2}$ as opposed to the $1/N$ scaling at criticality. The five-vertex model, which is a special asymmetric limit of the six-vertex model, can be viewed as an interacting domain wall system[8], the interacting dimers[9] and certain types of crystal surfaces.[7] In these applications, it is necessary to consider the asymmetric six-vertex (ASSV) model in which there are non-zero fields which break the arrow reversal symmetries. The quantum hamiltonian derived from the ASSV as an anisotropic limit is that of the asymmetric XXZ chain in which the two hopping rates are asymmetric.

The symmetric six-vertex (SSV) model (without the electrical fields) is critical in the parameter range $-1 < \Delta < 1$, with $\Delta$ being the standard model parameter, and its critical properties are described by the central charge $c$=1 conformal field theory[10] in the continuum limit. Since the finite-size scaling (FSS)[11] is an important tool in studying the critical phenomena, there have been many studies on the FSS properties of the SSV model[12], and the symmetric XXZ chain with and without the vertical field.[13,14,15] Also, the FSS for the five-vertex model is studied by Noh and Kim[8] in the critical region and by Gwa and Spohn[5] at a stochastic point. However, since most of these works are confined to special cases, it is necessary to extend them to obtain full understanding of the FSS behaviors of the general six-vertex model. In this work, we describe our recent results on the FSS properties of the ASSV model for both the critical region[16] and the region where the KPZ scaling holds.[17]

## 2. Bethe ansatz equation and the phase function

We consider the six-vertex model on the $N \times M$ square lattice under the periodic boundary condition. Let $w_i$ ($i$=1,...,6) be the six vertex-weights in the standard ordering of six vertices. Define four independent model parameters by

$$h = \frac{1}{4}\log\frac{w_1w_3}{w_2w_4}, \qquad v = \frac{1}{4}\log\frac{w_1w_4}{w_2w_3},$$
$$\Delta = \frac{w_1w_2 + w_3w_4 - w_5w_6}{2\sqrt{w_1w_2w_3w_4}} \quad \text{and} \quad \delta = \frac{1}{2}\log\frac{w_1w_2}{w_3w_4}. \tag{1}$$

Here $h$ ($v$) is the horizontal (vertical) field in units of $k_BT$ conjugate to the horizontal (vertical) polarization $1-2\rho_L$ ($1-2\rho_D$), $\rho_L$ ($\rho_D$) being the left (down) arrow density. For later convenience, also define two parameters;

$$\tilde{\Delta} = \frac{w_1w_2 + w_3w_4 - w_5w_6}{w_1w_3 + w_2w_4} = \frac{\Delta}{\cosh 2h} \quad \text{and} \quad s = \frac{w_1w_3 - w_2w_4}{w_1w_3 + w_2w_4} = \tanh 2h. \tag{2}$$

Let $Q$ denote the number of down-arrows in a row of $N$ vertical bonds and let $q \equiv Q/N$. The mean value of $q$ is $\rho_D$.

The row-to-row transfer matrix $\mathbf{T}$ can be diagonalized by the Bethe ansatz method in each sector, the subspace with fixed $Q$. The central equation in this method is the Bethe ansatz equation

$$Z_N(\alpha_j) = \frac{2\pi I_j}{N} \qquad (j = 1, \ldots, Q) \tag{3}$$

where $I_j$ are half-integers (integers) for $Q$ even (odd) and $Z_N(\alpha)$ is the *phase function* of the form

$$Z_N(\alpha) = p(\alpha) + 2hi - \frac{1}{N}\sum_{j=1}^{Q}\Theta(\alpha - \alpha_j). \tag{4}$$

The explicit forms of the functions $p(\alpha)$ and $\Theta(\alpha - \beta)$ depend on the range of $\Delta$ and are tabulated e.g. in Nolden[4] for $\Delta < 1$ and in Bukman and Shore[3] for $\Delta \geq 1$. The eigenvalue $\Lambda$ of $\mathbf{T}$ in sector $Q$ is of the form $\Lambda = \exp(-E^R) + \exp(-E^L)$ with the 'energy' $E^{R,L}$ given by

$$E^{R,L} = -v(N - 2Q) \mp \left(h + \frac{\delta}{2}\right) N - \sum_{j=1}^{Q}\Phi_{R,L}(\alpha_j) \tag{5}$$

where $\Phi_{R,L}(\alpha)$ are also tabulated in references.[3,4] Here, $\{\alpha_j\}$ are the distinct solutions of Eq. (3) and the upper (lower) sign corresponds to the case of R (L). Of the four model parameters, only $\Delta$ and $h$ (or $\tilde{\Delta}$ and $s$) enter the Bethe ansatz equation determining $\{\alpha_j\}$ while the $\Phi_{R,L}$ depends only on $\Delta$ and $\delta$. The vertical field $v$ simply adjusts $\rho_D$, the mean value of $q$.

The ground-state energy is obtained if $I_j = \phi_j \equiv -(Q+1)/2 + j$. An important class of levels is characterized by the '$m$-shifted states' obtained from the choice $I_j = \phi_j + m$. They correspond to the spin-wave excitations of the Gaussian model. Creating particles and holes at either ends of the Fermi sea from the $m$-shifted states, one can generate the whole class of excited states which scale as $1/N$ in the critical phase. These are called the scaling levels. For simplicity, we only consider the $m$-shifted states in section 3 and the first excited state from the $m = 0$ state in section 4.

In the thermodynamic limit $N, Q \to \infty$ with fixed $q = Q/N$, the roots for the ground state and the scaling levels form a continuous contour $\mathcal{C}$ in the complex-$\alpha$ plane with end points $A = -a + ib$ and $B = a + ib$. $A$ and $B$ are functions of $h$ and $q$ determined by the relations $Z_\infty(A) = -\pi q$ and $Z_\infty(B) = \pi q$.

### *2.1. Universal ratio of fluctuations*

The sums appearing in Eqs. (4) and (5) are of the form

$$\mathcal{S}[f] = \frac{1}{N}\sum_{j=1}^{Q} f(Z_N^{-1}(\frac{2\pi I_j}{N})) \tag{6}$$

where $f(\alpha)$ is an arbitrary function which does not depend explicitly on $h$ and $q$, $Z_N^{-1}$ is the inverse function of $Z_N$ and the set $\{I_j\}$ pertains to the particular level of interest.

The bulk contribution $\mathcal{S}_0[f]$ of the sum is

$$\mathcal{S}_0[f] = \frac{1}{2\pi}\int_A^B f(\alpha)R(\alpha)d\alpha \tag{7}$$

where $R(\alpha)$, the root density function, is given by the solution of the integral equation

$$R(\alpha)+\frac{1}{2\pi}\int_A^B \Theta'(\alpha-\beta)R(\beta)d\beta = p'(\alpha) \tag{8}$$

with $'$ denoting the derivative with respect to $\alpha$. $\mathcal{S}_0$ depends on $h$ and $q$ only through $A$, $B$ and $R(\alpha)$. Partial derivatives of $\mathcal{S}_0$ with respect to $h$ and $q$ can also be expressed in terms of the solution of integral equations of the type Eq. (8). Comparing those expressions, we find the following general relation exists between the second derivatives of $\mathcal{S}_0$;

$$\frac{\partial^2\mathcal{S}_0[f]}{\partial q^2} = -\left(\frac{\pi}{2D_0^2}\right)^2\frac{\partial^2\mathcal{S}_0[f]}{\partial h^2}. \tag{9}$$

Here, $D_0 \equiv D(A) = D(B)$, referred to as the dressed charge in Bogoliubov *et al.*[13], is a real constant depending on $h$, $q$ and $\Delta$ but not on $f$ where $D(\alpha)$ is the solution of

$$D(\alpha)+\frac{1}{2\pi}\int_A^B \Theta'(\alpha-\beta)D(\alpha)d\beta = 1. \tag{10}$$

Eq. (9) implies that even if both $\frac{\partial^2\mathcal{S}_0}{\partial q^2}$ and $\frac{\partial^2\mathcal{S}_0}{\partial h^2}$ are functionals of $f$, their ratio is independent of $f$ and hence is universal.

### *2.2. Critical phase*

The ASSV model is ferromagnetically ordered with $\rho_D$=0 or 1 when $\tilde{\Delta} > 1$ while it is antiferromagnetcially ordered with $\rho_D$=1/2 when $\tilde{\Delta} < \tilde{\Delta}_c$ with $\tilde{\Delta}_c$ given parametrically by

$$\pm\, h = \log\cosh\lambda - \lambda/2 - \sum_{n=1}^{\infty}\frac{(-1)^n}{n}e^{-2n\lambda}\tanh(\lambda n) \tag{11}$$

where $\Delta = -\cosh\lambda$. $\tilde{\Delta}_c$ decreases from $-1$ for $s$=0 to $-4$ for $s$=$\pm$1. In between, $\rho_D$ varies continuously as a function of $h$ and corresponds to the rounded portion of the ECS and to the incommensurate phase of the domain wall systems. This is the region where the system is critical. The critical region in the four-dimensional phase space is most conveniently visualized in the ($\tilde{\Delta}$,$s$,$q$) space for fixed $\delta$. It is bounded by the surfaces $\tilde{\Delta}$=1, $s = \pm 1$ and $1-2q = \pm 1$. Within this region, a plane of singularity defined by $q = 1/2$ and $\tilde{\Delta} < \tilde{\Delta}_c$ should be inserted. The surface $\tilde{\Delta}$=1 is where the KPZ scaling holds, and the transitions at $q$=0, 1/2 and 1 as $v$ is varied through the respective critical values are of the Pokrovsky–Talapov (PT) type.

## 3. FSS in the critical phase

When a two dimensional system is in a critical phase which possesses the conformal symmetry, the energy of the transfer matrix under the periodic boundary condition is expected to take the finite-size scaling properties of the form

$$-\log \Lambda_\alpha = N\bar{f} + i\theta + \frac{2\pi}{N}\zeta''\left(h_\alpha + \bar{h}_\alpha - \frac{c}{12}\right) - \frac{2\pi i}{N}\zeta'\left(h_\alpha - \bar{h}_\alpha\right) + o\left(\frac{1}{N}\right) \tag{12}$$

where $\alpha$ labels the energy levels, $\bar{f}$ is the bulk free energy (in units of $k_B T$), $\zeta'$ ($\zeta''$) is the real (imaginary) part of the complex anisotropy factor $\zeta = \zeta' + i\zeta''$, $c$ is the central charge, $(h_\alpha, \bar{h}_\alpha)$ is the conformal dimension associated with the level and $o(x)$ stands for terms smaller than $x$; $\lim_{x\to 0} o(x)/x = 0$. The $O(1)$ imaginary term $i\theta$ accounts for the phase factor which appears when the one lattice-unit translation by $\mathbf{T}$ is incommensurate with the periodicity of the underlying model.[18] To identify various parameters of Eq. (12) with the model parameters, one need to calculate $E_\alpha^{R,L}$ to $O(1/N)$. We discuss this for the $m$-shifted levels for simplicity.

### *3.1. Summation formula*

The general strategy here is to regard $Z_N^{-1}$ in Eq. (6) as known first and determine it self-consistently later. To proceed, we make the crucial assumption that $Z_N(\alpha)$ can be expanded in the form

$$Z_N(\alpha) = b_0(\alpha) + \frac{b_1(\alpha)}{N} + \frac{b_2(\alpha)}{N^2} + o\left(\frac{1}{N^2}\right) \tag{13}$$

where $b_0 \equiv Z_\infty$. Also we assume that the first derivative $Z'_\infty(\alpha)$ does not vanish anywhere along the contour $\mathcal{C}$ to ensure that $Z_N^{-1}$ also has the similar $1/N$ expansion on $\mathcal{C}$. The $1/N$ scaling of the energy gaps throughout the critical phase follows from these assumptions.

Applying the Euler-Maclaulin sum formula to Eq. (6) and using the Taylor expansion of $f(Z_N^{-1}(\phi))$ at $\phi = \pm\pi q$, one obtains $\mathcal{S}[f]$ in the form

$$\mathcal{S}[f] = \mathcal{S}_0[f] + \frac{\mathcal{S}_1[f]}{N} + \frac{\mathcal{S}_2[f]}{N^2} + o\left(\frac{1}{N^2}\right) \tag{14}$$

where $\mathcal{S}_n[f]$ are given as integrals of $f'(\alpha)b_n(\alpha)$ plus some surface terms. Applying this to Eq. (4) with $f(\beta) = -\Theta(\alpha - \beta)$ then gives a set of integral equations from which $b_n(\alpha)$ are determined.

On the other hand, the derivatives of the bulk term $\mathcal{S}_0$ with respect to $h$ and $q$ also produce similar expressions. Comparing them with the results for $\mathcal{S}_n[f]$, we find following relations between the finite-size corrections and the partial derivatives of $\mathcal{S}_0$;

$$\mathcal{S}_1[f] = m\int_A^B f'(\alpha)D(\alpha)d\alpha = i\pi m\frac{\partial \mathcal{S}_0}{\partial h}, \tag{15}$$

$$\mathcal{S}_2[f] = 2\pi i\left(\frac{D_0^2}{2}m^2 - \frac{1}{24}\right)(\zeta - \bar{\zeta}) \tag{16}$$

where

$$\zeta = \frac{1}{2}\left[\frac{\partial}{\partial q}\left(\frac{\partial S_0}{\partial h}\right) + \frac{\pi i}{2D_0^2}\frac{\partial^2 S_0}{\partial h^2}\right] = \frac{1}{2}\left[\frac{\partial}{\partial h}\left(\frac{\partial S_0}{\partial q}\right) - \frac{2iD_0^2}{\pi}\frac{\partial^2 S_0}{\partial q^2}\right] \tag{17}$$

and $\bar{\zeta}$ is the complex conjugate of $\zeta$. We use the notation $\zeta$ for the quantity on the right-hand side of Eq. (17) anticipating identification of it as the anisotropy factor. An analogous relation between the Fermi velocity and magnetic susceptibility has been obtained by Bogoliubov et al.[13] for the symmetric XXZ chain but our result is more general.

### *3.2. Finite-size corrections of the transfer-matrix spectra*

Now we apply the general result of the previous subsection to the energy eigenvalues Eq. (5) by choosing $f = \Phi_R$ or $f = \Phi_L$. For simplicity we only mention the case where $E^R < E^L$ . From Eq. (5), the bulk free energy of the sector $Q$ is $e(h,q) = -h - \delta/2 - S_0$ for all scaling levels. Since the partition function is obtained by summing over all levels in the sector $Q$ and then over $Q$, one makes the Taylor expansion of the summand around $\overline{Q} \equiv \rho_D N$ by inserting $q = \rho_D + (Q - \overline{Q})/N$ and assuming $Q - \overline{Q}$ is of $O(1)$. Using the relations Eq. (9) and (17), one then finds that the energy eigenvalue of the $m$-shifted level in the sector $Q$ is of the form of Eq. (12) with central charge $c$=1. The non-universal quantities appearing in Eq. (12) are explicitly related to the thermodynamic quantities; $\bar{f}(h,v) = \min_q\{e(h,q) - (1-2q)v\}$, $\theta = \pi m(1 + e_h) = \pi m(1 + \bar{f}_h) = 2\pi m \rho_L$, and $\zeta = -\frac{1}{2}e_{q,h} + iD_0^2 e_{q,q}/\pi = -\bar{f}_{v,h}/\bar{f}_{v,v} - 4iD_0^2/\pi \bar{f}_{v,v}$ where the subscripts denote partial derivatives. The conformal dimension of the level is $(\Delta_{m,n-\beta}, \overline{\Delta}_{m,n-\beta})$ where $\beta$ is the fractional part of $\overline{Q}$,

$$\Delta_{m,n} = \frac{1}{4}\left(\frac{m}{\sqrt{g}} + \sqrt{g}n\right)^2 \quad \text{and} \quad \overline{\Delta}_{m,n} = \frac{1}{4}\left(\frac{m}{\sqrt{g}} - \sqrt{g}n\right)^2 \tag{18}$$

with

$$g = \frac{1}{2D_0^2}. \tag{19}$$

Here the integer vorticity index $n$ is related to the sector index $Q$ by $Q - \overline{Q} = n - \beta$. Eq. (18) is exactly the conformal dimension of the Gaussian model operator with spin-wave index $m$ and vorticity index $n$ with $g$ the Gaussian coupling constant. Thus one sees that the critical phase of the ASSV model is indeed in the Gaussian-model universality class with $c = 1$ but that the operator content is modified by the $O(1)$ phase term $\theta$ and the shift of the vorticity index $n$ by $\beta$. Similar results have been obtained by Woynarovitch *et al.*[14] for the symmetric XXZ chain in vertical field. That the critical exponents are given by Eq. (19) is first derived by Izergin and Korepin[19] for the symmetric XXZ chain in field but it also holds for the ASSV model. Eq. (19)

reduces to the well-known result $g = 1 - \cos^{-1}(-\Delta)/\pi$ in the limit of $h$=$v$=0.

*3.3. Toroidal partition function*

The toroidal partition function is the $O(1)$ part of the full partition function $\sum_\alpha \Lambda_\alpha^M$ where the sum is over all levels. It contains whole information on the FSS of the scaling levels and is conveniently expressed in terms of the modular ratio $\tau = \frac{M}{N}\zeta$ and the nome $\mathsf{q} = \exp(2\pi i\tau)$. The modular ratio $\tau$ specifies the shape of the torus into which the $N \times M$ square lattice should be deformed to make the system isotropic in the continuum limit. Taking into account of all the excitations from the $m$-shifted levels, we find that the toroidal partition function of the ASSV is given by

$$\tilde{\mathcal{Z}} = \frac{1}{|\eta(\mathsf{q})|^2} \sum_{m,n\in\mathbf{Z}} e^{-2\pi i m\alpha} \mathsf{q}^{\Delta_{m,n-\beta}} \bar{\mathsf{q}}^{\bar{\Delta}_{m,n-\beta}} \tag{20}$$

where $\eta(\mathsf{q})$ is the Dedekind eta function $\eta(\mathsf{q}) = \mathsf{q}^{1/24}\prod_{n=1}^{\infty}(1 - \mathsf{q}^n)$ and $\alpha$ is the fractional part of $\rho_L M$. This is the Coulombic partition function of the $c = 1$ theory modified by the two mismatch parameters $\alpha$ and $\beta$, the fractional part of the mean number of left arrows per column and down arrows per row, respectively. The two mismatch parameters account for the incommensuration of arrow densities $\rho_L$ and $\rho_D$ with the system size $M$ and $N$, respectively. Eq. (20) satisfy the required modular covariant property under the transformation $\tau \to -1/\tau$ which rotates the square lattice by $90^0$ degree. More general toroidal partition function under the twisted boundary conditions can also be derived in a similar way.

*3.4. Curvature versus g*

Eqs. (9) and (19) relate the Hessian of the free energy to the Gaussian coupling constant $g$ by

$$H \equiv \bar{f}_{h,h}\bar{f}_{v,v} - \bar{f}_{h,v}\bar{f}_{v,h} = \left(\frac{2}{\pi g}\right)^2. \tag{21}$$

This implies that the geometric mean of the two principal curvature in the rounded portion of crystal surfaces[2] is directly related to the universal quantity $g$ via the relation

$$\kappa = \frac{2d_0^2}{\pi k_B T}\frac{1}{g}, \tag{22}$$

where $d_0$ is the distance between the crystal planes, *i.e.*, the step height. The values of $g$ varies from 0 at the $\tilde{\Delta} = 1$ surface, to 1/2 at the $q$=0 and 1 PT surfaces and also at the free fermion point ($\Delta = 0$), and to 2 at the $q = 1/2$ PT surface.

## 4. FSS near the stochastic line

At $\tilde{\Delta} = 1$, analysis of the previous section breaks down due to the fact that $Z'_\infty(\alpha)$

at the end points $A$ and $B$ vanishes. As a result, the energy gaps acquire new scaling behaviors belonging to the KPZ universality class. Recently one of us have developed a perturbative method from which systematic corrections to the bulk are obtained as a power series in $1/\sqrt{N}$ provided the scaling variable defined by

$$t = \sqrt{\frac{q(1-q)}{2\pi}} \left(\frac{1-\tilde{\Delta}}{s}\right) N^{3/2} \tag{23}$$

is kept finite.[17] The method is applied to the asymmetric XXZ hamiltonian which generates the time evolution of the probability distribution for the single-step, body-centered-solid-on-solid growth model. The hamiltonian is an extreme anisotropic limit of the ASSV model and the result can be applied to the ASSV model also.

### 4.1. *Isotropic-to-KPZ crossover scaling function*

We discuss here the energy gap of the first excited state in each sector $Q = qN$. The energy gap scales with $1/N$ for $\tilde{\Delta} < 1$ (isotropic scaling) and with $1/N^{3/2}$ at $\tilde{\Delta} = 1$ (anisotropic, KPZ scaling). The crossover behavior from the isotropic to the KPZ scaling is described by the universal scaling function $\mathcal{F}(t)$ given by

$$\mathcal{F}(t) = \frac{\sqrt{2}}{3}\pi^{3/2}\{Y_3(Y_1^{-1}(t)) - Y_3^0(Y_1^{0\ -1}(t))\} \tag{24}$$

where the real functions $Y_m^0$ and $Y_m$ are defined by

$$Y_m^0(y) = \mathrm{Re}\left(\frac{m+2iy}{m+2}(-i\sqrt{y+i})^m + \frac{1}{i}\int_0^\infty \frac{(-i\sqrt{y+i+t})^m - (-i\sqrt{y+i-t})^m}{e^{\pi t}-1}\,dt\right), \tag{25}$$

$$Y_m(y) = Y_m^0(y) + (-i\sqrt{y-i})^m - (-i\sqrt{y+i})^m. \tag{26}$$

$\mathcal{F}(t)$ increases monotonically from $\mathcal{F}(0)$=6.5091893... and behaves as $\sim t^{1/3}$ for large $t$.

The first energy gap of the ASSV model is given by

$$\log\frac{\Lambda_0}{\Lambda_1} = \frac{1-\gamma^2}{(1+\gamma m)^2}\left[-\frac{2\pi i}{N} + \frac{2\gamma(1-m^2)^{1/2}}{1+\gamma m}\frac{\mathcal{F}(t)}{N^{3/2}}\right] + O(\frac{1}{N^2}) \tag{27}$$

where $\gamma = \sinh 2h/(e^\delta - \cosh 2h)$ and $m = 1-2q$. The corresponding expression for the asymmetric XXZ hamiltonian is obtained from $\frac{s}{2}\frac{\partial}{\partial\gamma}\log\Lambda_0/\Lambda_1$ at $\gamma = 0$. In this case, $s$ and $m$ takes the meaning of the growth rate and the substrate slope, respectively. When $s = 1$, $m = 0$ (i.e. $q = 1/2$) and $t = 0$, one recovers the result of Gwa and Spohn[5].

### 4.2. *KPZ-to-EW crossover scaling function*

The result of previous subsection applies for $s \neq 0$. The limit $s = 0$ is the Edwards–Wilkinson point where the correlation functions satisfy diffusion equations and the energy gaps scale with $1/N^2$. The perturbative method allows one to obtain the KPZ-to-EW crossover scaling functions as a power series in $1/u$ where $u = s\sqrt{N}$ is the crossover scaling variable. For the asymmetric XXZ hamiltonian at $q = 1/2$, the energy gap takes the scaling form $\mathcal{G}(u)/N^2$ as $N \to \infty$ with $u$ fixed. We find that

$$\mathcal{G}(u) = \mathcal{F}(0)|u| + \frac{B_0'}{u^2} + \frac{B_1'}{u^4} + \cdots \tag{28}$$

where $B_0' = 16\pi^2\zeta(3)$, $\zeta(x)$ being the zeta function, and $B_1'$=$-3744.8012\ldots$.

## 5. Acknowledgments

This work is supported by Ministry of Education, Republic of Korea, by KOSEF through the Center for Theoretical Physics of Seoul National University, and also by NSF grant DMR-9205125. DK thanks Prof. M. den Nijs for hospitality during his visit to University of Washington where most of this work is carries out.

## 6. References


1. E. H. Lieb and F. Y. Wu, in *Phase Transitions and Critical Phenomena*, edited by C. Domb and M. S. Green, (Academic, London, 1972) Vol. 1.
2. H. van Beijeren, *Phys. Rev. Lett.* **38** (1977) 993; C. Jayaprakash, W. F. Saam and S. Teitel, *Phys. Rev. Lett.* **50** (1983) 2017; C. Jayaprakash and W. F. Saam, *Phys. Rev.* **B30** (1984) 3916.
3. J. D. Shore and D. J. Bukman, *Phys. Rev. Lett.* **72** (1994) 604; D. J. Bukman and J. D. Shore, *J. Stat. Phys.* **78** (1995) 1277.
4. I. M. Nolden, *J. Stat. Phys.* **67** (1992) 155.
5. L.-H. Gwa and H. Spohn, *Phys. Rev.* **A46** (1992) 844, *Phys. Rev. Lett.* **68** (1992) 725.
6. J. Neergaard and M. den Nijs , *Phys. Rev. Lett.* **74** (1995) 730.
7. M. Gulácsi, H. van Beijeren and A. C. Levi, *Phys. Rev.* **E47** (1993) 2473; M. Gulácsi, A. C. Levi and E. Tosatti, *Phys. Rev.* **E49** (1994) 3843.
8. J. D. Noh and D. Kim, *Phys. Rev.* **E49** (1994) 1943.
9. H. Y. Huang, F. Y. Wu, H. Kunz and D. Kim, *Interacting dimers on the honeycomb lattice: An exact solution*, to be published.
10. J. L. Cardy, in *Phase Transitions and Critical Phenomena*, edited by C. Domb and J. L. Lebowitz, (Academic, London, 1987) Vol. 11.
11. See, *e.g.*, J. L. Cardy, ed., *Finite-size Scaling*, (North-Holland, New York) 1988; V. Privman, ed., *Finite Size Scaling and Numerical Simulation of Statistical Systems*, (World Scientific, Singapore) 1990.
12. P. Di Francesco, H. Saleur and J.-B. Zuber, *Nucl. Phys.* **B285** (1987) 454, **B300** (1988) 300; V. Pasquier, *J. Phys.* **A20** (1987) L1229.

# QUANTUM IMPURITY PROBLEMS IN CONDENSED MATTER PHYSICS

Paul FENDLEY[(a)], Andreas W.W. LUDWIG[(b)], and Hubert SALEUR[(a)]

[(a)] *Physics Department, University of Southern California, Los Angeles, CA 90089-0484, USA*
[(b)] *Physics Department, University of California, Santa Barbara, CA 93106, USA*

ABSTRACT

A review for non-experts is given of the basic physical ideas underlying recent exact solutions of Quantum Impurity Problems. These are phrased in the language of "scale-invariant boundary conditions" and crossover phenomena between them. Special emphasis is given to the exact computation of transport properties through point contacts in Luttinger liquids and fractional Quantum Hall effect devices, from Bethe Ansatz integrability. These include exact results in linear response [conductance] as well as out of equilibrium [$I(V)$ curve and DC shot noise].

## 1.) INTRODUCTION: QUANTUM IMPURITY PROBLEMS

Quantum Impurity Problems have recently attracted much attention in Condensed Matter Physics. The reason for this is that many of these problems are, to a large extent, analytically tractable and are, at the same time, realized in many interesting physical situations. Perhaps the best known such problem is the Kondo problem, describing the effect of dilute magnetic impurities in normal metals [ 1]. In the early 30's, experiments found an unusual minimum in the temperature dependence of the electrical resistivity in such materials [ 2]. A first theoretical explanation of the occurance of the minimum in the resistivity was given by Kondo [ 3], using low order perturbation theory. In more modern language, Kondo's calculation revealed that the antiferromagnetic interaction of a quantum spin with a metal is asymptotically free. The resistivity minimum is caused by a logarithmic *increase* of the resistivity due to the interaction with the impurity spin, as the temperature is lowered, counteracting the conventional *decrease* of resistivity arising from interactions with the phonons of the lattice. It is the same type of logarithmic dependence that arises in perturbative expansions in all asymptotically free theories. However, in such theories all terms of perturbation theory diverge at low temperatures $T$; and so does Kondo's perturbative result. Thus, as $T \to 0$, this problem exhibits all the features that plague the many strongly interacting systems of condensed matter and particle physics. For this reason the low-temperature behavior of this system had remained a major puzzle until the Kondo model and its generalizations, were studied using renormalization group (RG) techniques, by Anderson [ 4], Wilson [ 5] and Nozières [ 6, 7], and others, and more recently using Bethe Ansatz integrability [ 8] and Conformal Field Theory [ 9, 10, 11, 12, 13, 14, 15]. Generalized "multi-channel" Kondo models [ 7,10] and other related quantum impurity models [ 16, 17, 18, 19] provide analytically tractable examples [ 20] of so-called "non-Fermi-liquid" behavior, which has attracted much attention recently in the context high-$T_C$ and heavy Fermion materials. Roughly, non-Fermi-liquids are metallic systems whose low energy excitations (those that are responsible for the behavior of the system at low temperatures) cannot be described by conventional independent-electron-type theories, built on the free Fermi gas ("Fermi-liquids"). The conceptual importance of the Conformal Field Theory (CFT) approach to these Quantum Impurity Problems is that it explains the nature of the "non-Fermi-liquid" state, occuring in these systems: it is an interacting CFT that replaces low-lying quasiparticle excitations near the Fermi-surface of a Fermi-liquid. Moreover, CFT enables one to compute exact static and dynamic correlation functions [ 11,13,14] in the non-Fermi-liquid state. Interesting work that might indicate how to bootstrap such non-Fermi-liquid states of Quantum Impurity Models (where the Fermi-surface is one point) to higher dimensions (where the Fermi-surface is a higher dimensional manifold) was recently initiated by F D M Haldane [ 21].

Perhaps it might be fair to say that the Kondo problem, and the large variety of other Quantum impurity models, have become paradigm models for strongly interacting systems in Solid State Physics, playing a role similar to the one played by the Ising model for classical Statistical Mechanics.

As in the Kondo effect, quantum field theories play an important role in understanding Quantum Impurity Problems in general.

Let us start by defining what we mean by a Quantum Impurity Problem. We have in mind a general class of systems which have the following features [ 17,18]. (i) There are extended gapless (critical) quantum mechanical degrees of freedom, which live in an infinite spatial volume, the "bulk". (ii) These interact with an impurity, localized at one point in position space. The impurity may carry quantum mechanical dynamical degrees of freedom.

To have an example in mind, consider the Kondo problem. (i): The extended degrees of freedom are those of the bulk metal. The presence of a Fermi-surface means that the metal sits at a RG fixed point(see e.g. [ 22]). Physically this is easily understood since the system of electrons has ("particle-hole") excitations of arbitrarily low energy about its Fermi-sea ground state, providing the critical degrees of freedom in the "bulk" of the metal. (ii): The impurity spin, located at one point in space (say the origin), is a dynamical quantum mechanical degree of freedom (the dynamical process is the spin-flip). In the Kondo model we also see another feature of quantum impurity problems: they are generically one-dimensional. Since the impurity spin sits only at one point in space, it is only the (orbital) s-wave wavefunctions of the metal electrons that can interact with the spin. Second-quantizing this s-wave theory, we get a (non-interacting) quantum field theory of 1-dimensional Fermions, defined on half-infinite (radial) position space (a half-infinite line), which interacts with the quantum spin at the end of the line. Considering a path-integral representation of the 1D theory of Fermions, we have a $(1+1)$-dimensional Lagrangian field theory, one dimension from the (half-infinite) radial space coordinate, and another dimension from the (say, imaginary) time coordinate. All interactions take place at one point in space, the end of the line, where the impurity is located. In the $(1+1)$ dimensional space-time picture, the impurity sits at the "boundary" of space-time, which can be viewed as the upper-complex plane, the "boundary" being the real axis. This will be the general picture that we use in the study of all Quantum Impurity Problems.

In the Kondo problem we see one way how a Quantum Impurity Problem can be realized experimentally: a bulk system (here the 3D metal) contains a finite but small concentration $x$ of quantum impurities. In the limit of very dilute impurities ($x << 1$) the impurities do not interact with each other (to lowest order in $x$), and the 1-impurity theory may be used to describe the physics of the bulk material in the presence of dilute impurities. Actually, experiments performed at very low concentrations are known to be in good agreement with the 1-impurity theory for the ordinary (1-channel) Kondo model.

A quite different realization of quantum impurity models occurs in the context of point contacts [ 23, 24, 25, 26, 27]. These are basically electronic devices: two leads (capable of transporting electrical current) are attached to a single quantum impurity. Each one of the leads is connected to a battery, so that electrical current is driven through the quantum impurity. One can then measure experimentally the electrical current $I$, flowing through the quantum impurity, as a function of the applied driving voltage $V$ (from the battery). The $I(V)$ curve, the differential (non-linear) conductance $G_{diff} = \partial I(V)/\partial V$, or the temperature dependence of the linear response conductance $G_{lr} = \lim_{V\to 0} I(V)/V$ are examples of experimental probes characteristic of the quantum impurity. Since the $I(V)$ curve as well as the differential conductance are non-equilibrium properties, the point-contact realizations of quantum impurity problems are theoretically more challenging than other realizations, in that much more than equilibrium statistical mechanics is involved to achieve a theoretical understanding of these quantities.

We will come back to such a realization of quantum impurity problem later on, when we discuss point contacts in fractional quantum Hall effect devices.

## 2.) DESCRIPTION OF QUANTUM IMPURITY MODELS BY BOUNDARY CONDITIONS

A general approach to quantum impurity models was developed in Ref. [ 10] (see Ref. [ 17,18] for a review). The crucial observation was the following: if the quantum impurity has internal degrees of freedom, interaction and renormalization effects turn the *dynamical* quantum impurity into a *boundary condition* on the extended "bulk" degrees of freedom, at large distances and low temperatures. At low temperatures the system may be in the strong coupling regime (this is where Kondo's result diverges). *The boundary condition is thus a way to think about the strongly interacting Quantum Impurity Model.* Nozières' physical picture of screening [ 6] illustrates how this works for the simplest case, the 1-channel Kondo model: the antiferromagnetic interaction of the impurity spin with the spin of the conduction electrons, which has renormalized to large values at low temperatures, causes complete screening of the impurity spin. A modified boundary condition on the electrons that are not involved in the screening, is left. This mechanism, however, appears to be much more general, and seems to apply to all quantum impurity problems.

The boundary conditions, generated in this process, may be highly non-trivial (see e.g. Ref. [ 10,14,16]). However, since the bulk is massless (critical), the induced boundary condition is scale invariant asymptotically at large distances and low temperatures. Actually, it is often even conformally invariant [ 28].

Since quantum impurity problems are intimately related to scale invariant boundary conditions, let us discuss critical behavior occuring at the boundary of a massless bulk theory from the point of view of the renormalization group (RG). A scale invariant boundary condition has a number of different aspects:

(i): A scale invariant boundary condition is a RG fixed point (recall that the "bulk" is always critical in the type of systems that we are considering here). As with any RG fixed point, there is a set of *relevant/marginal/irrelevant* 'boundary operators' (and couplings) associated with each scale-invariant boundary condition. These operators have support only 'at the boundary', i.e. at one point in position space (at the position of the impurity).

(ii): If no relevant boundary operators are allowed (e.g. by symmetry [ 12]), then the scale invariant boundary condition represents a stable fixed point [the zero temperature fixed point, describing the Kondo model at strong coupling, is an example]. Irrelevant boundary operators give perturbatively calculable corrections to physical properties evaluated *at* the RG fixed point. Many important physical features of the Kondo model are actually due to the effect of the leading (dominant) irrelevant boundary operator [ 10,13].

(iii): Adding a *relevant* boundary operator to the Hamiltonian describing a particular scale invariant boundary condition, *destroys* that boundary condition, and causes *crossover* to a new, scale invariant boundary condition at large distances and low temperatures (in the infrared). In other words, we have a (boundary) RG flow, describing the crossover from the initial scale invariant boundary condition (in the ultraviolet, i.e. at short distances or high temperatures) to a new scale invariant boundary condition (in the infrared, i.e. at large distances and low temperature). Some examples are given in [ 12,16]. Note that at any stage of this flow, the bulk remains always critical and unchanged; the only action is at the boundary. An interesting observation concerning general boundary RG flows was made in [ 29]: with each one of the two scale invariant boundary conditions [the ultraviolet (UV) and the infrared (IR) RG fixed point] one can associate a universal number, $S_{imp}^{UV}$ and $S_{imp}^{IR}$ respectively. The physical meaning of $S_{imp}$ is that of the zero-temperature entropy of the boundary. It is a property of a scale invariant boundary condition. In general, one always finds that

$$S_{imp}^{UV} > S_{imp}^{IR}, \qquad \text{(decrease of boundary entropy)}$$

This may be viewed as a boundary analogue of the well known $c$-theorem of bulk conformal field theory [ 30, 31]. (Note, however, that the universal numbers $S_{imp}^{UV}$ and $S_{imp}^{IR}$ do not seem to be obviously related to a dynamical quantity (as is the case for the central charge, which is related to the stress tensor of CFT [ 32])).

As an example consider the (1-channel) Kondo model. Initially, at weak coupling (at high temperatures, i.e. in the ultraviolet), we have a quantum mechanical spin *decoupled* from the electron degrees of freedom of the metal. An isolated ($s = 1/2$) spin has a zero-temperature entropy of $S_{imp}^{UV} = \ln 2$. At strong coupling (at low temperatures, i.e. in the infrared), this impurity spin is completely screened by the conduction electrons [ 6]. This means that *no* dynamical degrees of freedom are left, and thus we have $S_{imp}^{IR} = 0$.

Scale invariant boundary conditions are often also conformally invariant. Thus, the neighborhood of the RG boundary fixed points can often be solved exactly using methods of conformal field theory*. On the other hand, in crossovers between such boundary fixed points, scale invariance is lost. Conformal field theory methods are therefore in general not useful to understand entire RG crossover trajectories. So, different methods are in general needed to study crossovers exactly. A well known such method is Yang-Baxter integrability.

## 3.) YANG-BAXTER INTEGRABLE BOUNDARY CONDITIONS (BETHE ANSATZ)

Ever since the pioneering work by Yang and Yang [ 33], Baxter [ 34], Zamolodchikov and Zamolochikov [ 35], Korepin [ 36] and others, Yang-Baxter integrability has been used with much success to study crossover scaling in bulk 2D theories exactly. However, it is often objected that integrable models are not so relevant for experimentally observable physics for at least two reasons:

(a): In order to achieve integrability, extensive fine-tuning of parameters is often required. Therefore, it is often believed that exact predictions made by studying an exactly integrable model might often not be generic and therefore difficult to observe experimentally.

(b): A very important set of experimentally accessible observables are transport properties. Amongst those is the (linear response) conductance which is usually computed from (equilibrium) Green's functions using the Kubo-formula. In the past, it has not been possible to compute exact Green's functions, even when the system is (Bethe-Ansatz) integrable. Therefore, before the work to be discussed below, integrability was largely restricted to the computation of thermodynamic quantities, excluding transport properties at finite temperature.

*A main message of this talk is that this is radically different for integrable Quantum Impurity Problems. For these problems, exact transport properties can be computed (even out*

*Many examples are discussed, e.g. in [ 10,16,29,11,12,13,16,14,17,1819].

*of equilibrium) and integrability can answer directly experimentally important questions.* In particular:

(i): In order to achieve integrability in Quantum Impurity Problems, one often needs to adjust very few parameters – sometimes none! An example where no parameter has to be fine-tuned to achieve integrability are the point-contacts in fractional Quantum Hall Effect devices, to be discussed below in detail: these systems provide an experimental realization of an integrable systems without any fine-tuning (in addition to the earlier Kondo effect).

(ii): We have computed, for the first time [ 24], exact transport properties (at non-zero temperature) from integrability. The linear response conductance for the above mentioned quantum Hall point contact device agrees quantitatively with recent experiments by Milliken/Webb/Umbach [ 37].

(iii): We have not only computed linear response transport properties, but also *non-equilibrium* quantum transport properties [ 25], such as the exact $I(V)$ curve of the above mentioned quantum Hall point contact, from Yang-Baxter integrability. This is, as far as we know, the first exact calculation of non-equilibrium quantum transport for a system of electrons that cannot be transformed into a free theory. The so-obtained $I(V)$ curves show a universal non-equilibrium broadening, as the system is driven harder out of equilibrium, as well as a maximum in the differential conductance beyond a critical universal value of driving. This is a pure non-equilibrium phenomenon, occuring in the steady state at sufficiently strong driving.

(iv): In addition, we have computed [ 26] the exact DC-shot-noise from integrability for the Quantum Hall point contact, another non-equilibrium transport property. This is the first time exact shot noise has been computed in a system of interacting electrons. The shot-noise is particularly interesting for the physics of the fractional quantum Hall effect in that it provides an experimentally directly accessible signal of the fractional charge of the "Laughlin quasi-particles" of the highly correlated quantum Hall state.

The method that we use to obtain those exact results is quite unconventional: it relies crucially on a judicious choice of basis of the Hilbert space of the system. We use a basis that is natural from the point of view of integrability. It is simply the basis in which all the infinite number of conservation laws (that exist since we have an integrable system) are diagonal. This basis turns out to have a "Fock-space" like structure, i.e. it is spanned by "quasiparticles". Such bases are conceptually related to so-called "Yangian" bases of the long-range Haldane-Shastri spin-chains [ 38]. It is in this basis that the quantum impurity iteraction becomes tractable. In order to compute transport properties, we use a kinetic equation for those quasiparticles of the Bethe-Ansatz. This is highly non-trivial, since we are really describing a fully interacting system, where a single-particle (Fermi-liquid) concept such as a kinetic equation seems out of place, at first sight. The single particle kinetic equation would fail to produce exact results in interacting systems due to the existence of particle production processes in the single-particle basis. However, the quasiparticle basis dictated to us by integrability is precisely characterized by the absence of quasiparticle production (and "factorized scattering"). This particular and special feature of an integrable theory, allows us to use a kinetic equation to compute transport properties exactly from integrability. We will elaborate further on this crucial point later on.

## 4.) EXACT TRANSPORT THROUGH POINT CONTACTS IN FRACTIONAL QUANTUM HALL EFFECT DEVICES: NON-EQUILIBRIUM AND LINEAR RESPONSE

We will review in this section a particular Quantum Impurity Problem in which the experimentally relevant physics is encoded in the crossover between two scale invariant boundary conditions. It is experimentally realized in tunneling experiments through point contacts between fractional quantum Hall effect edge states, which measure the conductance of the point contact. We have computed this conductance exactly, using Yang-Baxter integrability (see below). The salient features of this calculation are the following:

(A): Quantitative agreement [ 24] of the exactly computed linear response conductance is obtained with recent experiments by Milliken/Webb/Umbach [ 37], of the same quantity. Here, exact results for an integrable model can be directly compared with experiments. The agreement is even more so important since the conductance is described by a *universal* scaling function. It has no adjustable parameter (apart from an overall scale).

(B): The exact calculation and the good agreement with experiments provide a strong confirmation of the edge-mode theory of the fractional quantum Hall effect (originally proposed by X. G. Wen [ 39]). The tunneling conductance through the point contact can be viewed as a "spectroscopic probe" of the quantum state in the edges of the strongly correlated quantum Hall fluid.

(C): Since the results of our calculation are also valid out of equilibrium, exact predictions are provided for future non-equilibrium transport experiments. These include predictions of the exact $I(V)$ curve [ 24,25] as well as DC shot noise [ 26], which are *universal* functions involving the relevant variables (temperature $T$, voltage $V$,...). Universality means that the exact curves contain *no* adjustable parameters (apart from an overall scale) and are independent of sample-specific details that might occur in the fabrication process of the point contact device. For that reason our results have high predictive power.

### 4a.) Quantum Hall Effect (reminder)

We will be discussing a system that is in the strongly correlated quantum Hall state [ 40] at fractional filling $\nu$. The Quantum Hall effect arises in a system of electrons that are constrained to move in two dimensions and are placed in a strong magnetic field perpendicular to the plane of motion. Crucial for the occurance of the Fractional Quantum Hall effect [FQHE] is the presence of disorder ('dirt') and electron interactions. The remarkable physics of the Quantum Hall effect consists in the universal behavior of this complex system. The transverse conductivity $\sigma_{xy}$ exhibits plateaux as a function of inverse magnetic field, which are quantized to extemely high accuracy: $\sigma_{xy} = \nu e^2/h$ [where $\nu$ is the "filling fraction"]. One may say that " dirt and complicated electron interactions give rise to universal physics". When the magnetic field is such that the transverse conductivity $\sigma_{xy}$ is locked to its plateau value, the longitudinal (dissipative) conductivity $\sigma_{xx}$ vanishes. This means that the bulk system is then an insulator. In this case the only gapless modes, that are able to transport electrical current, sit on the edge of the sample. On the other hand, only for those values of magnetic field, where the transverse conductance jumps from one plateau value to the other, is the longitudinal conductivity non-zero. It is only for those values of magnetic field that the bulk of the system is a conductor.

## 4b.) The Quantum Hall Device

We will now describe the point contact device that is used in the experiments by Milliken/Webb/Umbach [ 37]. A 2D electron gas is prepared in the $\nu = 1/3$ fractional Quantum Hall state, in the interior of a "Hall bar". A "Hall bar" is a strip with a finite width in the $y$-direction, which is very long (length $2l \to \infty$) in the $x$-direction. As discussed above, the bulk of the system (the interior of the strip) is an insulator. Current can only flow on the edge of the sample, which is here the upper and the lower edge of the Hall bar (i.e. of the strip). Due to the chirality selected by the magnetic field, the upper edge contains only left moving currents whilst the lower edge contains only right moving currents. The gapless (current carrying) degrees of freedom on the two edges are simply left (L) and right (R) moving free Bosons, $\phi_L$ and $\phi_R$, that live on the upper and lower edge respectively. Their dynamics is given by the usual Tomonaga-Luttinger Hamiltonian

$$H_0 = v_F \frac{\pi}{\nu} \int dx [j_L^2(x) + j_R^2(x)]$$

Here,

$$j_L(x) = \frac{-1}{4\pi}(\partial_t + \partial_x)\phi_L, \quad j_R(x) = \frac{1}{4\pi}(\partial_t - \partial_x)\phi_R$$

measure the density of charge carriers on the edges. The total charges on L- and R- edges are

$$eQ_L = e \int dx j_L(x), \quad eQ_R = e \int dx j_R(x)$$

To make an electrical current flow a bias voltage $V$ has to be applied. This is done in the following way: Recall that the charge carriers in the upper/lower edge move only to the left/right. So, the left (right) moving charge carriers are injected from the right (left) from a battery, with an equilibrium thermal distribution at chemical potentials $\mu_L(\mu_R)$. The applied bias voltage is

$$V = \mu_L - \mu_R$$

The voltage can be incorporated into the edge mode theory by adding a term

$$H_V = -(\frac{eV}{2})\Delta Q, \qquad \text{where} \quad \Delta Q = Q_L - Q_R$$

to the hamiltonian.

Note that the difference in chemical potentials corresponds to a voltage drop in the (short) $y$-direction of the Hall bar. In the absence of any interaction between the L- and R- moving edge modes, this voltage bias causes flow of a ("source-drain current") current in the (long) $x$-direction of the Hall bar,

$$I_0(V) = ev_F \frac{< Q_L >_V - < Q_R >_V}{2l} = ev_F \frac{< \Delta Q >_V}{2l}$$

where $2l$ is the length of the bar in the $x$-direction. A simple calculation shows that the conductance is (see e.g. [ 41])

$$G_0 = \frac{\partial I_0}{\partial V} = \nu e^2/h \tag{1}$$

This agrees with the transverse conductivity as it should, since current and voltage are perpendicular.

## 4c.) The Point Contact

In the Quantum Hall device discussed in Section 4b.), the L- and R- edges are spatially separated, and do therefore not interact with each other. In recent experiments, Milliken/Webb/Umbach introduce a point contact at one position, say at $x = 0$, along the long ($x$)- direction of the Hall bar. At this point ($x = 0$) the right and left moving edges are brought close to each other, so that they interact at one point. In the experiments this is done by applying an additional gate voltage $\mathcal{V}_g$ perpendicular to the Hall bar (in the $y$–direction) at $x = 0$, which depletes the electrons of the 2D electron gas, that participate in the formation of the $\nu = 1/3$ Quantum Hall state within the Hall bar, i.e. within the strip. The sole effect of this gate voltage is to bring R- and L- moving edges close to each other at the point contact. This causes processes where charge carriers hop from the L-edge to the R-edge and vice versa. [It is important not to confuse this gate-voltage $\mathcal{V}_g$, governing the strength of the tunneling at the point contact, with the voltage bias $V$ (from the battery) that drives current through the device.]

Generically, for non-zero gate voltage, there is an additional (negative) backscattering contribution $I_B$ to the current, arising form the hopping events at the point contact. Thus, the total ("source-drain current") will be

$$I(V) = I_0(V) + I_B(V)$$

where $I_0(V)$ is the "source-drain current" in the absence of the point contact.

What is measured in the experiments, performed so far [ 37], is the *linear response conductance* of the point contact,

$$G_{lr}[\mathcal{V}_g, T] \equiv \lim_{V \to 0} I(V)/V, \qquad \text{(linear response conductance)} \tag{2}$$

When one plots the measured values for $G_{lr}[\mathcal{V}_g, T]$ versus $\mathcal{V}_g$ at different temperatures $T$, one obtains a number of curves, one for each temperature. These curves exhibit sharp resonance peaks [ 37], at particular (resonance) values $\mathcal{V}_g^*$ of the point contact interaction (i.e. the gate voltage). These resonance peaks sharpen as the temperature is lowered. Ideally, on resonance, the conductance assumes its maximal value [ 41]

$$G_{lr}[\mathcal{V}_g^*, T] = \nu e^2/h, \qquad \text{(on resonance)}$$

independent of temperature. (In the experiments, $\nu = 1/3$.)

For any value of gate-voltage, which is not a resonance value, i.e. for $\mathcal{V}_g \neq \mathcal{V}_g^*$, the point contact becomes completely insulating at zero temperature,

$$\lim_{T\to 0} G_{lr}[\mathcal{V}_g \neq \mathcal{V}_g^*, T] = 0, \qquad \text{(off resonance)}$$

In other words, no current $I(V)$ flows at zero temperature.

Now focus attention on one of the resonance peaks. Such a peak is centered around some particular value of gate voltage $\mathcal{V}_g^*$. Based on scaling arguments [ 41] the vanishing of the conductance off resonance should be described by a universal scaling function

$$G_{lr}[\mathcal{V}_g, T]/(e^2/h) = g[A\frac{\delta\mathcal{V}_g}{T^{2/3}}], \quad \text{(scaling of the linear response conductance, } \nu = 1/3) \tag{3}$$

where $\delta\mathcal{V}_g = \mathcal{V}_g - \mathcal{V}_g^*$ is the deviation of the gate-voltage from the particular resonance value $\mathcal{V}_g^*$. Here $A$ is non-universal factor, and $g$ is a universal function. This scaling form implies that the experimental resonance curves, measured at different temperatures, should collapse onto a single (universal) curve. This is indeed the case [ 37, 42,24]. We will now review the basic ideas underlying the exact computation of this universal function. The so-obtained exact curve is in good agreement with the experimentally measured one [ 24].

## 4d.) Description of Point Contact using the Luttinger Edge State theory

In section 4b.) we discussed the Hamiltonian $H_0$ describing the free propagation of the edge modes, as well as the term $H_V$ describing the effect of the applied driving voltage. In the presence of tunneling processes between the two edges (at $x = 0$), the total Hamiltonian is

$$H = H_0 + H_B + H_V$$

where the backscattering interaction

$$H_B = \lambda[e^{i\phi_L(x=0)}e^{-i\phi_R(x=0)} + e^{-i\phi_L(x=0)}e^{i\phi_R(x=0)}] \tag{4}$$

describes hopping of charge carriers at the point contact. Note that the point contact hopping interaction preserves the global $U(1)$ symmetry corresponding to a simultaneous shift of the two chiral bosons $\phi_L$ and $\phi_R$, by the same phase. This symmetry amounts physically to conservation of the total charge $Q_L + Q_R$ at the point contact (preserved in each hopping event). However, one establishes by power counting (see e.g. [ 41]) that all other local operators (at $x = 0$) of the free boson theory, consistent with this

$U(1)$ symmetry, are irrelevant for $\nu > 1/4$ (in particular, for $\nu = 1/3$ relevant for the experiments). For $\nu = 1/3$ the only relevant coupling constant, $\lambda$ in the above equation, has scaling dimension $\Delta = 2/3$, again by power counting.

Setting the driving voltage $V = 0$ for a moment, we see that we are precisely in the situation of a boundary crossover flow, discussed in Section 2 above. The ultraviolet fixed point, corresponding to $\lambda = 0$, describes the completely conducting fixed point (on resonance, no tunneling between L- and R- moving edges), whereas the infrared fixed point describes the completely insulating point contact (off resonance, at $T = 0$). The coupling constant $\lambda$ is proportional to the deviation of the gate voltage from its resonance value,

$$\lambda \propto \delta V_g$$

Note that the cosine interaction in Eq.(4) is *untractable* in the boson basis (except when $\nu = 1/2$, in which case it can be solved by quadrature [ 41, 43,25]).

## 4e.) Exact Calculation of Transport through the Point Contact

The key observation that enables us to compute transport properties exactly (and also thermodynamic properties [ 44], even though we will not discuss those here) is that the point contact interaction $H_B$ is Yang-Baxter (Bethe Ansatz) integrable [ 45,44,24]. To reiterate: there is an infinite number of local operators allowed by $U(1)$ symmetry, that can occur at the position of the point contact $x = 0$. All of those will in general be present in the Hamiltonian, when there is a point contact in the Hall bar. However, for filling $\nu = 1/3$, only one of those operators (the one with coupling constant $\lambda \propto \delta V_g$) is relevant. *This single relevant operator is integrable.* So, the model is automatically integrable, without any fine-tuning of parameters at all.

The main idea to compute exact transport properties from integrability of the point contact interaction $H_B$, is to make use of a judiciously chosen basis of the edge mode theory (i.e. of the theory of free bosons, $\phi_L, \phi_R$). This basis is *not* the basis of oscillator and zero modes of the boson fields. In a conformal field theory, such as that of the edge modes, many bases of Hilbert space can be used. All of them, of course, describe the same physics (the dynamics of the edge modes). Here we chose a basis in which the point contact interaction is simple. A similar point of view, namely a choice of a suitable basis of the bulk degrees of freedom, has also been at the root of the conformal field theory solutions of quantum impurity model fixed points, such as e.g. the Kondo model [ 10,19]. The basis that is chosen for the quantum Hall point contact has a Fock-space like (quasiparticle) structure, and will be described in physical terms below. This basis is dictated by the integrability of the problem.

Having defined this quasiparticle basis, one may apply kinetic equations, which are conceptually identical to those used in the standard Landauer-Büttiker transport theory in free electron systems. The reason why the same type of equations can also be used in the interacting quantum Hall point contact problem is quite deep. It is related to the "factorized" scattering that occurs at the point contact in the quasiparticle basis discussed above, which is a consequence of the integrability of the theory. Scattering (at the point contact) looks different in different bases, even though the physics must be the same no matter what basis we use to describe scattering. Since the scattering matrix is basis *dependent*, it may look simple in one basis, but complicated in another. We will exploit this extensively in the following.

More specficially, to put the integrability of the point contact interaction to use, we will proceed in 3 steps:

STEP (i): Even/odd Boson decomposition of the edge mode theory

STEP (ii): Integrability of the odd-boson theory; factorized scattering and definition of the quasiparticle basis.

STEP (iii): Kinetic equation for the quasiparticles of the basis of the Bethe Ansatz.

We will discuss those steps now, in turn.

<u>STEP (i):</u>

Note that the backscattering point contact interaction $H_B$ only involves the combination

$$[\phi_L(x=0) - \phi_R(x=0)]$$

Therefore, instead of using the original bosonic edge modes $\phi_L(x,t), \phi_R(x,t)$, it will be convenient to work with the linear combinations

$$\phi_L^{e/o}(x+t) \equiv \frac{1}{\sqrt{2}}[\phi_L(x,t) \pm \phi_R(-x,t)], \qquad \text{(even/odd boson)}$$

Note that these are both left-moving, but are non-local. The non-locality will not be a problem, as we will see below, since the interaction sits only at one point. Since the interaction only involves the field $\phi^o$, the other combination $\phi^e$ will be unaffected by the point contact interaction and will remain free [ 46].

From the above definition we get the charge densities of the even/odd bosons

$$j_L^{e/o}(x+t) = \frac{1}{\sqrt{2}}[j_L(x,t) \pm j_R(-x,t)]$$

This implies the following simple but important relationship of the charges of the even/odd boson theory with the original charges of the left and right moving edge modes:

$$\Delta Q = Q_L - Q_R = \sqrt{2}Q^o, \qquad \text{(not conserved)}$$
$$Q_{tot} = Q_L + Q_R = \sqrt{2}Q^e, \qquad \text{(conserved)}$$

<u>STEP (ii):</u> In STEP (i) we have decomposed the degrees of freedom of the bosonic edge modes into those of a non-interacting *even* boson ($\phi^e$), and those of the *odd* boson, ($\phi^o$), that do interact with the point contact. As we will see below, it is also only the odd-boson

theory that is needed to compute the the conductance and the $I(V)$ curve. Therefore, we will focus now only on the interacting degrees of freedom($\phi^o$).

The interacting theory of the odd-boson, ($\phi^o$), *with* the point contact is integrable [ 45,44]. This means that the dynamics of the odd boson theory possesses an infinite number of conservation laws (which are mutually commuting). Clearly, the basis of the Hilbert space of the odd-boson theory, in which all those conservation laws are *diagonal*, should play a special role. This basis, and the eigenvalues of the infinite number of conservation laws in this basis, is completely fixed by the constraints of integrability. In particular, the eigenvalues of the (interacting) Hamiltonian for the odd-boson have the form

$$H = v_F \sum_i p_i$$

whereas the other conservation laws are

$$H^{(n)} = v_F \sum_i (p_i)^n, \qquad (n = 3, 5, 7, ...)$$

The form of the Hamiltonian itself already indicates the Fock-space like structure of this basis: the energy is a sum of individual contributions $v_F p_i$ each of which is identified with a quasiparticle moving with momentum $p_i$ ($v_F$ is a "Fermi-velocity"). The quasiparticles of this very special basis of the odd-boson Hilbert space, which is selected by the constraints of integrability, are non-interacting, i.e. form an ideal gas, in the sense that their energies are simply additive. However, even though the gas of quasiparticles is "ideal", the distribution functions, describing how the quasiparticles fill the single-quasiparticle momentum states $p_i$, are in general not those of the ideal Bose of Fermi gas (except when $\nu = 1/2$ [ 25]). These distribution functions are highly non-trivial scaling functions which can be computed exactly [ 44,24,25] from the Thermodynamic Bethe Ansatz (TBA).

Let us now describe physically the results of the solutions of the TBA equations. The Bethe ansatz is a well studied procedure to find the structure of the basis of Hilbert space the diagonalizes the interacting hamiltonian $H$ (as well as the higher conservation laws $H^{(n)}$) of an integrable theory such as the odd-boson theory. The solution of those equations gives us the following information. (a) In general there may be a (finite) number of species of quasiparticles, all of which can propagate with momentum $p$ (this information did not yet appear in the discussion of $H$ and $H^{(n)}$ above, since all species have the same eigenvalues of those operators). We label the species of possible quasiparticles by an index $j$. The information of how many species of quasiparticles there are comes out of the solutions of the Bethe ansatz equations. (b) The density of states per momentum interval, $n_j(p)dp$, available for quasiparticles of species $j$. (c) The "occupation number " function indicating the number $n_j(p)f_j(p)dp$ of states of momentum $p$ that are actually occupied by quasiparticles of species $j$.

Originally, the Bethe ansatz equations contained the information about there being an integrable point contact interaction at one point in space ($x = 0$). However, as the system size $2l$ becomes large, the point contact interaction can only give rise to corrections to the density of states and occupation functions, $n_j(p)$ and $f_j(p)$, which are subleading and vanish as the system size $l \to \infty$. Therefore, the density of states and the distribution functions can be obtained by solving the Bethe equations without any point contact interaction. In this case there is an additional conserved quantity, the odd-charge $\Delta Q = \sqrt{2}Q^o$. With the help of this odd charge the possible species of quasiparticles can be described as follows: for any value of filling fraction $\nu$ (or, equivalently, for any Luttinger liquid described by Bosons $\phi_L, \phi_R$) there are always two species of quasiparticles, one with charge $\Delta Q = +1$ (the "kink") and one with charge $\Delta Q = -1$ (the "antikink"). In addition, when $j-1 < 1/\nu \leq j$ there are $j-2$ additional species of quasiparticles with $\Delta Q = 0$, i.e. they are neutral. So, for filling fraction $\nu = 1/3$ there is only one neutral quasiparticle.

The actual distribution functions can be computed exactly [ 44,24,25] at temperature $T$, and applied bias voltage $V$ (we will refrain from displaying them here explicitly). The effect of the bias voltage is to shift the chemical potentials of kinks and antikinks by equal and opposite amounts $V/2$. Since these distribution functions are complicated (scaling) functions of momentum, voltage and temperature, this may seem to be a complicated way to think about the simple odd-boson theory in the absence of the point contact interaction. The real advantage of the so-constructed quasiparticle basis comes into play when we consider the point contact interaction. Consider the following scattering process: A state consisting of a set of quasiparticles (in the odd-boson Hilbert space) is incident on the point contact. An example of such a state could be $|+, p_1 > \otimes |+, p_2 > \otimes |0, p_3 > \otimes |-, p_4 >$, describing two kinks, one antikink, and a neutral quasiparticle, the state carrying total charge $\Delta Q = +1$. Scattering of any such state off the point contact is very simple, in this quasiparticle basis: First of all, due to the kinematic constraints arising from the existence of the extra conservation laws $H^{(n)}$ with $n > 1$, there is no quasiparticle production upon scattering off the point contact. I.e. the outgoing state, once all the quasiparticles have passed through the point contact, contains again the same number of quasiparticles (four, in the above example). All that can happen is that the internal species labels of the quasiparticles can have changed. Second, this scattering is "factorized". This means that the 4 quasiparticles of our example scatter "one-by-one", or "one-after-the-other". The scattering matrix of the a multi-quasiparticle state is the *product* of the scattering matrices of the single-quasiparticle states contained in it ("factorized scattering"). The single-quasiparticle scattering matrix has the form $S_{j,j'}(p/T_B)$ where $p$ is the momentum of the incoming quasiparticle which is preserved in the elastic scattering process, and $j,j'$ denote the species labels for the quasiparticle before and after scattering. Since the single-quasiparticle S-matrix (the scattering amplitude) is a number, it must depend on momentum measured in units of some scale. The only available scale is the "boundary crossover scale" $T_B = C\lambda^{1/(1-\nu)}$ ($C$ is a non-universal constant), where the point contact starts to become insulating. The scale $T_B$ is the analogue of the Kondo temperature in the point contact tunneling problem. This is the physical reason[†] why the single-quasiparticle S-matrix is a function of the dimensionless ratio $v_F p/k_B T_B$ (the factors of $v_F$ and $k_B$ may be included into the constant $C$, or otherwise suppressed). The single-quasiparticle scattering matrix can be computed exactly from the constraints of integrability. The result is displayed e.g. in [ 24]. An important feature of this scattering matrix is that $S_{\pm,0}(p/T_B) = S_{0,\pm}(p/T_B) = 0$. This means that a kink (+) can only scatter back into a kink (+) or into an antikink (−), or vice versa. These are the only scattering processes of the charged quasiparticles, that can occur, which is reflected in the unitarity of the kink-

† †Of course, this comes out directly by analyzing the constraints of integrability.

antikink scattering matrix, $|S_{\pm,\mp}(p/T_B)|^2 + |S_{\pm,\pm}(p/T_B)|^2 = 1$. Scattering of the neutral particle by the point contact is described by a momentum dependent phase shift(see e.g. [ 44]).

In this unconventional basis, the effect of the point contact interaction is completely transparent. In order to make contact with the transport of physical charge, one needs to recall that the charge of the kinks/antikinks is the physical charge $\Delta Q$, in units of the electron charge, transported by these quasiparticles which have a rather complicated relation with the original electrons* . This allows us to proceed to the last step of our calculation.

STEP (iii):

Using the factorized nature of scattering of the point contact in the quasiparticle basis of the Bethe ansatz, we can write a kinetic equation for the charge transport across the point contact.

More precisely, we will compute the backscattering current $I_B$, discussed above**. Consider first the situation without any backscattering ($\lambda = T_B = 0$). In this case the difference of charges on the two edges, $\Delta Q$, is conserved, and therefore independent of time. When we now turn on backscattering ($\lambda \neq 0$), charge carriers hop from the left edge to the right edge and vice versa. Such processes violate conservation of the $\Delta Q$ charge. In each such hopping event we have $\delta Q_L = -\delta Q_R$.

Imagine a carrier of charge $+q$ hopping from the left to the right edge at the point contact. In such a hopping event charge $q$ is lost from the source-drain current. Furthermore such a hopping event corresponds to a change $\delta(\Delta Q/2) = \delta(Q_L - Q_R)/2 = -q$. Similarly, if a carrier of charge $q$ hops from the right to the left edge at the point contact, charge $q$ is added to the source-drain current. Such a hopping event corresponds to a change $\delta(\Delta Q)/2 = \delta(Q_L - Q_R)/2 = q$. So, the backscattering current can be entirely expressed in terms of $\Delta Q$: the backscattering current is the rate at which $\Delta Q/2$ changes in time

$$I_B = \partial_t \Delta Q/2$$

In the steady state, maintained by applying a bias (driving) voltage $V$, this rate is constant. When (say) $V$ is positive, there are more kinks than antikinks, injected with a thermal distribution into the system from their respective (infinite) reservoirs. Hence, at the point contact, more kinks scatter into antikinks, than vice versa. Therefore, a backscattering current $I_B$ is flowing.

We now derive an analytic expression for the backscattering current using a kinetic rate equation for quasiparticles of the Bethe ansatz. It is possible to compute the rate of

*For $\nu = 1/2$ these Bethe ansatz quasiparticles are actually again free Fermions. So, for this value of $\nu$ the point contact tunneling problem is actually non-interacting and can be solved by quadrature [ 41,43]. This particular case is perhaps useful to gain some physical intuition about the nature of the special basis that we are using in the general, interacting case (see e.g. Ref. [ 25]).

**Recall that the total "source-drain" current is $I = I_0 + I_B$, where $I_0$ is the "source drain" current in the absence of backscattering.

change of $\Delta Q/2$ in the basis of the Bethe ansatz quasiparticles, since each scattering event of a kink (antikink) into an antikink (kink) changes the physical charge $\Delta Q/2$ by $-1$ (by $(+1)$). This kinetic equation is of course very familiar. However, in general there would be no reason why it should be applicable to an interacting system. But, it is exact in the case that we are considering, even though the system is interacting. The reason for this lies in the constraints of integrability: as discussed above, in the very special quasiparticle basis of the Bethe ansatz, these quasiparticles scatter off of the point contact independently ("one-by-one"), and all quasiparticle production processes are absent. This allows us to express the rate of change in $\Delta Q$, in terms of the following two (exactly known) quantities:

(a): the number of kinks and antikinks (carriers of charge $\Delta Q = \pm 1$) in the momentum range between $p$ and $p + dp$

$$n_V(p) f_\pm(p, V) d(v_F p)$$

(Here $v_F$ is a "Fermi-velocity").

(b): The transition probability for scattering a kink (antikink) into an antikink (kink), of momentum $p$ (recall that $p$ is conserved in each such scattering event)

$$|S_{+-}(p/T_B)|^2$$

The number of kinks of momentum $p$ that scatter into antikinks per unit time is

$$|S_{+-}(p/T_B)|^2 \rho_{+-} d(v_F p) \tag{5}$$

where $\rho_{+-}$ is the probability that the initial kink state is filled and the final antikink state is empty. For a system of free fermions we would have

$$\rho_{+-} = f_+[1 - f_-]$$

but in our interacting system we only have

$$\rho_{+-} = f_+ - f_{+-}$$

where $f_{+-}$ is the unkown probability that the *both*, the kink and the antikink states are filled. For the number of antikinks of momentum $p$ that scatter into kinks per unit time one finds a formula similar to Eq.(5), with $\rho_{+-} \to \rho_{-+}$. In the final rate equation, only the difference between these two probabilities

$$\rho_{+-} - \rho_{-+} = f_- - f_+$$

appears. Notice that the unknown $f_{+-} = f_{-+}$ has cancelled out. Therefore, the backscattering current is

$$I_B(V) = -e \int d(v_F p) n_V(p) |S_{+-}(p/T_B)|^2 [f_-(p, V) - f_+(p, V)] \tag{6}$$

This formula is our basic result. All ingredients in this formula are exactly known: the scattering matrix has a simple analytic form (e.g. [ 24]) and the occupation factors and densities of state are obtained exactly from the thermodynamic Bethe ansatz (TBA). Notice that this equation is valid for any value of the driving voltage $V$. It thus automatically describes non-equilibrium transport.

### 4f.) Discussion of the Results

First we discuss the exact linear response conductance, Eq.'s(2, 3) as computed from Eq.'s(1, 6). Fig.1 of [ 24] contains a plot of the conductance, Eq.(3), versus the parameter $(T_B/T)^{2/3} \propto \lambda/T^{2/3}$. This curve is compared with the Monte Carlo data by Moon et al. [ 42] and the experimental data by Milliken et al. [ 37]. The agreement with the Monte Carlo is excellent. The agreement with the experiments is good. There are some small discrepancies for low values of the conductance, in the tails. These might be due to the fact that the conductance peak, selected in the experiments, was actually smaller than $\frac{1}{3}e^2/h$ and was decreasing with temperature. This means that it was not a complete resonance. In order to achieve true resonance, two parameters [ 41] need to be adjusted, whereas only a single parameter, the gate voltage, has been varied in the experiment. We suggest that future experiments use the freedom to vary also the magnetic field on the $\nu = 1/3$ Hall plateau, to locate a resonance with maximal and temperature independent conductance on resonance. One would expect that this would still improve the agreement between experiment and theory. In any case, however, the curve shows a crossover from a $\propto \delta V_g/T^{2/3}$ powerlaw behavior at high temperatures to a $\propto T^4/(\delta V_g)^6$ powerlaw at low temperatures. This crossover of powerlaws is a highly non-trivial property of the universal scaling function, describing the linear response conductance. Coming back to the discussion in section 2, it is a consequence of the crossover between the completely conducting and the completely insulating scale invariant boundary conditions, describing the UV and the IR fixed points.

Second, the exact results for the differential conductance*, $\partial(I_0 + I_B)/\partial V$, which is a universal function of two independent ratios, $T_B/T$ and $V/T$, is plotted in Fig.'s 1 and 2 of Ref. [ 25]. (In Fig. 1 a different set of independent ratios, namely $T_B/V$ and $V/T$ was chosen.) Fig. 1 shows that the differential conductance curve develops a peak at a driving $V/T$ that is equal to a critical universal value of 7.18868.... This peak is pure non-equilibrium feature. Even for much smaller values of driving, $V/T$, the conductance already shows a significant universal " non-equilibrium broadening", displayed in Fig. 2 of Ref. [ 25]. It would seem this effect might be relatively easily visible experimentally.

### 4g.) DC Shot Noise

Using similar techniques we have also obtained [ 26] the DC shot noise at $T = 0$. We find a remarkably simple relationship between the DC noise $< I^2 >$ (which depends on

*Recall that the integral gives the $I(V)$ curve.

$T_B$ and $V$) and the $I(V)$ curve:

$$< I^2 >= \frac{\nu e}{2(1-\nu)}[V\frac{\partial I}{\partial V} - I]$$

This relation may be thought of as a non-equilibrium analogue of the fluctuation-dissipation theorem, for the Luttinger liquid (at $T = 0$).

Specializing this expression to the limit of weak backscattering, $\lambda$, and $T_B$ small, one finds the expression for shot noise of non-interacting particles of charge $\nu e$ (Laughlin 'quasi-particles')

$$< I^2 >\approx \nu e(I_0 - I), \qquad T_B \text{ small} \tag{7}$$

On the other hand, specializing to strong backscattering, $\lambda$, and $T_B$ large, our formula gives the expression for shot noise of particles of charge $e$ (electrons)

$$< I^2 >\approx eI, \qquad T_B \text{ large}$$

The DC shot noise thus gives a direct probe [ 47] of the fractional charge ($= \nu e$) of Laughlin's quasi-particles, underlying the theory of the FQHE. Our exact expression Eq.(7) interpolates between these two limits.

Acknowledgement:

One of us (A.W.W.L.) likes to thank Ian Affleck for a very fruitful collaboration on many of the results reviewed in sections 1 and 2. We thank M P A Fisher, S M Girvin, F D M Haldane, F Lesage, C Mak, F P Milliken, N P Warner for valuable discussions. This work was supported by the Packard Foundation, the National Young Investigator program (NSF-PHY-9357207), the DOE(DE-FG03-84ER40168), and the Alfred P. Sloan Foundation (A.W.W.L.).

# REFERENCES


1. For a recent review see e.g.: *The Kondo Problem to Heavy Fermions*, A C Hewson (Cambridge University Press, New York, 1993).
2. For a review of those early experiments, see e.g.: *The Theory of Metals*, A H Wilson (Cambridge University Press, Cambridge, 1953).
3. J Kondo, Prog. Theor. Phys. **32** (1964) 37.
4. P W Anderson, J. Phys. **C3** (1970) 2346.
5. K G Wilson, Rev. Mod. Phys. **47** (1975) 773.
6. P Nozières, J. Low Temp. Phys. **17** (1974) 31; *in* Proc 14th Int. Conf. on Low temperature physics, ed. M Krusius and M. Vuorio, Vol. 5 (North-Holland, Amsterdam, 1975) p.339.
7. P Nozières and A Blandin, J de Physique (Paris) **41** (1980) 193.
8. P B Wiegmann and A M Tsvelik, Pis'ma Zh. Exp. Teor. Fiz **38** (1983) 489 [JETP Lett. **38**, 591]; Adv. Phys. **32** (1983) 453; N Andrei, K Furuya and J Lowenstein, Rev. Mod. Phys. **55** (1983) 331; N Andrei and C Destri, Phys. Rev. Lett. **52** (1984) 364.
9. I Affleck, Nucl Phys **B336** (1990) 517.
10. I Affleck and A W W Ludwig, Nucl. Phys. **B 352** (1991) 849; *ibid* **B 360** (1991) 641.
11. A W W Ludwig and I Affleck, Phys. Rev. Lett. **67** (1991) 3160.
12. I Affleck, A W W Ludwig, H-B Pang, D L Cox, Phys. Rev. **B 45** (1992) 7918.
13. I Affleck and A W W Ludwig, Phys. Rev. **B 48** (1993) 7297.
14. A W W Ludwig and I Affleck, Nucl. Phys. **B 428** (1994) 545.
15. A W W Ludwig, in Proceedings of the ICTP Summer School on *Low Dimensional Quantum Field Theories for Condensed Matter Physicists*, Trieste (Italy), Sept. 1992, Editors: S Lundqvist, G Morandi and Yu Lu (World Scientific, New Jersey, 1995).
16. I Affleck and A W W Ludwig, Phys. Rev. Lett. **68** (1992) 1046; I Affleck, A W W Ludwig, B Jones, Phys. Rev. **B52** (1995) 9528.
17. A W W Ludwig, Int. J. Mod. Phys. **B 8** (1994) 347.
18. I Affleck, in *Correlation Effects in Low-Dimensional Electron Systems*; Eds.: A Okiji and N Kawakami (Springer-Verlag, Berlin, 1994).
19. A W W Ludwig, "Quantum Impurity Problems: Exact Methods" (proceedings of the Conference on *Quantum Field Theory and Statistical Mechanics*, Univ. of Southern California, Los Angeles, April 1993; World Scientific, in press).

20. A W W Ludwig, Physica **B 199&200** (1994) 406. (Proceedings of the International Conference on Strongly Correlated Electron Systems, San Diego, 1993).

21. F D M Haldane, Lectures given at the *International School of Physics, Enrico Fermi*, Varenna, Italy, July 1992.

22. R Shankar, Rev Mod Phys **66** (1994) 129.

23. D C Ralph, A W W Ludwig, J v.Delft, R A Buhrman, Phys. Rev. Lett. **72** (1994) 1064.

24. P Fendley, A W W Ludwig, H Saleur, Phys Rev Lett **74** (1995) 3005.

25. P Fendley, A W W Ludwig, H Saleur, Phys Rev **B 52** (1995) 8934.

26. P Fendley, A W W Ludwig, H Saleur, Phys Rev Lett **75** (1995) 2196.

27. P Fendley, A W W Ludwig, H Saleur, *" Exactly Solvable Field Theory of Tunneling Between Edge States in the Fractional Quantum Hall Effect"*, Proceedings of the ICTP-Trieste Conference on 'Quantum Field Theory and Statistical Mechanics', April 1995, Trieste, Italy (World Scientific, in press).

28. Conformally invariant boundary conditions were first discussed from an abstract point of view by Cardy (see e.g. J L Cardy, Nucl. Phys. **B324** (1989) 581, and references therein).

29. I Affleck and A W W Ludwig, Phys. Rev. Lett. **67** (1991) 161.

30. A B Zamolodchikov, JETP Lett **43** (1986) 730.

31. A W W Ludwig and J L Cardy, Nucl Phys **B285** (1987) 687.

32. A A Belavin, A M Polyakov, A B Zamolodchikov, Nucl Phys **B241** (1984) 333.

33. C N Yang and C CP Yang, J Math Phys **10** (1969) 1115.

34. R J Baxter, *" Exactly Solved Models in Statistical Mechanics"* (Academic Press, New York, 1982).

35. A B Zamolodchikov and AL B Zamolodchikov, Ann Phys (N.Y.) **120** (1979) 253; Al B Zamolodchikov, Nucl Phys **B342** (1991) 695.

36. "*Quantum Inverse Scattering Method and Correlation Functions*", V E Korepin, N M Bogoliubov, A G Izergin (Cambridge Univ. Press, Cambridge, 1993).

37. F P Milliken, C P Umbach and R A Webb, Solid State Communications (1995) [in press].

38. F D M Haldane, Z N C Ha, J C Talstra, D Bernard and V Pasquier, Phys Rev Lett **69** (1992) 2021.

39. X G Wen, Phys Rev **B 41** (1990) 12838; **43** (1991) 11025; **44** (1991) 5708.

40. see e.g.: " *The Quantum Hall Effect*", Editors: R E Prange and S M Girvin (Springer Verlag, New York, 1990).

41. C L Kane and M P A Fisher, Phys Rev **B46** (1992) 15233; **B46** (1992) 7268.

42. K Moon, H Yi, C L Kane, S M Girvin, M P A Fisher, Phys Rev. Lett **71** (1993) 4381.

43. F Guinea, Phys Rev **B32** (1985) 7518.

44. P Fendley, H Saleur, N P Warner, Nucl Phys **B430** (1994) 577.

45. S Goshal and A B Zamolodchikov, Int. J. Mod. Phys. **A9** (1994) 3841.

46. This decomposition has enabled the solution of the X-ray singularity problem in a Luttinger liquid: I Affleck and A W W Ludwig, J Phys **A27** (1994) 5375.

47. C L Kane and M P A Fisher, Phys Rev Lett **72** (1994) 724.

# EXPERIMENTAL EVIDENCES OF DIFFERENCES BETWEEN DISCRETE AND CONTINUUM PERCOLATION MODELS

Sasuke Miyazima, Kaneyasu Maruyama,
Koichi Okumura and Akihiko Okazaki

*Department of Engineering Physics, Chubu University Kasugai, Aichi 487, Japan*

A fact that discrete and continuum percolation models belong to the different universality class is proved by experiments and simulation which are related to the electric conductivity, the permeability, the elasticity and the normal vibrational mode energy.

## 1. Introduction

There are many systems which indicate characteristic properties of percolation. Most of percolation systems in nature are seemed to be a continuum percolation medium which consists of several different kinds of materials with different physical quantities such as random mixtures of conductive and non-conductive subsystems, elastic and nonelastic matters and so on. Generally the systems show the phase transition e.g. from conductive to nonconductive phase or from elastic to nonelastic system, depending on the concentration of constituents. Critical properties of these systems are very interesting from the point of theoretical and applicative purposes.

In 1957, Broadbent and Hammerseley introduced so called percolation problem on the discrete lattice, for the first time.[1] This is a very simple and basic model which shows a prominent change in various physical quantities as well as Ising spin model. Most of physicists thought that continuum percolation system can be approximated by introducing an infinitely fine lattice. However, some differences are noticed by examining natural materials which are considered as continuum percolation problems. Halperin *et al.* pointed out that critical exponents of the percolation problems might be different between discrete and continuum percolation models.[2] The main conclusion is listed in Table 1 for the electric conductivity, elastic constant and permeability in the two- and three-dimensions.

We have some variations of original percolation problem. In Figs. 1(a) and 1(b), two kinds of continuum percolation models are shown. The model shown

Table 1. The critical exponents for electric conductivity, elastic constant and permeability of the discrete, the continuum Swiss-cheese and the continuum inverted Swiss-cheese percolation models in the two- and three dimensions.[2]

| Model | | Electric Conductivity | Elastic Constant | Permeability |
|---|---|---|---|---|
| Two Dimension | Discrete | 1.3 | 3.7 | 1.3 |
| | Swiss | 1.3 | 5.2 | 2.8 |
| | Inverted Sw. | 1.3 | 3.7 | 1.3 |
| Three Dimension | Discrete | 1.9 | 3.6 | 1.9 |
| | Swiss | 2.4 | 6.1 | 4.4 |
| | Inverted Sw. | 1.9 | 4.1 | 2.4 |

in Fig. 1(a) is called as Swiss-cheese model because of the structure which is similar to Swiss-cheese and the other one shown in Fig. 1(b) is inverted Swiss-cheese because of inverse relation of medium and hole. The exponents of the electric conductivity for discrete, Swiss-cheese and inverted Swiss-cheese models are the same in the two-dimension. However, the exponents of the elastic constant are different. Several years ago, Benguigui realized an experiment of the above mentioned continuum percolation problem,[3] and the electric conductivity and elastic constant of a piece of aluminum foil with randomly distributed holes, which is shown in Fig. 2. His conclusion was not definite unfortunately. Therefore three-dimensional experiments which are extensions of Benguigui's two-dimensional experiment have been required. However, although it is easy to punch off holes in the two-dimensional system such as the aluminum sheet shown in Fig. 2, it is difficult to make cavities inside the three-dimensional material and take it off to outside without leaving any damage inside the object.

Here we propose a three-dimensional experiment of Swiss-cheese percolation with a simple idea, which gets rid of the above mentioned difficulty. At first, we set electrically conductive fluid such as mercury or pure water in a container. If we push rubber balls into the fluid, the electrically conductive fluid spills over from the container by the same volume as the rubber balls. The electric conductivity decreases as the number of balls which are pushed into the fluid increases. The conductivity is reduced to zero at a threshold volume ratio of fluid component.

In the next section, we discuss the exponent of electrical conductivity and permeability of three-dimensional Swiss-cheese model, and we can show the

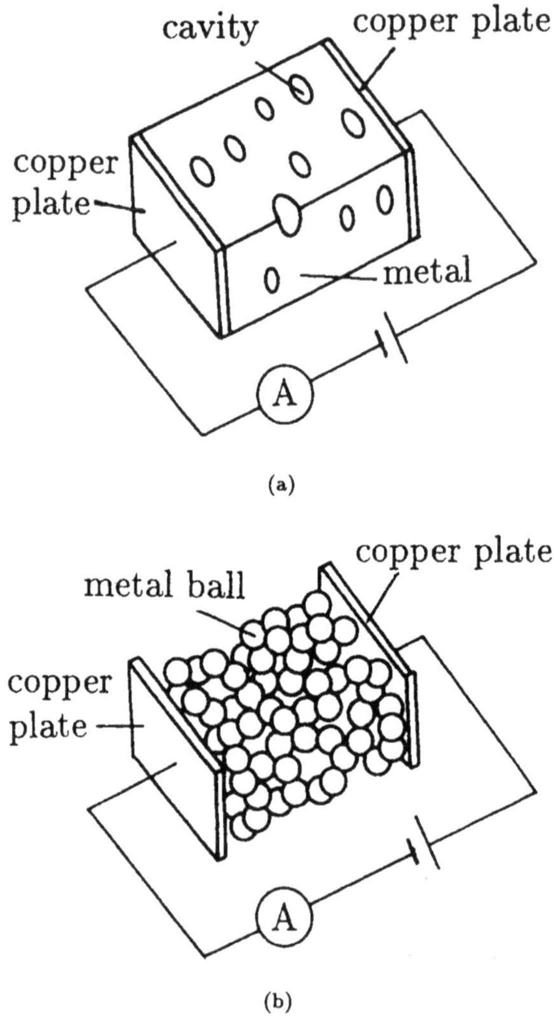


Fig. 1. (a) Swiss-cheese model. (b) Inverted Swiss-cheese model.

exponents are 2.4 and 4.2, respectively, which are equivalent values to the theoretical prediction. In Sec. 3, we discuss an inverted Swiss-cheese model in which the elastic constant is an object of investigations.

There are many important physical quantities in most of materials such vibration, heat conductivity additionally to electrical conductivity, elasticity

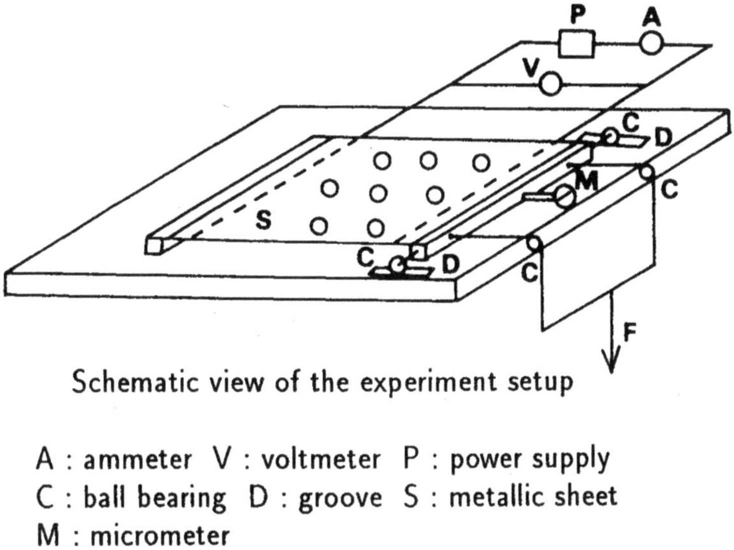


Fig. 2. Two-dimensional experiment of continuum percolation model. The electric conductivity and the elastic constant of aluminum foil are measured.

and permeability which are discussed by Halperin et al. Especially here we are interesting of differences between full materials in the two- and three-dimension and corresponding porous materials such as Swiss-cheese.

## 2. Exponent of Swiss-cheese Percolation Model

The electric conductivity of lime stone was measured and gave a clue to understanding of differences between the discrete and continuum percolation models. An intermediate model between discrete and continuum percolation models which will be discussed in Sec. 4 also leaves some evidences for the difference. The experiment by Benguigui is another trial to make the difference clear.

In 1991, we showed an experimental evidence which revealed the difference between the discrete and the continuum percolation models, getting rid of the above mentioned difficulty by introducing electrically conductive fluid instead of solid material.[4] Although it is very difficult to make cavities in the solid material and to take it off from the material, we can simulate the situation by inserting rubber balls which play role of cavities into fluid.

Our apparatus is shown in Fig. 3, where a container is an acrylic tube of diameter 10 cm and contains 360 rubber balls (the diameter is 16 mm). The

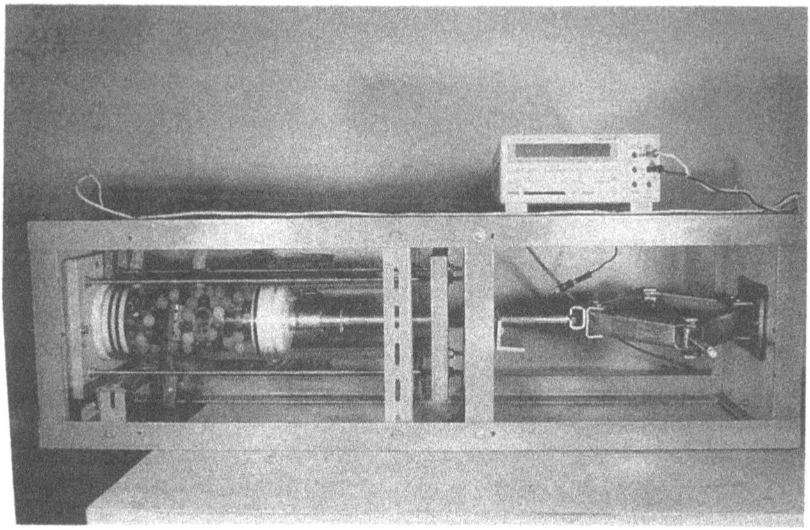

Fig. 3. Experimental apparatus for electric conductivity in the three-dimensional Swiss-cheese model.

piston at the right hand side is used to control the space among the rubber balls filled by the electrically conductive fluid, i.e. if one presses the piston, the gap becomes narrow and therefore the electric conductivity of fluid between both the end of acrylic tube decreases. Only one trial gave the exponent of electric conductivity 2.4 ± 0.1 together with very good reproducibility.

By using similar apparatus shown in Fig. 4 we can measure the permeability of fluid through the porous object made by balls pressed by piston. In this case we obtained the exponent of fluid permeability 4.2 ± B0.3,[5] which should be compared with the predicted value 4.4 shown in Table 1.

## 3. Exponent of Inverted Swiss-cheese Percolation Model[6]

Here we consider an elastic system which consists of many rubber balls. When the rubber balls form a zigzag chain from one side to other side of a container by adding one by one (see Fig. 5), then the chain shows an elastic property against pressing the piston at a side of the container. One of the difficulties for realization of this inverted Swiss-cheese model is that we have to lift up balls in the container without any support. Here we introduce a randomly netted pile of very thin copper wire (so called enamel wire). Many balls are hung up

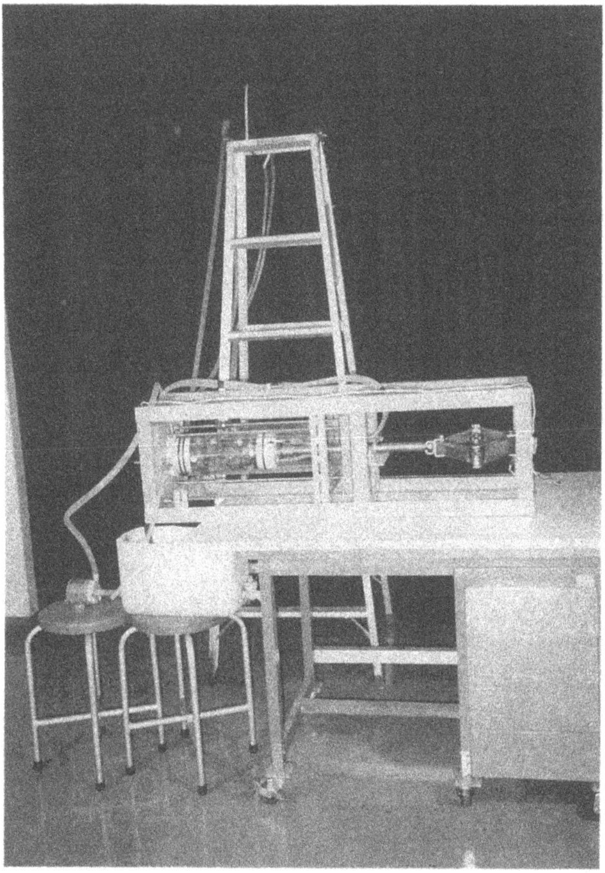

Fig. 4. Experimental apparatus for permeability in the three-dimensional Swiss-cheese model.

with the wire, which has a negligibly small elastic constant. This is shown in Fig. 5.

At first there is no sequence of balls from a side to another side of container. When we press a side into the container, the rubber balls get contact each other gradually and form a sequence or zigzag chain. As soon as the sequence is formed, if we press a side a little, stress can be measured at the other side with load transducer. Calculating the elastic constant from the data and plotting them by a log-log plot, we obtain the exponent 4.15, which is comparable with the predicted exponent 4.1 in Table 1.

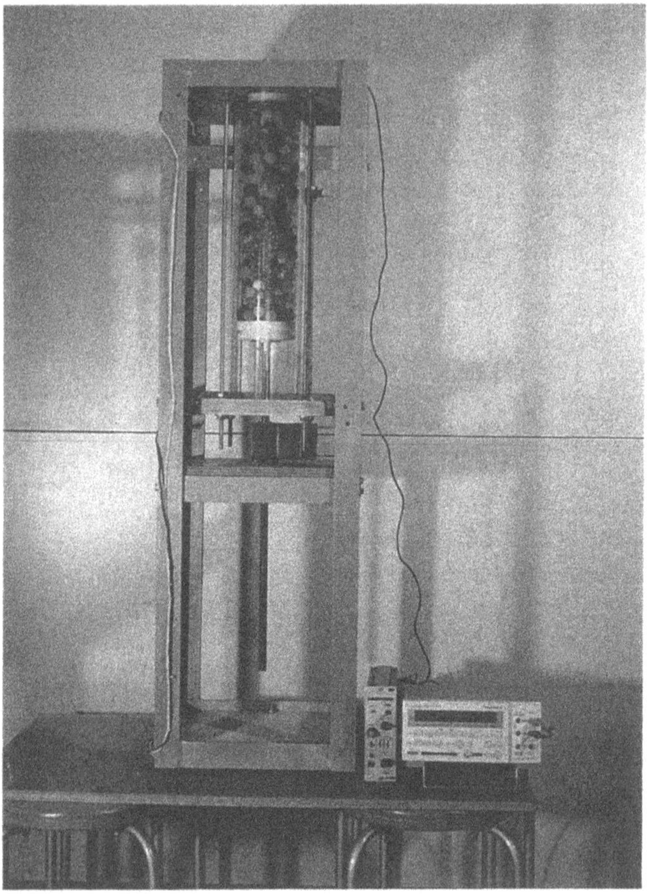

Fig. 5. Experimental apparatus for elasticity in the three-dimensional inverted Swiss-cheese model.

## 4. Summary and Discussion

Up to here we consider the electric conductivity and permeability in Swiss-cheese model and the elasticity in inverted Swiss-cheese model. It is clearly shown in these experiments that there are differences between the discrete and the continuum percolation models. However, if one looks at Table 1, everyone easily notices that there are still many exponents which are not experimentally checked.

Furthermore there are still more physical quantities in solids such as vibration, specific heat, and so on. These quantities also change when the porosity of solids changes. For example, the lowest energy of a basic vibration mode of aluminum plate is measured as the porosity is changed. We already noticed decrease of vibrational energy with porosity and it is also clear that we have a threshold value that the mode energy becomes zero. In a similar way, every physical quantities depend on the porosity and there should exist the threshold at somewhere of the porosity.

Coupled with theoretical and experimental investigations, nowadays simulation is also another very important method for research. Simulation study of the continuum percolation model is not difficult in principle, but very difficult in practice even in two-dimensional space. Here we discuss the connectivity of random circle holes of various radius. By using finite size scaling technique, the exponent of correlation length are obtained as $n = 1.65$, which is different from 4/3 for the discrete percolation model.[7]

Finally, we discuss an intermediate model between the discrete and the continuum percolation models which consists of conductive and nonconductive rigid balls. If these rigid balls occupy the regular sites, then this model is a discrete model. However, when we pack these rigid balls randomly, they occupy almost regular sites locally, but partially they are not regular but they include many defects like in crystal. They have no long range order any more, and the position of center of the balls becomes continuous variables. Therefore this model can be considered between the discrete and continuum model. The experimental result for the threshold value and the exponent in this model are 0.284 and 3.93, respectively.[8] These results are worthy to pay attention because the model itself is an intermediate one between the discrete and the continuum percolation model, the threshold value is very near to that in the discrete model and the exponent is much larger than both the models.

Thus we have been able to show the difference between the discrete and the continuum percolation models in its critical exponents of the electric conductivity, the permeability and the elasticity. In the discrete and the continuum percolation model, there are many interesting and stimulus points of study and there remain a lot of subjects which are not clear.

## References

[1] S. R. Broadbent and J. M. Hammersley, Proc. Cambridge Philos. Soc. **53** (1957) 629.

[2] B. I. Halperin, S. Feng and P. N. Sen, Phys. Rev. Letters **54** (1985) 2391.

[3] L. Benguigui, Phys. Rev. Letters **53** (1984) 2028.

[4] S. Miyazima, K. Maruyama and K. Okumura, J. Phys. Soc. Jpn. **60** (1991) 2805.
[5] K. Maruyama, K. Okumura and S. Miyazima, Physica **A191** (1992) 313.
[6] K. Maruyama, K. Okumura and S. Miyazima, Fractals **1** (1993) 904.
[7] A. Okazaki, K. Maruyama, K. Okumura and S. Miyazima, to be published in Fractals **3** (1995).
[8] K. Maruyama, K. Okumura and S. Miyazima, to be published in Fractals **3** (1995).

# CHARGE FRUSTRATION, BICONTINUITY AND INTERFACES

DAVID CHANDLER
*Department of Chemistry, University of California, Berkeley, CA 94720-1460 USA*
E-mail: chandler@cchem.berkeley.edu

ABSTRACT

This paper reviews and discusses implications of the results obtained with the charge frustrated Ising model. It is a model of self-assembly for oil-water-surfactant mixtures. It applies to other systems as well.

## Oil, water and surfactants

At normal temperatures, oil and water don't mix. On large length scales and away from criticality, the only structure that exists in this system is the oil-water interface. It's an Ising system. The addition of some surfactant, however, changes the physics. With the surfactant, the system self assembles, forming very many mesoscopic structures. The change in physics is due to frustration. The frustration follows from the constraint of stoichiometry. This constraint forces oil-like and water-like species to exist within a microscopic distance – the length of a surfactant molecule. On larger length scales, the entropic effects of this constraint are the same as the energetic effects of electroneutrality in a charge particle system. Stillinger and others have noted this apparent mapping.[1–3] We have used it to motivate the charge frustrated Ising model for oil-water-surfactant mixtures.[4,5]

The charge frustrated Ising model is characterized by one energy and two lengths. The energy is essentially the low temperature oil-water surface tension in the absence of surfactant. The two lengths are the correlation length of the oil-water system in the absence of surfactant, and the mean length of a surfactant molecule. Analysis of this minimalist model and comparison with experiment is reviewed in this paper. Since the reviewed material is well described in the primary literature,[4–7] what I say now is appropriately brief.

## Charge frustrated Ising model

The model has two binary variables: $s_i = \pm 1$, indicating whether lattice site $i$ is occupied by a polar or hydrophobic species, and $t_i = 1, 0$, indicating whether that species is or is not part of a surfactant (if it is not, the species corresponds to water or oil). One may imagine, for example, a cubic lattice with lattice spacing $a$. The symmetry of the lattice should be irrelevant, but the microscopic length scale it introduces is significant. It corresponds to the correlation length of the oil or water phases in the absence of surfactant. It is a small length of the order of 2 to 3 Å.

The Hamiltonian of the charge frustrated model is

$$\mathcal{H} = -J \sum_{ij}{}' s_i s_j + \frac{1}{2} q^2 \sum_{i \neq j} v_{ij}\, s_i s_j t_i t_j - \sum_i (h s_i + \mu t_i) \; . \tag{1}$$

The primed double sum is over nearest neighbor pairs, and the coupling $J$ is positive. The second double sum is the Coulomb term. Namely, $v_{ij}$ is the inverse Fourier transform of $4\pi/k^2$. Its strength is governed by the size of the charge $q$. This charge is given by Stillinger's formula [1]

$$q^2 = d/4\pi\beta\rho\Delta^2 \; , \tag{2}$$

where $d$ refers to dimensionality, and $\beta$ is the reciprocal temperature $1/k_{\mathrm{B}}T$. The length $\Delta$ is essentially the root mean square separation of polar and hydrophobic groups within a surfactant molecule, and $\rho$ is the average bulk density of surfactant. The quantities $h$ and $\mu$ are the chemical potentials that govern the relative amounts of oil and water, and the concentration of surfactant. The charge depends upon surfactant concentration, i.e., $\rho$. The model must therefore be treated self consistently.

The size of the ferromagnetic coupling, $J$, can be estimated from the oil-water surface tension, $\sigma$. Far from oil-water criticality,

$$J \approx \sigma a^2/2 \; , \tag{3}$$

and at room temperature, $\sigma \approx 0.1 k_{\mathrm{B}}T\text{Å}^{-2}$. Thus, the parameters characterizing the charge frustrated Ising model of oil-water-surfactant systems are all reasonably well known. This knowledge is of significance when using the model to explain experimental observations.

The Coulomb interactions favor anti parallel spins, the ferromagnetic interactions favor parallel spins. The charges will organize so that electroneutrality is maintained in volumes much larger than those circumscribed by a screening length. Debye's formula for the screening length in this model is $r_{\mathrm{D}} = \Delta/\sqrt{6}$. If the amphiphilic surfactants are significantly longer than the lattice spacing, $a$, the competing interactions will cause frustration. Spatially modulated phases may therefore follow.

In general, constraints have a corresponding electrostatic representation. The Coulomb interaction is the longest wavelength manifestation of the constraint.

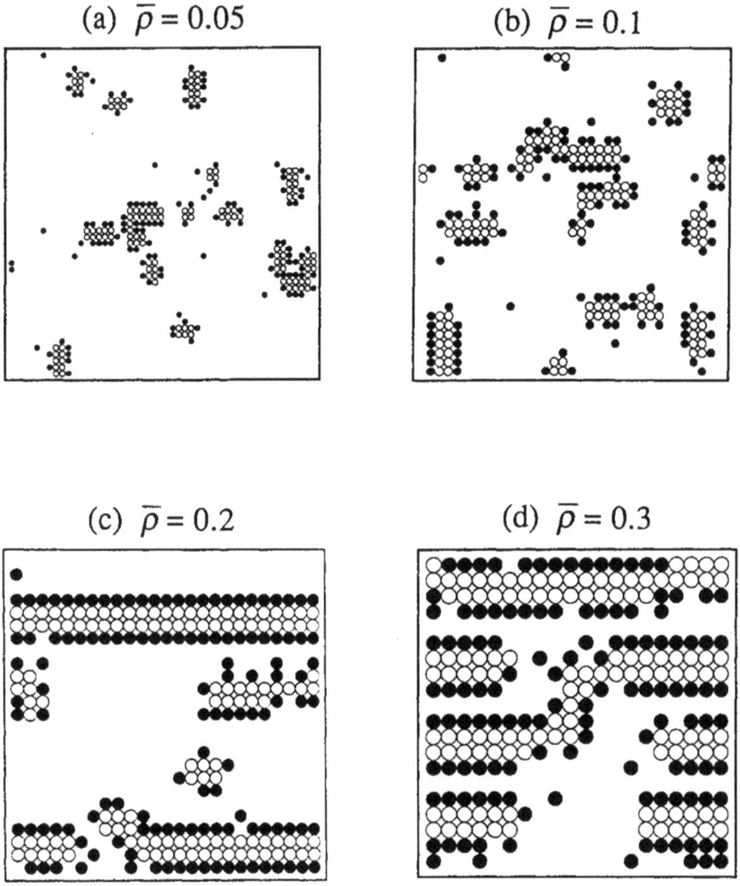


Figure 1: Snapshots of some typical configurations of the charge frustrated Ising system. The open spheres (∘) represent the -1 spin (hydrophobic); the black spheres (•) represent the +1 spin (polar). For clarity the solvent spins (+1 spins without a charge) are not drawn. Values of $\bar{\rho} = \rho a^2$ indicate different concentrations of charges (i.e., surfactant). This figure is taken from Ref. 4.

## Monte Carlo

Computer simulation provides an assessment of the possible structures that result from Coulomb frustration of the ferromagnetic Ising system. Figures 1 and 2 show typical configurations of equilibrated $d = 2$ Monte Carlo simulations. The pictured cases correspond to surfactant added to either pure water or pure oil. Thus in these cases, the self assembled structures are built from

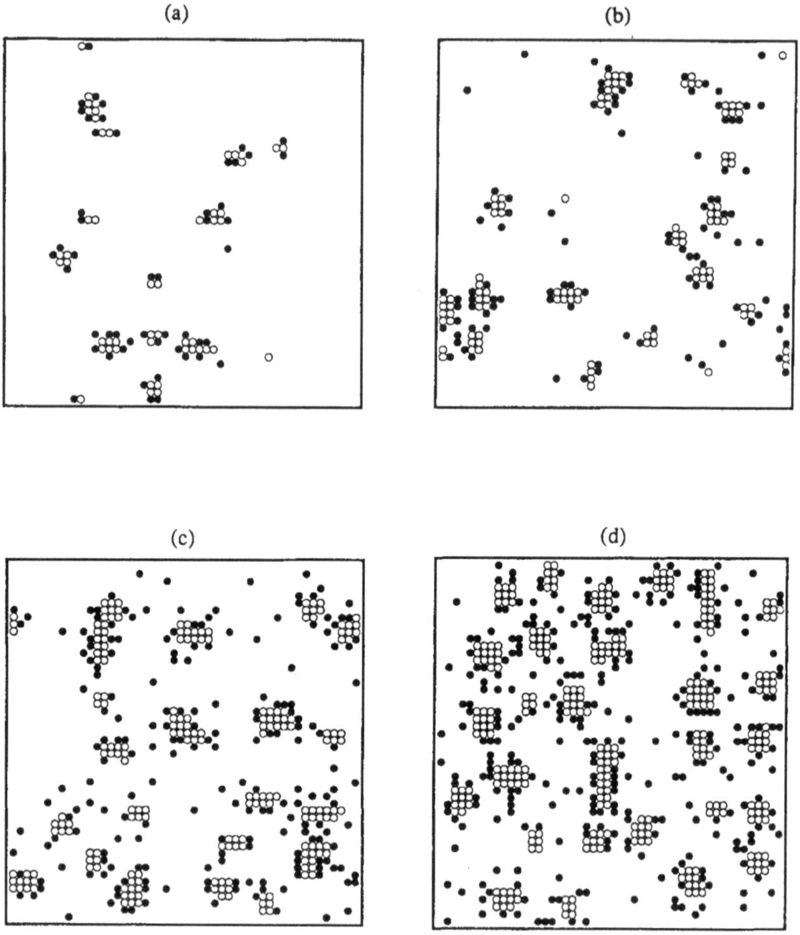


Figure 2: Snapshots of some typical configurations of the charge-frustrated Ising model for various surfactant concentrations, at a higher temperature than those in Fig. 1. The concentration of charged particles is in (a) $\rho = 0.02/a^2$; in figure (b) $\rho = 0.04/a^2$; in figure (c) $\rho = 0.08/a^2$; and in figure (d) $\rho = 0.12/a^2$. This figure is taken from Ref. 4.

micelles or bilayers. The arrangements of micelles and bilayers look very much like complex fluid assemblies observed in Nature, including dense micellar liquids and oriented lamellar phases.

The Monte Carlo equilibration of these phases is not trivial. Standard single spin moves fail in this regard. The frustrated system quickly organizes, but it then fails to move further unless the clustered spins are displaced collectively.

Standard Monte Carlo converges to a quenched configuration, not a sequence of equilibrated fluid configurations. The pictures given in Figs. 1 and 2 were obtained from a Monte Carlo algorithm with collective moves designed specifically to simulate fluid self assembly.[5]

## Bicontinuous microemulsions

In some situations, analytical analysis of the charge frustrated Ising model is particularly simple. A bicontinuous microemulsion is one such case. On a macroscopic scale, this type of material is a homogeneous mixtures of oil and water. On the microscopic scale, however, surfactant divides mesoscopic domains of oil and water. The separated domains percolate throughout the system. Surfactant interfaces are therefore ubiquitous.

The analysis of interface Hamiltonians is standard for treating systems that are extensive in interfaces.[7] This approach, however, does not address issues such as the stability of the assembled interface and the role of molecular parameters such as surfactant length.

A field theoretic formulation of the charge frustrated Ising model has been used to treat bicontinuity in oil-water-surfactant mixtures.[5] Here, the variables $\{t_i\}$ can be integrated out in a Gaussian approximation leaving a reduced Hamiltonian involving only the spins, $\{s_i\}$. For the case of equal volume fractions of oil and water, the reduced Hamiltonian is

$$\bar{\mathcal{H}} = -J \sum_{ij}{}' s_i s_j + \frac{1}{2} q^2 \sum_{i \neq j} L_{ij}\, s_i s_j \tag{4}$$

where $L_{ij}$ is the screened Coulomb interaction. It is the inverse Fourier transform of

$$\hat{L}(k) = \frac{fa/\Delta}{1 + \Delta^2 k^2/6} \tag{5}$$

where $f = 2\rho a^2 \Delta$ is essentially the volume fraction of surfactants. Widom's pioneering lattice model of microemulsions[9] can be obtained as an approximation to Hamiltonian (4). Specifically, Widom's model coincides to a particular truncation of $L_{ij}$ and a particular choice of the parameters $q$, $\Delta$ and $f$. The restrictions that lead to Widom's model, however, are without a molecular basis.

Teubner and others have used a random wave model to describe the structure of bicontinuous microemulsions.[10–12] This model provides a picture from which the partition function for Hamiltonian (4) can be treated to a good approximation. In particular, imagine expanding about a Gaussian reference Hamiltonian

$$\mathcal{H}_{\rm G} \approx (k_{\rm B}T/2)\sum_{i,j}\psi_i G_{ij}\psi_j \tag{6}$$

where $s_i$ is determined by $\psi_i$ through a local transformation

$$s_i = c(\psi_i) \tag{7}$$

One encounters such transformations, for example, in the field theoretic formulation of the Ising model.[13] In that case, $c(\psi) = \tanh(\psi)$, which is the specific form we use. In the random wave model, interfaces are constructed from a level clipping of a Gaussian field. This construction corresponds to taking $c(\psi)$ as a step function. On length scales that are large compared to interfacial widths, differences between continuous and discontinuous transformations are physically inconsequential. For the purposes of relating the Gaussian model to the actual Hamiltonian, however, a step function is not acceptable as it is not invertible. An invertible transformation is required, however, to construct the Jacobian that relates the metric of $\{s_i\}$ to that of $\{\psi_i\}$. Provided the Jacobian exists, the partition function for Hamiltonian (4) can be estimated by perturbation theory expanding about the Gaussian reference.

A form of Hartree self consistent field theory can be carried out in this way. For quantitative results, it is important to impose upon the calculation the average effect of the hard spin condition

$$\langle s_i s_j\rangle = 1 \text{ for } i = j\ . \tag{8}$$

This imposition gives a mean spherical approximation. Its evaluation for the charge frustrated Ising model is remarkably simple, giving the following small wave vector expression for the spin-spin structure factor:[5]

$$S_k = \frac{1}{\bar{a}^2 + c_1 k^2 + c_2 k^4} \tag{9}$$

with

$$\bar{a}_2 = \frac{c^2}{4}\left(\frac{a^6}{16\pi^2 c_2^2} - \frac{c_1}{c_2}\right)^2 \tag{10}$$

$$c_1 = \beta\sigma a^4/2 - fa\Delta/6 \tag{11}$$

$$c_2 = fa\Delta^3/36 \tag{12}$$

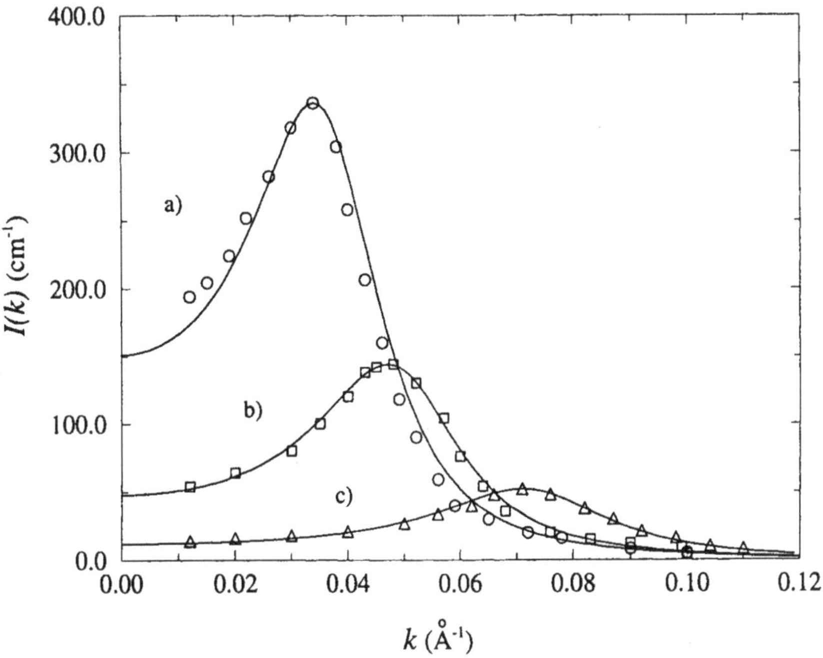


Figure 3: SANS spectra of $D_2O$-n-octane-$C_{12}E_5$ systems at varying surfactant volume fraction, $f$. Data is from Ref. 20 a) $f = 0.0566$, b) $f = 0.121$, and c) $f = 0.200$. The corresponding solid lines are the result of Eqs.(9)-(12), with $\sigma \approx 0.1 k_B T$ and $\Delta \approx 11$Å. This figure is taken from Ref. 5.

Equation (9) is valid provided $k$ is small in comparison with $2\pi/\Delta$. This condition holds for most of the range of wave vectors probed in small angle neutron and x-ray scattering.[14] The algebraic form of Eq.(9) was introduced phenomenologically by Teubner and Strey.[15] The charge frustrated Ising model provides the molecular interpretation of the phenomenological coefficients. Comparison of theory and experiment is successful, as illustrated in Fig. 3. It is remarkable that frustration described by just two microscopic lengths, $a$ and $\Delta$, leads to separated domains of oil and water many times larger in size.

For wave vectors beyond the range considered in Fig. 3, or for describing interfacial fluctuations, Eq.(9) does not suffice. The transformed Gaussian field, however, remains useful. For the purpose of visualization, for example, a configuration of the Gaussian field can be generated by computer. The level surface of this field can then be rendered with computer graphics giving a

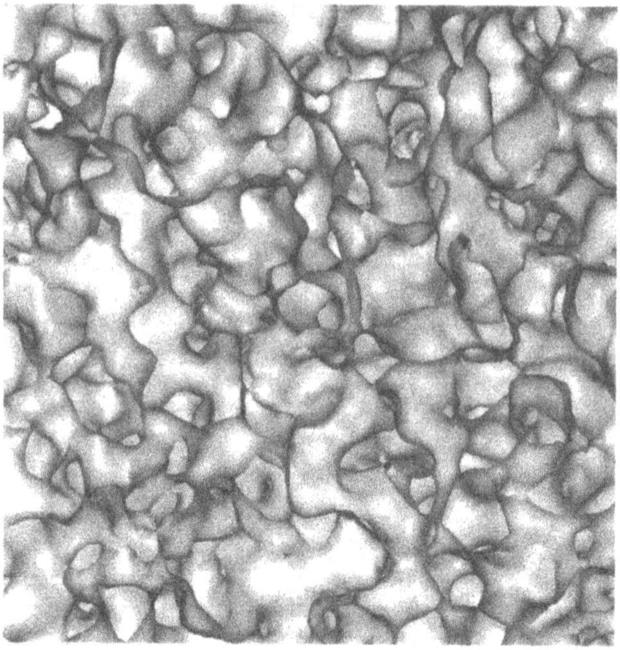

Figure 4: The three-dimensional level surface indicating the interfacial structure of the bicontinuous microemulsion studied in Ref. 21. The view is along the $z$ axis of a 64 x 64 x 64 grid with 30 Å resolution. This figure is taken from Ref. 7.

picture of a bicontinuous phase. This is the type of rendering done by Chen and coworkers. [12] An example taken from our work is shown in Fig. 4.

Beyond the main peak, the structure factor of the bicontinuous microemulsion decays as $k^{-4}$. See Eq.(9). This power law is observed whenever narrow random interfaces fill a system.[16] The converse is not necessarily true, however. Namely, while Eq.(9) holds, it is possible that interfaces can be broad, with widths comparable to the domain size. In other words, with enough surfactant, a microemulsion can be formed with largely diffuse regions of surfactant dividing oil and water. Such a material would not be microphase separated, in the traditional sense of the terminology. Calculations employing the charge frustrated model predict that some bicontinuous microemulsions exist in this way, with diffuse interfaces.[7]

## Metallic and insulating behaviors in liquids

The fact that charge frustration can lead to bicontinuous structure allows us to understand how this structure appears in a system that, at first thought, seems very different than oil-water-surfactant systems.

By computer simulation, Klein and coworkers have found a phase of metal-ammonia solutions that is bicontinuous.[17] According to their simulation, electrons are microphase separated from the solution of ammonia molecules and metal ions. This bicontinuous structure characterizes the conducting phase of these solutions.

The connection between this result and charge frustration begins with an observation made by Alavi and Frenkel.[18] They examined a version of Falicov and Kimball's itinerant electron model.[19] At high enough density, Alavi and Frenkel found that a gas of ideal fermions will phase separate from a fluid of classical blockers. In essence, this phase separation occurs when the Fermi pressure overwhelms the pressure of the blocks, thus squeezing out the blocks. At lower fermion chemical potential, the fermions and blocks are mixed. In the mixed phase, the fermions are at low concentration, localized in voids. The mixed phase is insulating, and the separated phase is conducting.

Electrons are born into liquid ammonia by dissolving metal, like Li or Na. The electrons leave the metal atoms because the solvation energy of a metal ion out weighs the ionization potential of the metal atom. The itinerant electrons thus formed experience a random field due to the disordered arrangement of ammonia and metal ions in the liquid. The field has regions of high potential (like the location of blockers) and regions of low potential (like the voids in the sea of blockers). At low concentration, the electrons will be localized in the largest regions of low potential energy. There are relatively few of these regions. Hence, due to the Pauli exclusion principle, only a few electrons can be accommodated in this way. As a result, phase separation would seem to occur at a high enough electron chemical potential, provided this chemical potential is below the threshold to extended states. This threshold is generally very high for electrons in polar liquids. Thus, the Alavi-Frenkel phase separation should be relevant.

Metal-ammonia solutions are complicated, however, by Coulomb interactions. Macroscopic separation of electrons from ions cannot occur as it would violate electroneutrality. Coulomb interactions will therefore frustrate the Alavi-Frenkel transition. Given the bicontinuous structure that is produced when Coulomb frustration alters the Ising transition, the bicontinuous structure discovered by Klein and his coworkers seems understandable. It is a possibility that merits quantitative study.

### Acknowledgments

I am grateful to my coworkers listed in References 4-7. My research on complex fluids is supported by the U.S. National Science Foundation and the Office of Naval Research.


### References


1. F. H. Stillinger, J. Chem. Phys. **78**, 4654 (1983).
2. L. Leibler, Macromolecules, **13**, 1602 (1980).
3. T. Ohta and K. Kawasaki, Macromolecules, **19**, 2621 (1986).
4. D. Wu, D. Chandler, and B. Smit, J. Phys. Chem. **96**, 4077 (1992).
5. H.-J. Woo, C. Carraro and D. Chandler, submitted to Phys. Rev. E (1995).
6. M. W. Deem and D. Chandler, Phys. Rev. E **49**, 4268 (1994).
7. M. W. Deem and D. Chandler, Phys. Rev. E **49**, 4276 (1994).
8. W. Helfrich, Z. Naturforsch. **28c**, 693 (1973).
9. B. Widom, J. Chem. Phys. **84**, 6943 (1986).
10. M. Teubner, Europhys. Lett. **14**, 403 (1991).
11. S. Marčelja, J. Phys. Chem. **94**, 7259 (1990); P. Pieruschka and S. Marčelja, J. Phys. II France **2**, 235 (1992); P. Pieruschka and S. A. Safran, J. Phys. Condens. Matt. **6**, A357 (1994).
12. S. H. Chen, S. L. Chang, R. Strey, J. Samseth and K. Mortensen, J. Phys. Chem. **95**, 7427 (1991).
13. See, for example, G. Parisi, *Statistical Field Theory*, (Addison-Wesley, Redwood City, CA, 1988), p. 209-213.
14. M. Kahlweit, R. Strey, D.Hasse, H. Kunieda, T. Schmeling, B. Faulhaber, L. Magid, O. Söderman, P. Stilbs, J. Winkler, A. Dittrich, and W. Jahn, J. Colloid Interface Sci. **118**, 436 (1987).
15. M. Teubner and R. Strey, J. Chem. Phys. **87**, 3195 (1987).
16. G. Porod, Kolloid Z. **124**, 83 (1951); **125**, 51 (1952).
17. Z. Deng, G.J. Martyna and M.L. Klein, Phys. Rev. Lett. **71**, 267 (1993); Z. Deng, G.J. Martyna and M.L. Klein, J. Chem. Soc. Faraday Trans. **90**, 2009 (1994).
18. A. Alavi and D. Frenkel, J. Chem. Phys. **97**, 9249 (1992).
19. L. M. Falicov and J. C. Kimball, Phys. Rev. Lett. **22**, 957 (1967); J. F. Freericks and L. M. Falicov, Phys. Rev. B **41**, 2163 (1990).

20. M. Kotlarchyk, S.-H. Chen, J. S. Huang, and M. W. Kim, Phys. Rev. Lett. **53**, 941 (1984).
21. L. Auvray, J.-P. Cotton, R. Ober and C. Taupin, J. Phys. Chem. **88**, 4586 (1984).

# CRYSTALS IN RANDOM MEDIA

DAVID A HUSE
*AT&T Bell Labs, Murray Hill NJ 07974 USA*
*huse@physics.att.com*

Vortex lattices in type-II superconductors, charge-density waves, spin-density waves, and electron Wigner lattices are examples of systems that can order into a state that is spatially periodic, with a period incommensurate with the underlying atomic crystal. In ideal samples without imperfections, these systems show true long-range order. However, pinning by randomly placed chemical or structural imperfections in the sample induces divergent displacements, destroying the positional long-range order, as first argued by Larkin[1]. Very strong pinning will produce an amorphous ground state, that is in no sense crystalline or periodic. Intermediate between the long-range ordered state in the absence of pinning and the fully amorphous state for strong pinning, there is the possibility of a distinct thermodynamic phase that is topologically ordered (dislocation-free) at large scales, although the positional long-range order has been degraded by the divergent displacements[2-4].

To investigate this question, we[5] have studied the simplest model with the essence of this physics, which we argue is the random-field XY model. We have simulated this model in 2 and 3 dimensions, looking at the density of static topological defects and at correlation functions. In 2 dimensions it appears that topological defects are present at low density even at weak disorder, so there is no indication of the proposed defect-free intermediate phase. However, for 3 dimensions the density of defects vanishes much more rapidly as the random field is reduced. The simulation data are consistent with a transition at a nonzero random field to the proposed defect-free phase, which would then exist over a finite range of weak random field in this 3-dimensional system.

References

1. A I Larkin, Zh. Eksp. Teor. Fiz. 58, 1466 (1970) [Sov. Phys. JETP 31, 784 (1970)].
2. D A Huse and L R Radzihovsky, in Fundamental Problems in Statistical Mechanics VIII, H van Beijeren and M H Ernst, eds. (Elsevier, Amsterdam, 1994) p. 49.
3. D A Huse, Physica B 197, 540 (1994).
4. T Giamarchi and P le Doussal, Phys. Rev. Lett. 72, 1530 (1994).
5. M J P Gingras and D A Huse (preprint).

# TYPE-I SUPERCONDUCTORS: A NEW PLAYGROUND FOR WETTING PHENOMENA

J.O. INDEKEU

*Laboratorium voor Vaste-Stoffysica en Magnetisme, Katholieke Universiteit Leuven, B-3001 Leuven, Belgium*

E-mail: joseph.indekeu@fys.kuleuven.ac.be

and

J.M.J. VAN LEEUWEN

*Instituut-Lorentz, Rijksuniversiteit Leiden, 2311 SB Leiden, the Netherlands*

ABSTRACT

The Ginzburg-Landau theory, which is known to be quantitatively correct for classical superconductors, is used to show that a type-I superconductor with *enhanced* order parameter $|\psi|^2$ at the surface displays an interface delocalization or "wetting" transition. In this transition, which takes place in magnetic fields $H_c(T)$ at which normal and superconducting phases coexist, a *macroscopically* thick layer of the superconducting phase intrudes from the surface into the bulk normal phase. Surprisingly, the order of the transition is controlled by a *bulk* material constant, the Ginzburg-Landau parameter $\kappa$. This is most unusual, since in other systems (fluids, Ising magnets, etc.) the order of the wetting transition depends on surface parameters that are difficult to determine or control. For superconductors, *first-order* wetting is predicted for $0 \leq \kappa < 0.374$, and *critical* wetting for $0.374 < \kappa < 1/\sqrt{2}$. The possibility of critical wetting in superconductors is especially exciting in view of the fact that this phenomenon has largely eluded experimental verification in any system until now. Type-I superconductors with $\kappa > 0.374$, such as Pb for example, are thus likely to be an ideal test case for observing critical wetting. Furthermore, superconductors provide a realization of wetting in systems with exponentially decaying wall-interface interactions (so-called "short-range" forces). This is different from most other systems where algebraically decaying ("long-range") forces prevail, and has important consequences for critical wetting.

## 1 Introduction

For classical superconductors the Ginzburg-Landau (GL) theory has been a successful guide for studying a rich variety of phenomena, in particular phase transitions and critical phenomena in bulk. In this report we aim at demonstrating that the GL theory also displays an almost equally rich behavior in surface and interfacial phase transitions. It is standard practice to discuss these transitions in the context of wetting theory, which stems from the description of the behavior of liquids in contact with a substrate [1]. Wetting theory goes far beyond the classical situation of a structureless substrate in contact with a liquid-vapor system at coexistence. For example, also magnetic systems can display wetting phenomena, which are usually referred to as interface delocalization transitions. It has also been long known that superconductivity exhibits typically surface phenomena, such as the nucleation of a superconducting sheath at a surface under circumstances where bulk superconductivity does not exist [2]. To our knowledge the two phenomena, wetting and superconductivity, have rarely been considered from a single viewpoint. In this paper we aim at formulating surface superconductivity in terms of the wetting language, thereby indicating that superconductors may show certain realizations of wetting theory which have hardly been seen in standard wetting experiments.

The essential parameter in wetting phenomena is the surface field which is a measure of the preferential adsorption of one of the coexisting bulk phases at a surface. For adsorbed fluids it is the surface chemical potential increment $\Delta\mu_1$. For magnets as described by, e.g., the Ising model, it is the surface magnetic field $H_1$. We show that the corresponding quantity for superconductors is the surface extrapolation length $b$. It measures the distance from the surface where the superconducting order parameter $\psi$ would vanish upon a linear extrapolation of $\psi$ away from the surface. Normally $b > 0$, which means that the order parameter is *suppressed* at the surface and $b \to 0$ when one considers a surface between a superconductor (SC) and a ferromagnet which causes $\psi$ to vanish at the surface. For a vacuum/SC surface $b = \infty$, which means that the gradient of $\psi$ normal to the surface vanishes. Less familiar are negative values of $b$ which imply that $\psi$ is *enhanced* at the surface. This occurs for instance when a thin superconducting layer with higher $T_c$ is present at the surface [3]. It occurs also near twinning planes inside the bulk SC's Sn, In, Nb, Re and Tl and at surfaces where grinding-induced twinning planes accumulate [4]. It is for $b < 0$ that the interesting new "wetting" phenomena occur.

The basic thermodynamic arguments pertaining to the wetting transition are applied to type-I superconductors as follows. Consider bulk two-phase coexistence between normal (N) and superconducting (SC) states, at the so-called critical field $H_c(T)$. We recall that the Ginzburg-Landau parameter $\kappa$ is the ratio of the magnetic penetration depth $\lambda$ to the superconducting coherence length $\xi$. For type-I superconductors, that is, for $\kappa < \kappa_c = 1/\sqrt{2}$, the interfacial tension $\gamma_{SC,N}$ of the SC/N interface is positive [5]. This is essential in the discussion of the wetting transition, and sets a natural limit, $\kappa = 1/\sqrt{2}$, to the validity of most of the phenomena we will describe. The value $\kappa = 1/\sqrt{2}$ corresponds to a bulk multicritical point.

For a semi-infinite system with a surface or wall (W) one can consider two surface free energies, $\gamma_{W,N}$ and $\gamma_{W,SC}$, depending on whether the bulk boundary condition (infinitely far from the wall) is the N or the SC phase. The premise of the wetting problem is that one of the two bulk phases, N or SC, is preferred by the surface. Our calculations show that the SC phase is preferred for all $\kappa$ when $b \leq 0$, and for sufficiently large $\kappa$ when $b > 0$. Therefore, we consider $\gamma_{W,SC} < \gamma_{W,N}$. If we now impose the disfavored phase, N, as the bulk boundary condition, the question arises whether the state near the wall is also predominantly N, with perhaps a *thin* superconducting sheath, or whether a *macroscopic* superconducting layer can occur, which separates the N phase from the wall. The thermodynamic equilibrium condition

$$\gamma_{W,N} \leq \gamma_{W,SC} + \gamma_{SC,N} \tag{1}$$

is realized as a strict inequality in the former case (generally denoted by "partial wetting") and as an equality in the latter case ("complete wetting").

We shall demonstrate that, for surfaces with $b < 0$, all type-I superconductors undergo the wetting transition at some temperature $T_D$ strictly below $T_c$. At $T_D$ the inequality (1) first becomes an equality, and remains so for $T > T_D$. Like for magnetic systems, the wetting transition in superconductors is suitably denoted by "interface delocalization" transition, because, as we shall show, the transition is between a state in which a SC/N interface is either absent or *localized* near the wall, to a state in which the SC/N interface is *delocalized*, or *unbound*, infinitely far away from the wall.

## 2 Ginzburg-Landau Theory and Dynamical Analog

The Ginzburg-Landau surface free energy functional, including the surface contribution, reads

$$\gamma[\psi, \mathbf{A}] = \frac{\hbar^2}{2mb} \mid \psi(0) \mid^2 \tag{2}$$

$$+ \int_0^\infty dx \left[\alpha \mid \psi \mid^2 + \frac{\beta}{2} \mid \psi \mid^4 + \frac{1}{2m} \mid (\frac{\hbar}{i}\nabla - q\mathbf{A})\psi \mid^2 + \frac{\mid \nabla \times \mathbf{A} - \mu_0 \mathbf{H} \mid^2}{2\mu_0}\right]$$

Here, $x$ is the coordinate perpendicular to the surface, which is at $x = 0$. $\mathbf{A}$ is the vector potential and the applied field $\mathbf{H} = H\mathbf{e}_z$ is parallel to the surface. Furthermore, $\alpha \propto T - T_c$, where $T_c$ is the bulk critical temperature. We choose the gauge so that $\mathbf{A} = (0, A(x), 0)$. Due to the translational invariance along $y$ and $z$, and by redefining $A(x)$, it is possible to work with real functions $\psi(x)$. $A$ and $\psi$ are determined by the Ginzburg-Landau equations.

The role of $b$ becomes clear when the free energy is minimized with respect to $\psi(0)$. This leads to the familiar boundary condition [5]

$$\dot{\psi}(0) = b^{-1}\psi(0), \tag{3}$$

where the dot means $d/dx$.

The determination of the delocalization transition is greatly facilitated by constructing a "phase portrait" in which $\dot{\psi}(0)$ is plotted versus $\psi(0)$, for *arbitrary* $b$. Fig.1 illustrates this for two representative cases, $\kappa = 0.3$ and $\kappa = 0.5$, for bulk two-phase coexistence ($H = H_c$). The bulk phase is fixed to N. For $\kappa = 0.3$ (solid lines) the concave branch, from $O$ to $Y$, corresponds to

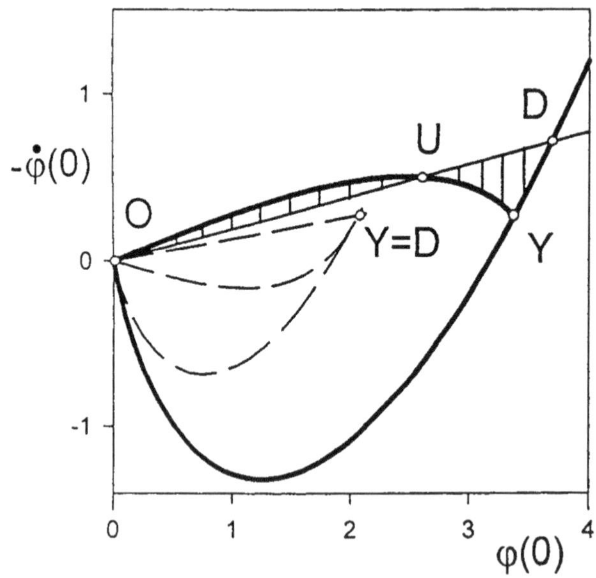


Figure 1: Phase portraits $-\dot{\varphi}(0)$ versus $\varphi(0)$, for $\kappa = 0.3$ (solid lines) and $\kappa = 0.5$ (dashed lines). Here, $\varphi \equiv \psi\xi/\kappa$ and the dot means $\xi d/dx$. For $\kappa = 0.3$ an equal-areas construction determines the location of the first-order interface delocalization transition. For $\kappa = 0.5$ the meeting point $Y$ determines the location of the critical delocalization transition.

profiles with a superconducting surface sheath. As $Y$ is approached, the sheath thickness $l$ diverges. The convex branch corresponds to W/SC/N profiles, which start from the surface and approach the SC phase for $x \to \infty$. An SC/N interface is added (at $x = \infty$), such that finally the N phase is reached, across an infinite SC layer. The straight line represents the boundary condition (3). Intersections of this line with the branches correspond to extrema of the free energy. There is a local minimum at the origin (absence of superconductivity), a local minimum at $D$ (macroscopic SC layer and delocalized SC/N interface), and a maximum at $U$ (thin SC sheath and localized SC/N interface). In order to locate the phase transition, where the two minima exchange stability, we substitute into (2) the $A$ and $\psi$ that satisfy the Ginzburg-Landau equations and the boundary condition $\dot{A}(0) = \mu_0 H$, but with $\psi(0)$ *fixed*. This leads to the surface free energy function $\tilde{\gamma}$, given by

$$\tilde{\gamma}(\psi(0)) = \frac{\hbar^2}{2m}\left[\frac{\psi(0)^2}{b} + \sigma(\psi(0))\right] \tag{4}$$

By varying $\sigma$ with $\psi(0)$ we obtain $d\sigma(\psi)/d\psi = -2\dot{\psi}_0(\psi)$, where $\dot{\psi}_0 \equiv \dot{\psi}(0)$. Combined with (4) this leads to

$$\tilde{\gamma}(\psi(0)) = \tilde{\gamma}(0) + \frac{\hbar^2}{m}\int_0^{\psi(0)} d\psi \left[\frac{\psi}{b} - \dot{\psi}_0(\psi)\right] \tag{5}$$

from which one easily derives an equal-areas rule. Thus, a *first-order* phase transition between a state with $\psi = 0$ and a macroscopic SC layer takes place when $b$ is adjusted so that the hatched areas are equal.

For $\kappa = 0.5$ (dashed lines in Fig.1) the phase portrait is qualitatively different. Both branches are convex and, depending on $b$, we have either a stable finite sheath or an infinite sheath. The sheath thickness $l$ diverges continuously approaching $Y$. A *critical* delocalization transition occurs when the straight line passes through $Y = D$. The precise location of $D$ is obtained by studying the asymptotic behavior of $A$ and $\psi$, for $x \to \infty$, for W/SC profiles.

## 3 Wetting Phase Diagram at Bulk Coexistence

Fig.2 presents the global wetting phase diagram in the $(\kappa, \xi/b)$-plane, at bulk two-phase coexistence. For a given bulk material, changing $T$ and $H$, such that $H = H_c(T)$, corresponds to moving horizontally. Increasing $T$ towards $T_c$ means increasing $\xi/\mid b \mid$ towards $\infty$, since $\xi \propto \mid T - T_c \mid^{-1/2}$. For $0 \leq \kappa < 0.374$ the delocalization transition (line $FD$) is of first order. In the limit $\kappa \to 0$ we have calculated analytically that it occurs at $\xi/b = -0.602$. For $\kappa > 0.374$, the delocalization transition (line $CD$) is critical. The sheath thickness diverges as $l \propto \log(1/(T_{CD} - T))$. The nucleation of the sheath takes place on the critical line $CN$. The continuation of this line, which is the metastability limit of the normal surface state, is indicated by $ML$ (dashed line).

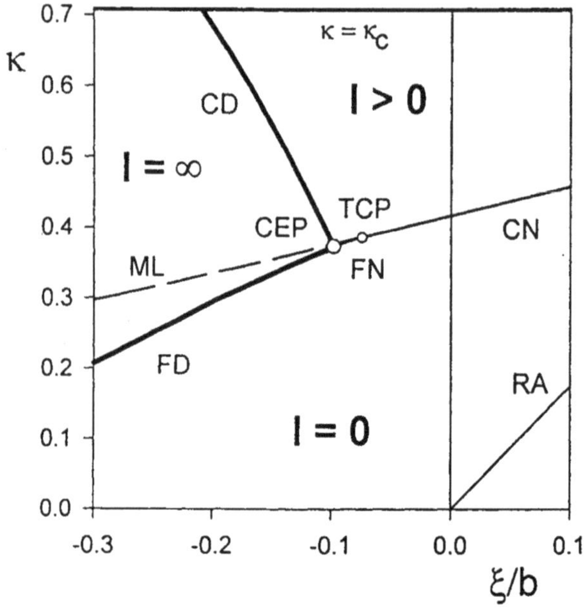


Figure 2: Global phase diagram of interface delocalization transitions for type-I superconductors. The phases are labeled by the thickness $l$ of the superconducting surface sheath: $l = 0$ (no superconductivity), $l > 0$ (SC/N interface localized at the surface), $l = \infty$ (delocalized SC/N interface). The transitions are explained in the text.

The lines $CD$ and $FD$ meet in a *critical endpoint* $CEP$, at $\kappa = 0.374$. On the other hand, the nucleation transition changes from second to first order at a *tricritical point* $TCP$, at $\kappa = 0.388$. There, the line $CN$ changes (with common tangents) into the first-order nucleation line $FN$. On this line, between $TCP$ and $CEP$, nucleation proceeds with a thickness jump, from $l = 0$ to $l = l_{FN}$, and $l_{FN}$ diverges when $CEP$ is approached. Note that for $b < 0$ all type-I SC's are bound to undergo an interface delocalization transition at some $T_D < T_c$. For completeness, we draw attention to the line $RA$, on which $\gamma_{W,N} - \gamma_{W,SC}$ passes through zero, corresponding to reversal of preferential adsorption. To the left of this line the SC phase is preferred by the surface.

## 4 $H - T$ Phase Diagrams

From the point of view of wetting phenomena as well as that of superconductivity it is important to study what happens as a function of the magnetic field, especially for $H > H_c(T)$. *First-order* wetting transitions have extensions into the bulk one-phase region, called *prewetting lines* [1]. In superconductivity, critical nucleation of a SC surface sheath occurs at a field $H_{c3}$ [2], which is a strongly increasing function of $\xi/\mid b\mid$ for $b < 0$ [3].

The connection between prewetting and critical nucleation is demonstrated in Fig.3, for $\kappa = 0.2$ and $b < 0$. The temperature variable is $t \equiv (T-T_c)/(T_c - T_D)$, such that the first-order delocalization transition occurs at $t_D = -1$.

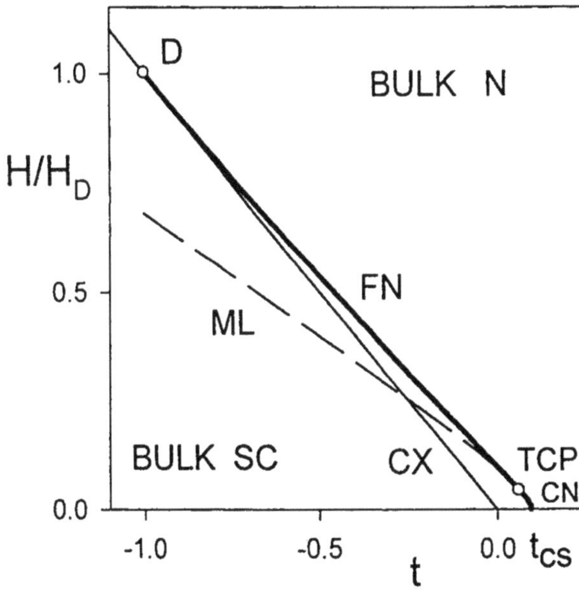


Figure 3: $(H,T)$ surface phase diagram for $\kappa = 0.2$ and $b < 0$. The line $CX$ denotes bulk two-phase coexistence. The surface superconductivity transitions can be interpreted as first-order (line $FN$), tricritical (point $TCP$), and critical (line $CN$) prewetting transitions, associated with the first-order delocalization transition at $D$.

Similarly, the magnetic field is expressed in units of $H_D$. The connection with measurable quantities is most easily made by locating the critical point of *surface* superconductivity in zero field, at $T_{cs}$. Indeed, for $b < 0$, $T_{cs} > T_c$ and one can show analytically that $\xi(T_{cs})/b = -1$. Since the function $\xi(T)$ is known from the microscopic theory, measuring $T_{cs}$ allows to estimate $b$. This in turn allows to locate $T_D$, because $t_{cs} = \xi(T_D)^2/b^2$ depends *only* on $\kappa$, and is obtained directly from our calculations. For $\kappa = 0.2$, $\xi(T_D)/b = -0.308$.

The first-order delocalization transition, at $D$, has an extension along the first-order nucleation or "prewetting" line $FN$. This line, which is tangential to the bulk coexistence line $CX$ at $D$, changes into a critical nucleation line $CN$, at a tricritical point $TCP$. Note that this is different from prewetting in fluids or magnets, where the prewetting line stops at a simple critical point. At $TCP$, $FN$ and $CN$ meet with common tangents, and to the left of $TCP$, $CN$ continues as the metastability limit $ML$ (dashed line) of the normal surface phase. To the right of $TCP$, the line $CN$ terminates at the surface critical temperature $T_{cs}$. An analytic calculation shows that $H_{c3} \propto (T_{cs} - T)^{1/2}$, near $T_{cs}$. For the thickness $l$ of the surface sheath when bulk two-phase coexistence is approached at fixed $T$, we find a divergence $l \propto \log(1/(H - H_c))$, for $T_D < T < T_c$. This is the approach to "complete wetting". In contrast, for $T < T_D$, $l = 0$.

The $(H, T)$ surface phase diagram for a type-I SC with $\kappa > 0.374$ and $b < 0$, which displays critical delocalization, is qualitatively different, as Fig.4 illustrates for $\kappa = 0.5$. The line $FN$ is absent, and the line $CN$ continues

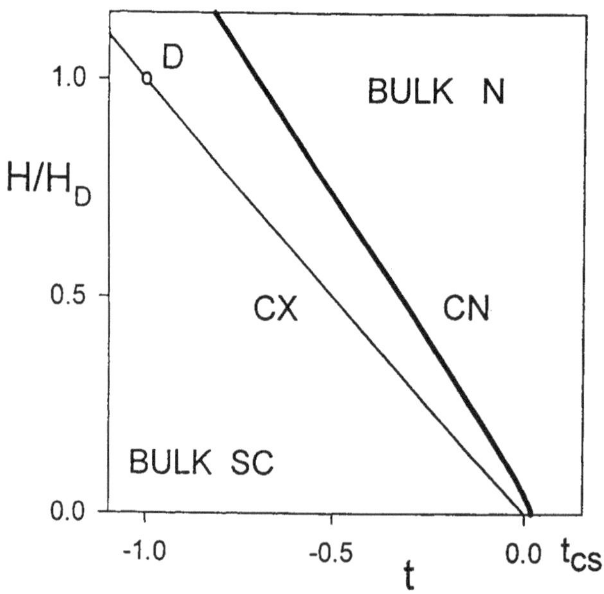


Figure 4: $(H, T)$ surface phase diagram for $\kappa = 0.5$ and $b < 0$. The line $CX$ denotes bulk two-phase coexistence. The surface superconductivity transitions are critical (line $CN$). They are unrelated to and disconnected from the critical delocalization transition at $D$.

down to temperatures $T < T_D$. The transition point $D$, at bulk coexistence, is then isolated. Experimentally, $D$ could be detected by monitoring $l$ versus $T$. Also, $T_D$ can be estimated indirectly, as explained above, by measuring the zero-field surface transition temperature $T_{cs}$.

## 5 Conclusions

Our main conclusions and final remarks are the following.
A) For all type-I superconductors with enhanced surface order parameter ($b < 0$), there exists a genuine "wetting" temperature $T_D$, which marks an interface delocalization phase transition (Fig.2). This transition occurs at bulk two-phase coexistence, so that $H_D = H_c(T_D)$. The existence of these transitions demonstrates that, contrary to what is commonly believed, the thickness $l$ of the superconducting surface sheath can be much larger than the bulk coherence length $\xi$.
B) The complete phenomenology of wetting is realized in type-I superconductors. Surprisingly, unlike in all other wetting systems that we know of, the order of the wetting transition is controlled by a *bulk* material constant, the GL parameter $\kappa$. First-order wetting takes place for $0 \leq \kappa < 0.374$, and critical wetting for $0.374 < \kappa < 1/\sqrt{2}$. Consequently, e.g., Al ($\kappa \approx 0.01$) and Sn ($\kappa \approx 0.1$) are predicted to display first-order wetting, whereas Pb ($\kappa \approx 0.6$) is a candidate for critical wetting. Thus there is new physics associated with wetting in superconductors, because in other systems the order of the transition depends on *surface* parameters which are difficult to determine or control.
C) The prediction of critical wetting is especially important from an experimental viewpoint. Up to now, detailed experiments on wetting in fluids have provided unambiguous evidence for first-order wetting and prewetting [6]. However, there is to our knowledge no clear experimental verification yet of critical wetting in any system. Thus, type-I superconductors with $\kappa > 0.374$ may be an ideal testing ground for observing this phenomenon.
D) Wetting in superconductors is also new in that it provides a realization of systems with exponentially decaying wall-interface interactions (with characteristic length scales $\lambda$ or $\xi$), usually referred to as short-range forces. This leads to a *logarithmic* divergence of the sheath thickness when approaching complete wetting ($T > T_D$) and critical wetting. In contrast, in other systems power-law interactions (van der Waals or other) prevail, which are referred to as long-range forces. Especially for critical wetting, theory has since long predicted interesting anomalies for exponentially decaying interactions [7].
E) We give a new interpretation and make further predictions for twinning-plane superconductivity (TPS). The $(H, T)$ surface phase diagram (Fig.3) is

similar to that of TPS, at the same $\kappa$ [4]. But, first-order TPS transitions are now interpreted as prewetting, and we thus predict a first-order wetting transition in which the SC/N interface delocalizes from the twinning plane, at the point $D$ at bulk coexistence. Neither this nor the possibility of critical wetting has to our knowledge been predicted yet for TPS.
F) The case of a (classical) superconductor is known to be a fairly unique example of a system for which the Ginzburg-Landau theory is quantitatively correct. The experiments on TPS confirm this once more. For other systems (fluids, magnets, etc.) the corresponding mean-field theory is at best qualitative and often a poor approximation. Therefore, the mean-field wetting phase diagram of type-I superconductors (Fig.2) is exceptional in that it provides a quantitative guide for the experimentalist.

Publication of this research is in progress [8].

**Acknowledgements**
We thank R. Blossey, A.I. Buzdin, P.-G. de Gennes, V.V. Moshchalkov and C. Strunk for discussions. J.O.I. is Research Director of the Belgian National Fund for Scientific Research. J.M.J.vL. is in part supported by Human Capital and Mobility Program "Materials Research at KULeuven" of the EC. This research is supported by the Belgian Inter-University Attraction Poles (IUAP) and the Concerted Action Research Programme (GOA).

## References

1. For a review on wetting, see, e.g., S. Dietrich, in *Phase Transitions and Critical Phenomena*, eds. C. Domb and J.L. Lebowitz (Academic, London 1988), vol. 12, p.1.
2. D. Saint-James and P.-G. de Gennes, *Phys. Lett.* **7**, 306 (1963).
3. H.J. Fink and W.C.H. Joiner, *Phys. Rev. Lett.* **23**, 120 (1969).
4. I.N. Khlyustikov and A.I. Buzdin, *Adv. Phys.* **36**, 271 (1987).
5. See, e.g., P.-G. de Gennes, *Superconductivity of Metals and Alloys* (Addison-Wesley Publ.) 1966.
6. For ordinary fluids, see, e.g., H. Kellay, D. Bonn, and J. Meunier, *Phys. Rev. Lett.* **71**, 2607 (1993); for quantum fluids, see, e.g., K.S. Ketola, S. Wang, and R.B. Hallock, *Phys. Rev. Lett.* **68**, 201 (1992); J.E. Rutledge and P. Taborek, *Phys. Rev. Lett.* **69**, 937 (1992).
7. R. Lipowski, D.M. Kroll, and R.K.P. Zia, *Phys. Rev.* **B27**, 4499 (1983); E. Brézin, B.I. Halperin, and S. Leibler, *Phys. Rev. Lett.* **50**, 1387 (1983); M.E. Fisher and A.J. Jin, *Phys. Rev. Lett.* **69**, 792 (1992).
8. J.O. Indekeu and J.M.J. van Leeuwen, *Phys. Rev. Lett.* (to appear).

# NONEQUILIBRIUM SURFACE TENSION DRIVEN PHENOMENA

MANUEL G. VELARDE
*Instituto Pluridisciplinar, Universidad Complutense de Madrid, Paseo Juan XXIII, n.1, 28040-Madrid, Spain*

ABSTRACT

At the free surface of a liquid, the pressure at the liquid side of the surface depends on the local surface tension and the local curvature of the surface, and the shear stress depends on the local gradient of surface tension. Accordingly, if the surface tension is nonuniform, equilibrium is not possible and motion is induced in the liquid (Marangoni effect). Up to the sign, surface/interfacial tension is to the surface what pressure is to the bulk of the liquid. Flow is the nonequilibrium (mechanical) response of a liquid to a pressure or surface tension gradient. An overview is here provided of recent analytical, numerical and experimental results, concerning steady and time-dependent motions driven by nonuniform, nonequilibrium surface tension. First, I consider the case of a liquid drop immersed in another immiscible liquid when heat and/or mass transfer mechanisms act at the drop surface leading to drop motion, with drag or thrust, due to instability induced by the Marangoni effect. Then I consider a few aspects of Bénard steady convective patterns in a liquid layer open to the ambient air and heated from the liquid side. Finally, I describe a few findings concerning time-dependent motions, waves at the open surface in a Bénard layer heated from the air side.

## 1. Introduction

When there is an open surface or an interface exists between two liquids, the interfacial tension accounts for the jump in normal stresses proportional to the surface curvature across the interface, hence affects its shape and stability. Gravity competes with the Laplace forces in accomodating equipotential levels with curvature. Their balance defines, for instance, the stable equilibrium of spherical drops or bubbles. If the surface tension varies with temperature or composition, and consequently, with position along an interface, its change takes care of a jump in the tangential stresses. Hence its gradient acts like a shear stress applied by the interface on the adjoining bulk liquid (Marangoni stress), and thereby generates flow or alters an existing one. Surface tension gradient-driven flows are known to affect the evolution of growing fronts, and measurements of transport phenomena. The variation of surface tension along an interface may be due to the existence of a thermal gradient along the interface or perpendicular to it. In the former case we have instantaneous convection while in the latter, flow occurs past an instability threshold [1-5].

To qualitatively assess the role of gravity relative to forces induced by surface tension or its gradient, some dimensionless groups help. There is the (static) Bond number $Bo=gl^2\Delta\rho/\sigma$, where $\Delta\rho$ is the density difference between two fluids (e.g. a drop immersed in another liquid), $\sigma$ is the interfacial tension and l is the space scale involved in the problem. There is also the modified (dynamic) Bond number $Bo^*=gl^2\Delta\rho/\Delta\sigma$ and the Marangoni number $Ma=l\Delta\sigma/\eta\kappa=PrRe_\sigma$, where $\eta$ is the dynamic viscosity ($\eta=\rho\nu$) and $\kappa$ is the thermal diffusivity. $Pr=\nu/\kappa$ is the Prandtl number. $Re_\sigma$ denotes a surface Reynolds number, with the velocity scale $V=\Delta\sigma/\eta$ due to $\Delta\sigma=(d\sigma/dT)\Delta T$. Hence Ma is a Péclet number, $Pe=Vl/\kappa$.

Surface tension gradient-driven flows have been observed as laminar flows, with or without cellular patterns, and only high enough Marangoni numbers turbulent flows occur [4-7]. For flows driven by capillary forces the Péclet/Marangoni number appears in the heat equation. Then for high Prandtl number fluids the velocity field is slaved by the temperature while for low Prandtl number fluids the later slaves the former. Hence the interest in proceeding to high Péclet/Marangoni number flows (with high Pr fluids) to explore features of strongly dissipative space-time chaos at variance with the Reynolds-Kolmogorov turbulence, which is mostly inertial.

At an open surface the (simplest) dimensional tangential stress balance is $\eta\partial u/\partial z=-\partial\sigma/\partial T=-(\partial\sigma/\partial x)(dT/dx)$. It is here that the Marangoni number appears in the engine bringing the flow (Marangoni-driven flow). Let $ar=d/l$ be an aspect-ratio, with d the depth of a layer of horizontal extent l. If $[ar^2Re_\sigma]\ll1$, inertia can be neglected, and the effects of surface tension penetrate by viscosity all throughout the bulk of the fluid (viscous flow). With $X=x/l$, $Z=z/d$ and $U=u/V^*$ we have $\partial U/\partial Z=-[(\partial\sigma/\partial T)\Delta T/\eta V^*](d/l)(d\Theta/dX)$, where $V^*=[(\partial\sigma/\partial T)\Delta T/\eta](d/l)$ and $\Theta=(T-T_c)/\Delta T$. If, however, $[ar^2Re_\sigma]\gg1$, the effect of surface tension remains near the interface in a boundary layer, $\delta$, much thinner than the liquid depth, $\delta\ll d$ (boundary layer flow). Then $\delta$ rather than d is the scale for vertical displacements. $\delta/l=ar[Re_\sigma]^{-1/2}$. Hence $V^*=[\nu(\partial\sigma/\partial T)^2\Delta T^2/\eta^2 l]^{1/3}$. That $[ar^2Re_\sigma]\ll1$ does not imply that the Reynolds number is small. Indeed, $Re_\sigma$ can be greater than unity provided the aspect-ratio is small enough. $[ar^2Re_\sigma]\gg1$ does not preclude the flow from being laminar for large enough aspect-ratio. In flows with very low Re, inertia can be neglected in the flow (e.g. Stokes creeping flow, at zero Reynolds number) of very viscous fluids like the silicone oils, where however possible high values of Ma exist.

Surface deformation and surface curvature appear in the balances of tangential and normal stresses, respectively. The former balance yields a capillary number $Ca=\eta V^*/\sigma=(Bo/Bo^*)ar$. For boundary layer flows the Weber number appears from the same normal balance, $We=lV^{*2}\Delta\rho/\sigma$. Consequently, both Ca and We account, in their appropriate flow, accounting for the relevance of surface deformation effects in a surface tension gradient-driven flow. Strictly speaking the only new parameter appearing is the contact angle at a boundary.

## 2. Drops and bubbles

I shall discuss in some detail the case of drops or bubbles for it is illustrative of one and many problems and quite a number of recent earth-based and low-g, space experiments have been conducted [1,2,8-10]. A liquid drop is a model for many natural systems as disparate as planets, nuclei, etc. Drop dynamics provides the test ground for theoretical analysis (e.g. matched asymptotic expansions), experimental techniques and numerical methods (which on ocassion perform better than real experiments). The study of drops also allows to sequentially proceed from simple to complex. Clearly, drop dynamics is not just of mere fluid physics interest. It is of relevance in atmospheric sciences, chemical engineering, materials science, etc.

Let us see how the hydrodynamic force on a drop has been estimated since the pioneering work of Newton and Stokes. From Newton's experiments in 1710, and later observations, the magnitude of the drag force on solid drops/spheres in steady motion of a viscous fluid was given as

$$F_D = 0.22\pi R^2 \rho U^2 \tag{1}$$

where $U$ is the relative velocity between particle and fluid,$R$ is the particle radius, and $\rho$ is the fluid density. This relation is for "large" values of $U$, for which inertial effects are important.

Stokes, in 1850, suggested that at very low relative velocities the inertial effects are so small that they can be omitted from the Navier-Stokes equations. Under these conditions, the total drag on a sphere is

$$F_D = 6\pi R\eta U \tag{2}$$

Oseen pointed out that at a great distance from the sphere the inertia terms become more important than the viscous terms, and suggested a possible improvement of the Stokes law (2) by taking inertial-force terms partly into consideration. His drag force is

$$F_D = 6\pi R\eta U\left(1+\frac{3}{16}R_e\right) \tag{3}$$

where here $R_e = \dfrac{(2R)U}{\nu}$. Today we know that neither Stokes' nor Oseen's laws are uniformly valid, and the latter is not really an improvement of the former. Rather Stokes' analysis is valid in a small enough neighborhood of the sphere (creeping flow) and Oseen's analysis though valid for matching at the flow velocity values far from the sphere is not valid when approaching it. The Oseen approximation although incorporates inertial terms is still a linear theory and as Stokes' is a steady state theory. Several authors did attempt, not always successfully, an improvement upon the Oseen and Stokes analyses.

The mathematical problems solved by Stokes and Oseen come from different approximations to the Navier-Stokes equations [8],

$$\frac{\partial \mathrm{v}}{\partial t} + \mathrm{v}\cdot\nabla\mathrm{v} = -(\nabla p)/\rho + \nu\nabla^2\mathrm{v} \tag{4}$$

together with appropriate i.c. and b.c. Using the vorticity $\omega$=**rot v**, Eqs (4) become

$$\frac{\partial}{\partial t}\omega + \mathrm{v}.\nabla\omega = \omega.\nabla\mathrm{v} + \nu\nabla^2\omega \tag{5}$$

where $\omega.\nabla\mathrm{v}$ is the source of vorticity in the balance equation (yields stretching of vortex lines with consequent increase of vorticity), and $\nu$ is its diffusivity.

For the axisymmetric motion of a sphere there is always a stream function, i.e. a velocity potential $\Psi(r,\theta,\phi)$ which is constant along a streamline ($\phi$ is not relevant in the axisymmetric case). The existence of $\Psi$ is neither limited to steady motions nor to ideal, viscous-free flows. The existence of $\Psi$ for axisymmetric motions depends entirely on the *kinematical* asumption of incompressibility, hence exists for both ideal and viscous flows for those two motions differ only in their *dynamical* properties.

In the axisymmetric case with (r, θ, φ) coordinates we have:

$$v_r = -\frac{1}{r^2 \sin\theta}\,,\quad v_\theta = \frac{1}{r\sin\theta}\frac{\partial\Psi}{\partial r}\,,\quad v_\phi = 0$$

Then Eqs. (4) or (5) reduce to

$$\frac{\partial}{\partial t}(E^2\Psi)+\frac{1}{r^2\sin\theta}\frac{\partial\left(\Psi,E^2\Psi\right)}{\partial(r,\theta)}$$
$$-2\frac{E^2\Psi}{r^2\sin^2\theta}\left(\frac{\partial\Psi}{\partial r}\cos\theta-\frac{1}{r}\frac{\partial\Psi}{\partial\theta}\sin\theta\right)=\nu E^4\Psi \tag{6}$$

with $\frac{\partial(f,g)}{\partial(r,\theta)}=\begin{pmatrix}\frac{\partial f}{\partial r}\frac{\partial f}{\partial\theta}\\ \frac{\partial g}{\partial r}\frac{\partial g}{\partial\theta}\end{pmatrix}$ and $E^2=\frac{\partial^2}{\partial r^2}+\frac{\sin\theta}{r^2}\frac{\partial}{\partial\theta}\left(\frac{1}{\sin\theta}\frac{\partial}{\partial\theta}\right)$. Consequently, the Stokes creeping flow aproximation is the reduction of Eqs. (6) to

$$E^4\Psi=0 \tag{7}$$

Eq. (7) with the appropriate, non-penetration, no-slip/stick b.c. and suitable asymptotic behavior for large r, yields

$$\Psi(r,\theta)=(-\frac{1}{4}U\frac{R^3}{r}+\frac{3}{4}URr-\frac{1}{2}Ur^2)\sin^2\theta \tag{8}$$

hence the hydrodynamic drag force (2). Note that from (8) follows that the disturbance of the sphere extends to infinity as 1/r, and the presence of a boundary or another drop can modify the flow appreciably even when placed at a distance of many diameters from the drop.

Eq. (7) [see also Eq. (5)] shows that vorticity obeys the Laplace equation $\nabla^2\omega=0$. The flow is due solely to steady diffusion of vorticity to infinity in all directions; the drop is a source of vorticity due to the no-slip b.c. at its surface. Vorticity decreases as $1/r^2$ as originated from a dipolar source that produces for each component of $\omega$ equal positive and negative quantities at the surface of the sphere.

Oseen's apparent improvement of Stokes creeping flow approximation was a consideration of inertia terms linearized around the far field velocity hence a heuristic extension of the Stokes problem. It was not until the fifties that Stokes' and Oseen's results were properly put in context and generalized [11-13].

In 1911-12 another line of thought was developped. Rybczynski and Hadamard, independently, solved the Stokes problem for a liquid drop with flows outside and inside [1,2,8]. Their extension of Eq. (2) is

$$F_D=6\pi\eta_0 UR[1+\frac{2\mu/3}{1+\mu}] \tag{9}$$

with $\mu=\eta_i/\eta_o$ ($\eta_i=\eta_{drop}$). Clearly, the limit $\eta_i$ going to infinity yields Stokes law (2), while $\eta_0 >> \eta_i$ yields the corresponding law for a bubble, with 4 rather than 6 in Eq. (2).

In 1957 appeared a remarkable paper by Proudman and Pearson [13] where earlier ideas proposed by Kaplun on matched asymptotic expansions were applied to the flow dynamics around the sphere. The basic concept was to consider Stokes solution as a local (or inner) solution of the problem and Oseen's as a regular (or outer) solution rather than considering Oseen's as an improvement upon Stokes'.The local (inner) solution was assumed to be valid in a spherical region of radius $1/Re$ around the sphere while the outer solution was valid from infinity down to the $1/Re$ neighborhood. In the overlapping zone both solutions were accepted as valid, hence the need for appropriately matching them.

Proudman and Pearson found that for nonvanishing although low Reynolds number flows (Re<<1) the hydrodynamic drag on the sphere is

$$F_D = 6\pi\eta UR\left(1+\frac{3}{16}Re+Re^2+\frac{9}{160}Re^2\ln Re+...\right) \tag{10}$$

which shows the non-analytic form of the expansion. Extensions in various directions were made by different people. For instance, Acrivos and Taylor [14] using the Peclet number $\frac{(2R)U}{\kappa}$ considered the convective terms in the heat equation while using the Stokes flow in Eq. (6). They obtained a Peclet number expansion of the (convective) Nusselt number,

$$Nu = 2+\frac{1}{2}Pe+0.03404Pe^2+\frac{1}{4}Pe^2\ln Pe+\frac{1}{16}Pe^3\ln Pe+... \tag{11}$$

which also shows a non-analytic form. Subramanian [15] used matched asymptotic expansions to obtain the hydrodynamic force on a drop including convective terms in the heat equation while maintaining the Stokes approximation for the velocity field. His series expansion in terms of the Péclet/Marangoni number, and subsequent improvement by his collaborator Merritt, did not show any logarithmic term [16]. This was due to the way the outer solution was treated, which was different from that of Acrivos and Taylor. Taylor and Acrivos [17] also showed that for the deformation of the sphere to be noticeable the capillary number must be non-negligible. For a slightly deformable bubble they obtained the radius in term of the angle θ

$$R(\theta) = R\left[1-\frac{5}{96}ReCa\left(3\cos^2\theta-1\right)\right] \tag{12}$$

for $Re$<<1 and $Ca$<<1. At $Re$=0 (i.e. in the creeping flow approximation) a drop or a bubble remain spherical irrespective of the low or high value of the (constant) surface tension. However, deformation may be relevant even when inertial effects are ignored if as earlier noted the surface tension $\sigma$ is not constant. If over the spherical surface $\Delta\sigma$ is small compared to the value of $\sigma$, the capillary number will be small, and the drop or bubble may be asummed spherical with negligible error. Different authors have also extended the theory to account for various non-Newtonian fluids.

Young, Goldstein and Block [18] were the first to realize the possibility of levitating a drop or a bubble by means of capillary forces (Marangoni stresses).They showed that a drop or a bubble placed in another fluid where a temperature gradient exists instantaneously tends to move towards the hotter point. This is the motion of the drop relative to the flow induced along its surface by the lowering of surface tension at its leading pole (hotter than the rear pole). Using the Stokes-Rybczynski-Hadamard approximation they also computed the terminal velocity of a drop or a bubble in the field of gravity (Re<<1, Pe<<1), and experimentally checked the theoretical prediction within reasonable accuracy (within 20%) with an experiment using rising bubbles in a liquid layer heated from below (diameters 2R=$10^{-3}$-22x$10^{-3}$cm; dT/dz=10-90 K/cm). Using neutrally buoyant liquid water at 4°C, Bratukhin, Evdokimova, Pschenichnikov, Briskman and Zuev [19] did a similar experiment with rising bubbles in a laterally heated vertical liquid layer .

For the Young, Goldstein and Block problem the balance between capillary, buoyancy and hydrodynamic forces is

$$F_\sigma + F_g = AU \tag{13}$$

with the (Marangoni) capillary force

$$F_\sigma = \frac{4\pi R^2}{(1+\mu)(1+\delta)}\left(\frac{d\sigma}{dT}\right)(\nabla T)_\infty \tag{14}$$

and the buoyancy force

$$F_g = \frac{4}{3}\pi R^3 g(\rho_i - \rho_o) \tag{15}$$

As

$$A = 4\pi\eta_o \frac{1+\frac{3}{2}\mu}{1+u} R > 0 \tag{16}$$

the hydrodynamic force $AU$ represents drag. $\delta = \lambda_i/\lambda_o$ is the ratio of thermal conductivities (drop to surrounding fluid). Clearly, $AU$ is an extension of both the Stokes law $(\eta_i \to \infty, \mu \to \infty)$ and the Rybzcynski-Hadamard law $\left([\nabla t]_\infty = 0\right)$.

In their experiment, Young et al. used a liquid bridge. An improvement eliminating possible capillary convection at the open sides was carried out by Hardy. He used a closed cavity with silicone oil and air bubbles ($2R=5\text{-}25\times10^{-3}$ cm; dT/dz=40-140 K/cm). Hardy [20] clearly pointed out the role of contamination at the surface of the bubble, as long ago suggested by Levich [1,2]. Further improvement came with an experiment by Merritt and Subramanian [16]. Experimentalists started using drops rather than bubbles. Barton and Subramanian [21] used neutrally buoyant drops (2R=20-600mm, dT/dz=2.4 K/mm). Recent earth-based and low-g work by Braun *et al* [22] on thermocapillary migration of drops provided the most accurate verification of theprediction by Young *et al.*. Their Péclet/Marangoni numbers were in the range $10^{-5}\text{-}10^{-6}$, but with liquids having the non standard surface tension gradient behavior, i.e. surface tension increasing with the increase of temperature (2 butoxyethanol-water mixture with liquid/liquid phase separation at 61.14 $^0$C on the lower branch of the closed miscibility gap; 2R=11μm, dT/dz=36.9K/m, $d\sigma/dT>0$) .

Worth mentioning are also experiments carried out by Neuhaus and Feuerbacher [23] who found disagreement with Young, Goldstein and Block prediction but who invoked that the solution of the difficulty was in augmenting the theory with a surface dilational viscosity, in agreement with a conjecture already put forward long ago by Boussinesq [1,2,8] and later on by Lucassen [24]. Another interesting experiment was made by Hähnel, Delitzsch and Eckelmann [25] using a Plateau tank. They observed the motion of two water drops in butyl benzoate (2R=1.7 mm and 0.6 mm; dT/dz=2.07 K/cm). The curious thing about their experiment is that at first the bigger drop moved faster than the smaller one (first 479 s) and then the smaller drop overtook the larger which (at 689 s) lagged well behind the other. Their Péclet number was of order unity.These and a plethora of related studies are still demanding further theoretical work and feedback from theory to go beyond 'weak' forces.

The work on drops and bubbles so far recalled, refers to "passive" drops. Now I turn to some recent findings about "active" drops [26-31]. By an "active" drop or bubble I mean a drop or a bubble with internal volume heat sources, with a surface where chemical reactions may occur, etc. Take e.g. a drop at rest in a homogeneous medium and assume that there is (uniform) internal heat generation or a surface chemical reaction. The state of rest can be unstable. Consider, for instance, the latter case with given uniform composition far off the drop. A composition fluctuation at the surface of the drop promotes the Marangoni effect which yields flows inside and outside the drop which can be sustained if the Marangoni effet is strong enough, i.e. past an instability threshold. Indeed, as the drop moves the flow brings to the leading pole the higher solute at the surface, it makes the concentration far off the drop larger than that in its vicinity; hence the motion sustained past a certain threshold. Alternatively, if a velocity fluctuation tending to move the drop in certain direction spontaneously occurs, it breaks the initial spherical symmetry in composition, hence Marangoni stresses which in turn can help sustaining the velocity fluctuation past an instability threshold. If the initial state is that of a uniform, constant drop velocity or there is an externally imposed temperature or composition gradient, then instability is also possible leading to a different drop motion. These are not the only possible instabilities as we also have the effects due to deformability and time-dependence.

To illustrate how we have proceeded along the path set by earlier mentioned authors dealing with passive drops now I consider the case of a spherical drop moving with constant velocity in a temperature and composition homogeneous, infinitely extended surrounding fluid. Both the inner and outer fluids are taken immiscible. The surface tension is assumed to vary linearly with temperature. The outer fluid is assumed to have a uniform concentration of a solute which is allowed to react exo- or endothermally at the surface of the drop. Far off the drop both temperature and the concentration of solute are constants, hence, I insist, no external gradient. Stefan flow is negligible (convective flow of the reacting components in a direction normal to the surface where the reaction is taking place; it is generally a small effect for most chemical reactions and is normally important only in the presence of strong ablation or condensation). In the low Reynolds and Péclet numbers approximation with however $MaPe = 1$. In dimensionless form the steady Eq. (6) becomes

$$\frac{Re}{\nu^{*}r^{2}}\left\{\frac{\partial\left(\psi_i,E^2\psi_i\right)}{\partial(r,\delta)}+2E^2L_r\psi_i\right\}=E^4\psi_i \quad (17)$$

which is nonlinear and has to be considered together with the corresponding heat and mass diffusion equations, and appropriate b.c. Here $Re$ is defined with the constant drop velocity, $\delta=\cos\theta$, $\nu*$ refers to (kinematic) viscosity (i=1 outer fluid, $\nu*=1$; i=2 drop, $\nu*=\nu$), and $L_r$ corresponds to the operator appearing in the third term of Eq. (6).

The linear solution of Eq. (17) yields the hydrodynamic force

$$F = -4\pi\eta_1 rAU \quad (18)$$

If A is negative we have drag while if A is positive there is thrust, hence selfpropulsion, and possible autonomous motion of the drop in a medium originally uniform. We have

$$A = -[1 + (3/2)\mu + 3m] / (1 + \mu + m) \quad (19)$$

with $\mu=\eta_1/\eta_2$ inverse of (9), and m a suitably scaled Marangoni number accounting for a balance between $d\sigma/dT$, the chemical reaction rate, and (viscous and heat) dissipation. We see that at $m=m_1=-1/3-\mu/2$ we have A=0, whereas at $m=m_2=-1-\mu$, A diverges to infinity. Clearly, the study of both cases demands nonlinear analysis. Before referring to this we note that for A=3, the Marangoni effect combined with the chemical reaction yields the possibility of thrust, while for A=-1 we have drag, as well as for A=-3/2, and even more, for A=-3 there is enhanced drag due to the secondary backflow around the drop.

The weakly nonlinear result for $(m-m_1)$ (>0, <<1) provides the value of the hydrodynamic force

$$F = -4\pi\eta_1 R[A + B(RU/\nu_1)]U \quad (20)$$

with

$$B = -(1/4)A^2 + (1/2)Pr[m(6 + 3L - AL)/(1 + \mu + m)] \quad (21)$$

where here Pr= $\nu_1/\kappa_1$, and L=$\kappa_1/D$ is the (inverse) Lewis number of the homogeneous surrounding medium. A representation of F versus U straightforwardly shows that for ($m-m_1$)<0 (A>0) there are three possible values of U for zero hydrodynamic force, hence possible autonomous motion of the drop. The addition of an external force field like buoyancy, or an externally imposed thermal gradient like in the Young *et al.* experiment, provides the possibility of three genuinely different non-zero velocities for a given force field strength. As already mentioned we have multiplicity of steady states of motion which are not all actually realizable. An estimate of the autonomous motion velocity yields about U=0.5Rm/s with threshold at $\Delta$T=0.02 K (respectively, $\Delta$T=0.2K) for R=1 mm (R=0.1 mm).

The other case yields that for $m<m_2$, there are three possible values of the hydrodynamic force for zero velocity, hence three possible coexisting levitation levels of which one is not stable. Levitation or motion is a consequence of the nonlinearity in the flow coupled to the Marangoni stress, and not a consequence of some external thermal gradient as in the case treated by Young *et al.*[18].

Finally, addition of a time-varying gravity field as it occurs in space (g-jitter) and in some earth-based experiments leads to a time-dependent, weakly nonlinear vector-form Landau equation for the velocity of the drop [31]. This equation can be used not only to find the stationary regimes and analyze their stability, but also to consider time varying motions. In the simplified case of a small amplitude buoyancy force changing sinusoidally with time the result found is that an active drop capable of autonomous motion actually tends to move in a direction orthogonal to the time-varying force. Thus an *active* drop shows fascinating new features worth exploring with low-g levels.

## 3. Liquid layers. Convective patterns

The onset of patterned convective motion in heated fluid layers with a free upper surface has been extensively studied since the original experiments by Henri Bénard [16,7,32]. Depending on the depth of the layer one distinguishes two basic mechanisms of instability. In sufficiently deep cells, or in cells in which the fluid is confined between rigid horizontal plates, the convective motion settles when buoyancy forces overcome viscous forces (Rayleigh-Bénard problem) [32,33]. Alternatively, in sufficiently shallow layers with an open surface, inhomogeneities in the surface-tension are responsible for the onset of motion (Bénard-Marangoni problem). In both cases, the characteristic wavelength of the convective structure may be comparable to the depth of the cell or much larger, depending on whether or not the horizontal boundaries are good thermal conductors.

Close to the instability threshold the system may be described by amplitude equations having a universal form. The coefficients of these amplitude equations are specific to each instability, i.e., they depend on the dimensionless numbers of the problem containing fluid properties, boundary conditions and the external forcing. Here, for convenience, we write

$$\mathrm{Ma} = -\frac{\partial\sigma}{\partial T}\frac{d\Delta T}{\kappa\eta} \ ; \ \mathrm{Bi} = \frac{hd}{k} \tag{22}$$

where $\eta$, $\kappa$, $\sigma$ refer to the liquid, d is the thickness of the layer, $\Delta$T is the temperature diference between the open surface and the heated bottom plate, h is the thermal surface conductance and k is the heat conductivity of the fluid layer.The Biot number is a characteristic of the heat transfer at the boundaries; an infinite value of the Biot number corresponds to a perfectly conducting boundary while a zero value corresponds to a poorly conducting surface.

To the Navier-Stokes, continuity and energy equations, in the dimensionless Boussinesq approximation [34,35], we add the boundary conditions:
at the lower uniformly heated rigid plate:

$$\begin{aligned} &v = 0 \\ &\partial_z T - Bi\, T = \text{constant}. \end{aligned} \tag{23.a}$$

and at the top undeformable open surface:

$$\begin{aligned} &\omega = 0 \\ &\partial_z \sigma = \eta \partial_z u \\ &\partial_y \sigma = \eta \partial_z v \\ &\partial_z T + Bi\, T = \text{cte}. \end{aligned} \tag{23.b}$$

Table 1. Bénard cells

| $\Delta$ = **Ma-Ma$_c$** | **Stable patterns** |
|---|---|
| $\Delta<\Delta_c$ | Motionless diffusion |
| $\Delta_c<\Delta<0$ | Motionless diffusion, Hexagons |
| $0<\Delta<\Delta_1$ | Hexagons |
| $\Delta_1<\Delta<\Delta_2$ | Hexagons, Rolls |
| $\Delta>\Delta_2$ | Rolls |

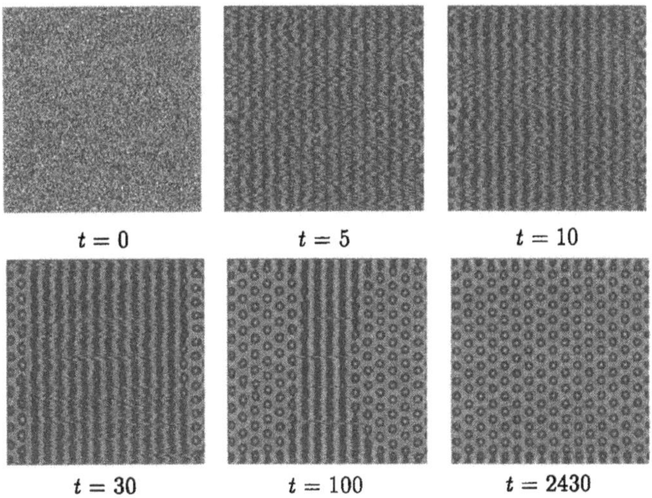


Fig. 1. Bénard-Marangoni convection. Evolution of a random initial condition towards the hexagonal planform

To study the transition between the motionless diffusion state and the convective state, and the dynamics of the structures that are expected to appear in this convective state, we use a multiple scale perturbation theory in the vicinity of the onset of convection. A small parameter is used to separate the fast variables that describe the instability and the slow ones that are useful to describe the pattern dynamics. The temperature is:

$$T = T(z)\,[A_1(X,Y,\tau)\exp(iK^{(1)}\cdot r) + A_2\exp(ik^{(2)}\cdot r) + A_3\exp(ik^{(3)}\cdot r) + c.c] \quad (24)$$

where $k^{(i)}$ denotes three linearly critical wave vectors oriented at 120 degrees in the horizontal plane. The amplitude equations in the horizontal plane are ( e.g. for $A_1$):

$$\begin{aligned}
\alpha_t \partial_t A_1 &= \alpha_l \Delta A_1 + \alpha_q A_2^* A_3^* \\
&- \alpha_{cs} A_1 |A_1|^2 - \alpha_{ci} A_1 (|A_2|^2 + |A_3|^2) \\
&+ \alpha_d (k^{(1)}\cdot \nabla_x)^2 A_1 + i\,\beta_1 (k^{(1)}\cdot \nabla_x)(A_2^* A_3^*) \\
&+ i\beta_2 [A_2^* (k^{(2)}\cdot \nabla_x) A_3^* + A_3^* (k^{(3)}\cdot \nabla_x) A_2^*] \\
&+ i\beta_3 [A_3^* (k^{(2)}\cdot \nabla_x) A_2^* + A_2^* (k^{(3)}\cdot \nabla_x) A_3^*]
\end{aligned} \quad (25)$$

where

$$\begin{aligned}
\alpha_l &= 0.0038 \\
\alpha_t &= 0.05 + 0.013\,Pr^{-1} \\
\alpha_q &= 0.0203 - 0.0046\,Pr^{-1} \\
\alpha_{cs} &= 0.016 + 0.0049\,Pr^{-1} + 0.00077\,Pr^{-2} \\
\alpha_{ci} &= 0.0217 + 0.003\,Pr^{-1} + 0.0018\,Pr^{-2} \\
\alpha_d &= 0.021 \\
\beta_1 &= \beta_2 = \beta_3 = \beta = 0.016 - 0.0041\,Pr^{-1} \\
\Delta &= Ma - Ma_c
\end{aligned} \quad (26)$$

Similar equations appear for $A_2$ and $A_3$ (with circular permutation of the indices) [36,37]. We give the numerical value of the coefficient in the specific case of a poor conducting upper surface and good conducting lower plate. Other thermal conditions at the top and bottom change the numerical values of the coefficients. These equations are generalized Ginzburg-Landau equations with genuinely new advective terms with $\beta$ coefficients. For these equations there is no Lyapunov functional hence for some value of the $\beta$ we may observe non steady behaviour. In the computations to be described attention is paid to the two parameters $\Delta$ and $\beta$ which measure, respectively, the (sub/supercritical) distance to the threshold and the strength of the advective terms.

The values of the constants for $Pr \rightarrow \infty$ are

$$\Delta_c = -\frac{\alpha_q^2}{4\alpha_1(\alpha_{cs}+2\alpha_{ci})} \approx -0.456$$

$$\Delta_1 = \frac{\alpha_q^2\alpha_{cs}}{\alpha_1(\alpha_{cs}-\alpha_{ci})^2} \approx 53.4$$

$$\Delta_2 = \frac{\alpha_q^2(2\alpha_{cs}+\alpha_{ci})}{\alpha_1(\alpha_{cs}-\alpha_{ci})^2} \approx 179.2 \qquad (27)$$

The study of Eq. (25) without spatial terms in *infinite* horizontal extent is known. Results of linear stability analysis of rolls and hexagons are summarized in Table 1.Two hysteresis cycles, hence coexistence and bistability appears in the intervals $[\Delta_c, 0]$ and $[\Delta_1, \Delta_2]$, respectively.

Busse´s formulæ [38] for space independent amplitude equations in *infinite* geometry provide a starting point for the exploration of the complete equation in *finite* geometry. Let us list the results of five numerical studies of the amplitude equations (5) in a square geometry, starting from random initial conditios (cases i-iv) or rolls (v). To solve the equations a finite difference method using a semi-implicit splitting-type scheme was applied. To know *when* the system attains a steady state it was used a method based on the evaluation of a norm, providing the rate of change of the distance between two successive states of the system, due to both the saturation and to the phase evolution of the pattern [39,40]:

i. ($\Delta$=50,$\beta$=0.1; Fig. 1). The system evolves to a stationary hexagonal pattern. The large value of $\beta$ does not affect much the pattern. Computations with $\beta$=0 confirm that an increase of $\beta$ only slightly distorts the pattern. Thus the advective terms in the amplitude equations do not play an important role in the case of an hexagonal structure. The fluid rises in the center of the cells in accordance with experimental observations. No defects are observed.

ii. ($\Delta$=75,$\beta$=0). The system is in the bistable region. Both hexagons and rolls are possible. The simulation displays the coexistence of these two structures.

iii. ($\Delta$=150,$\beta$=0). The rolls are the preferred structure. At the boundary, the rolls tend to be perpendicular to the sidewalls.

iv. ($\Delta$=150,$\beta$=0.1). Increasing the parameter $\beta$ to 0.02 no apprecialbe changes with respect to case III were observed. The system does not reach a steady state. Besides the roll structure, defects appear moving through the system.

v. ($\Delta$=150,$\beta$=0.1). Here the computation is with the same parameter values as in iv but with rolls plus slight superposed noise added as initial condition. First, the structure evolves to rolls without defects, but as time goes on the rolls start to bend leading to a'zig-zag'-like instability [41]. The system does not show evolution to a steady state.

## 4. Liquid layers. Waves and dissipative ('aging'/imperfect) solitons

Let us consider now the liquid layer heated from the air side or open to suitable mass adsorption from a vapor phase above, with subsequent absorption in the bulk hence creating a thermal gradient inside and focus on the deformability of the interface. Contrary to the case of a layer heated from the liquid side the problem refers to oscillatory motions, waves, and not to steady patterns (Bénard cells).

Both mass absorption and desorption, and heat transfer experiments have been carried out [42-45]. For the case of heat transfer (liquid depths from 0.3 to 0.8 cm) liquid octane was poured in square or cylindrical vessel, and in annular channel (1.5 and 2.0 cm inner and outer radii, respectively). The bottom was cooled by air or water at 20 C and the cover made of quartz placed at 0.3 cm above the liquid was heated hence establishing in the liquid layer a temperature gradient. For values of this gradient ranging from 10 to 200K/cm, solitary waves and periodic wave trains have been observed which show behavior in head-on and oblique collisions similar to results from mass transfer experiments. For mass transfer with vapor absorption by the liquid, Fig. 2 depicts a set-up used. In vessel A either we have a cylindrical container or, by using a suitable ring, an annular channel filled with liquid (liquid depth 1.8 cm). The vessels $B_1$ and $B_2$ are also filled with another liquid. With pentane in $B_1$ and $B_2$ either xylene, nonane, trichloroethylene or benzene were used as absorbing liquid in A, while with toluene as absorbing liquid in A either hexane, pentane, acetone or diethylether in $B_1$ and $B_2$ were used. In all cases the results were qualitatively the same. The glass cover C is placed on top of the vessels $B_1$ and then when the vessel C is full of hexane vapor, say, we place it on top of A thus allowing the absorption of hexane by the toluene liquid in A.The absorption process is a strong one, hence creating Marangoni surface stresses strong enough to trigger interfacial instability. During the whole duration of the experiment, hexane vapor is also allowed to diffuse from the two vessels $B_2$ to A.

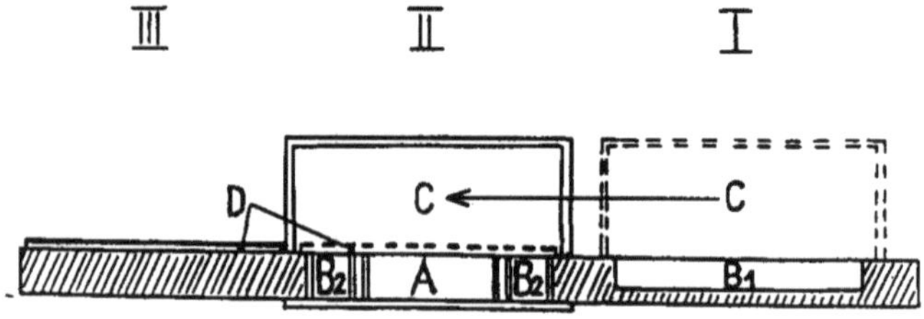


Fig. 2. Bénard-Marangoni set-up for mass transfer.

Observation and recording with a CCD camera was made by shadowgraph from the top with pointlike illumination from the bottom up (work is in progress with more sophisticated means). For instance, with cylindrical or annular cylindrical containers at first rather violent chaotic motions occur along the surface in A with waves moving in practically all directions but finally after about one minute when most of the vapor in C is absorbed, a dramatic self-organization occurs leading to regular wave motion. Long time lasting, synchronically colliding counterrotating periodic wave trains have been observed for about 50 to 200 seconds while the single (periodic) wave train with either clockwise or counterclockwise rotation remained up to 450 seconds. As the Marangoni stresses decay the number of crests in each train increases with corresponding wavelength decrease. Subsequently, a single set of either clockwise or counterclockwise (periodic) rotating waves remains (Fig. 3) until finally when the constraint fades away all convective motions disappear. Typical mean wave velocities at the outer wall of the annular channel before and after collision are, respectively, 2.7 and 1.7 cm/s (corresponding to angular speeds of 71.4

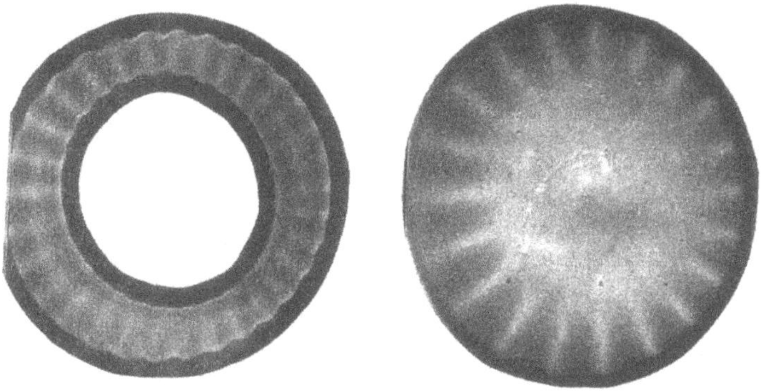

(a) cylindrical annular channel (b) cylindrical container.

Fig. 3. Periodic wave trains in Bénard-Marangoni oscillatory convection with mass transfer.

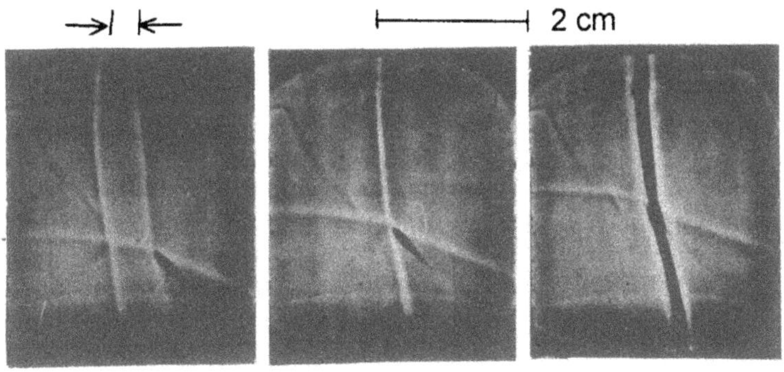


Fig. 4. Head-on collision of two waves in Bénard-Marangoni oscillatory convection.

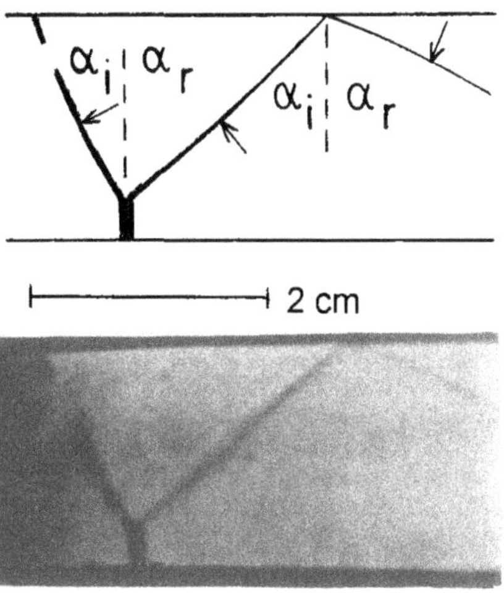


Fig. 5. Regular and Mach reflection of waves in Bénard-Marangoni oscillatory convection.

degrees/s and 45.7 degrees/s, respectively). Thus the mean wave velocity right after collision is about 64% (with less than 2% error) the mean wave speed measured before collision. About 0.2 seconds after collision the original wave speed is recovered.The phase shift shows the same sign as in the case of head-on collisions of 'solitary' waves in rectangular vessels (illustrated in Fig. 4). A couple of reflections at walls also illustrate the solitonic/shock behavior of the waves (Fig. 5) which occur with and without the formation of a (phase locked, third wave) Mach stem according to the angle of incidence.

Based upon the knowledge earlier described, dissipation-modified, shallow layer Korteweg-de Vries (KdV) and (proper) Boussinesq equations have been derived [45-52]. Such equations describe long wave oscillatory instability in problems where an input-output energy balance exists besides the usual (local) KdV balance between nonlinearity and dispersion. Hence they are natural extensions to real fluids of the original dissipation-free, inviscid problem considered by Boussinesq, Rayleigh, Korteweg and de Vries, and others. Some of the solutions of these equations have been studied, such as solitary waves, cnoidal waves (periodic wave trains) as well as the (head-on and oblique) interaction of these waves.

For simplicity I consider the particular case of one-sided (left to right) moving waves describable by a dissipation-modified KdV equation [46,49]

$$h_t + (h^2)_y + h_{yyy} + \delta[h_{yy} + h_{yyyy} + D(h^2)_{yy} + \alpha h] = 0 \tag{28}$$

where h(y,t) is a scaled elevation of the surface in the study of one-side steadly propagating waves. The coefficient D can be either positive or negative, while $\alpha$ is always nonnegative. We consider the dissipation as a perturbation upon the viscous-free case, hence $\delta \ll 1$. Note that besides the addition of dissipative terms in Bénard-Marangoni problem Eq. (28) also contains a generalization of the surface tension effets already correctly introduced by Korteweg and de Vries and later on discussed by Benjamin and others.

To illustrate the energy balance sustaining, say a solitay wave, I multiply Eq. (28) by h, and integrate with respect to y along the available space. The mean value of h along the layer is zero (mass conservation). The time evolution of energy

$$E = \frac{1}{2}\int h^2\,dy \tag{29}$$

is governed by the balance

$$\frac{dE}{dt} = \delta\left(\int h_x^2\,dx - \int h_{xx}^2\,dx + 2D\int hh_x^2\,dx - \alpha\int h^2\,dx\right) \tag{30}$$

whose value vanishes for a steadily travelling wave. The first term on the right hand side of (30) describes the energy input at rather long wavelengths due to instability, the second and fourth terms describe energy dissipation on short and long wave lengths, respectively, and the third term accounts for nonlinear (feedback) correction to long wave energy input (for h positive, positive if D is positive and negative otherwise). Note that in the absence of dissipation and continuous energy supply Eq. (28) reduces to the standard KdV equation and solitary waves or cnoidal waves (periodic wave trains) are still possible. When dissipation plays negligible role in experiment they can be excited from appropriate initial conditions either numerically or in water tanks where viscosity plays negligible role. In the present case the situation is different. Indeed, at variance with the (integrable, dissipation-less) KdV equation where a one-parameter family of solutions exists, hence as a consequence of initial conditions all possible amplitudes and corresponding velocities exist, here the input-output energy balance selects a single wave or a single amplitude periodic wave train, a bound state or an erratic/chaotic wave train [47,48]. On the other hand, when considering the three-dimensional problem [49] phase shifts following collision or reflection at walls depend upon the incident angle (e.g. measured front to front or twice the value front to wall, i.e. by $\pi/2-\alpha_i$, Fig. 5; a reflection is like a collision with a mirror image wave). At the approximate value of $\pi/2$ no phase shift is expected while for lower collision angles the phase shift has the sign of phase shifts upon head-on collisions. Higher values than $\pi/2$ (or $\alpha_i<\pi/4$) lead to a change of sign in the phase shift and the formation of a Mach stem, i.e. a third wave evolving phase locked with the post collision or reflected front. The phase shift sign in such case is the same as the sign in the overtaking collisions discussed by Zabusky and Kruskal [53]. The formation of bound states has been numerically observed. Starting with e.g. an initial condition of two nearby 'solitary' pulses the system evolves according to Eq. (28) to a wave train with unequally spaced maxima. All maxima have the same value, hence the same velocity dictated by the energy balance (30) in the steady state.

## Acknowledgments

The results reported here come from years of collaboration with H. Linde, C.I. Christov, Yu. S. Ryazantsev, V. I. Nekorkin, A. A. Nepomnyashchy, X.-L. Chu, A. N. Garazo, J. Bragard, J. Pontes, A Ye. Rednikov, V.N. Kurdyumov, A. Maksimov, G.-X. Huang and P. D. Weidman. Work sponsored by DGICYT (Spain) under Grant PB 93-81, by the European Union under Network Grants ERBCHRXCT 93-106 and 93-107, and by the Fundación "Ramón Areces" (Spain).

## References

1. B. G. Levich, *Physicochemical Hydrodynamics*, Prentice-Hall, Englewood Cliffs, N.J. (1965).
2. B. G. Levich, B.G. and V.S. Krylov, Ann. Revs. Fluid Mech. 1 (1969) 293-316, and references therein.
3. L. E. Scriven and C. V. Sternling, Nature 187 (1960) 186.
4. S. Ostrach, Ann. Revs. Fluid Mech. 14 (1982) 313, and references therein.
5. S. H. Davis, Ann. Revs. Fluid Mech. 19 (1987) 403, and references therein.
6. E. L. Koschmieder, *Bénard Cells and Taylor Vortices*, Cambridge U. P., Cambridge, (1993)
7. M. G. Velarde and C. Normand, Sci. Amer. 243 (1980) 92-108.
8. J. Happel and H. Brenner, *Low Reynolds Number Hydrodynamics*, Prentice-Hall, Englewood Cliffs, N.J. (1965), and references therein.
9. G. Wozniak, J. Siekmann and J. Srulijes, Z. Flugwiss. Weltraumforsch.,12 (1988) 137, and references therein.
10. R. S. Subramanian, in *Transport Processes in Bubbles, Drops and Particles* (editors: R. P. Chhabra and D. de Kee), Hemisphere, N.Y. (1992), and references therein.
11. S. Kaplun, ZAMP 5 (1954) 111. See also S. Kaplun and P.A. Lagerstrom, J. Math. Mech. 6 (1957) 585.
12. P. A. Lagerstrom and J.D. Cole, J. Rat. Mech. Anal. 4 (1955) 817.
13. I. Proudman and J. R. A. Pearson, J. Fluid Mech. 2 (1957) 237.
14. A. Acrivos and T. D. Taylor, Phys. Fluids 5 (1962) 387.
15. R. S. Subramanian, AIChE. J, 27 (1981) 646.
16. R. M. Merritt and R. S. Subramanian, J. Coll. Interf. Sci. 125 (1988) 333.
17. T. D. Taylor and A. Acrivos, J. Fluid. Mech. 18 (1964) 466.
18. N. O. Young, J. S. Goldstein and M.J. Block, J. Fluid Mech. 6 (1959) 350.

19. Yu. K. Bratukhin, O.A. Evdokimova, A.F. Pschenichnikov, Izvestia Akad. Nauk USSR, 1979, n. 5. See also Yu. K. Bratukhin, V.A. Briskman, A.L. Zuev, A.F. Pschenichnikov and V. Ya. Rivkind, in Hydrodynamics and Heat-Mass Transfer in Weightlessness, Nauka, Moscow (1982)
20. S. C. Hardy, J. Coll. Interf. Sci. 69 (1979) 157.
21. K.D. Barton and R.S. Subramanian, J. Coll. Interf. Sci. 133 (1989) 211.
22. B. Braun, Ch. Ikier, H. Klein and D. Woermann, in *Rocket Experiments in Fluid Science and Materials Sciences*, ESA SP-1132, Paris (1994), vol. 4, 119.
23. D. Neuhaus and B. Feuerbacher, in *Procs. 6th. European Symposium on Materials Sciences under Microgravity Conditions*, ESA SP-256, Paris (1987), 112
24. J. Lucassen, Trans. Faraday Soc. 64 (1968) 2221
25. M. Hahnel, V. Delitzsch and H. Eckelman, Phys. Fluids A1 (1989) 1460.
26. A.Ye. Rednikov, Yu.S. Ryazantsev and M.G. Velarde, Phys. Fluids 6 (1994) 451.
27. A.Ye. Rednikov, Yu.S. Ryazantsev and M.G. Velarde, J. Coll. Interf. Sci. 164 (1994) 168.
28. A.Ye. Rednikov, Yu.S. Ryazantsev and M.G. Velarde, Int. J. Heat. Mass Transf. 37, Supp. 1 (1994) 361.
29. A.Ye. Rednikov, Yu. S. Ryazantsev and M.G. Velarde, J. Non-Equilib. Thermodyn. 19 (1994) 95-113
30. A.Ye. Rednikov, Yu.S. Ryazantsev and M.G. Velarde, Phys. Scripta T55 (1994) 115.
31. A.Ye. Rednikov, V.N. Kurdyumov, Yu. S. Ryazantsev and M.G. Velarde, Phys. Fluids 7 (1995), to appear
32. Ch. Normand, Y. Pomeau and M.G. Velarde, Rev. Mod. Phys. 49 (1977) 581, and references therein.
33. M. C. Cross and P.C. Hohenberg, Rev. Mod. Phys. 65 (1993) 851
34. R. Perez-Cordon and M. G. Velarde, J. Phys. (Paris) 36 (1975) 591
35. M. G. Velarde and R. Perez-Cordon, J. Phys. (Paris) 37 (1976) 176
36. J. Bragard, J. Pontes and M. G. Velarde, Int. J. Bif. Chaos (1995) to appear
36. J. Bragard and M. G. Velarde, (1995), manuscript in preparation.
38. F. H. Busse, J. Fluid Mech. 30 (1967) 625.
39. C. I. Christov, J. Pontes and M.G. Velarde, in *Advanced Concepts and Techniques in Thermal Modelling, Procs. Eurotherm 36,* Editions Européennes, Paris (1995), to appear.
40. J. Pontes, C. I. Christov and M.G. Velarde, Int. J. Bif. Chaos (1995) to appear
41. P. Manneville, *Dissipative Structures and Weak Turbulence*, Academic Press, San Diego (1990)
42. P. D. Weidman, H. Linde and M.G. Velarde, Phys. Fluids A4 (1992) 921.
43. H. Linde, X.-L. Chu and M.G. Velarde, Phys. Fluids A5 (1993) 1068
44. H. Linde, X.-L. Chu, M.G. Velarde and W. Waldhelm, Phys. Fluids A5 (1993) 3162
45. X.-L. Chu and M.G. Velarde, Phys. Rev. A43 (1991) 1094
46. A. N. Garazo and M.G. Velarde, Phys. Fluids A3 (1991) 2295
47.C. I. Christov and M.G. Velarde, Physica D (1995) to appear
48. V. I. Nekorkin and M.G. Velarde, Int. J. Bif. Chaos 4 (1994) 1135
49. A. A. Nepomnyashchy and M.G. Velarde, Phys. Fluids 6 (1994) 187
50. M. G. Velarde, V.I. Nekorkin and A. Maksimov, Int. J. Bif. Chaos (1995) to appear
51. A. Ye. Rednikov, M.G. Velarde, Yu. S. Ryazantsev, A.A. Nepomnyashchy and V. N. Kurdyumov, Acta Appl. Math. 39 (1995) 457
52. M. G. Velarde and C. I. Christov (editors), *Fluid Physics*, World Scientific, Singapore, 1995
53. N. J. Zabusky and M. D. Kruskal, Phys. Rev. Lett. 15 (1965) 57

# SIMULATED ANNEALING METHODS UNDER ANALYTICAL CONTROL

M.A. VIRASORO
*International Centre for Theoretical Physics ,*
*I-34100 Trieste, Italy*

**Using the replica method on constrained systems we show that Simulated Annealing can be understood and its accuracy controlled. For certain systems we show that it will give wrong results. In those cases we show how to improve it.**

## 1 Introduction

Both by the way it was discovered[1], as well as by the breadth of its applications, the sheer existence of the method of Simulated Annealing (SA), as a general purpose, non specific method for optimization, is a strong piece of evidence in favour of the idea that different complex systems share non trivial common features. Therefore it comforts the idea of building a theory of them.

However, up to now this incipient 'theory of complex systems' has been unable to say much about the deep reasons behind the method. Useful insight has been acquired on its range of applicability but it is fair to say that it was derived mainly empirically and at a pre-theoretical level. As a consequence, its formulation is still very general, but vague, as it suits the present level of understanding.

The present contribution wants to assess whether the time has come to try:

- to delimit the definition so as to give a more precise and better controlled formulation of the method
- to investigate how the details in the protocol procedure may affect both the computing time and the outcome
- to control the error

Towards this goal we will use analytical technics that have been developped in the context of the mean field theory of Spin Glasses. It is clear that only a limited subset of problems is tractable with this method. However this is more a practical limitation than a fundamental one. An analytical treatment leads to a deeper understanding that unveils characteristics of the system responsible for a specific behavior and this is, by definition, exportable to other systems.

On the other hand we could also obtain analytically a numerical estimate for the difference in cost/energy between the local minimum found following our precription and the absolute minimum and this is relevant only for systems that we can analytically solve.

Our main conclusion will be to show how the efficiency of SA depends on the relative variance of the cost/energy surface roughness at different scales. The two same classes of Complex Systems recently distinguished by their off equilibrium dynamics will reappear here [2,3].

An interesting byproduct of our analysis is that for those cases for which we prove that SA does not work, we will be able to propose a modified version that is more costly computation wise but that leads to a good local minimum.

The methods of Statistical Physics apply if one assumes that every instance of an optimization problem belongs to a class defined by the statistical properties of the parameters (the quenched variables). Our results will, of course, apply to the generic instance in that class. They do not describe the rare instances, nor the worst instance.

## 2 Simulated Annealing as Hierarchical Search

At zero temperature a system's evolution is governed by the energy surface. At finite temperature the latter has to be replaced by the **free energy**. In the mean field approximation the free energy is a surface in a $kN + 1$ dimensional space where $N$ is the number of sites and $k$ is the number of parameters required to specify the single site probability distribution of the state. This space does not depend on the temperature and therefore as the temperature varies one generates a family of surfaces.

With increasing temperature the free energy surface becomes smoother. The entropy substracts from the energy and decreases the roughness starting from the shorter scales on.

Annealing is a way of probing successively surfaces with increasing roughness. In the free energy surface at a definite temperature $T$ valleys and hills will be broader and smoother than in the corresponding surface at a lower temperature. Many systems undergo a phase transition at a critical temperature $T_c$ above which roughness is not more effective and the system behaves as having a single accessible global minimum. The latter multifurcates as the temperature lowers and this process repeats itself in a continuous fashion all the way down to zero temperature.

We can thus say that Simulated Annealing is basically a way of:

- exploring a free energy surface at a certain temperature $T_1$ and there choosing a "good" valley;

- decreasing the temperature one step and then limiting the choice of the new valley among those that were generated by multifurcation from the previous choice;
- iterating this last step until zero temperature is reached.

Therefore by definition SA is an example of **Hierarchical Search**. In artificial databases if one has to retrieve an item among $2^N$ ones, an efficient method consists in classifying all of the items in a tree with $K$ generations and $P$-furcations of each branch at each level (so that $K^P = 2^N$). One then introduces a criterium **at each level** that leads to a choice of a particular branch among the $P$ ones. At the end one has discriminated among $PK$ cases instead of $K^P$ ones.

By construction Hierarchical Search is efficient and artificial databases are built so that it be exact. In natural systems (and in generic optimization problems) it will still be efficient but its accuracy will have to be assesed.

First of all one has to identify the natural classification tree. SA adopts the tree constructed through temperature multifurcation but this is not the only possibility. In systems with long range interactions where the mean field approximation can be justified the powerful replica method introduced by Edwards and Anderson and Parisi's replica symmetry breaking (RSB) scheme imply the existence of a natural classification tree as a consequence of ultrametricity [4]. This tree is different from the temperature tree.

The existence of different classification trees poses the problem of how to compare and grade the efficiencies of the hierarchical search methods based on them [a]. For the sake of this comparison we can imagine just two levels trees. We will call the valleys that appear at the higher, ancestor, level the h-valleys while the descendants will be called l-valleys.

In SA the higher temperature is chosen so that the energy is slightly above the barriers between h-valleys. Still when the system is in a particular configuration the energy will be such that:

$$\frac{dS_{hv}}{dE_{hv}} = \frac{1}{T} \tag{1}$$

where the entropy $S_{hv}$ refers to the partial entropy corresponding to a particular h-valley. In going from one h-valley to another, the system may climb or

[a] We should here stress that we expect this kind of methods to be only heuristics. A better procedure leads to a lower local minimum. A generalized "chaos in temperature" [5] phenomenon, implies in general that the overlap between the configuration reached by any of these methods and the global unique minimum is minimal.

dive in energy according to eq(1). The amount of time spent in each h-valley will be proportional to:

$$\exp\left(S_{hv} - \beta E_{hv}\right) \tag{2}$$

We see the partial free energy of the h-valley appearing. The natural criterium at this level will be to choose the h-valley with minimum free energy. Why should we expect that the minimal energy configuration lies inside it? An explicit analysis of this has never been made. One possible answer would assume a linear extrapolation from energies above the barriers to the bottom of the h-valley. In fact the minimal energy configuration is defined by the equation:

$$S_{hv}(E_{min}) = 0 \tag{3}$$

Therefore:

$$S_{hv}(E_{hv}) + \frac{dS}{dE}(E_{min} - E_{hv}) = 0 \tag{4}$$

so that

$$E_{min} = E_{hv} - TS_{hv} \tag{5}$$

Unfortunately, we do not know of any argument to justify the neglect of the higher order terms in eq(5) [b] . Therefore we have to conclude that it is not possible to control such an extrapolation.

On the other hand the ultrametric tree is constructed at a fixed temperature by grouping together configurations that contribute to the partition function. The different levels in the tree correspond to different distances within which configurations are grouped in clusters. There is no extrapolation from high energy to low energy. In a two level tree at the h-level there are clusters of configurations with radius $d_h$ containing l-clusters of radius $d_l$. The formalism assigns to each h-cluster a quantity that resembles a free energy:

$$\frac{\exp(-\beta G_{hv})}{\sum_{hv'} \exp(-\beta G_{hv'})} = \frac{\sum_{lv \in hv} \exp(-\beta G_{lv})}{\sum_{hv'} \sum_{lv' \in hv'} \exp(-\beta G_{lv'})} \tag{6}$$

The criterion at the higher level is again to select the h-cluster with minimum $G_{hv}$ but now the replica analysis proves that for $G_{lv}$ near $G_{hv}$ the average number of l-valleys with free energy $G_{lv}$ is given by:

$$d\mathcal{N}_{lv}(G_{lv}) = \exp(\beta\ x(d_l)\ (G_{lv} - G_{hv}))dG_{lv} \tag{7}$$

where $x$ can be analytically computed and is related to the inverse function of Parisi's order parameter $q(x)$ at the overlap value $q$ corresponding to the

[b] The 2nd order term in eq(5) is equal to $\frac{TS_{hv}^2}{2C}$ where $C$ is the specific heat

distance $d_l$. In a direct comparison between two randomly chosen h-valleys with free energies $G_1$ and $G_2$ such that:

$$G_1 - G_2 = O(N) > 0 \tag{8}$$

one finds with probability one that the h-valley 2 contains lower free energy l-valleys.

However, before reaching any further conclusion we have to take into account that the number of h-valleys also grows exponentially with $G_{hv}$. In fact the replica method also implies that

$$d\mathcal{N}_{hv}(G_{hv}) \propto exp(\beta\ x(d_h)\ G_{hv})dG_{hv} \tag{9}$$

and if

$$x(d_h) > x(d_l) \tag{10}$$

the growth of the number of h-valleys overcompensates the decrease in probability so that the h-valley containing the lowest l-valley will be a **high free energy** one. In such a case hierarchical search will lead us astray.

This possibility was overlooked for quite a long time because Parisi's probability interpretation requires that $x(d)$ be a monotonous decreasing function of $d$. In fact, even if it was known that for some spin glasses the continuous Parisi function $q(x)$ had the wrong slope, in all such cases that solution was unstable and the Replica Symmetry Breaking scheme that gave the correct $Z$ function was parametrized by a one single stage of symmetry breaking with a corresponding discontinuous $q(x)$ function. In the next section we will discuss the physical interpretation of eq(10).

## 3 Surface Roughness as a Function of the Length Scale

### *3.1 The Generalized Random Energy Model*

In the previous section we have discussed at least two ways of constructing a family of free energy surfaces with gradual smoothing from the shorter length scales. Substracting successive surfaces we obtain different spectral components of the surface fluctuations. The variance of the roughness as a function of the scale is the crucial property that distinguishes different models.

We can compute for instance (following Derrida[6]) the average probability that configurations $\mu$, $\nu$ have energies $E_\mu$, $E_\nu$:

$$\mathcal{P}(E_\mu, E_\nu) \sim \exp(-\frac{(E_\mu - E_\nu)^2}{2Nv(d)}) \tag{11}$$

where v(d) is, in most models, an increasing function of the distance between configurations $d$. The positive correlation between the energies of two near configurations decreases monotonously as the distance grows. But at the same time the number of pairs of configurations also grows with $d$ (like $\exp(\alpha(d))$).

In Derrida's Generalized Random Energy Model[6] both effects are present and its solution depends on the behavior of a function $x(d)$ that explicitely depends on the parameters of the model (the functions $v(d), \alpha(d)$). If $x(d)$ is monotonously decreasing with $d$ then one can explicitely check that the lowest energy l-valley lies in the lowest energy h- valley. On the other hand if $x(d)$ has a different behavior then the lowest energy l-valley is not contained in the lowest energy h-valley. In the model one can explicitely calculate how high in energy one has to go to choose the right h-valley. It turns out, as it could be expected, that it depends on how steep is the growth of $x$ with $d$. In one limit, one recovers the Random Energy Model where no hierarchical search strategy is possible. On the other extreme if $x(d)$ is almost flat the lowest l-valley will be in an h-valley that although is not the lowest it lies near it.

### *3.2 The Partition Function of Constrained Systems and the 2nd Replica Trick*

It would be extremely interesting to calculate in more realistic models how high one has to go in energy at the h-level to find the sought l-valley but to do that one has to go beyond equilibrium.

In fact we need two basic quantities;

- the distribution of h-valleys with free energy:

$$d\mathcal{N}_{hv}(G) = \rho_{hv}(G_h)dG_h \tag{12}$$

- the distribution of l-valleys in free energy $G_l$ inside an h-valley with free energy $G_h$:

$$d\mathcal{N}_{lv}(G_l, G_h) = \rho_{lv}(G_l, G_h)dG_l \tag{13}$$

In collaboration with R. Baviera we propose to derive these two quantities by computing the partition function of a constrained system [7] made out of $R_h$ groups of $R_l$ configurations such that: 1) two configurations belonging to the same group lie at a distance $d_l$ of each other; 2) two configurations belonging to different groups are at a distance $d_h$.

Then the partition function can be written:

$$Z_J(R_h, R_l, d_h, d_l) = \sum_{\substack{\{C_{aa'}; a=1,\cdots,R_h; a'=1,\cdots,R_l\} \\ |C_{aa'} - C_{bb'}| = d_h \\ |C_{aa'} - C_{ab'}| = d_l}} e^{-\beta \sum_{a,a'} H_J[C_{aa'}]} \tag{14}$$

The index $J$ is a reminder that the $Z_J$ so defined still depends on the quenched variables. As usual, one has to average over $J$ the $\log Z_J$.

From the ultrametric hypothesis we can deduce that all $R_h$ groups must be inside the same h-valley. Therefore:

$$Z_J = \int d\hat{G}\ \rho_{hv}(\hat{G}) e^{-\beta R_h \hat{G}} \tag{15}$$

where $e^{-\beta \hat{G}}$ is the statistical weight of the h-valley occupied by a single group of $R_l$ configurations. $\hat{G}$ is therefore not the free energy of the h-valley but if the constrain is loose (see later the 'technical equations' 19) then:

$$\hat{G} = R_l G_h \tag{16}$$

Finally:

$$e^{-\beta R_l G_h} = \int dG_l\ \rho_{lv}(G_l, G_h) e^{-\beta R_l G_l} \tag{17}$$

In the thermodynamic limit $G_h$, $G_l$, $\log \rho_{hv}$, $\log \rho_{lv}$ are all extensive, proportional to $N$, quantities. Therefore the equations (15), (17) are dominated by either saddle points or end points contributions.

If we can calculate and **analytically continue** (thus the name 2nd replica trick) in $R_h, R_l$ the average over $J$:

$$\overline{\log Z_J(R_h, R_l, d_h, d_l)} \tag{18}$$

we can immediately retrieve through Legendre transformations the desired quantities $\rho_{lv}$, $\rho_{hv}$.

We must still choose the distances $d_h$, $d_l$ that define the branching levels of the 2-tiers tree. We found out that in order to perform the analytical continuation in $R_h, R_l$, the parameters $d_h$, $d_l$ have to be chosen saddle points of $\overline{\log Z_J}$. This adds two 'technical' equations:

$$\frac{d\overline{\log Z_J}}{dd_h} = \frac{d\overline{\log Z_J}}{dd_l} = 0 \tag{19}$$

We performed all the calculations in the spherical p-spin model[c]. For this model many properties are known:

- the continuous replica symmetry breaking solution $x(d)$ is monotonically increasing with $d$ and does not contribute to the partition function $Z$[9,10].
- Simulated Annealing does not work.
- Dynamical relaxation off equilibrium does not converge to the equilibrium state[2].
- the distribution of states in free energy is known by virtue of a calculation of the number of TAP equations solutions[11].

Our calculations are consistent with all of these results. In particular, by choosing $R_l = 1$ and $d_h$ corresponding to the Edwards Anderson $q_{EA}$ we rederived Crisanti-Sommers calculation[11].

For those readers that are expert in replica calculations we would like to comment the close relationship between the saddle points dominating the quantity $\overline{\log Z_J(R_h, R_l, d_h, d_l)}$ and certain solutions to a subset of the saddle point equations that appeared already in the unconstrained problem. Thus if we consider saddle points corresponding to two levels of RSB with parameters $(x_1, x_0)$ fixed at arbitrary, non stationary values then we generate in this way all the solutions necessary to calculate 18. We are aware that something similar (although at the 1RSB level) was already apparent in the dynamical analysis of the same p-spin model[2].

## 4 Conclusion. How to Improve on Simulated Annealing

In a 2-tiers tree with $P$ and $P'$ branchings:

$$\begin{aligned} P &= e^{\alpha N} \\ P' &= e^{\alpha' N} \end{aligned} \tag{20}$$

hierarchical search examines (discriminates among) $e^{\alpha N} + e^{\alpha' N}$ instead of $e^{(\alpha+\alpha')N}$.

[c] The details can be found in Baviera's thesis report[8]

We have shown that for certain systems the minimum energy configuration has to be searched in higher than minimum h-valleys. The number of such h-valleys will be $e^{\gamma N}$ with $\gamma < \alpha$. Therefore the total number of exams will be:

$$e^{\frac{\alpha}{\gamma}N} + e^{(\alpha'+\gamma)N} \tag{21}$$

which is still smaller than $e^{(\alpha+\alpha')N}$.

We can now use the calculations of $\rho_{hv}$ and $\rho_{lv}$ to give a precise estimate of how high in free energy one has to go.

$\rho_{hv}$ is the average total number of h-valleys in the system. When it becomes smaller than one, it means that not all $J$'s define systems with such an h-valley. Analogously $\rho_{lv}(G_h, G_l) \times \rho_{hv}(G_h)$ is the average total number of l-valleys and again cannot be smaller than 1. Instead $\rho_{lv}$ is the average number of l-valleys in a h-valley. When it is smaller than one, it means that not all h-valleys have an l-valley in that interval. This is consistent. We can choose an h-valley but the $J$'s have to remain generic. Therefore

$$\begin{aligned} \rho_{hv}(G_h) &\geq 1 \\ \rho_{lv}(G_h, G_l)\rho_{hv}(G_h) &\geq 1 \end{aligned} \tag{22}$$

while $\rho_{lv}$ can be greater or smaller than 1.

The hierarchical search leads to the equations:

$$\begin{aligned} \rho_{hv}(G_h^0) &= 1 \\ \rho_{lv}(G_h^0, G_l^0) &= 1 \end{aligned} \tag{23}$$

The solutions $G_h^0, G_l^0$ will be stable if:

$$\frac{d\rho_{lv}(G_h^0, G_l^0)}{dG_l^0} > \frac{d\rho_{hv}(G_h^0)}{dG_h^0} \tag{24}$$

while when these conditions are not satisfied the new equations are:

$$\begin{aligned} \rho_{lv}(G_h^1, G_l^1)\rho_{hv}(G_h^1) &= 1 \\ \frac{d(\rho_{lv}(G_h^1, G_l^1)\rho_{hv}(G_h^1))}{dG_h^1} &= 0 \end{aligned} \tag{25}$$

valid only if eq(22)are satisfied.

In conclusion we would like to make two comments:

1. Both solutions $G_h^0, G_l^0$ and $G_h^1, G_l^1$ when valid are accurate in the leading order in $N$.

2. Larger number of levels in the classification tree will make the method more efficient. The soutions will still be accurate in leading order in $N$ but we expect them to be worst in non-leading orders.

A numerical verification of these methods is in our future program.

## Acknowledgments

We acknowledge useful conversations and comments from M. Ferrero, J. Kurchan, R. Monasson and G. Parisi.

## References

1. S. Kirkpatrick, C.D. Gelatt Jr and M.P. Vecchi, Science **220** (1983) 671.
2. L.F. Cugliandolo and J. Kurchan, Phys.Rev.Lett.**71** (1993) 173;
3. L.F. Cugliandolo and J. Kurchan, J.Phys A **27**, 5749 (1994)
4. M. Mézard, G. Parisi, and M.A. Virasoro, "Spin glass theory and beyond", World Scientific (Singapore 1987);
5. I.Kondor, J.Phys. A **22** (1989) L163, I.Kondor, A. Végsö, J.Phys. A **26** (1993) L641
6. B. Derrida, J. Physique Lett, **46** (1985) 401. B. Derrida and E. Gardner, J. Phys. C: Solid State Phys. **19** (1986) 2253. B. Derrida and E. Gardner, J. Phys. C: Solid State Phys. **19** (1986) 5783.
7. S.Franz, G.Parisi, M.A. Virasoro, J.Physique I **2** (1992) 1969, Europhys.Lett. **22** (1993) 405
8. R. Baviera "Metodo della Cavità nei vetri di spin soggetti a vincoli" Tesi di Laurea in Fisica (1994-1995), Università di Roma "La Sapienza".
9. A. Crisanti, H. Horner and H-J Sommers, Z.Phys. **B92** (1993) 257; A. Crisanti, H.J.Sommers, Z. Phys. B **87** (1992) 341.
10. J.Kurchan, G.Parisi, M.A. Virasoro, J.Physique I **3** (1993) 1819
11. A.Crisanti, H.Sommers, preprint cond-mat/9406051
12. G.Parisi preprints cond-mat/9411115, cond- mat/9412004

# QUANTUM CHAOS AND LOCALIZATION: BAND RANDOM MATRIX APPROACH

F.M.IZRAILEV
*Budker Institute of Nuclear Physics*
*Novosibirsk 630090, Russia*
E-mail: izrailev@physics.spa.umn.edu

**Abstract**

Short review of recent developments of the author and his cooworkers in modern random matrix approach to complex dynamical and disordered systems is given. The distinctive peculiarity of this approach is a band-like structure of random matrices which is associated with generic properties of realistic quantum systems.

# 1 Introduction

It is well known that the main approach in describing statistical properties of spectra of complex quantum systems is related to random matrices. They have been introduced by Wigner about fifty years ago (see, in [1]) in application to nuclear physics and since that time the random matrix theory (RMT) has found many other applications. Recently, the interest to the RMT has been significantly increased due to quantum chaos problems. It was found that maximal statistical properties both spectra and eigenstates of quantum dynamical (without any random parameters) systems which exhibit strong chaos in the classical limit, are well described by an ensemble of random matrices. The situation where random matrices can be used, is characterized by two conditions. First, classical motion has to be completely chaotic; second, the influence of quantum localization effects should be sufficiently small. In this extreme limit all eigenstates are random gaussian functions and spectrum statistics is universal (the so-called Wigner-Dyson statistics). The well known examples of such systems are quantum billiards [2], kicked rotator model [3, 4] and the kicked tops [5] in the semiclassical region.

In spite of very success of the RMT in the description of statistical properties of complex systems like many-electron atoms and heavy nuclei, its application is restricted due to the absence of free parameters which can be associated with specific properties of given quantum systems. It is clear that full random matrices can be used to describe only local properties of spectra when any dependence on the energy is neglected. Also, such quantum effects as localization of eigenstates are washed out in the conventional RMT due to the assumption of infinite range of the interaction between all unperturbed states.

At the same time, there are many examples when quantum chaos is not well developed, however, statistical approach is valid. The similar situation occurs in the physics of disordered solids when describing the effects of strong localization. To take into account localization effects, in [6] a new type of random matrices has been proposed which are characterized by band structure. This suggestion was made on the base of the detailed study of the kicked rotator model in the region when classical chaos is strong, however, quantum chaos may be weak due to the influence of localization.

In what follows, we discuss the properties of some ensembles of structured random matrices and their applications in physics.

# 2 Standard Band Random Matrices

The ensemble of standard real symmetrical Band Random Matrices (BRM) is specified by the band width $2b$ and by the size of matrices $N$. Inside the band, all matrix elements $H_{nm}$ are distributed according to the Gaussian distribution with the zero mean and the fixed variance,

$$< H_{nm} > = 0; < H_{nm}^2 > = \frac{(N+1)(1+\delta_{nm})}{b(2N-b+1)}. \qquad (1)$$

With this normalization, all eigenvalues (in the limit of large $N \rightarrow \infty$) are located in the interval $(-2, 2)$, independently of the value $b$. The delta-symbol ($\delta_{nm} = 0$ for $n \neq m$ and $\delta_{nm} = 1$ for $n = m$ ) is introduced to assure the normalization of Gaussian orthogonal ensemble (GOE) in the limit of full matrices ($b = N$). Analytical results have been obtained in the limit $N \gg b \gg 1$ [7, 8, 9, 10] and for general case of a sufficiently fast decay of matrix elements for large $|n - m| \gg b$.

An increasing interest to the matrices of this kind is related to their broad applications. In particular, much developments have been done in connection to the quantum chaos (see, e.g. [11, 4]). It was shown that generic structure of Hamiltonian matrices in some basis is of band-like, which reflects a finite range of a perturbation. If the corresponding classical motion is chaotic, one can assume the pseudo-random character of matrix elements inside the band (see details in [4]).

Another important application is related to 1D and quasi-1D solid state models with random potentials. In the 1D geometry, the size of the band

$b \gg 1$ has the meaning of the range for random hoppings in the tight-binding representation. In the quasi-1D case, parameter $b$ plays a role of the number of transfer channels; this analogy has recently been proved rigorously by making use of non-linear sigma model [7].

For infinite BRM ($N \to \infty$) with a fixed band $b \gg 1$ all eigenstates are found to be exponentially localized [4, 6, 12, 7]. This means that amplitudes $\varphi_n$ of eigenstates decay exponentially fast away from the center of localizations, $\varphi_n \sim \exp(-|n - n_0|/l_\infty)$. The quantity $l_\infty$ known as the localization length, is of great importance in describing the global structure of eigenstates. Based on scaling arguments, in [4, 6] it was predicted that the localization length $l_\infty$ is proportional to $b^2$. An exact expression was obtained in [7]:

$$l_\infty = c\rho^2(E)\, b^2, \tag{2}$$

where $\rho(E)$ is the density of states given by the semicircle law,

$$\rho(E) = \frac{2}{\pi R^2}\sqrt{R^2 - E^2}\,; \quad R^2 = 8bv^2; \quad v^2 \equiv < H_{nm}^2 > . \tag{3}$$

and $c$ is the numerical factor which depends on the form of the decay of matrix elements when moving from the main diagonal. One can see that in the limit $b = N$ the case of full random matrices of GOE is recovered.

To study the structure of eigenstates in the finite basis, one can use the definition of localization length based on the so-called information entropy of eigenstates, $\mathcal{H} = -\sum_{n=1}^{N} w_n \ln w_n; \quad w_n = |\varphi_n^2|$ where $\varphi_n$ is the component of an eigenstate in a given basis. Since the maximal value of $\mathcal{H}$ is bounded by the size $N$, in [13, 4] it was suggested to define the localization length $l_N^{(1)}$ as

$$l_N^{(1)} = N \exp(< \mathcal{H}_N > - \mathcal{H}_{GOE}). \tag{4}$$

In this definition, the localization length is normalized in such a way that for the most extended eigenstates which appear in the limit $b = N$, the value of $l_N^{(1)}$ is equal to the size of the basis, $l_N^{(1)} = N$. Therefore, the factor $\mathcal{H}_{GOE} \approx \ln(N/2.07)$ has the meaning of the entropy $\mathcal{H}$ for random eigenstates of GOE (see details in [4]). In Eq. (4) the ensemble average $< .... >$ is introduced over a number of eigenstates of the same energy. The entropy localization length $l_N^{(1)}$ defined by Eq. (4) varies from 1 to $N$ depending on the degree of localization and hence becomes a convenient measure to characterize quantitatively properties of eigenstates.

The essential point of studies [4, 6, 14, 8, 9] is that the properties both of eigenstates and spectra depend on the ratio of localization length $l_\infty$ to the size of the basis $N$. Since for any energy the localization length $l_\infty$ is proportional to $b^2$, the scaling parameter $\lambda = b^2/N$ in the limit $N \gg b \gg 1$ was suggested. In particular, in [6] it was numerically found that the localization length $l_N^{(1)}$ scales with $\lambda$ as follows:

$$\beta^{(1)} \equiv \frac{l_N^{(1)}}{N} = \frac{c_1\lambda}{1+c_1\lambda}, \tag{5}$$

where the coefficient $c_1$ is estimated as $c_1 \approx 1.4$. Detailed analytical study [10] has shown that the above scaling expression (5) is very close to the correct one. For the comparison with the results of [10] it is convenient to use another variables, $y = \beta^{(1)}/(1-\beta^{(1)})$ and $a = \ln c_1$, in which the scaling (5) reads as $\ln y = \ln \lambda + a$. In [9] the coefficient $a$ was shown to have different values in two limits, $a(-\infty) \approx 1.58$ for strongly localized states ($\lambda \ll 1$) and $a(+\infty) \approx 1.79$ for extended states ($\lambda \gg 1$). Recent numerical data [15] for the scaling are presented in Fig. 1. Although the data confirm the existence of the two slopes $a(-\infty), a(+\infty)$, however, the shift $\Delta a_{\rm exp} = a(+\infty) - a(-\infty) \approx 0.12$ is about as twice as less than the theoretical prediction. This discrepancy for the shift $\Delta a$ is still not understood, it may be related to not sufficiently large/small values of $\beta^{(1)}$ (one should stress that the band size has to be large enough, $b \gg 1$, in order to expect the correct limits).

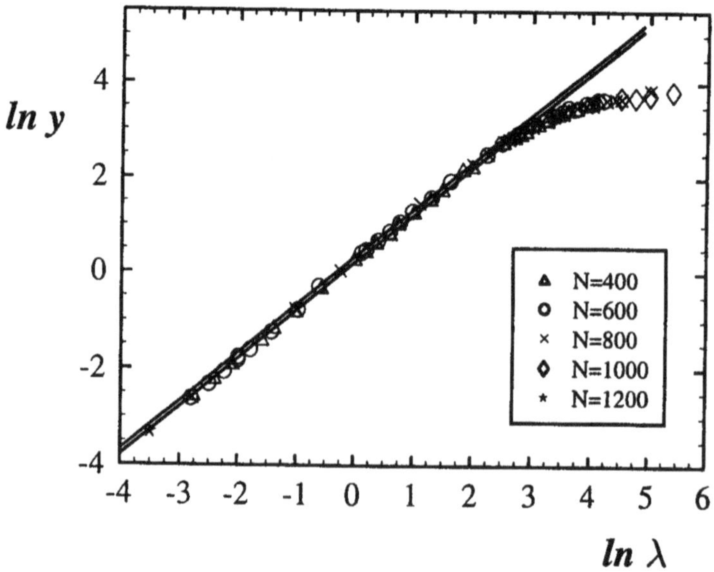


Fig. 1. Scaling properties of the localization length (4–5) on the parameter $\lambda = b^2/N$.

In [9] it was found that the scaling of the form (5) is exact for another definition of the localization length, which is based on the so-called inverse participation ratio $P_2$,

$$P_2 = \sum_{n=1}^{N} (w_n)^2 \; ; \quad w_n = |\varphi_n|^2. \tag{6}$$

By normalizing the inverse participation ratio to its value for full random matrices, one can introduce the rescaled localization length $l_N^{(2)}$,

$$\beta_2 \equiv \frac{l_N^{(2)}}{N} = (\frac{< P_2 >}{P_2(GOE)})^{-1} = \frac{3/N}{< P_2 >} \tag{7}$$

According to the theoretical results [9], the scaling $\ln y = \ln \lambda + a$ for the quantity $\beta_2$ is exact.

The above scaling expression (5) for the rescaled localization length appears to be universal since it was also found for other models like kicked rotator model [4, 16], 1D Anderson and Lloyd models [17], dimer model [18]. This scaling is expected to be explained by renormalization theory, but this question is still open for rigorous analysis.

It is well known that conductance of finite samples generically show anomalous sample-to-sample fluctuations. In particular, in metallic regime the so-called universal conductance fluctuations have been discovered, the remarkable property of which is that the ratio of the variance of conductance to its mean value does not depend on the size of the sample. In this regime the fluctuations of the conductance are Gaussian. In other limit case of strong localization, the fluctuations are of very specific form, namely, the log-norm distribution for the conductance holds if the localization length $l_\infty$ is much less than the size of samples. Since these properties of conductance are directly related to the structure of eigenstates, it is important to know fluctuation properties of eigenstates.

First investigation of the distribution of the entropy localization length $l_N^{(1)}$ of eigenvectors for standard BRM was described in [19]. In papers [10, 20] many other characteristics of eigenstates are studied for standard BRM both in localized $\lambda \ll N$ and delocalized $\lambda \gg N$ regimes. Scaling properties of the inverse participation ratio $P_2$ can be seen from the relation between the mean square deviation $\delta = (< P_2^2 > - < P_2 >^2)^{1/2}$ and the mean value $< P_2 >$. As is known in the theory of disordered solids such a relation is of very important characteristics of the type of fluctuations. Dependence of $\delta$ on $< P_2 >$ was recently studied both analytically [20] and numerically [15]. In Fig. 2 the numerical data [15] are shown for the wide range of the localization

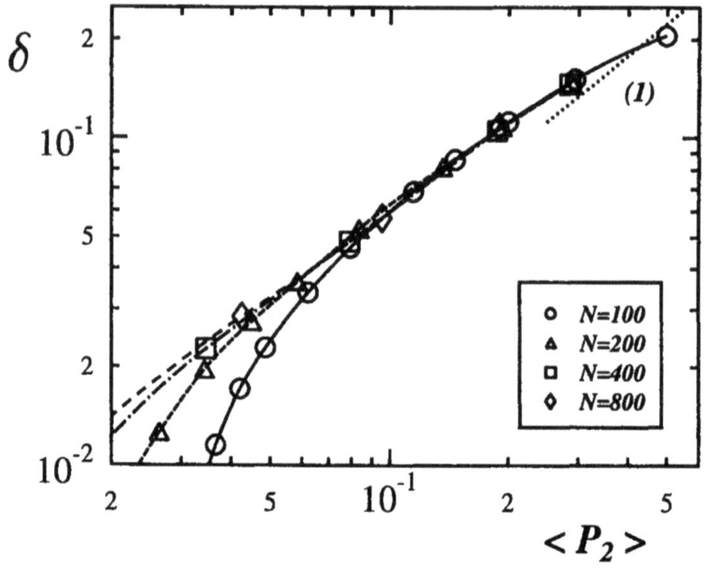


Fig. 2. Mean deviation $\delta$ versus mean value of the inverse participation ratio $\langle P_2 \rangle$.

parameter $\lambda$. In [10, 20] it was shown that in the limit of strong localization the ratio $\delta / < P_2 >$ for the non-symmetrical hermitian BRM approaches a constant value $1/\sqrt{5}$. In the Fig. 2 this dependence is represented by the dotted straight line (1). It is seen that the scaling holds only in the strongly localized regime, for $< P_2 > \geq 0.1$. The constant ratio of $\delta / < P_2 >$ for localized states reflects extremely large fluctuations in the width of eigenstates (see also, [10, 20]); such strong fluctuations have been also found in the kicked rotator model for the eigenstates, localized in the momentum space due to dynamical localization [4]. The detailed numerical study [15] has shown that only for extremely large values of $\lambda^{-1} \geq 20$ the ratio $\delta / < P_2 >$ approaches the limit value.

Another important question arises for the spectrum statistics. The commonly used quantity is the distribution $P(s)$ of spacings between neighboring levels in the spectrum. The spacings $s$ are assumed to be normalized to the mean local spacing $< s >$ which is defined by $\rho(E)$. It is well established that the level spacing distribution $P(s)$ can be generically served as a measure of randomness (chaos) of a quantum system. In the limit case of integrable (regular) systems, the distribution $P(s)$ is very close to Poissonian law, $P(s) \sim \exp(-s)$ and in the other limit of fully chaotic systems the well known Wigner-Dyson surmise

$$P(s) = A\, s^{\beta} \exp(-Bs^2) \tag{8}$$

appears which is predicted by the RMT with $\beta = 1, 2, 4$ depending on symmetry of the system.

For band random matrices of infinite size, $N \to \infty$, the level spacing distribution is close to the Poisson, due to exponential localization of all eigenstates in the infinite basis (see discussion is [4]). For finite matrices, the distribution $P(s)$ turns out to be intermediate between the Poisson and the Wigner Dyson (8). To describe the intermediate statistics for $P(s)$, in [14] some phenomenological expression has been proposed,

$$P(s) = As^{\beta}(1 + B\beta s)^{f(\beta)} \exp\left[-\frac{\pi^2}{16}\beta s^2 - \frac{\pi}{2}(1 - \frac{\beta}{2})s\right] \tag{9}$$

where $A$ and $B$ are normalizing parameters and $f(\beta) = \beta^{-1}2^{\beta}(1 - \frac{\beta}{2}) - 0.16874$ is a fitting function (see details in [14]) and $\beta$ is the only free parameter. Expression (9) is constructed in such a form that it gives very good correspondence to the exact results of the RMT for all three known ensembles of full random matrices characterized by the repulsion parameter $\beta = 1, 2$ and 4, also, for $\beta = 0$ the form of Eq.(9) is the Poisson. This distribution even closer to the exact one than the Wigner-Dyson surmise (8) itself.

In numerical experiments [14] for different values of $b$ and $N$ of standard band random matrices it was found that the repulsion parameter $\beta$ is a smooth function of the scaling parameter $\lambda = b^2/N$. For strongly localized states, $\lambda \ll 1$, the distribution $P(s)$ is close to the Poissonian law and in the other limit of the extended states, it is close to the Wigner-Dyson expression. Similar situation is known for the kicked rotator model where this intermediate statistics has been found in the case of strong classical chaos and strong localization (see details in [4]).

# 3 Wigner Band Random Matrices

Another important type of random matrices which have band structure, has been introduced by Wigner in his early papers on random matrices [21]. It is interesting that random matrices of this type have been well forgotten for a long time, in favor of full random matrices which enjoy the rotational invariance and, therefore, are much easier for the rigorous analysis. Only recently, due to attempts of quantum chaos theory in describing statistical properties of conservative systems with few degrees of freedom, the interest to this ensemble of band random matrices has been raised significantly.

This type of the BRM is characterized by the main diagonal with increasing matrix elements, in addition to the band structure described in the previous section. As a result, the ensemble of such matrices is defined by real symmetrical Hamiltonians of the type

$$\tilde{H}_{nm} = \alpha n \delta_{nm} + H_{nm} \tag{10}$$

where $H_{nm}$ is defined by the expression (1) for standard band random matrices and $\alpha$ is the spacing between energies $E_n^{(0)} = \alpha n$ of the unperturbed ($H_{nm} = 0$) Hamiltonian. In fact, generic properties of spectra and eigenfunctions in the parameter range under consideration are similar to that for generic case when the spacings are uncorrelated with the mean value $\alpha$. Therefore, the unperturbed part can be associated with an integrable Hamiltonian which has the poissonian distribution for the energy level spacings $P(s)$. As a result, the properties of the model are specified by three parameters, the mean level spacing $\alpha$, the variance of the matrix elements of the perturbation $v^2 = < H_{nm}^2 >$ and the band width $2b$. In [21] the simplest case of matrix elements with random signs $v_{mn} = \pm v$ was chosen.

The above model can be treated as the generic Hamiltonian of an integrable conservative system with complex perturbation which couples finite large number $b \gg 1$ of unperturbed states. Typical examples are Hamiltonians of heavy atoms and nuclei in shell model representation [22, 23]. Recently, very detailed study of compound states in the chaotic $Ce$ atom has been performed in [22] where it was revealed that this class of the BRM can be used rather efficiently to describe such a physical system. In solid state physics, the band random matrices (10) with a leading diagonal can be associated with quasi-1D disordered Hamiltonian in the presence of an external electric field determined by the value of $\alpha$ [20].

Quite close ideas were developed recently in the paper [24] where the model for an integrable Hamiltonian $H_0$ perturbed by some generic perturbation $\epsilon V$ was considered. It was argued that in the basis of semiclassical eigenstates of the unperturbed part $H_0$, the perturbation matrix $\langle m \mid \epsilon V \mid n \rangle$ has a form of sparse band. It is also known that in shell models of atoms and nuclei, many off-diagonal matrix elements are zeros due to existence of additional quantum numbers. For this reason in [25] the above Wigner model of band random matrices has been generalized by introducing additional parameter $S$ of sparsity which defines the relative number of non-zero elements in each row of perturbation matrix $H_{nm}$. Below, we discuss the statistical properties of Wigner Band

Random Matrices (WBRM) of infinite size and without sparsity (the case of finite matrices has been studied in [26, 27].

The most important quantity which characterizes many essential properties of the model is the so-called *local spectral density of states* (in the literature, LDOS). This function, also known in nuclear physics as *strength function*, is defined by

$$\rho(E,n) = \sum_{\nu} \mid \varphi_{\nu}(n) \mid^2 \delta(E - E_{\nu}) \tag{11}$$

where $\varphi_{\nu}(n)$ is $n^{th}$ component of the eigenvector $\varphi_{\nu}$ corresponding to the energy level $E_{\nu}$, $n = 1, 2, ...N$. After averaging over disorder, the LDOS depends (in the limit $N \to \infty$) only on the combination $\epsilon_n \equiv E - \alpha n$, therefore, $\overline{\rho(E,n)} = \rho_L(\epsilon_n)$. As a consequence, the global density of states for the large matrix is independent of the energy, $\overline{\rho(E)} = \frac{1}{N}\sum_{n=1}^{N} \rho_L(\epsilon_n) \to \frac{1}{\alpha N}$ provided the parameter $\alpha$ is small enough for the sum being replaced by the integral [28].

One of important questions is the form of LDOS in dependence on the parameters of the model. It is clear that in one limit of very small $\alpha$, the form of $\rho(E,n)$ is close to the semicircle (3) with the radius $R = v\sqrt{8b}$. On the other hand, in some region of parameters the form of LDOS was predicted [21] to have the Lorentzian shape

$$\rho_{BW}(E,n) = \frac{\Gamma/2\pi}{(E - \alpha n)^2 + \Gamma^2/4} \tag{12}$$

with an effective width $\Gamma$. Such a form is well known in nuclear physics as the Breit-Wigner law. It is considered to be well in agreement with experimental data for nuclei [29] and complex atoms [22]. This fact makes the WBRM to be much more attractive as a model for physical systems as compared with the full random matrices.

The detailed analytical and numerical consideration of the LDOS for general case of sparse WBRM has been recently performed in [25]. Below, for simplicity we discuss the results when sparsity is absent. It was found that the transition from semicircle to Breit-Wigner form is governed by the Wigner parameter $q$ which is defined as

$$q = \frac{v^2}{b\alpha^2} \tag{13}$$

The region of the semicircle corresponds to the limit case $q \gg 1$, in other limit $q \ll 1$ the Breit-Wigner form (12) can be shown to appear, with the width $\Gamma \approx 2\pi v^2/\alpha$. One should stress however, that in the latter case the value of $q$ has to be large enough, $q \gg b^{-1}$, otherwise the number of unperturbed states

$\Gamma/\alpha$ covered by perturbation, is much less than 1 and standard perturbation theory is applicable. Therefore, the Breit-Wigner regime is expected in the region $b^{-1} \ll q \ll 1$.

In Fig. 3 the transition from the semicircle to the Breit-Wigner law is shown for large matrices of size $N = 1000$ with the fixed band size $b = 10$ [25]. For such parameters, the influence of finite $N$ is small due to relation $b^2 \ll N$; on the other hand, there is a small range of $q$ -values where the Breit-Wigner shape for the LDOS is expected. The smooth curves in Fig. 3 correspond to the solution of the analytical expression derived in [25] and histograms are obtained by the direct computations of the LDOS (11) from eigenvectors of a number of matrices with the same parameters. From this figure, the whole transition is seen in detail. Unexpected peculiarity of these data is that in the critical point (for $q \approx 0.4$) there is a local minimum in the center of the LDOS. For smaller values of $q \ll b^{-1}$ the form of the LDOS is different from the Breit-Wigner law and seems to be close to the gaussian form, in agreement with other numerical data [28].

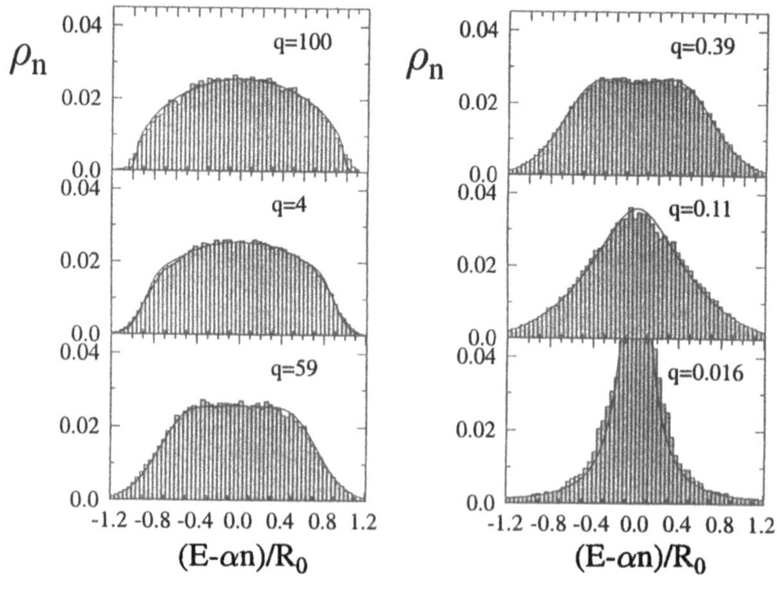


Fig. 3. Dependence of spectral local density $\rho_n$ on the Wigner parameter q.

Specific question arises about the form of tails of the LDOS in the Breit-Wigner regime. In the paper [22] on *Ce* atom, an exponential Wigner's expression [21] for the tails has been corrected (see discussion in [30]) However, the difference can hardly be seen, due to numerical experiments performed in [25].

Another important question concerns the structure of eigenstates in the two above regimes. Recent analysis [27] of old numerical data [12] together with new numerical experiments [30] have revealed unexpected scaling properties of eigenstates and spectra. The goal of the approach suggested in [27, 30] is based on the concept of maximal localization length $l_{\max}$ of eigenstates in the energy space. This localization length is defined by the ratio of the width of LDOS to the mean spacing $\alpha$ between unperturbed energy levels, $l_{\max} \approx \Delta E/\alpha$. In the semicircle region ($q \gg 1$) the energy width is defined by $\Delta E \approx 2R \approx 4v\sqrt{2b}$ and in the Breit-Wigner regime it is given by $\Delta E \approx \Gamma \approx 2\pi v^2/\alpha$. According to [27, 30], statistical properties both eigenstates and spectrum depends on the ergodicity parameter $\beta_l$ which is the ratio of the actual localization length $l$ in the unperturbed basis to its maximal value $l_{\max}$ ,

$$\beta_l = \frac{l}{l_{\max}} = \beta_l(\Lambda) \leq 1; \quad \Lambda = \frac{l_\infty}{l_{\max}} \tag{14}$$

where the new scaling parameter $\Lambda$ is introduced through the localization length $l_\infty \sim b^2$ of the standard band random matrices (for $\alpha = 0$). In the regime when the LDOS is semicircle, the scaling parameter has the dependence $\Lambda \propto \alpha b^{3/2}$; in the Breit-Wigner regime one gets $\Lambda \propto (\alpha\beta)^2$ [25]. The actual localization length $l$ can be defined by the common definition $l_N^{(2)}$ (see Eq.(7)) through to inverse participation ratio $P_2$ (6).

The analysis of numerical data has confirmed the above suggestion. In particular, it was found that in the Breit-Wigner regime, $q \ll 1$, the ergodicity parameter is maximal, $\beta_l \simeq 1$. This means that all eigenstates are ergodic in an available energy space which results in the limiting Wigner-Dyson statistics for the level spacing distribution $P(s)$. The important fact is that this result is independent of the value of the other parameter $q$, therefore, the Wigner-Dyson law for $P(s)$ is directly related to the Breit-Wigner form of the LDOS. This remarkable fact is far from being trivial since in the basis of unperturbed basis of the Hamiltonian (10) all eigenstates are localized. The explanation is due to the fact that eigenstates with close energies are always strongly overlapped in the unperturbed basis.

Unexpected situation occurs in other limit case of the semicircle form of the LDOS. As it was discovered in [30], in this case the ergodicity parameter can have any value depending on the model parameters, which means the possibility of strong localization of eigenstates in the energy space. As a result, for small values $\beta_l \ll 1$ the Poisson law for the level spacing distribution $P(s)$ occurs, and for $\beta_l \simeq 1$ ,again, the Wigner-Dyson law for $P(s)$ arises.

Another remarkable result discovered in [27, 30] is that the ergodicity parameter turns out to be very close (within the 2% accuracy) to the repulsion parameter $\beta$ in the phenomenological expression (9), according to the numerical data. This intriguing fact still has no any explanation; one should stress that repulsion parameter is basis independent quantity, however, the ergodicity parameter $\beta_l$ is obtained through the localization length which is basis dependent.

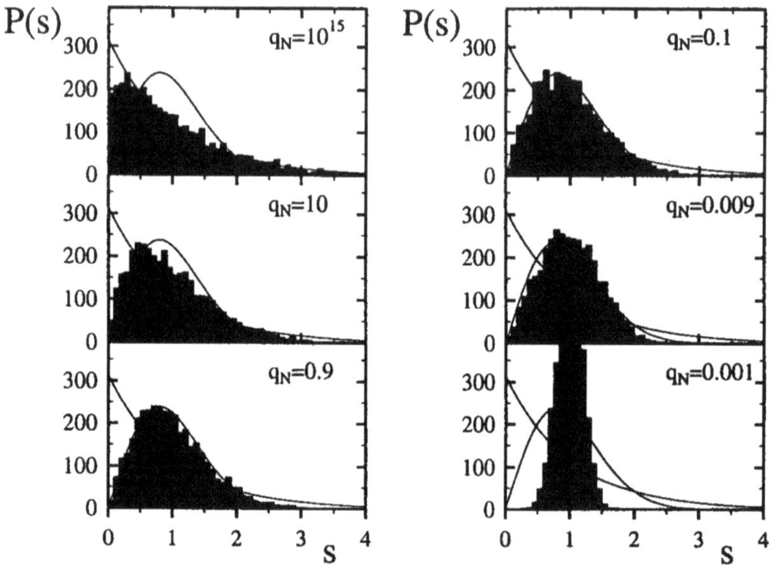


Fig. 4. Dependence of level spacing distribution P(S) on the Wigner parameter q.

To show how the level spacing distribution changes for the fixed band size $b$ with variation of the ratio between perturbation $V$ and unperturbed spacing $\alpha$, in Fig. 4 the data for $P(s)$ are presented as a dependence on the Wigner parameter $q_N$ for finite matrices of size $N = 1000$ and fixed values $b = 10$, $V = 0.1$ (the parameter $\alpha$ changes only). It is clearly seen the transition from the form closed to Poisson, to that of Wigner-Dyson (smooth curves in the figure) and further, to the non-generic one for very small values of $q_N$ corresponding to the perturbative region. In the extreme limit $q_N \to 0$, the dependence $P(s)$ approaches delta-function. It is interesting, that for the case of uncorrelated unperturbed spacings, the poissonian dependence of $P(s)$ arises in the two limits of very large and very small values of $q_N$, with the Wigner-Dyson one in between.

# 4 Two-Body Random Matrices

As was mentioned above, in the paper [22] the direct computation of Hamiltonian matrix for $Ce$ atom has shown that the WBRM ensemble discussed in the precious section has much in common. In particular, the LDOS and spectrum statistics are quite similar in the region of parameters corresponding to the $Ce$ atom. However, many other properties can not be explained in the frame of the WBRM. The most serious problem is that in random matrix approach the two-body character of interaction is not reflected. Also, such important physical quantities as density matrix, temperature and occupation numbers are not possible to introduce in the conventional band random matrices discussed above. For this reason, in recent paper [31] an "intermediate" approach has been suggested which is based on a simple mathematical model with random interaction and takes into account the most important features of many-body systems such as single-particle orbitals, two-body interaction and the Pauli principle.

In this model of Random Two-Body Interaction (RTBI) it is assumed that the system consists of $n_0$ Fermi-particles which can occupy $m_0$ orbitals with no double occupancy (extension to Bose-systems is straightforward). As a generic example, the energies of the orbitals has been taken in the form

$$\epsilon_k = d(k + \frac{1}{k}), \qquad k = 1, 2, ..., m_0 \tag{15}$$

where the second term is introduced in order to avoid the degeneracy of many-particle states. Matrix elements of two-body interaction are chosen as random variables distributed according to the Gaussian law with the zero mean and the variance $V_0^2$. Therefore, the random matrix model has three independent parameters: $n_0, m_0$ and $V_0/d$ .

The basis in which matrices are defined, corresponds to the products of single-particle orbitals ordered by increasing many-body unperturbed energies. Therefore, without interaction the matrix is diagonal with increasing elements. The size $N$ of RTBI-matrix is given by $n_0, m_0$. $S = N = C_{n_0}^{m_0} = m_0!/(n_0!(m_0 - n_0))!)$ and is exponentially large for large $n_0$ and $m_0$. The matrix elements of the two-body interaction are zero between basis states which differ by more than two orbitals. Therefore, the Hamiltonian matrix has the sparsity (ratio of total number $N^2$ to the number of non-zero elements ) defined by $n_0, m_0$. At large $n_0, m_0$ the sparsity is exponentially large, $S \sim \exp(n_0)$. Actually, the number $N_2 = m_0^2(m_0 - 1)^2/4$ of independent variables (number of pair interactions) is even smaller than the number of non-zero matrix elements.

There are three types of interactions between many-body states: diagonal interaction which contains $k_{int} = n_0(n_0 - 1)/2$ two-body terms, interaction between states which differ by one orbital with $k_{int} = n_0 - 1$ terms and interaction between states which differ by two orbitals with $k_{int} = 1$. Finally, distant many-body states which differ by more than two orbitals have zero matrix elements. Thus, the Hamiltonian matrix $H_{ij}$ has large and increasing diagonal, plus sparse band-like structure with a decrease of off-diagonal elements as a function of the distance $|i-j|$ from the diagonal.

In [22] a particular case has been considered which is close to $Ce$ atom: $m_0 = 11, n_0 = 4$ (therefore, $N = 330$), $d = 1, V_0 = 0.12$. According to [32], the density of states (DOS) tends to Gaussian function when $m_0 \gg n_0 \gg 1$. For the case of $m_0 = 11, n_0 = 4$ the DOS turns out to be quite close to the Gaussian form with the mean $< E >= 25.1$ and $\sigma = 5.7$ [31].

A quantity of relevant interest is the LDOS defined by (11). According to the numerical results [31], in the center of the spectrum ($n \approx N/2$) the spreading width of the LDOS, which has the Breit-Wigner form (12) is equal to $\Gamma_c \approx 1.0$ One should note that close to the edges $n = 1, N$, the symmetric form of the distribution $\rho(E,n)$ is strongly distorted. However, the width of the distribution itself changes slightly. What is more important, the function $\rho(E,n)$ in the tails decays much faster than in the Breit-Wigner law. This is the consequence of the band-like structure of the Hamiltonian matrix. The value of $\Gamma_c$, as well as the strong decay of tails, are found to be in agreement with *Ce*-atom calculations [22]. The ground state turns out to be quite simple containing $l_N^{(1)} \approx 1$ of basis components. In fact, this can be explained by the low density of states near the ground state. The localization length $l_N^{(1)}$ is maximal in the center $< E >$ of the energy band and, according to the numerical data, can be well described by the Gaussian function: $l_N^{(1)} = A \exp(-(E- < E >)^2/(2\sigma^2))$ with $A \approx 135$ and $\sigma \approx 5.45$ .Thus, in the center of the spectrum the number of principle components is about 100 which is very close to the results of direct calculations for Ce-atom [22].

As is known, quantum statistical laws are derived for systems with infinite number of degrees of freedom, or for systems in a thermostat. From this point of view, it is of importance to study how statistical laws appear in systems with finite number of particles. In the RTBI model one can introduce a reasonably accurate statistical description based on such macroscopic characteristics as the temperature $T$, the chemical potential $\mu$ and average occupations numbers $n_i$ for the orbitals. In the mean field approximation these quantities can be obtained from the following set of equations:

$$\sum_{i=1}^{m_0} n_i = n_0 \tag{16}$$

$$\sum_{i=1}^{m_0} \epsilon_i n_i + < \sum_{i \geq k}^{m_0} V_{ik} n_i n_k > = E \tag{17}$$

$$n(\epsilon_i) = (1 + \exp(\tilde{\epsilon}_i - \mu)/T)^{-1} \tag{18}$$

$$\tilde{\epsilon}_i = \epsilon_i + < \sum_{k=1}^{m_0} V_{ik} n_k > \tag{19}$$

In the RTBI model, the average interaction $< V_{ik} >$ is zero, therefore, one can omit interaction term in (17) and (19). On the other hand, there exist other important effects of the interaction which appear beyond the mean field approximation. Namely, the interaction leads to the spreading width for the basis states and for the single-particle orbitals ($\Gamma_i$). It also results in the shift of average energies, $\tilde{\epsilon}_i = \epsilon_i + \delta\epsilon_i$. According to the numerical data [31], the magnitudes $\delta\epsilon_i$ are smaller than $\Gamma_i$ and vanish in the mean ($\delta\epsilon_i < 0$ for low orbitals, $\delta\epsilon_i > 0$ for high orbitals and $\delta\epsilon_i \approx 0$ near the center). For this reason, one can take into account the effect of the spreading width $\Gamma_i$ only.

One can now introduce the occupation numbers $n_i$ which are obtained by averaging the Fermi-Dirac distribution $n(\epsilon)$ in Eq. (18) over the interval $\Gamma_i$:

$$n_i = \int_{\epsilon_i - \Gamma_i/2}^{\epsilon_i + \Gamma_i/2} n(\epsilon) \frac{d\epsilon}{\Gamma_i} = 1 - \frac{T}{\Gamma_i} \ln \left[ \frac{1 + \exp \frac{(\epsilon_i + \frac{\Gamma_i}{2} - \mu)}{2T}}{1 + \exp \frac{(\epsilon_i - \frac{\Gamma_i}{2} - \mu)}{2T}} \right] \tag{20}$$

This function takes into account the finite spreading width of "quasi-particle" and in the limit $\Gamma_i = 0$ it gives $n_i = n(\epsilon_i)$. The numerical solution [31] of equations (16,17,20) is given in Fig. 5. To study the influence of the spreading width $\Gamma_i$, two theoretical curves are given for $\Gamma_i = 0$ and $\Gamma_i = 3.0$ (dotted and solid lines). The latter value was deliberately taken larger than the spreading width of basic components $\Gamma_c \approx 1.0$ in order to elucidate the effect of $\Gamma_i$. The temperature $T$ and chemical potential $\mu$ are calculated from the above equations (16,17,20) at fixed excitation energy $\delta E = E - E_{min}$. It was found that for small $\delta E$ the values of temperature $T$ are quite different for different values of $\Gamma_i$. Surprisingly, in spite of this, the two "theoretical" curves practically coincide. This means that the temperature imitates the statistical effect of the interaction, the phenomenon far from being trivial. Direct computations of average occupation numbers from exact eigenstates are shown in Fig. 5 with circles (for comparison, the occupation numbers are shown by squares for the case of $V_0 = 0$). Notice that even for small number of degrees of freedom (four particles) there is a reasonable agreement between the theory and the data of numerical experiments.

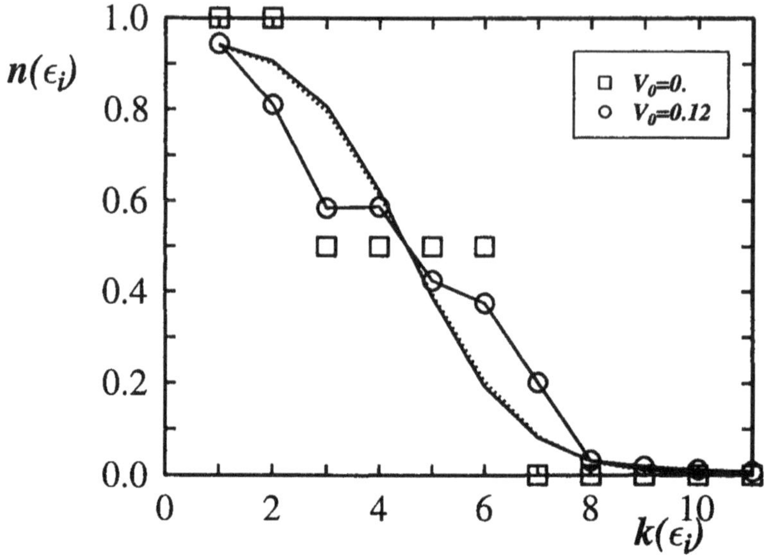


Fig. 5. Distribution of occupation numbers for the excitation energy $\delta E = 1.63$.

We have to stress that equilibrium distribution, or "thermalization", in this few–particle system is due to the interaction which "chaotically" mixes neighboring basis states with different occupation numbers. To demonstrate this, in Fig. 5 the occupations numbers for the same system with no interaction ($V_0 = 0$) are plotted. Their distribution is not smooth and for some values of $\delta E$ even not monotonous.

The main problem in the compound state theory is the calculation of effects of an external perturbation. The TBRI model also allows to study matrix elements of any single-particle operator expressed in terms of elementary density matrix operators $\rho_{ik} = a_i^\dagger a_k$ which transfer the particle from the orbital $k$ to the orbital $i$. In Ref. [22] the following equations for variance of the matrix elements of $\rho_{ik}$ between compound states $< 1|$ and $< 2|$ have been derived:

$$\overline{|\rho_{ik}^{(21)}|^2} = \overline{\langle 2|\hat{n}_i(1-\hat{n}_k)|2\rangle} \sum_r w_1(E_r + \omega_{ik}) w_2(E_r) \tag{21}$$

where $\omega_{ik} = \epsilon_i - \epsilon_k$ and $w(E) =< |\varphi_n^2| >$. The sum is taken over many-body basis components with $w_1, w_2$ being the weights of these components in the states $< 1|$ and $< 2|$. If we know the shape of the strength function $w$, this sum can be replaced by the integral and explicitly evaluated. The result can be expressed as $D_1 \tilde{\delta}(E_2 - E_1 + \omega_{ik})$ where $D_1$ is average level spacing near the state $< 1|$ and $\tilde{\delta}$ is a "spreaded" $\delta-$function ($\int \tilde{\delta} dE = 1$). For example, for the Breit-Wigner shape of $w$, the function $\tilde{\delta}$ has the Breit-Wigner shape too, with the width $\Gamma = \Gamma_1 + \Gamma_2$ where $\Gamma_1, \Gamma_2$ are the widths of $w_1$ and $w_2$. The quantity $\tilde{\delta}$ reflects an energy conservation for quasi-stationary basis states.

The results of the direct numerical study of density matrix in the RTBI model have been compared in [31] with the theoretical predictions (21). To reduce fluctuations, an average over subintervals was performed (each subinterval in $k1, k2$ contains 10 compound states). Additional average was taken over 5 realizations of Hamiltonian matrices (function $\tilde{\delta}$). The typical shape of matrix elements of the density matrix is shown in Fig. 6 [31]. The band structure of the density matrix (21) originates from the approximate energy conservation for the transitions between quasi-stationary basis states. The data have shown that the shape of density matrix, presented in Fig. 2, is quite close to the theoretical estimate (21). However, a more detailed analysis have revealed that near the maximum, theoretical values are approximately 1.3 time smaller than the "experimental" ones. On the other hand, for small values of matrix elements (far from the maximum), theoretical values are above the "experimental" ones. This fact needs further investigation and possible improvement of statistical theory.

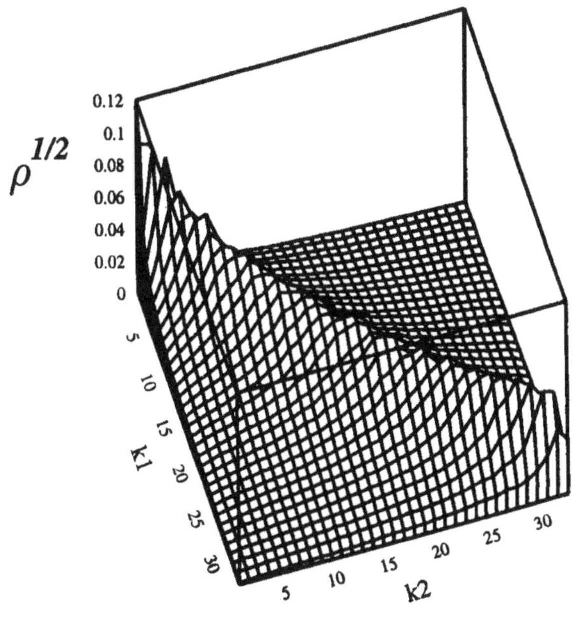


Fig. 6. Numerical data for matrix elements of density matrix operator.

The discussed here the TBRI model seems to be extremely effective and promising for the study many physical problems. As one example, in [31] the problem of the enhancement of weak perturbation in chaotic many-body systems was discussed in the frame of the TBRI. In particular it was shown that even for small number of particles ($n_0$), noticeable enhancement of the weak perturbation is possible.

# 5 Acknowledgements

The author is very thankful to his collaborators on papers reported in this talk. He also would like to thank the organizers of the 19th IUAP International Conference on Statistical Physics for the invitation to contribute the Conference, and for the financial support during his stay in Xiamen. The financial support of Grant No RB7000 from the International Science Foundation is gratefully acknowledged.

# References

[1] *Statistical Theories of Spectra: Fluctuations*, ed. C.E.Porter, Academic Press, (1965).

[2] O.Bohigas and M.-J. Giannoni, *Lect. Notes Phys.* **209** (1984) 1.

[3] B.V.Chirikov, F.M.Izrailev and D.L.Shepelyansky, *Physica* **D33** (1988) 77.

[4] F.M.Izrailev, *Phys. Rep.* **196** (1990) 299.

[5] F.Haake, *Quantum Signatures of Chaos*, (Springer, Berlin, 1991).

[6] G.Casati, L.Molinari and F.M.Izrailev, *Phys. Rev. Lett.* **64** (1990) 1851.

[7] Y.V.Fyodorov and A.D.Mirlin, *Phys. Rev. Lett.* **67** (1991) 2405.

[8] Y.V.Fyodorov and A.D.Mirlin, *Phys. Rev. Lett.* **69** (1992) 1093.

[9] A.D.Mirlin and Y.V.Fyodorov, *J. Phys.* **A26** (1993) L551.

[10] Y.V.Fyodorov and A.D.Mirlin, *Phys. Rev. Lett.* **71** (1993) 412.

[11] T.H.Seligman, J.J.M.Verbaarschot and M.R.Zirnbauer, *Phys. Rev. Lett.* **53** (1985) 215; M.Feingold and A.Peres, *Phys. Rev.* **A34** (1986) 591; M.Feingold, D.M.Leitner and O.Piro, *Phys. Rev.* **A39** (1989) 6507; M.Feingold, D.Leitner and M.Wilkinson, *Phys. Rev. Lett.* **66** (1991) 986.

[12] M.Wilkinson, M.Feingold, and D.M.Leitner, *J.Phys.* **A24** (1991) 175.

[13] F.M.Izrailev, *J. Phys.* **A22** (1989) 865.

[14] G.Casati, F.M.Izrailev and L.Molinari, *J. Phys.* **A24** (1991) 4755.

[15] F.M.Izrailev, L.Molinari and K.Życzkowski, (1995) to be published.

[16] G.Casati, I.Guarneri, F.M.Izrailev and R.Scharf, *Phys. Rev. Lett.* **64** (1990) 5.

[17] G.Casati, S.Fishman, I.Guarneri, F.M.Izrailev and L.Molinari, *J. Phys.* **C4** (1992) 149.

[18] F.M.Izrailev, T.Kottos and G.Tsironis, (1995) to be published.

[19] K.Życzkowski, M.Lewenstein, M.Kuś and F.M.Izrailev, *Phys. Rev.* **A44** (1991) 73.

[20] Y.V.Fyodorov and A.D.Mirlin, *Int. J. Mod. Phys.* **B8** (1994) 3795.

[21] E.Wigner, *Ann. Math.* **62** (1955) 548; *ibid* **65** (1957) 203.

[22] V.V.Flambaum, A.A.Gribakina, G.F.Gribakin and M.G.Kozlov, *Phys. Rev.* **A50** (1994) 267.

[23] V.Zelevinsky, M.Horoi and B.A.Brown, Preprint MSUCL-942, (1994); M.Horoi, B.A.Brown and V.Zelevinsky, Preprint MSUCL-953 (1994).

[24] T.Prozen and M.Robnik, *J. Phys.* **A26** (1994) 1105.

[25] Y.V.Fyodorov, O.Chubikalo, F.M.Izrailev and G.Casati, (1995) to be published.

[26] M.Feingold, A.Gioletta, F.M.Izrailev and L.Molinari, *Phys. Rev. Let.* **70** (1993) 2936.

[27] G.Casati, B.Chirikov, I.Guarneri and F.M.Izrailev, *Phys. Rev.* **E48** (1993) R1613.

[28] M.Feingold and D.Leitner, *J. Phys.* **A26** (1993) 7367.

[29] A.Bohr and B.Mottelson, in *Nuclear Structure*, v.1 (Benjamin) 1968.

[30] G.Casati, B.V.Chirikov, I.Guarneri and F.M.Izrailev, (1995) to be published.

[31] V.V.Flambaum, F.M.Izrailev and G.Casati, (1995) to be published.

[32] T.A.Brody, J.Flores, J.B.French, P.A.Mello A.Pandey and S.S.M.Wong, *Rev. Mod. Phys.* **53** (1981) 385.

# APPLICATION OF CLOSED TIME PATH GREEN'S FUNCTIONS TO QUANTUM TRANSPORT THEORY

ZHAO-BIN SU
*Institute of Theoretical Physics, Academia Sinica*
*P.O. Box 2735, Beijing, 100080, China*
E-mail: zbs@itp.ac.cn

ABSTRACT

The Kubo formula is of fundamental importance for the quantum transport theory. However, its applicability has been discussed for a long time. Applying the Closed Time Path Green's Function approach, we show explicitly that the Kubo formula can be derived from the formal solution of the linearized quantum transport equation of Kadanoff-Baym-Mahan-Hansch. Moreover, using the collective coordinate quantization method, we show the generic equivalence between the force-force correlation description for the resistivity and the current-current correlation description for the conductivity within the framework of adiabatic switching ansatz.

## 1. Introduction

Because of its practical importance, the investigation of electronic transport in solids received much attention. A general approach for the linear transport process, the Kubo current-current correlation formula [1] is widely regarded as an exact formalism for quantum many-body systems, in the sense that no assumption is made about the rate at which the charge carrier elections are scattered. The other general method, e.g, the Boltzmann equation, however, has often been treated on the classical or semiclassical levels [2]. Mahan and Hansch [3] proposed a set of quantum Boltzmann equations (QBE) for the linear transport of many-electron systems. Their QBE is derived form the Dyson equation of nonequilibrium Green's function [4–9]. In the Born approximation, the Mahan-Hansch QBE produces the same results for dc conductivity of electron-phonon-impurity systems as those derived from the Kubo formula in the ladder diagram approximation. We notice that, traditionally, the equivalence between the Kubo formula and Boltzmann equation has also been examined only on the level of weak scattering limit. Then it is natural to raise the question as whether these two theories are exactly equivalent or not. Although the equivalence between them could be physically expected, it is not actually obvious because of their quite different forms. The Kubo formula relates the conductivity to an equilibrium current-current correlation function which is a kind of two-particle (four-point) Green's function, while the QBE provides a kinetic equation for the nonequilibrium quantum distribution function which is a kind of one-particle (two-point) Green's function. In this paper, we review the exact equivalence between these two theories [10].

However, it seems that neither of these two methods is a powerful tool for practical calculations in the case of a large scattering rate. If the Kubo conductivity $\sigma$ is expanded in the strength of the scattering potential $\lambda$ , the lowest order term diverges as $\sigma \sim \lambda^{-2}$. Therefore, even in the limit of weak scattering the correct evaluation of the linear response conductivity requires the summation of an infinite series of divergent terms [11]. Because of this difficulty, a number of attempts have been made to find explicit formula for the resistivity $\rho = \sigma^{-1}$ [12,13], rather than the conductivity $\sigma$, with the hope of being easier to evaluate for small $\lambda$. Because $\rho$ is of order $\lambda^2$ and thus, it has been expected to be easier to evaluate the first nonvanishing term in the expansion of $\rho$ in powers of $\lambda$. However, it is found that the summation of an infinite series of suitable terms is also required even for small $\lambda$ [14]. With the similar motivation, many approaches have been proposed following the force-balance (FB) theory for both the linear and nonlinear resistivity $\rho$ [15]. The resulted impurity resistivity often has the expression as $\rho \sim < 1/\tau >$ [13,14]. Here, $\tau$ is the scattering relaxation time. If we compare this result with the conductivity obtained from the Kubo formula or the Boltzmann equation, $\sigma \sim < \tau >$, we find $< 1/\tau > \neq 1/ < \tau >$. This discrepancy has caused a number of contentions and has been resolved in certain sense by re-summing the divergent terms [14].

In this paper we review a new formulation [16] for the FB theory along another line to maintain the equivalence between the FB theory and the dynamic-response (DR) theory. In this approach, we apply the collective coordinate quantization method [16] within the frame work of the adiabatic switching Ansatz. In the DR theory one calculates the transport current as a nonlinear response of an applied electric field, while in the FB theory one expresses the electric field as a function of steady current. The relation between them is precisely a general conjugate relation. We show transparently that, by using a functional Legendre transformation, the FB theory and DR theory are mutually equivalent and complementary. In particular, in the weak-field limit the resistivity obtained from the FB theory is exactly the inverse of the Kubo conductivity formula.

## 2. Equivalence between Kubo Formula and Quantum Boltzmann Equation

In the presence of a uniform constant electric field $\mathbf{E}$, the Hamiltonian can be written as

$$H_T = H_A + H_{int} = \frac{(\mathbf{p} - e\mathbf{A})^2}{2m} + H_{int}, \tag{1}$$

where $H_{int}$ is the interaction between electrons and phonons or impurities, etc., and $\mathbf{A}(t) = -\mathbf{E}t$ is the vector potential. This system is typically nonequilibrium for which the nonequilibrium Green's function (nGF) provides a convenient description [4–9]. We start our discussion from the following Dyson equation:

$$\int dz[\Gamma_{0A}(x_1,z)_{\alpha\mu} - \Sigma_A(x_1,z)_{\alpha\mu}]\eta_\mu G_A(z,x_2)_{\mu\beta} = \delta_{\alpha\beta}(x_1 - x_2), \tag{2}$$

$$\Gamma_{0A}(x_1, x_2)_{\alpha\beta} = [i\partial_{t_1} - H_A(t_1)]\delta_{\alpha\beta}(x_1 - x_2). \quad (3)$$

In Eq. (2) the nGF $G_{\alpha\beta} = -i < T_p\psi_\alpha\psi_\beta^+ >$, $\alpha(\beta) = +$ or $-$, indicating that they are defined on the closed-time path, which runs from $-\infty$ to $+\infty$ (positive "+" branch) and comes back from $+\infty$ to $-\infty$ (negative "-" branch), with $T_p$ being the generalized time-ordering operator. It is easy to verify $G_r = 1/2\xi_\alpha\eta_\beta G_{\alpha\beta}$, $G_a = 1/2\eta_\alpha\xi_\beta G_{\alpha\beta}$, here and in the following context the repeated time branch indices are summed over "+" and "-" branches with $\xi_+ = \xi_- = 1$ and $\eta_+ = -\eta_- = 1$. Similarly, for the generalized $\delta$ function, $\delta_{++} = -\delta_{--} = \delta$, and $\delta_{+-} = \delta_{-+} = 0$. Because the interaction $H_{int}$ is independent of the vector potential, the self energy $\Sigma_A$ depends on the field through its functional dependence upon the $G_A$, $\Sigma_A = \Sigma[G_A]$. To avoid confusion, we use $G$, $\Sigma$ to denote the equilibrium GF to distinguish them from the nGF.

We now derive the QBE straightforwardly from the Dyson equation. To get rid of the gauge-dependent translational non-invariance, we make use of the following transformation [3]:

$$\begin{aligned} G_A(x_1, x_2) &= \exp[-ie\mathbf{E}(t_1 + t_2) \cdot (\mathbf{x}_1 - \mathbf{x}_2)/2]\tilde{g}(x_1, x_2), \\ \Sigma_A(x_1, x_2) &= \exp[-ie\mathbf{E}(t_1 + t_2) \cdot (\mathbf{x}_1 - \mathbf{x}_2)/2]\tilde{\sigma}(x_1, x_2). \end{aligned} \quad (4)$$

Here $\tilde{g}$ and $\tilde{\sigma}$ are introduced as the reduced Green's function and self-energy.

For linear transport, $\tilde{g}_{\alpha\beta} = G_{\alpha\beta} + \delta g_{\alpha\beta}$. Utilizing the above transformation and the expansion to linear order in $\mathbf{E}$, we obtain the following QBE:

$$\begin{aligned} [i\frac{\partial}{\partial t_1} & - \frac{\hat{p}_1^2}{2m}]\tilde{g}_{\alpha\beta}(x_1, x_2) - \int dz \tilde{\sigma}_{\alpha\mu}(x_1, z)\eta_\mu \tilde{g}_{\mu\beta}(z, x_2) \\ & = \delta_{\alpha\beta}(x_1 - x_2) - \frac{1}{2}e\mathbf{E} \cdot \left[(\mathbf{x_1} - \mathbf{x_2}) + \frac{p_1}{m}(t_1 - t_2)\right] G_{\alpha\beta}(x_1 - x_2) \\ & - \frac{i}{2}e\mathbf{E} \cdot \int dx'[(\mathbf{x_1} - \mathbf{x}')(t' - t_2) - (\mathbf{x}' - \mathbf{x_2})(t_1 - t')] \\ & \times \Sigma_{\alpha\nu}(x_1 - x')\eta_\nu G_{\nu\beta}(x' - x_2), \end{aligned} \quad (5)$$

where the self-energy $\tilde{\sigma} = \Sigma + \delta\sigma$, and $\delta\sigma = \Sigma[G + \delta g] - \Sigma[G] + \delta_A\sigma$, in which

$$\begin{aligned} \delta_A\sigma_{\alpha\beta}(x_1 - x_2) &= \int dy_1 dy_2 \frac{\delta\Sigma_{\alpha\beta}(x_1, x_2)}{\delta G_{\mu\nu}(y_1, y_2)}\eta_\mu\eta_\nu \frac{i}{2}e\mathbf{E}(t_{y_1} + t_{y_2} - t_{x_1} - t_{x_2}) \\ & \cdot(\mathbf{y}_1 - \mathbf{y}_2)G_{\mu\nu}(y_1, y_2). \end{aligned} \quad (6)$$

This is a new term describing the $\mathbf{E}$ field modification of the scattering effect, which has been ignored by Mahan and Hansch [3]. This correction term vanishes only in the Born approximation where the self-energy $\Sigma$ is linear function of $G$, and then it agrees exactly with the Mahan-Hansch equation for linear quantum transport. In order to show the exact equivalence between the QBE and the Kubo formula, it is necessary to include this correlation term.

For the linear response term $\delta g$, we can derive the following integral equation from Eq. (6):

$$\delta g_{\alpha\beta}(x_1,x_2)-\int dy_1dy_2\Pi_{\alpha\beta\mu\nu}(x_1,x_2,y_1,y_2)\eta_\mu\eta_\nu\delta g_{\mu\nu}(y_1,y_2)$$
$$=e\mathbf{E}[\int dzG_{\alpha\mu}(x_1,z)\eta_\mu G_{\mu\beta}(z,x_2)\frac{\hat{p}_z}{2m}(2t_z-t_{x_1}-t_{x_2})$$
$$-\int dy_1dy_2\Pi_{\alpha\beta\mu\nu}(x_1,x_2,y_1,y_2)(t_{x_1}+t_{x_2}-t_{_1}-t_{y_2})$$
$$\times\frac{i}{2}(\mathbf{y}_1-\mathbf{y}_2)\eta_\mu\eta_\nu G_{\mu\nu}(y_1,y_2)] \tag{7}$$

$$\Pi_{\alpha\beta\mu\nu}(x_1,x_2,z_1,z_2) = \int dy_1dy_2G_{\alpha\delta}(x_1,y_1)\eta_\delta\frac{\delta\Sigma_{\delta\gamma}(y_1,y_2)}{\delta G_{\mu\nu}(z_1,z_2)}$$
$$\eta_\gamma G_{\gamma\beta}(y_2,x_2), \tag{8}$$

where the momentum operator $\hat{\mathbf{p}}=(\vec{\nabla}-\bar{\nabla})/2i$. The formal solution of this equation can be expressed in terms of the two-particle Green's function as follows:

$$\delta g_{\alpha\beta}(x_1,x_2)=e\mathbf{E}\cdot\int dzG^{(2)}_{\alpha\beta\mu\mu}(x_1,x_2,z,z)\eta_\mu\frac{\hat{p}_z}{2m}(2t_z-t_{x_1}-t_{x_2}), \tag{9}$$

where

$$G^{(2)}_{\alpha\beta\mu\nu}(x_1,x_2,x_3,x_4)=<T_p\psi_\alpha(x_1)\psi^+_\beta(x_2)\psi_\mu(x_3)\psi_\nu(x_4)>. \tag{10}$$

The proof of Eq. (9) being the solution of Eq. (7) can be achieved either by the direct manipulation with the help of an integral equation satisfied by $G^{(2)}$,

$$G^{(2)}_{\alpha\beta\mu\nu}(x_1,x_2,x_3,x_4)-\int dz_1dz_2\Pi_{\alpha\beta\delta\gamma}(x_1,x_2,z_1,z_2)\eta_\delta\eta_\gamma G^{(2)}_{\delta\gamma\mu\nu}(z_1,z_2,x_3,x_4)$$
$$=G_{\alpha\nu}(x_1,x_4)G_{\beta\mu}(x_2,x_3), \tag{11}$$

or by the Feynman diagram expansion method.

According to the transformation in Eq. (4), the electric current $\mathbf{j}=(-ie/m)(\hat{\mathbf{p}}_x-e\mathbf{A})G_{A+-}(x,x)$ can be expressed as $(-ie/m)\hat{\mathbf{p}}_x\delta g_{+-(x,x)}$. ¿From Eq. (9) and after some management concerning the closed time path time ordering definition, the conductivity $\sigma$ ($\mathbf{j}=\sigma\cdot\mathbf{E}$) takes the following form

$$\sigma = -ie^2\int dz(t_z-t_x)\frac{\hat{p}_z}{m}\frac{\hat{p}_x}{m}G^{(2)}_{+-\mu\mu}(x,x,z,z)\eta_\mu$$
$$= \int dz(-i)\Theta(x-z)\left\langle[\psi^+(x)\frac{e\hat{p}_x}{m}\psi(x),\psi^+(z)\frac{e\hat{p}_z}{m}\psi(z)]\right\rangle. \tag{12}$$

Obviously, our result for dc conductivity, Eq. (12) derived from QBE does exactly coincide with the Kubo formula. Therefore, we prove the equivalence between the quantum kinetic equation approach (QBE) and the Kubo formula for linear transport in the static uniform electric field.

## 3. Equivalence between Kubo Formula and Force-Balance Equation

As mentioned in Sec. 1, many attempts based on the idea of the FB approach have been proposed and encountered some troubles when compared with the Kubo theory. Since the impurities are randomly distributed, we consider first a fixed impurity configuration and take the configuration average at a later time. We introduce the following generating functional:

$$Z[E_+, E_-] = \mathrm{tr}\left[U(\infty, -\infty; E_+)\hat{\rho}_0 U(\infty, -\infty; E_-)\right], \tag{13}$$

where

$$U(t, t_0; E) = T\exp\left[-i\int_{t_0}^{t}[H + H'(t_1)]dt_1\right], \tag{14}$$

with $T$ the time ordering operator, $H$ the total Hamiltonian of the system with and $H'$ the interaction of the system with an external electric field $\mathbf{E}$. In Eq. (13), $\hat{\rho}_0$ is the normalized initial density matrix so that $Z[E_+, E_-]|_{E_+=E_-} = 1$. Note that we have also distinguished formally $E_+(t)$ from $E_-(t)$, while the physical expectation values and correlation functions can be produced from the generating functional by setting $E_+(t) = E_-(t) = E(t)$ at the end of calculations.

Now we isolate the center of mass (c.m.) coordinates of the electrons. This can be done most easily in the path integral formalism. Firstly, $U$ and $U^+$ in (14) can be expressed as an unified evolution operator defined on a closed time path. It has a branch index denoted by $\alpha = \pm$. Secondly, we apply the standard collective coordinate quantization method [16] to extract the center of mass coordinate $q(t)$ and the total momentum $P(t)$ of the electrons as a collective coordinate and its conjugate momentum. We then express the electron wave field in terms of the c.m. wave field, $\tilde{\psi}(x, t)$ as

$$\psi(x, t) = \tilde{\psi}(x - q(t), t)\exp\{ip(t)(x - q(t)\},$$

in order to make $q$ and $P$ unconstrained. Insert further the functional identities

$$\int \mathcal{D}P(t)\delta[P(t) - \int dx\psi^+(x)(-i\nabla)\psi(x)] = I,$$

$$\int \mathcal{D}q(t)\delta[Q[\tilde{\psi}^+, \tilde{\psi}]]|\frac{\partial Q[\tilde{\psi}^+, \tilde{\psi}]}{\partial q}|^{-1} = I,$$

into the path integral where $Q[\tilde{\psi}^+, \tilde{\psi}] = \int dx\tilde{\psi}^+(x, t)x\tilde{\psi}(x, t)$ and $P(t) = Np(t)$. After some calculations, including the integrating out of $P(t)$, we obtain the following expression for the partition function:

$$Z[E_+, E_-] = \int \cdots \int \Pi_\alpha Dq_\alpha(t)\mathcal{F}(q_+, q_-)\exp[i\int dt[\frac{1}{2}M\dot{q}_+^2(t) - NeE_+(t)q_+(t) - \frac{1}{2}M\dot{q}_-^2(t) + NeE_-(t)q_-(t)]] \tag{15}$$

where $M = Nm$, $m$ being the electron mass, and $N$ the coral electron number, $N = \int dx\psi^+(x,t)\psi(x,t) = \int dx\tilde{\psi}^+(x,t)\tilde{\psi}(x,t)$, and

$$\mathcal{F}(q_+, q_-) = \int \cdots \int \Pi_\alpha D\tilde{\psi}_\alpha^+ D\tilde{\psi}_\alpha D\phi_\alpha \langle +(t=-\infty)|\hat{\rho}_0| - (t=-\infty)\rangle \times\delta\left[\int dx\tilde{\psi}_\alpha^+(x,t)x\tilde{\psi}_\alpha(x,t)\right]\delta\left[\int dx\tilde{\psi}_\alpha^+(x,t)(-i\nabla)\tilde{\psi}_\alpha(x,t)\right] \times\exp\left(iI[\tilde{\psi}_+^+, \tilde{\psi}_+, \phi_+; q_+; V] - iI[\tilde{\psi}_-^+, \tilde{\psi}_-, \phi_-; q_-; V]\right), \tag{16}$$

where $I[\tilde{\psi}^+, \tilde{\psi}, \phi; q; V]$ is the action functional of the electron-phonon-impurity system, where $\psi(x,t)$ is replaced by $\tilde{\psi}(x,t)$, the phonon field $\phi(x,t)$ by $\phi(x+q(t),t)$, and the impurity potential $V(x)$ by $V(x+q(t))$.

It is important to note that in the expression of $\mathcal{F}[q_+, q_-]$ the time derivative of $q_\pm(t)$ does not appear so that in the corresponding operator expression $q$ and $\dot{q}$ obey the canonical commutation relation. It is known that as long as the quantization of coordinates and the constraints are treated correctly, the infrared divergence associated with the translational mode of the collective coordinate does not appear [17].

In fact, the physically observable quantities are the random-potential average over the ensemble average of the observable. For example, the observable current at time t has the expression $<\dot{q}(t)>= \overline{(d/dt)\mathrm{tr}[q\hat{\rho}(t)]} = \overline{\mathrm{tr}[\hat{\rho}\dot{q}(t)]}$, where $\mathrm{tr}(\hat{\rho})$ is the ensemble average of Heisenberg picture operators for a fixed impurity configuration, the overbar denotes the random-potential average, and $< >$ is for the combination of the two. Since we have a description of two branches of time evolution which connects directly to the observable quantities, we may take the random-potential average directly on the generating functional [7]. Thus we introduce

$$\overline{Z[E_+, E_-]} \equiv \exp(iW[E_+, E_-]) \tag{17}$$

Because of the normalization of density operator we have $W[E_+, E_-]|_{E_+=E_-=0} = 0$. Next we define

$$Neq_\alpha(t) = -\alpha\frac{\delta W[E_+, E_-]}{\delta E_\alpha(t)} \qquad (\alpha = +, -). \tag{18}$$

Using the unitarity and composition property of the evolution operator we can easily prove $< q(t) >\equiv q(t)$ that

$$q(t) = q_+(t)|_{E_+=E_-} = q_-(t)|_{E_+=E_-} \equiv R(t, [E]), \tag{19}$$

where $[E]$ symbolizes its functional dependence on the applied field $E(t)$. Eq. (19) together with Eqs. (17) and (18) provide a description of the c.m. coordinates' motion as a nonlinear response of the applied electric field, which is in fact the general DR theory. By taking successive functional derivatives with respect to $E(t)$ one can easily verify that $R(t,[E])$, as a response functional of $E(t)$, is purely retarded and has the correct causality behavior. As an example let us look at the lowest-order expansion: the linear response. Since our choice of $\rho$ is such that the current expectation value is zero in the zero-field limit, by taking the functional derivative of Eq. (19) with respect to $E(t)$ once, we derive $\dot{q}(t) = \int dt'\Delta_r(t-t')E(t')$ with

$$\begin{aligned}\Delta_r(t-t') &= \frac{d}{dt}\frac{\delta R(t,[E])}{\delta E(t')}|_{E=0} \\ &= -iNe < \Theta(t-t')[\dot{q}(t), q(t')] > . \end{aligned} \tag{20}$$

Consequently, for the case of dc applied field, we derive the Kubo formula of conductivity: $Ne\dot{q} = \sigma E$,

$$\sigma = Ne\Delta_r(\omega=0) = N^2e^2[ImD_r(\omega)/\omega|]_{\omega=0}, \tag{21}$$

where $\Delta_r(\omega)$ is the Fourier transform of $\Delta_r(t)$, while $ImD_r$ is the imaginary part of the Fourier transform of the retarded current-current correlation function $D_r(t) = -i < \Theta(t)[\dot{q}(t),\dot{q}(0)] >$. In the derivation of (21) the real part of current-current correlation function was canceled by the equal-time commutator $[q,\dot{q}] = iI/M$ which appears in the difference of $\omega\Delta_r(\omega)$ and $D_r(\omega)$.

The calculation of the generating functional gives an expression for $q(t)$ as a functional of the applied field $E(t)$, i.e., Eq. (18). Now we develop a calculational method that provides $E(t)$ as a functional of $q(t)$. For this purpose we use the method of effective action [18], which is essentially a Legendre transformation method in field theory. In our case the transformation is given by

$$\Gamma[q_+, q_-] = Ne\int dt[E_+(t)q_+(t) - E_-(t)Q_-(t)] + W[E_+, E_-]. \tag{22}$$

Accordingly

$$NeE(t) = \frac{\delta\Gamma[q_+,q_-]}{\delta q_+(t)}|_{q_+=q_-} = \gamma(t,[q]), \tag{23}$$

where $[q]$ means a gain the functional dependence on $q(t)$.

From the general property of Legendre transformation, we know that Eq. (23) is nothing but the force-balance equation. For the steady-state achieved by the applied dc field, $\gamma(t,[q])|_{\dot{q}=v_D}$ is right the friction force acting on the c.m. coordinate.

This discussion shows that the DR theory and the FB theory are two mutually complemental descriptions of the nonlinear transport of many-electron systems. As far as the adiabatic switching Ansatz is valid, we have shown in the case of spatially homogeneous applied electric field that the two approaches are equivalent and are connected by a Legendre transformation.

As an illustration we show explicitly the equivalence of the two approaches in the weak-field limit. In this case we keep only the linear term of $q(t)$ in the functional of $q(t)$ in Eq. (23):

$$NeE(t) = \int dt \Gamma_r(t-t')q(t'), \tag{24}$$

where $\Gamma_r(t-t') = [\delta\gamma(t,[q])/\delta q(t)]|_{q=0}$. For the steady-state transport we have $q(t) = v_D(t-t_0)$, where $t_0$ is arbitrary. It is interesting to note that Eq. (24) is consistent with time translational invariance only when $\int dt\Gamma_r(t) = \Gamma_r(\omega = 0)$ is valid, and the latter is the consequence of the space translational invariance. Let us call this relation the consistency condition. Using this in the Fourier representation of Eq. (24) we obtain $E = \rho \cdot Nev_D$ with the resistivity expression

$$\rho = (\frac{1}{Ne})^2[\mathrm{Im}\Gamma_r(\omega)/\omega]_{\omega=0}. \tag{25}$$

Since the FB theory is the Legendre transformation of the DR theory, their linear-approximation coefficients, the resistivity, and the conductivity must be related to each other. Using Eqs. (18), (19) and (23) we derive

$$\frac{1}{Ne}\int d\tau \frac{\delta\gamma(t,[q])}{\delta q(\tau)}\frac{\delta R(\tau,[E])}{\delta E(t')} = \delta(t-t'). \tag{26}$$

In this expression we have set $E = 0$. Taking its Fourier transform, and using the consistency condition, we derive straightforwardly

$$\rho \cdot \sigma = 1. \tag{27}$$

That is, the resistivity obtained from the FB theory is the inverse of the conductivity formula of Kubo's linear response theory.

In conclusion, we would like to emphasize that, despite of great interests in quantum transport in recent years, the corresponding theory must be developed quite carefully, i.e., it must be consistent with the widely accepted theories in the linear response regime. In the present discussion, we renew the interest of old problems and show the exact equivalence between the linearized QBE (Mahan-Hansch) and the Kubo formula, the force-force correlation expression for resistivity and current-current correlation expression for conductivity.

## 4. Acknowledgements

The work covered by this review was carried out in pleasant collaboration with

Professors Liao-Yuan Chen and B. Sakita.

## 5. References


1. R. Kubo, *Can. J. Phys.* **34** (1956) 1274; R. Kubo, S. J. Muyake, and N. Hashitsume, in *Solid State Physics*, ed. H. Ehrenreich and D. Turnbull (Academic, New York, 1965), Vol. 17, P.269.
2. J. M. Ziman, *Principles of the Theory of Solids* (Cambridge, Londen, 1960); *Electrons and Phonons* (Clarendon, Oxford, 1967); W. Kohn and J. M. Luttinger, *Phys. Rev.* **108** (1957) 590.
3. G. D. Mahan, *Phys. Rep.* **145** (1987) 251; W. Hansch and G. D. Mahan, *Phys. Rev.* **B28** (1983) 1902.
4. J. Schwinger, *J. Math. Phys.* **2** (1961) 407.
5. L. V. Keldysh, *Zh. Eksp. Teor. Fiz.* **47**, (1964) 1515 [*Sov. Phys. JETP* **20** (1965) 1018].
6. L. P. Kadanoff and G. Baym, *Quantum Statistical Mechanics* (Benjamin, New York, 1962).
7. K. C. Chou, Z. B. Su, B. L. Hao and L. Yu, *Phys. Rep.* **118**, (1985) 1.
8. Z. B. Su, L. Y. Chen, X. T. Yu and K. C. Chou, *Phys. Rev.* **B37** (1988) 9810.
9. J. Rammer and H. Smith, *Rev. Mod. Phys.* **58** (1986) 323.
10. Liao-Yuan Chen and Zhao-Bin Su, *Phys. Rev.* **B40** (1989) 9309.
11. G. V. Chester and A. Thellung, *Proc. Phys. Soc.* **73** (1959) 745.
12. J. S. Rousseau, J. C. Stoddart and N. H. March, *J. Phys.* **C5** (1972) L175.
13. V. Kenkre and M. Dresden, *Phys. Rev.* **A6** (1972) 769; V. Kenkre, *ibid.* **7** (1973) 772.
14. See M. Huberman and G. V. Chester, *Adv. Phys.* **24** (1975) 489, and references therein.
15. e.g., X. L. Lei and C. S. Ting, *Phys. Rev.* **B32** (1985) 1112.
16. Zhao-Bin Su and B. Sakita, *Phys. Rev.* **B40** (1989) 9959.
17. J. L. Gervais and B. Sakita, *Phys. Rev.* **D11** (1975) 2943; see also B. Sakita, *Quantum Theory of Many Variable Systems and Fields* (World Scientific, Singapore, 1985).
18. C. De Dominicis and P. C. Martin, *J. Math. Phys.* **5** (1964) 14; **5** (1964) 31; R. Jackiw, *Phys. Rev.* **D9** (1974) 1686.

# Systems out of equilibrium: some exactly soluble models

B. Derrida

Laboratoire de Physique Statistique, ENS, 24 rue Lhomond, 75005 Paris, France
and Service de Physique Théorique, CE Saclay, F91191 Gif sur Yvette, France.

## ABSTRACT

A series of exact results have been obtained recently for the one dimensional asymmetric exclusion model, a model of particles which hop to their right at random times, on a one dimensional lattice, provided that the site they jump to is empty. The steady state properties can be calculated exactly using a matrix formulation. It consists in writing the steady state weights of a system of $N$ sites as the matrix element of a product of $N$ matrices: in this product, each matrix can take two possible forms $D$ or $E$ depending on whether the corresponding site is occupied or empty.

Several extensions of the model have been solved exactly using this matrix method, allowing the calculation of the diffusion constant of a tagged particle, of the profile of a shock and of the steady state properties of systems with two species of particles.

# 1 Introduction

In equilibrium statistical mechanics, one usually associates an energy function $E(C)$ to every possible configuration $C$ of a system and the weight of that configuration is proportional to $\exp[-E(C)/T]$ where $T$ is the temperature. The problem is then to calculate equilibrium properties by averaging with respect to these weights.

In out of equilibrium systems, there is usually no energy function and the system is only defined by its dynamical rules. The weigths of the configurations (i.e. the probability of appearance of each configuration in the steady state) have to be determined before starting the calculation of averages. Calculating these steady state weights is in general not an easy task. The goal of this lecture is to describe a matrix approach which gives the weights of the configurations in the steady state of a particularly simple one dimensional driven diffusive system [1], the asymmetric exclusion model [2, 3, 4].

The asymmetric exclusion model can be defined as follows: each site of a one dimensional lattice is either occupied by one particle or empty. A configuration of the system can therefore be characterized by binary variables $\{\tau_i\}$ where $\tau_i = 1$ if site $i$ is occupied by a particle and $\tau_i = 0$ if site $i$ is empty. During an infinitesimal time interval $dt$, each particle hops with probability $dt$ to its right if this site is empty (and does not move otherwise).

# 2 The steady state for open boundary conditions

Here we will first consider a finite system of $N$ sites with open boundary conditions [5, 6, 7], where in time $dt$ a particle may enter the lattice at site 1 with probability $\alpha dt$ (if this site is empty) and a particle at site $N$ may leave the lattice with probability $\beta dt$ (if this site is occupied).

The steady state properties of this asymmetric exclusion model with open boundary conditions can be calculated exactly [6, 7, 8] for any $N, \alpha$ and $\beta$. I will describe here the solution developped in collaboration with M. Evans, V. Hakim and V. Pasquier [7] where the weight of each configuration is written as the matrix element of a product of $N$ matrices. This type of approach, initially introduced in connection with the Bethe ansatz [9, 10], has been used to solve several problems in statistical mechanics in particular the problem of directed lattice animals [11].

The main idea is to try to write the weight $f_N(\tau_1 \dots \tau_N)$ of each configuration in the steady state as

$$f_N(\tau_1 \dots \tau_N) = \langle W| \prod_{i=1}^{N} [\tau_i D + (1-\tau_i)E] |V\rangle \ , \tag{1}$$

where $D, E$ are matrices, $\langle W|$, $|V\rangle$ are vectors and $\tau_i$ are the occupation variables. Thus, in the product (1) we use matrix $D$ whenever a site is occupied ($\tau_i = 1$) and $E$ whenever a site is empty ($\tau_i = 0$). Since $D$ and $E$ do not commute in general, the weights $f_N(\tau_1 \dots \tau_N)$ are complicated functions of the configuration $\{\tau_1 \dots \tau_N\}$. As the weights $f_N(\tau_1 \dots \tau_N)$ given by (1) are not necessarily normalized, the probability $p_N(\tau_1, \dots \tau_N)$ of a configuration $\{\tau_1, \dots \tau_N\}$ in the steady state is

$$p_N(\tau_1, \dots \tau_N) = \frac{f_N(\tau_1, \dots \tau_N)}{\sum_{\tau_1=1,0} \cdots \sum_{\tau_N=1,0} f_N(\tau_1, \dots \tau_N)} \ . \tag{2}$$

In [7] it was shown that (1-2) do give the weights in the steady state of the asymmetric exclusion model when the matrices $D$ and $E$ and the vectors $\langle W|$ and $|V\rangle$ satisfy the following algebraic rules

$$DE = D + E \tag{3}$$

$$\langle W|E = \frac{1}{\alpha}\langle W| \tag{4}$$

$$D|V\rangle = \frac{1}{\beta}|V\rangle \tag{5}$$

Proofs that this is so are not complicated and have already been published [7, 12]. Let us just explain it on a simple example. Consider a configuration of the form

$$\underbrace{0\,0\cdots0}_{p}\ \underbrace{1\,1\cdots1}_{N-p} \quad . \tag{6}$$

During each time interval $dt$, the probability of entering this configuration is

$$\frac{\langle W|E^{p-1}DED^{N-p-1}|V\rangle}{\langle W|(D+E)^N|V\rangle}dt \tag{7}$$

by coming from configuration

$$\underbrace{0\,0\cdots0}_{p-1}\ 1\ 0\ \underbrace{1\,1\cdots1}_{N-p-1}\ . \tag{8}$$

The probability of leaving this configuration is given by

$$(\alpha+\beta)\frac{\langle W|E^pD^{N-p}|V\rangle}{\langle W|(D+E)^N|V\rangle}dt \tag{9}$$

since one particle may enter at site 1 or the particle at site $N$ may leave. In the steady state these two expressions should be equal to ensure that the weight of configuration (6) remains unchanged

$$\langle W|E^{p-1}DED^{N-p-1}|V\rangle = (\alpha+\beta)\langle W|E^pD^{N-p}|V\rangle \tag{10}$$

and this equality is easy to check using (3-5). A generalization of this reasoning to arbitrary configurations is given in [12].

Once the weights are given by (1,2), one can calculate all the equal time steady state properties. For example the average occupation $\langle\tau_i\rangle$ of site $i$ is given by

$$\langle\tau_i\rangle = \frac{\sum\limits_{\tau_1=0,1}\cdots\sum\limits_{\tau_N=0,1}\tau_i\, f_N(\tau_1,\ldots\tau_N)}{\sum\limits_{\tau_1=0,1}\cdots\sum\limits_{\tau_N=0,1} f_N(\tau_1,\ldots\tau_N)} \quad , \tag{11}$$

which becomes in terms of the matrices $D$ and $E$ and the vectors $\langle W|$ and $|V\rangle$

$$\langle\tau_i\rangle = \frac{\langle W|(D+E)^{i-1}D(D+E)^{N-i}|V\rangle}{\langle W|(D+E)^N|V\rangle} \quad . \tag{12}$$

We see that it is necessary to calculate matrix elements of the form

$$\langle W|D^{m_1}E^{n_1}D^{m_2}E^{n_2}\cdots|V\rangle. \tag{13}$$

One way to do it [7, 12] is to construct matrices and vectors which satisfy (3-5). However the calculation of these matrix elements can be done directly from the algebraic rules (3-5). Clearly, from (4-5) some matrix elements are very easy to calculate :

$$\langle W|E^p D^q|V\rangle = \frac{1}{\alpha^p \beta^q}\langle W|V\rangle \ . \tag{14}$$

It is also easy to see, using $DE = D + E$, that any more complicated matrix element of the form (13) can be reduced to calculating simple matrix elements of the form (14). In particular, one can obtain that way the expression of the denominator $\langle W|(D+E)^N|V\rangle$ for all $N$

$$\frac{\langle W|(D+E)^N|V\rangle}{\langle W|V\rangle} = \sum_{p=1}^{N} \frac{p\ (2N-1-p)!}{N!\ (N-p)!}\ \frac{\beta^{-p-1}-\alpha^{-p-1}}{\beta^{-1}-\alpha^{-1}} \ . \tag{15}$$

In the large $N$ limit, one can calculate in particular the average occupation $\rho = \langle\tau_i\rangle$ for a site $i$ far from the boundaries and one finds the phase diagram shown in Figure 1 with the three following phases:

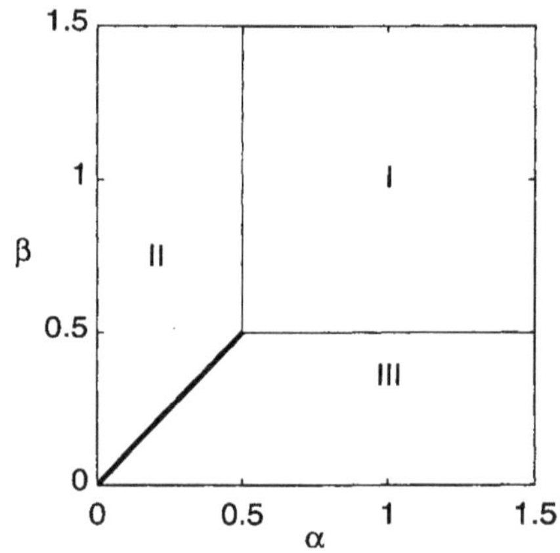


Figure 1: The phase diagram.

**Phase I** For $\alpha > \frac{1}{2}$ and $\beta > \frac{1}{2}$ ,

$$\rho = \frac{1}{2} \ , \tag{16}$$

**Phase II** For $\alpha < \frac{1}{2}$ and $\alpha < \beta$ ,

$$\rho = \alpha \ , \tag{17}$$

**Phase III** For $\beta < \frac{1}{2}$ and $\beta < \alpha$ ,

$$\rho = 1 - \beta \ . \tag{18}$$

Phase I is the maximal current phase, phase II is a low density phase and phase III is a high density phase. The lines $\alpha = 1/2$, $\beta > 1/2$ and $\beta = 1/2$, $\alpha > 1/2$ are second order transition lines (as $\rho$ is continuous) whereas the line $\alpha = \beta < 1/2$ is a first order transition line (discontinuity in $\rho$).

The first order transition can be seen in Figure 2 where the average occupation of the central site $\rho = \langle \tau_{(N+1)/2} \rangle$ is plotted versus $\alpha$ (at $\beta = .2$ fixed) for two system sizes $N = 61$ and $N = 121$. As $N$ increases, one observes that $\rho$ becomes discontinuous at $\alpha = \beta$.

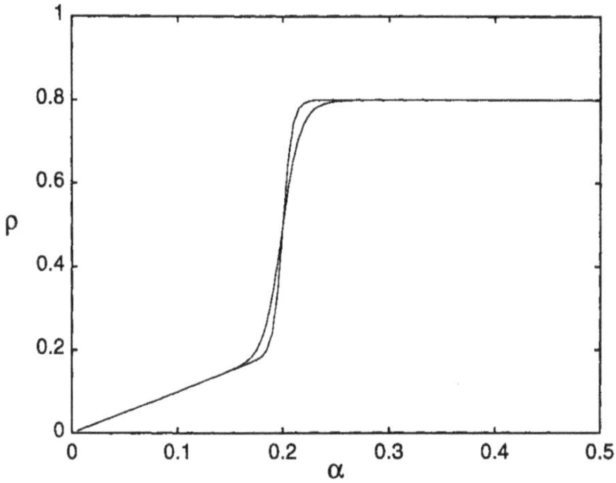


Figure 2: The average occupation $\rho = \langle \tau_{(N+1)/2} \rangle$ of the central site versus $\alpha$ for $N = 61$ and $N = 121$ when $\beta = .2$ .

# 3 The diffusion constant

We have seen that for open boundary conditions, one can calculate all equal time correlations using matrices. Other steady state properties related to unequal time correlation functions can also be calculated using the matrix approach [13, 14]. One such quantity is the diffusion constant of a tagged particle [15, 16]. In order to keep the discussion as simple as possible, we consider the case of $M$ particles on a ring of $N$ sites for which the steady state is known to be particularly simple: all configurations (with exactly $M$ particles) have equal weight [17].

If we tag one of the particles (without in anyway changing its dynamics) and call $Y_t$ the number of hops the particle has made up to time $t$, one expects that in the long time limit:

$$\lim_{t\to\infty} \frac{\langle Y_t \rangle}{t} = v \ ; \quad \lim_{t\to\infty} \frac{\langle Y_t^2 \rangle - \langle Y_t \rangle^2}{t} = \Delta \ . \tag{19}$$

The exact expressions for the velocity $v$ and the diffusion constant $\Delta$ are given by [13]

$$v \ = \ \frac{N-M}{N-1} \ , \tag{20}$$

$$\Delta = \frac{(2N-3)!}{(2M-1)!\,(2N-2M-1)!}\left[\frac{(M-1)!\,(N-M)!}{(N-1)!}\right]^2 . \tag{21}$$

The expression for $v$ is a simple consequence of the fact that in the steady state all configurations with $M$ particles have equal probability [17]. The expression for $\Delta$ on the other hand is harder to obtain and was derived in [13] using a generalization of the matrix approach.

An interesting limit is that of a given density $\rho$ of particles in an infinite system ($M = N\rho$ as $N \to \infty$)

$$\Delta \simeq \frac{\sqrt{\pi}}{2}\left[\frac{(1-\rho)^{3/2}}{\rho^{1/2}}\right]\frac{1}{N^{1/2}} . \tag{22}$$

The fact that $\Delta$ vanishes for $N \to \infty$ indicates that in the infinite system for fixed initial conditions the fluctuations in the distance travelled by a tagged particle are subdiffusive [18].

Another interesting limit is $N \to \infty$ with $M$ fixed

$$\Delta \to \frac{4^{M-1}\,[(M-1)!]^2}{(2M-1)!} \tag{23}$$

showing that $\Delta$ depends on the number $M$ of particles even in the limit of an infinitly large ring.

# 4 Two species of particle on a ring

The matrix method can be extended to obtain the exact steady state of systems with two species of particle [19, 20]. Consider a ring of $N$ sites with two species of particle represented by 1 and 2, and holes represented by 0 and in which the hopping rates are

$$\begin{array}{lcll} 1\ 0 & \to & 0\ 1 & \text{with rate } 1\ , \\ 2\ 0 & \to & 0\ 2 & \text{with rate } \alpha\ , \\ 1\ 2 & \to & 2\ 1 & \text{with rate } \beta\ . \end{array} \tag{24}$$

One can show [19] that the steady state weights may be written as

$$\text{trace}(X_1 X_2 \cdots X_N)\ , \tag{25}$$

where $X_i = D$ if site $i$ is occupied by a 1 particle, $X_i = A$ if it is occupied by a 2 particle and $X_i = E$ if it is empty, provided that the matrices $D, A$ and $E$ satisfy the following algebra

$$DE = D + E \quad ; \quad \beta DA = A \quad ; \quad \alpha AE = A\ . \tag{26}$$

The second two of these equations are satisfied when $A$ is given by

$$A = |V\rangle\langle W| \tag{27}$$

and

$$D|V\rangle = \frac{1}{\beta}|V\rangle \quad ; \quad \langle W|E = \frac{1}{\alpha}\langle W|\ . \tag{28}$$

So one can use the same matrices $D, E$ as those of section 2 for the case of open boundary conditions, and construct matrix $A$ from the vectors $\langle W|$ and $|V\rangle$.

A case of the two species problem of particular interest and which has been used to study shocks is that of first and second class particles [21, 22, 23]. This corresponds to $\alpha = \beta = 1$ so that all hopping rates are 1 and both first and second class particles hop forward when they have a hole to their right, but when a first class particle has a second class particle to its right the two particles interchange positions. Thus a second class particle tends to move backwards in an environment of a high density of first class particles and tends to move forwards in an environment of a low density of first class particles so that it gets trapped by a shock. The matrix method was used in [19] to calculate exactly the profile of a shock as seen from a second class particle.

To illustrate a situation with two species, let us consider on a ring of $N$ sites a single particle 2 and $M$ particles 1. The parameters $\alpha$ and $\beta$ which appear in (24) are chosen such that

$$\beta < 1 - \alpha \, .$$

This implies in particular that $\alpha < 1$ and $\beta < 1$, so that the particle 2 is slower than particles 1 and as $\beta < 1$, it plays the role of a moving obstacle. In other words, particles 1 can be thought as cars and the particle 2 as a single slow truck.

Using the algebraic rules (26), it is possible to calculate $Y(N, M)$ defined by

$$Y(N,M) = \frac{1}{\langle W|V\rangle} \sum_{\tau_1=0,1} \cdots \sum_{\tau_{N-1}=0,1} \delta_{\tau_1+\cdots\tau_{N-1},M} \, \langle W| \prod_{i=1}^{N-1} [\tau_i D + (1-\tau_i)E] \, |V\rangle \tag{29}$$

and show that for $1 \leq M \leq N-2$

$$Y(N,M) = \sum_{p=0}^{N-M-1} \sum_{q=0}^{M} \frac{1}{\alpha^p \beta^q} \, \frac{(N-2-p)! \, (N-2-q)! \, [p(M-q) + q(N-1-M)]}{M! \, (N-1-M)! \, (M-q)! \, (N-1-M-p)!} \tag{30}$$

whereas $Y(N,0) = \alpha^{-(N-1)}$ and $Y(N,N-1) = \beta^{-(N-1)}$. Then the average velocity $v_2$ of particle 2 (the truck) is given by

$$v_2 = \frac{Y(N-1,M) - Y(N-1,M-1)}{Y(N,M)} \tag{31}$$

whereas the average velocity $v_1$ of particles 1 (cars) is given by

$$v_1 = v_2 + \frac{N}{M} \, \frac{Y(N-1,M-1)}{Y(N,M)} \, . \tag{32}$$

These velocities are shown in figure 3 as a function of the density $\rho = M/N$ of particles 1 for two system sizes ($N = 100$ and $N = 200$). For large $N$, we see that as $\rho$ varies, $v_2$ presents a plateau for $1 - \alpha < \rho < \beta$ where $v_2 = \alpha - \beta$. Along this plateau, the ring consists of two macroscopic regions: a region of high density $\rho_+ = 1 - \alpha$ following the truck and a region of low density $\beta$ ahead of the truck. These two regions are of length $L_+ = N(\rho - \beta)/(1 - \alpha - \beta)$ and $L_- = N(1 - \alpha - \rho)/(1 - \alpha - \beta)$ so that the total density is

$$\rho = \frac{L_+\rho_+ \; + \; L_-\rho_-}{N} .$$

Apart from (32), the calculation of the average velocity $v_1$ of the particles 1 can also be obtained in this coexistence region, by

$$v_1 = \frac{L_+\rho_+v_+ \; + L_-\rho_-v_-}{M}$$

with $v_\pm = 1 - \rho_\pm$ i.e.

$$v_1 = \alpha - \beta + \frac{(1-\alpha)\beta}{\rho}$$

For a very large system, three phases can be observed as the density of particles 1 $\rho = M/N$ varies, where the average velocities $v_1$ of the particles 1 and $v_2$ of the particle 2 have the following expressions (which can be obtained from the asymptotic behavior of (30)):

For $\rho < \beta$, $\quad v_1 = 1 - \rho \quad ; \quad v_2 = \alpha - \rho$

For $\beta < \rho < 1 - \alpha$, $\quad v_1 = \alpha - \beta + (1-\alpha)\beta/\rho \quad ; \quad v_2 = \alpha - \beta$

For $1 - \alpha < \rho$, $\quad v_1 = 1 - \rho \quad ; \quad v_2 = 1 - \beta - \rho$

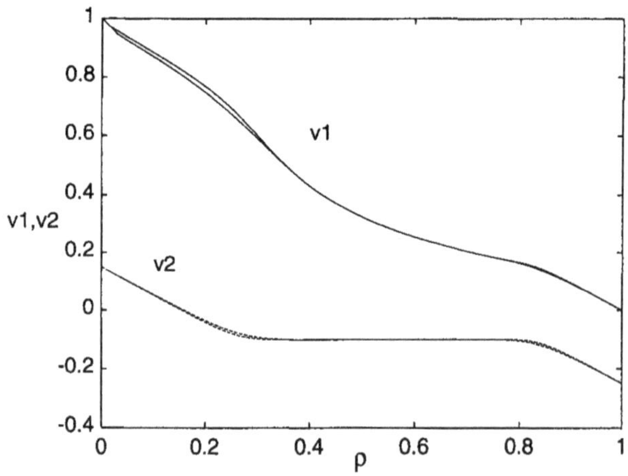


Figure 3: The velocities $v_1$ and $v_2$ of the particles 1 and 2 versus the density $\rho$ of first class particles for two rings of 100 and 200 sites when $\alpha = .15$ and $\beta = .25$. Finite size effects are small enough to make the curves for 100 and 200 hard to distinguish.

It is interesting to notice that in the range $\beta < \rho < 1 - \alpha$, the truck (particle 2) has a constant velocity, whereas in that range, cars (particles 1) have an average velocity slower than $1 - \rho$ (the velocity in absence of the truck). In this whole range there is coexistence of two macroscopic regions, one of low density $\beta$ and one of high density $1 - \alpha$, very similar to the coexistence between the liquid and the gas for a fluid.

# 5 Conclusion

The asymmetric exclusion model, beyond its own interest as a problem in probability theory [2, 3, 4, 20, 21, 22] has connections with several physical problems: superionic conductors [24, 25], traffic flow models [26, 27], polymer dynamics [28, 29, 30], interfacial growth and directed polymers in a random medium [31, 32, 33, 34, 35].

Several extensions of the results presented above have been published recently: they concern partial asymmetry [7, 36, 37, 38], the diffusion constant in systems with open boundary conditions [14], transient properties [39, 40], steady state in an infinite system with two species of particles [12, 19]. However there remains a number of simple generalizations of the asymmetric exclusion model such as the case of a fixed blockage [41], two species of particles with open boundaries [42], more than two species of particles [43], calculation of general unequal time correlations [44], more general reaction diffusion models [45], for which exact solutions are still lacking. One possibility to better understand when the matrix approach discussed here should work would be to investigate its connections to the Bethe ansatz [46, 47, 43].

**Acknowledgements**
This talk is based on works done in collaboration with E. Domany, M.R. Evans, V. Hakim, S. Janowsky, J.L. Lebowitz, K. Mallick, D. Mukamel, V. Pasquier, E.R. Speer. I would like to thank them as well as V. Rittenberg, G. Schütz, H. Spohn and R. Stinchcombe for very useful discussions.

# References

[1] B. Schmittmann and R.K.P. Zia, *Statistical Mechanics of Driven Diffusive Systems*, in *Phase Transitions and Critical Phenomena* vol. 17, C. Domb and J.L. Lebowitz, eds. (Academic, New York) (1995)

[2] F. Spitzer, *Adv. Math.* **5**, 246 (1970).

[3] T.M. Liggett, *Interacting Particle Systems* (Springer-Verlag, New York, 1985).

[4] H. Spohn, *Large Scale Dynamics of Interacting Particles* (Springer-Verlag, New York, 1991).

[5] J. Krug, *Phys. Rev. Lett.* **67**, 1882 (1991).

[6] B. Derrida, E. Domany and D. Mukamel, *J. Stat. Phys.* **69**, 667 (1992).

[7] B. Derrida, M.R. Evans, V. Hakim and V. Pasquier, *J. Phys.* A **26**, 1493 (1993).

[8] G. Schütz and E. Domany, *J. Stat. Phys.* **72**, 277 (1993).

[9] L.D. Faddeev, *Sov. Sci. Rev.* C**1**, 107 (1980).

[10] R.J. Baxter, *Exactly Solved Models in Statistical Mechanics* (Academic, New York, 1982).

[11] V. Hakim and J.P. Nadal, *J. Phys.* A**16**, L213 (1983).

[12] B. Derrida, M.R. Evans, *The asymmetric exclusion model: exact results through a matrix approach*, in *Nonequilibrium Statistical Mechanics in One Dimension* V. Privman, ed. (Cambridge University Press) in print

[13] B. Derrida, M.R. Evans and D. Mukamel, *J. Phys.* A **26**, 4911 (1993).

[14] B. Derrida, M.R. Evans and K. Mallick, *J. Stat. Phys.* **79**, 833 (1995).

[15] A. DeMasi A. and P.A. Ferrari, *J. Stat. Phys.* **38**, 603 (1985)

[16] R. Kutner and H. van Beijeren H, *J. Stat. Phys.* **39**, 317 (1985)

[17] P. Meakin, P. Ramanlal, L.M. Sander and R.C. Ball, *Phys. Rev.* A **34**, 5091 (1986).

[18] S.N. Majumdar and M. Barma *Phys. Rev.* B **44**, 5306 (1991).

[19] B. Derrida, S.A. Janowsky, J.L. Lebowitz and E.R. Speer, *Europhys. Lett.* **22**, 651 (1993); *J. Stat. Phys.* **73**, 813 (1993).

[20] E.R. Speer, *The Two Species Totally Asymmetric Exclusion Process*, in *Micro, Meso and Macroscopic Approaches in Physics*, M. Fannes, C. Maes and A. Verbeure eds. (Plenum, 1994).

[21] E.D. Andjel, M. Bramson and T.M. Liggett, *Prob. Theory Rel. Fields* **78**, 231 (1988).

[22] P.A. Ferrari, C. Kipnis and E. Saada, *Ann. Prob.* **19**, 226 (1991).

[23] C. Boldrighini, G. Cosimi, S. Frigio and M.G. Nuñes, *J. Stat. Phys.* **55**, 611 (1989).

[24] P.M. Richards, *Phys. Rev.* B **16**, 1393 (1977).

[25] W. Dieterich, P. Fulde and I.Peshel, *Adv. Phys.* **29**, 527 (1980).

[26] A. Schadschneider and M. Schreckenberg, *J. Phys.* A **26**, L679 (1993).

[27] T. Nagatani, *J. Phys.* A **26**, L781 (1993); *J. Phys.* A **26**, 6625 (1993).

[28] M. Rubinstein, *Phys. Rev. Lett.* **59**, 1946 (1987).

[29] T.A.J. Duke, *Phys. Rev. Lett.* **62**, 2877 (1989).

[30] J.M.J. van Leeuwen and A.Kooiman, *Physica* A **184**, 79 (1992).

[31] M. Kardar, G. Parisi and Y.-C. Zhang, *Phys. Rev. Lett.* **56**, 889 (1986).

[32] J. Kertész and D.E. Wolf, *Phys. Rev. Lett.* **62**, 2571 (1989).

[33] T. Halpin-Healy and Y.-C. Zhang, *Phys. Rep.* **254**, 215 (1994).

[34] J. Krug, P. Meakin and T. Halpin-Healy, *Phys. Rev.* A **45**, 638 (1992).

[35] J. Krug and L.-H. Tang, *Phys. Rev.* E **50**, 104 (1994).

[36] S. Sandow, *Phys. Rev.* **50**, 2660 (1994).

[37] S. Sandow and G.M. Schütz, *Europhys. Lett.* **26**, 7 (1994).

[38] F.H.L. Essler and V. Rittenberg, preprint 95

[39] R.B. Stinchcombe and G.M. Schütz, *Europhys. Lett.* **29**, 663 (1995);

[40] R.B. Stinchcombe and G.M. Schütz, preprint 95

[41] S.A. Janowsky S A and J.L. Lebowitz, *Phys. Rev.* A **45**, 618 (1992)

[42] M.R. Evans, D.P. Foster, C. Godrèche and D. Mukamel, *Phys. Rev. Lett.* **74**, 208 (1995)

[43] K. Mallick et al, work in progress

[44] G. Schütz, *Phys. Rev.* E **47**, 4265 (1993).

[45] F.C. Alcaraz, M. Droz, M. Henkel and V. Rittenberg, *Ann. Phys. (N.Y.)* **230**, 250 (1994).

[46] D. Dhar, *Phase Transitions* **9**, 51 (1987).

[47] L.H. Gwa and H. Spohn, *Phys. Rev.* A **46**, 844 (1992).

# Dynamics of Slip-and-Stick Fluctuations in a Random Growing Interface

C. K. Chan
*Institute of Physics, Academia Sinica, Taipei Taiwan*

Experiments and MC simulations are carried out to study the dynamics of the slip-and-stick motion produced by a 2D air-water interface moving in a random medium. We find that the slip-and-stick motions of the interface can be characterized by exponential probability distributions which are very different from those of the slip-and-stick models of earthquake. Futhermore, the dynamics of the slip-and-stick motion of the interface is found to produce a $1/f^2$ power spectrum.

## 1 Introduction

It is well known that the slidings of solids on solid surfaces are characterized by slip-and-stick motions. Two solid surfaces will stick together when the driving force is not large enough to overcome the static friction. However, when the driving force is large enough to overcome the static friction, slipping motions occur once the two surfaces start to slide because dynamic frictions are usually smaller. One could imagine that interesting dynamics will result when individual slip-and-stick systems are coupled together to form a larger system. In 1967, Knopoff et al [1] considered such a system which consists of solid blocks connected by springs driven to slide on a rough surface to mimic earthquakes. In this simple model, the slip-and-stick motions of the individual blocks are coupled to form stick (slow build up) and slip (sudden release) fluctuations in the energies stored in the springs of the system which give statistics similar to real earthquakes.

However, it is interesting to note that Knopoff et al 's model is also a model of growing or driven interfaces in a random medium. Figure 1a shows the proposed experiment of Ref.1 in which point masses are connected by springs driven to slide on a rough surface. Here the rough surface is moving with a constant velocity $V$. Figure 1a is similar to the situation of an interface driven to move through a random medium when one thinks of the springs and the point masses as the interface and the rough surface as the random medium. The problem of a driven interface moving through a random medium has received a lot of attentions after the work of Kardar, Parisi and Zhang (KPZ) [2] who have proposed the famous KPZ equation as a model for interfacial growth. Most of the studies have been mainly on the scaling properties of the roughness of the interface [3,4,5] and the depinning transition [6,7,8] which occurs when the interface is driven to move in the presence of quenched noises. The

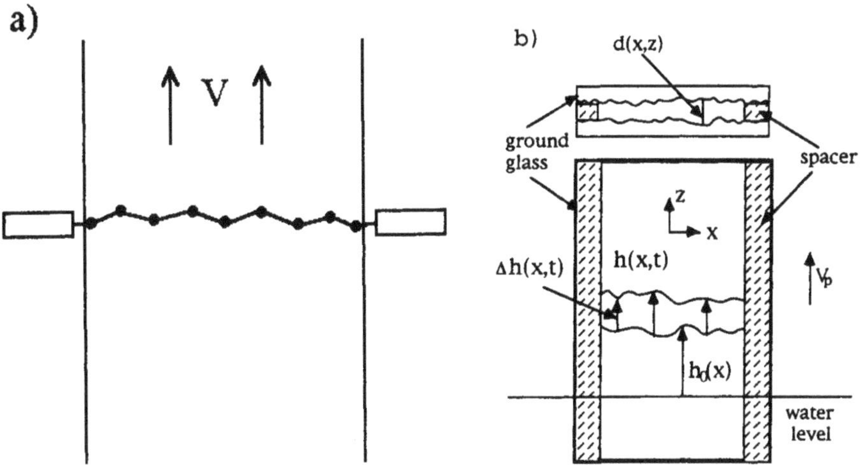


Figure 1: a) proposed experiment of Knopoff and b) experimental setup and the sample cell.

scaling properties of the driven interface is characterized by the roughness exponent $\chi$ which describes how the width of the interface scales with the sample size $L$ as : $w(L) \sim L^{\chi}$ where $w$ is the width of the interface defined as the rms of the fluctuation of the height $h(x,t)$ of the advancing interface, i.e. $w^2 = \langle h(x,t) - \langle h(x,t) \rangle^2 \rangle$. The depinning transition is characterized by an exponent $\theta$ which describes how the velocity of the depinned interface scales with the driving force. However, very little attention has been paid to the slip-and-stick dynamics of the driven interface.

In this short article, we report the results of an experiment and a MC simulation to study the slip-and-stick fluctuations of an air-water interface driven to move through a random medium. We find that the slip-and-stick motions of the interface can be characterized by exponential probability distributions which are very different from those of the slip-and-stick models of earthquake. Futhermore, the dynamics of the slip-and-stick motion of the interface is found to produce a $1/f^2$ power spectrum. The nature of the depinning transition and the effects of different pulling velocities can be found elsewhere [9].

## 2 Experiment and Simulation Model

Our experimental setup is shown in Figure 1b. The sample consists of two pieces of ground glasses with dimensions $12cm \times 5cm$ and two spacers which are mounted on a Plexiglass holder to form a narrow channel of dimension $12cm \times 5cm \times d_o$ where $d_o$= 400 $\mu m$ is the thickness of the spacers. Because of the roughness of the glass surfaces, the distance $d(x, z)$ between the two glass surfaces is not constant but a function of position $(x, z)$ as shown in Figure 1b. The average variations of the rough surfaces of the glasses is 25 $\mu m$ and therefore the average gap width between the two glass surfaces is $d_o + 50\mu m$. No special care has been exercised to clean the surfaces of the glasses. They are cleaned by laboratory glass cleaning fluid, acetone and alcohol.

When the sample is partially immersed under water, there is a position dependent capillary rise $z = h_o(x)$ because of the position dependent gap width $d(x, z)$. Presumably $h_o(x)$ is determined by the balance of gravitational pressure with the capillary pressure and the minimization of the surface and gravitational energies. Therefore $h_o(x)$ might be different in different experiments. But we find that we can consistently get $\langle h_o(x) \rangle = 1.6$ cm when all the glass surfaces have been prewetted first. Now, the situations would be very similar to that shown in Figure 1a when the sample is driven to move upward with a velocity $V$. The interface $h(x, t)$ will stick to the glass at places where the capillary forces can support a pressure difference of $\rho g h(x, t)$ where $\rho$ and $g$ are the density of water and gravitational acceleration respectively. However, the interface will not be pinned to the glass forever as the $\rho g h(x, t)$ at the pinned points will increase linearly with time and finally overcome the capillary force at that point and slipping motions will occur. These slip-and-stick motions of the interface are recorded by a $512 \times 480$ pixel$^2$ video system. As the motions of the interface is quite slow, we are able to measure the positions of the interface every 1/30 sec and detail analysis of dynamics is possible.

To simulate the above experimental situations, we have considered a one-dimensional discrete lattice growth model. Let $h_i$ be the discrete height of the interface at discrete lattice points $i$. The following rules and periodic boundary conditions are used to update the heights in these discrete points:

1) $h_i = 0$ and assign random capillary forces, $0 < \gamma_i < 1$ to all $i$.
2) $h_i = h_i + \delta h$
3) if $\alpha_i + \beta_i - \gamma_i > 0$ then $h_i = h_i - \delta h'$ and new $\gamma_i$ are assigned
   where $\alpha_i = (2h_i - h_{i-1} - h_{i+1})$ and $\beta_i = h_i$
4) adjust $h_i = h_i - \delta h'$ to satisfy the solid-on-solid condition for all i
5) goto setp 2

Step 1 means we start with a flat interface. Step 2 simulates the constant pulling velocity $V$ and all the points on the interface are pinned to the glass. The increment $\delta h$ in height in every step is simply $V\delta t$ which is the displacement produced by the constant pulling in a time step $\delta t$. In the results reported below, we have used $\delta h = 0.001$. Step 3 is to determine the stability of the interface by considering the balance of capillary with gravitational pressure and the local shape of the interface. Here the $\alpha$ term is the local shape and the

$\beta$ term is the gravitational energy. If the capillary forces cannot support the gravitational pressure, there will be slipping motions. One can see that this is an uniform infinite growth model if Step 3 is omitted. With Step 3, there will be fluctuations in the growth of $h_i$ and there will also be a steady state with a maximum average height $h_{\max} = < h_i >$, similar to the experiment. The discrete slipping distance $\delta h'$, which indicates how far the interface will slip, is a function of the roughness of the ground glass. In the simulations, $\delta h'$ is chosen to be 0.01 which is ten times larger than the step increment $\delta h$. The choice of these simulation parameters is to best mimic our experimental results. Detail discussions of the different effects of these parameters can be found in Ref.10.

The simulation model described above is in some aspects similar to the depinning model of Sneppen [11]. The main difference between our model and the Sneppen model is that : in our model, the balance of forces determine which points on the interface will slip and therefore more than one point of the interface might slip at the same time. However, in the Sneppen model, depinning only occurs at the global minimum. An other important feature of our model is that the depinning forces are self-adjusted to drive the interface to move with an average constant velocity. In most other simulations and experiments, a constant driving force is used.

## 3 Results and Discussions

To look at how the interface $h(x,t)$ evolves when the pulling is started, let's focus on the behaviour of a particular point on the interface $x_0$ chosen close to the center of the sample. Figure 2a is the time dependence of the $h(x_o,t)$ when the sample is being pulled by a velocity of 0.029 cm/sec. Note that the motion of the interface can be divided into three regimes. In the first regime, there are very little depinnings, the interface moves nearly with the velocity of the glass. However, at latter time, when the potential energy of the pinned interface increases, there are more slipping motions. Finally, at even later times, when the interface attains its final maximum height, there are very strong fluctuations in the motion of the interface. At this steady state, the interface is moving with an average constant speed $V$, relative to the glass. In other words, the depinning velocity of the interface in the final steady always equal to the pulling velocity $V$.

Although the interface is driven by a constant pulling velocity, there are very strong velocity fluctuations in the depinned interface. Figure 2b shows the depinning velocity of the interface in these three regimes. At early time, the interface is nearly pinned to the glass and therefore has nearly zero depinning

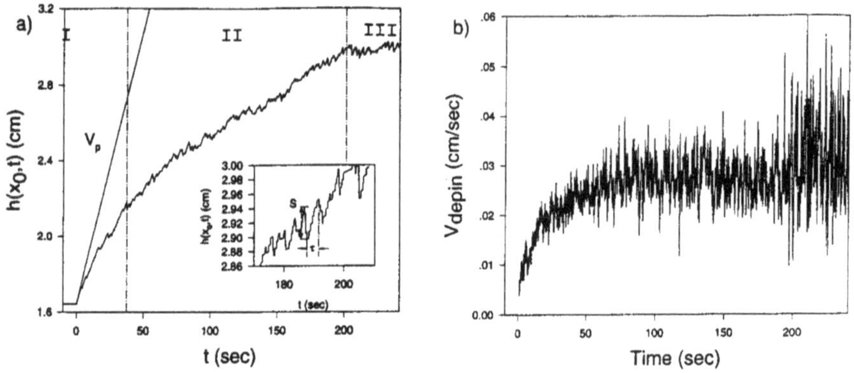


Figure 2: a) Time dependence of the capillary rise $h(x_o, t)$ at a particular point when the sample is pulled by a speed of 0.029 cm/sec.b) the corresponding velocity of the interface relative to the glass. The inset of a) shows the details of the stick-and-slip motions in the steady state.

velocity. The depinning velocity increases with time because of the increase in the pressure difference $\rho g h(x, t)$ which is the driving force of the depinning. When the interface reaches the final steady state, there is a sudden increase in the fluctuations in velocity. This is probably due to the fact that, all the energy put in by the pulling is now being fed to the fluctuations.

In the inset of figure 2b, we have also shown the details of the slip-and-stick motions of the interface. These slip-and-stick motions are similar to those reported in Ref.1 and are characterized by the waiting time $\tau$ and the slipping size $S$. One of the simplest way to analyze these fluctuations is to look at the power spectrum of the data shown in the inset of figure 2. Figure 3 shows the power spectrum of these fluctuations in the steady state. One can see from Figure 3 that the slip-and-stick motions do not give a $1/f$ noise spectrum but rather a $1/f^2$ spectrum. To compare the experimental results with simulations, the results from the Sneppen model and our simulation model are also shown in Figure 3. It is clear that our experimental results and the simulations all show similar form. This $1/f^2$ behaviour can also be found in the the droplet coalescence in breath figures[13]. However, we believe that this $1/f^2$ form comes from the sharp edges and the sudden jump in the data.

We have also performed the analysis on the probability distribution of $S$ and $\tau$. Note that $S$ here is not the same as quake size described in most earthquake models [12]. Usually, quake size is defined as the total release of energy or sites updated during an avalanche event such as in the Sneppen model. However, as mentioned above, in our simulations and experiments,

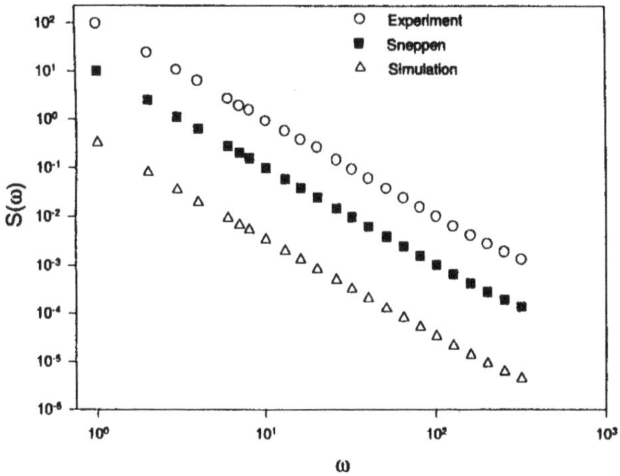


Figure 3: Power spectrum of the stick-and-slip motions in the steady state for a pulling speed of 0.029 cm/sec. Note that units of $\omega$ of the simulations have been rescaled to match with the data from experiments. Also, all $S(\omega)$ are normalized to one at $\omega = 1$ but they are displaced vertically for reason of clarity.

more than one point of the interface can move at the same time. It is difficult to identify a single avalanche event. Therefore, in our analysis of $S$ and $\tau$, we simple collect all the statistics of the $S$ and $\tau$ from different points of the interface. We find that the data collected this way have the probability distributions $P(S)$ and $P(\tau)$ in the the form as shown in Figure 4 with $P(S) \sim e^{-(\frac{S}{S_o})^2}$ and $P(\tau) \sim e^{-(\frac{\tau}{\tau_o})}$ where $S_o = 350\mu m$ and $\tau_o = 1.7sec$. The $S_o$ and $\tau_o$ are sample dependent and are related to the surface roughness of the ground glass [10]. To compare with the simulations, results from our model and the Sneppen model are also shown in Figure 4. Again, the experimental results and simulation results show similar form except for the Sneppen model when $S$ is large. This difference comes mainly from the global minimum search in the Sneppen model which has a correlation length of the size of the system.

Finally, to compare our experiment with other experiments, we have also measured the roughness of the interface at two different $V$ as shown in Figure 5. We find that the two pulling speeds give the values of $\chi = 0.57$ for $V = 0.11cm/sec$ and $\chi = 0.60$ for $V = 0.029cm/sec$. These measured values are close to values reported by other experiments and are in between the PKZ value of 0.5 and the 0.63 in the Sneppen model. From the above discussions, it is clear that the slip-and-stick dynamics of the air-water interface produced in our experiment can be qualitatively described by our simulation model which

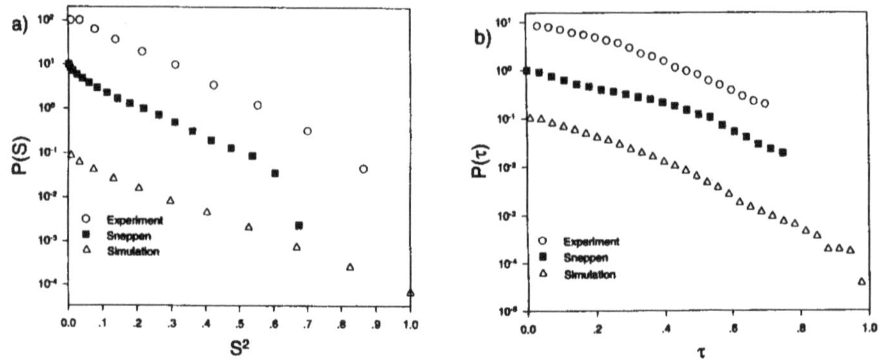


Figure 4: Probability distributions of the stick-and-slip motions. a) $P(S)$ and b) $P(\tau)$ for a pulling speed of $V$ =0.11cm. Note that units of $S$ and $\tau$ from experiments have been rescaled to compare with simulation data. Also, $P(S)$ and $P(\tau)$ are all normalized to one but they are displaced vertically for reason of clarity.

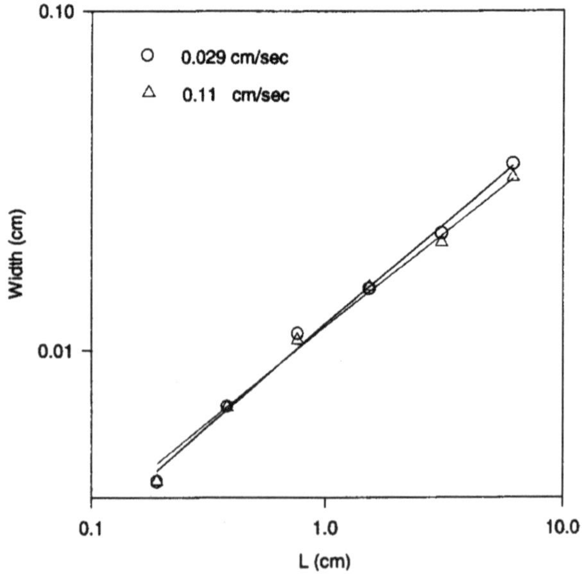


Figure 5: The width of the interface as a function of length scales.

is close to the Sneppen model. The origin of the $1/f^2$ behavior observed in the experiments and the simulation models is still not clear. It is important to know if this $1/f^2$ is a generic characteristic of slip-and-stick dynamics.

**Acknowledgments**

This work is supported by the grant NSC-85-2112-M-001-038 from the National Science Council of ROC.

**References**

1. R. Burridge and L. Knopoff, Bull. Seis. Soc. Am, 57, 341, (1967).
2. M. Kardar, G. Parisi and Y.-C. Zhang, Phys. Rev. Lett. 56, 889 (1986)
3. M. A. Rubio, C. A. Edwards, A. Dougherty, and J. P. Gollub, Phys. Rev. Lett. 63, 1685 (1989)
4. S. V. Buldyrev, A.-L. Barabasi, F. Caserta, S. Havlin, H. E. Stanley, and T. Vicsek, Phys. Rev. A , 45, R8313 (1992)
5. S. He, G. L. M. K. S. Kahanda, and P.-Z. Wong, Phys. Rev. Lett. 69, 3731 (1992)
6. T. Nattermann, S. Stepanow, L.-H. Tang and H. Leschhorn, J. Phys. II, (Paris) 2, 1483 (1992)
7. O. Narayan D. S. Fisher, Phys. Rev. *B*48, 7030 (1993)
8. L.-H. Tang, Ann. Rev. Comp. Phys. II, 1 (1995)
9. N. Y. Liang and C. K. Chan (preprint 1995)
10. N. Y. Liang and C. K. Chan (in preparation).
11. K. Sneppen, Phys. Rev. Lett. 69, 3539 (1992)
12. J. M. carlson and J. S. Langer, Phys. Rev. Lett. 62, 2632 (1989)
13. A. Steyer, P. Guenoun and D. Beysens, Phys. Rev. Lett., 68, 1869 (1992).

# STATISTICAL MECHANICS OF RIVER NETWORKS

A. MARITAN[1]

[1] *International School for Advanced Studies and Istituto Nazionale di Fisica Nucleare INFN, via Beirut 2-4, 34014 Trieste, ITALY*

and

J. R. BANAVAR[2], M. CIEPLAK[3], F. COLAIORI[1], A. FLAMMINI[1], A. GIACOMETTI[1], A. RINALDO[4] and I. RODRIGUEZ-ITURBE[5]

[2] *Department of Physics and Center for Materials Physics, 104 Davey Laboratory, University Park, PA, 16802, USA*

[3] *Polish Academy of Science, 02-668 Warsaw, POLAND*

[4] *Istituto di Idraulica "G. Poleni", Universita' di Padova, via Loredan 20, I-35131 Padova, ITALY*

[5] *Department of Civil Engineering, Texas A & M University, College Station, Texas 77843, USA*

**Abstract**

We will describe a class of models of river networks obtained by an energy minimization principle and show that over a range of parameter values, only two distinct universality classes are obtained for homogeneous basins. We show that even an infinitesimal heterogeneity is relevant and leads to qualitatively new behaviour. We calculate the exponents exactly for all three classes of behaviour. We also present the results of a model of river networks when the erosional properties of the surface soil are strongly heterogeneous. It yields fractal aggregation patterns following non-trivial power laws in experimentally relevant distributions. Scaling arguments will be presented that allow all the exponents to be determined from the fractal dimension.

## 1 Introduction

Experimental analyses of river networks[1] have shown clear examples of behaviour analogous to critical phenomena characterized by the absence of a single well-defined length scale reflected in a power-law behaviour of various quantities. A fundamental question that arises from these observations is whether, in analogy with conventional critical phenomena, one may fruitfully classify this behaviour into universality classes that are characterized by different sets of exponents. A related point is whether there exist scaling relationships between the various exponents of a given universality class. Another vital issue is the elucidation of simple models amenable to analysis that nevertheless capture some of the features of river networks.

River patterns in the large scale limit can be fractal or self-affine. We will show using scaling arguments that in both cases the geometry of the rivers, as defined by the fractal dimension in the former case or by the wandering exponent in the latter, determines uniquely a variety of other exponents.

We then introduce optimal channel networks (OCN)[2, 3] and discuss some previously introduced models. The relevance of hetereogeneity in river basins will then be shown in the context of a simple model.[4] In the limit of strong hetereogeneity we obtain a new universality class for the fractal river basin.[5] A stochastic differential equation approach to model landscape evolution is presented at the end of this paper.[6]

## 2 Definitions and Scaling Laws

A river network is a spanning tree defined in a lattice of arbitrary size and shape. In the following we will consider a $L \times L$ square lattice. A network is a set of drainage directions. A river basin is a scalar field of elevations consistent with the mapping of drainage directions defined by the corresponding network. Drainage directions are usually identified by steepest descent, i.e. the local gradient of the elevation field.

In a river basin, the area of a point of the substrate is defined as the total contributing area draining into the point through drainage directions, or the number of sites upstream of the point connected by the network. From the computational point of view, it can also be regarded as a measure of the flowrate if a unit weight is assigned to each source thus simulating a unit constant injection of mass uniformly over the basin. In natural basins these areas can be investigated through experimental analysis of digital elevation maps (DEM's) e.g.[1] The area, $s_i$, associated with any bond $i$ is defined as the number of sites upstream of $i$ connected by the network. In a real river basin a site is assumed to coincide with a pixel (about $30 \times 30 m^2$) in the DEM. The equation for the area at bond $i$, $s_i$, is

$$s_i = \sum_j s_j + R_i \tag{1}$$

where the sum over $j$ is carried out over bonds that input into $i$ and $R_i$ denotes the rainfall (precipitation) at the $i$-th bond. $R_i$ is commonly assumed to be independent of $i$ and equal to 1. Sometimes it will be convenient to associate the area $s_i$ of the $i$-th bond to the output site of the bond.

It is experimentally found in river basins that the probability that $s_i$ at any site exceeds a value ($s$) scales as $P(s) \propto s^{1-\tau}$.[1] The probability density $p(s, L)$ of a site having area $s$ in a system of size $L$, is expected[7, 8] to have the scaling form

$$p(s,L) = s^{-\tau} F(\frac{s}{L^{\phi}}) \tag{2}$$

where $F(x)$ is a scaling function which takes into account finite size effects and $\phi$ is the finite size exponent: $L^{\phi}$ is the characteristic area of the basin.

The distribution of upstream lengths has been measured experimentally. In our model, at a given site the areas (see eqn. (1)) of the nearest-neighbours are checked. The site with the largest value leads to the outlet. The site with the next-largest value is defined to be an upstream site – it indicates the longest path towards the source. If two (or more) equal areas are encountered, one is randomly selected.

The distribution of the lengths is also found to display a universal scaling form:

$$\pi(l,L) = l^{-\gamma} f(\frac{l}{L^{d_l}}) \tag{3}$$

where $L^{d_l}$ is the characteristic length in the basin.

In natural basins the drainage area $s$ and the stream length $l$ are related by Hack's law[9]:

$$s \sim l^{1/h} \tag{4}$$

The sub-basin from any site defined as all the upstream sites connected to it is characterized by typical longitudinal and transverse lengths $\xi_{\|}$ and $\xi_{\perp}$. For self-affine river networks $\xi_{\perp} \sim \xi_{\|}^{H}$ with $H \leq 1$. For fractal river networks (in which each rivulet originating from any site and proceeding to the global outlet is a fractal characterized by the same fractal dimension $d_l$) the wandering exponent $H = 1$, i.e. $\xi_{\perp} \propto \xi_{\|}$.

Thus for fractal river basins we get:[5]

$$H = 1, \phi = 2, \tau = 2 - \frac{d_l}{2}, h = \frac{d_l}{2}, \gamma = \frac{2}{d_l} \tag{5}$$

whereas for self-affine river networks ($H \leq 1$):[8]

$$d_l = 1, \phi = 1 + H, \tau = \frac{1+2H}{1+H}, h = \frac{1}{1+H}, \gamma = 1 + H. \tag{6}$$

Note that for the limiting case $H = d_l = 1$, eqs. (5) and (6) coincide and are the same as in mean field theory.[10]

## 3 Optimal Channel Networks

Recently[2, 3] a model based on the principle of minimum energy dissipation has been formulated: the potential energy associated with the precipitation is dissipated to maintain water flow. (As a result the water velocity is almost constant over the whole basin.[11]) Erosion caused by the flow sculpts the landscape and drives it into scale-free self-organized structure so that the energy expenditure is minimized resulting in a final stage in which erosion ceases and the landscape evolution stops. The energy dissipated by the network is then written as

$$E = \sum_i K_i(\delta h)_i Q_i \tag{7}$$

where $Q_i$ is the mean (annual) discharge in the bond $i$, $(\delta h)_i$ is the height drop along the drainage direction (thus $(\delta h)_i Q_i$ represents the dissipated power) and $K_i$ is a quantity (assumed to be independent of $i$ and equal to 1, without loss of generality for a homogeneous basin) related to the soil properties such as the erodability. Field observations indicated the empirical relationship $(\delta h)_i \sim Q_i^{\gamma-1}$[11, 12] with a range of values for $\gamma$ centered around 0.5. Based on the minimum energy expenditure and within the framework of simple solvable geometries, it is possible to show that $\gamma = 0.5$.[2, 3] A surrogate of $Q_i$ is the cumulative area defined above i.e. $Q_i \propto s_i$. Thus Eq.(7) becomes (apart from an irrelevent multiplicative constant)

$$E = \sum_i K_i s_i^{\gamma}. \tag{8}$$

Extensive numerical investigations of the networks minimizing the functional (8) have been carried out by two groups[2, 3] and[7, 13]. In references [2,3] the $\gamma = 1/2$ case has been mainly investigated. Their studies were restricted to the statistics of local minima and yielded exponents in excellent agreement to observational data (see table) besides many other features of landscape shape. In references [7,13] the interest was in the behavior of the global minimum of (8). Various $\gamma$ were studied in the interval (0,1) and all the data are consistent with scaling relations (5) and (6) with $H = 1$ and $d_l = 1$, i.e. the mean field values $\phi = 2$, $\tau = 3/2$ etc. are obtained.

Recently we have worked out an exact proof [4] of the above findings. Indeed we were able to show that the global minimum of (8) in the large size limit scales as Min $E \sim$ Max $(L^2, L^{1+2\gamma})$ for $0 \leq \gamma \leq 1$. Assuming scaling (2), one deduces that $\phi = 2$ and $\tau = 3/2$ (Eq. (6) holds with $H = 1$) for all $1/2 \leq \gamma < 1$. Logarithmic corrections are also expected for Min $E$ when $\gamma = 1/2$. Even though we are unable at present to extend our analysis to the $0 < \gamma < 1/2$ case, the evidence suggests that mean field behavior is expected in the whole range $0 < \gamma < 1$. The $\gamma = 1$ deserves special treatment. It is not difficult to prove[14] that in this case the minimum of (8) is highly degenerate: all directed river networks – each bond of the network oriented in the flow direction has a positive projection along the diagonal (passing through the global outlet) towards the outlet, have the same energy $E \sim L^3$ which is the minimum obtainable. This corresponds to the Scheiddeger model of river networks.[15] It can be solved exactly[10]: the exponents are given by (6) with $H = 1/2$.

We conclude this section by observing that it is intriguing that homogeneous OCN yield results in agreement with observations when a statistics based on local minima is performed.[2, 3] This has suggested[16] the concept of feasible optimality which states that nature is unable to reach the true ground state for complex optimization problems and therefore the relevant scaling ought to be obtained within the ensemble of accessible local minima.

In the following two sections we will be concerned only with the scaling properties of the global minimum.

## 4 Heterogeneous River Basins

Natural river basins are known to be heterogeneous. The simplest heterogeniety that one may consider is random precipation with $R_i$ no longer being uniform in (1). Sites with large $s_i$ have contributions from the $R_i$ of all the upstream sites. Thus, fluctuations in $R_i$ that do not have a long tail in their distribution would be expected to average out to yield the homogeneous result when minimizing (8). Indeed, an explicit analysis[10] within the framework of the Scheidegger model, shows that the universality class is unchanged on adding this kind of heterogeneity.

We have shown[4], within the $\gamma = 1$ context, that the incorporation of even an infinitesimal disorder or heterogeneity in the $K_i$ in Eq. (8) leads to a new universality class. We calculate exactly the exponents characterizing the new behavior. Eq. (6) holds with $H = 2/3$. Strikingly, the exponents of the heterogeneous model are found to be in good accord with observational data (see table). Our results underscore the necessity of including disorder in the modelling of river networks.

## 5 Disorder Dominated River Basins

In this section we will be considering an extreme case of the heterogeneity[5] defined in the previous section. To each bond $i$ of the lattice, independent random numbers $P_i$, e.g. in the range (0,1), are assigned representing the erodability of the surface soil of the bond $i$. Water flow proceeds in an invasive way[17]: The weakest link is selected and assumed to be a part of the network. The second-ranking weakest link is then selected and so on. The process is iterated in the ensemble of the remaining links until all sites are connected, i.e. they all have a route to the outlet. Loops need to be excluded from the process due to the fact that once a given route is selected, alternative routes formed due to the presence of a loop would be energetically unfavorable. Operationally, one obtains the network by incorporating the regions in order of increasing strength so that no loops are formed, yet all sites on the lattice are connected to the outlet.

We note that the resulting network is a union of the invasion percolation paths from each of the sites to the outlet. One can imagine that aggregation of new bonds to the evolving pattern occurs on quite different times scales depending on the strength of $P_i's$. The resulting pattern is found to be fractal, i.e. it is characterized by scaling relations (5) with $d_l = 1.21 \pm 0.02$. The exponents are shown in the table.

We note that this model could be alternately viewed as an optimization problem that selects the spanning tree that minimizes $\sum_i P_i$ where $P_i$ denotes the erodability of the $i$-th bond, the sum over $i$ runs over all the bonds of the tree and the $P_i's$ are assumed to be widely separated in magnitude, so that the sum of the $P_i's$ along any path is essentially equal to the largest $P_i$ encountered. On identifying $P_i$ as the energy dissipation in the $i$th bond, the model then becomes a special case of OCN.

In conclusion we have shown that homogeneous OCN with $\frac{1}{2} \leq \gamma \leq 1$ lead to two distinct universality classes: the Scheidegger class[10, 15] for $\gamma = 1$ and the mean field one for the other $\gamma's$.[4] Heterogeneous OCN on the other hand lead to new universality classes: we have given examples of both self-affine and self-similar river basins satisfying scaling relations (6) and (5) respectively.

## 6 Continuum Model for River Network Evolution

The model is a field theory for the soil height $h(\vec{x}, t)$ and takes into account the effects of random precipitation, erosion and the avalanching of soil.[6] An initially smooth landscape evolves into a nontrivial spatially self-organized state in which Hack's law and the algebraic distribution of drainage basin areas are obtained.

The evolution equation may be written in the compact form:

$$\frac{\partial}{\partial t} h(\vec{x}, t) = D \cdot h(\vec{x}, t) - k\nabla^4 h(\vec{x}, t) + \eta(\vec{x}, t) \tag{9}$$

where $D\cdot \equiv D_1 \partial_y^2 + \tilde{D}_2(|\nabla h|)\partial_x^2$ and $\vec{x} \equiv (x, y)$. The coefficient $\tilde{D}_2(|\nabla h|)$ is defined as:

$$\tilde{D}_2[|\nabla h|] = \begin{cases} D_1 > 0 & if \quad |\nabla h| > M \\ D_2 < 0 & if \quad |\nabla h| > M \end{cases} \tag{10}$$

and the random noise $\eta(\vec{x}, t) \equiv \epsilon r(\vec{x}, t)$ is Gaussianly distributed with zero average and correlation

$$< \eta(\vec{x}, t)\eta(\vec{x}', t') >= \epsilon^2 \delta^d(\vec{x} - \vec{x}')\delta(t - t') \tag{11}$$

The equation describes the temporal evolution of a two dimensional landscape with periodic boundary conditions in the $x$ direction and a dominant water flow in the $y$ directions due to an incline. The equation allows for ordinary diffusion in the $y$ direction and a stabilizing $\nabla^4 h$ term working as an ultraviolet regulator. In the $x$ direction, a diffusion term is operational as long as $|\nabla h| > M$. These diffusional processes mimic the avalanching effect discussed by Leheny and Nagel[18]: when neighbouring sites have large differences in heights, avalanching of the soil occurs leading to a smoothing effect. The erosional processes are captured by the negative diffusion coefficient $D_2$ when $|\nabla h| < M$. Such a term accentuates the height difference and captures the physics of erosion-the water flowing in the shallower parts of the landscape erode the soil further, *increasing* the height difference. In the simple version of our model, the erosional processes are blind – the erosion is uncorrelated with the water flow. The noise $\eta$ mimics the erosional effects of random precipitation.

Our equation is a generalization of equations developed in other contexts. In the limit that $D_1 = D_2 > 0$, the equation reduces to the Edwards-Wilkinson equation for linear growth processes.[19] When $D$ and $D_2$ are both positive but different, the equation is akin to Barenblatt's equation[20] describing the pressure in an elasto-plastic porous medium that allows for the contraction and expansion of the medium due to the flow. The novel ingredient in Eq.(9) is the possibility of a negative diffusion coefficient.

We have discretized Eq.(9) and followed by numerical integration the evolution of a initial random landscape. As diffusion and erosion proceed (the $\nabla^4 h$ term smooths the landscape on short distances) one observes a gradual self-organization of the landscape. The river patterns, obtained as explained in sec. II, have a scaling behavior consistent with the Scheidegger universality class, i.e. $H = 1/2$ in Eq.(6).

We have also monitored the temporal evolution of the roughness of the landscape. The roughness $W$ is defined as:

$$W(t) = (\frac{1}{L^2} \sum_{\vec{x}} < (h(\vec{x}, t) - \bar{h}(t))^2 >)^{1/2} \qquad (12)$$

where $\bar{h}(t) = \sum_{\vec{x}} h(\vec{x}, t)/L$. The average was performed over many samples corresponding to different realizations of the noise. We find an intermediate regime with an algebraic growth $W(t) \sim t^{\beta}$ with an exponent $\beta = 0.21 \pm 0.02$. This exponent value is different from the exactly solvable Edwards-Wilkinson model[19] where $\beta = 0$, i.e. $W(t) \sim \sqrt{\ln t}$.

The model captures some of the key ingredients of Glock's theory of the evolution of river networks[21] – small tributaries are eliminated as the main rivers swell in size. Larger aggressive stream captures the flow of a neighboring stream. Such effects had also been shown to occur by Leheny and Nagel[18] in their lattice model for river networks.

A more realistic model than the one presented here, would couple $h(\vec{x}, t)$ to the flow. This would lead to non-local effects like the one found in real basins.

This work was supported by KBN, NSF, NATO, ONR grants.

**TABLE**

| | Scheidegger | Mean Field | Random Erosion | Strong Disorder | River Basins |
|---|---|---|---|---|---|
| $d_l$ | 1 | 1 | 1 | $1.22 \pm 0.04$ | $1.1 \pm 0.02$ |
| $H$ | 1/2 | 1 | 2/3 | 1 | 0.75 - 0.80 |
| $\tau$ | 4/3 | 3/2 | 7/5 | $1.38 \pm 0.03$ | $1.43 \pm 0.02$ |
| $h$ | 2/3 | 1/2 | 3/5 | $0.62 \pm 0.02$ | 0.57 - 0.60 |
| $\gamma$ | 3/2 | 2 | 5/3 | $1.60 \pm 0.05$ | 1.8 - 1.9 |
| $d_F$ | 1 | 1 | 1 | $1.21 \pm 0.02$ | |

The exponents predicted by the scaling arguments, measured in our simulations and for river basins. $d_F$ is the fractal dimension of the river basin boundary. The measured values in the strong disorder simulations are consistent with the scaling predictions (5) on assuming that $d_l = 1.21 \pm 0.02$.

# References


[1] I. Rodriguez-Iturbe, R. L. Bras, E. Ijjasz-Vasquez, D. G. Tarboton, *Water Resour. Res.* **28**, 988 (1992); D. G. Tarboton, R. L. Bras and I. Rodriguez-Iturbe, *Water Resour. Res.* **24**, 1317 (1988): **26**, 2243 (1990); P. La Barbera and R. Rosso, *Water Resour. Res.* **25**, 735 (1989); S. P. Breyer and R. S. Snow, *Geomorphology* **5**, 143 (1992); D. R. Montgomery and W. E. Dietrich, *Nature* **336**, 232 (1988); *Science* **255**, 826 (1992); D. G. Tarbonton, R. L. Bras and I. Rodriguez-Iturbe, *Water Resour. Res.* **25**, 2037 (1989).

[2] I. Rodriguez-Iturbe, A. Rinaldo, R. Rigon, R. L. Bras, E. Ijjasz-Vasquez, *Water Resour Res.* **28**, 1095 (1992); *Geophys. Res. Lett.* **19**, 889 (1992); A. Rinaldo *et al*, *Water Resour. Res.* **28**, 2183 (1992) and reference therein for previous relates work.

[3] A. Rinaldo, I. Rodriguez-Iturbe, R. Rigon, E. Ijjasz-Vasquez and R. L. Bras, *Phys. Rev. Lett.* **70**, 822 (1993); R. Rigon, I. Rodriguez-Iturbe and A. Rinaldo, *J. Geophys. Res.* **99**, 11971 (1994).

[4] A. Maritan, M. Cieplak and J. R. Banavar, preprint.

[5] M. Cieplak, A. Giacommetti, A. Maritan, A. Rinaldo, I. Rodriquez-Iturbe and J. R. Banavar, preprint.

[6] A. Giacometti, A. Maritan and J. R. Banavar, *Phys. Rev. Lett.* (in press).

[7] T. Sun, P. Meakin and T. J$\phi$ssang, *Phys. Rev. E* **49**, 4865 (1994).

[8] A. Maritan, A. Rinaldo, R. Rigon, I. Rodriguez-Iturbe and A. Giacometti, submitted to *Phys. Rev. E* (1995).

[9] J. T. Hack, *U.S. Geol. Surr. Prof.* Paper **294-B**, 1 (1957).

[10] H. Takayasu, M. Takayasu, A. Provata and G. Huber, *J. Stat. Phys.* **65**, 725 (1991).

[11] L. B. Leopold, *Am. J. Sci.* **251**, 606 (1953).

[12] L. B. Leopold and T. Maddock, Jr., *U.S. Geol. Surv. Prof. Paper* **252**, 56 (1953); M. G. Wolman, *U.S.Geol. Surv. Prof. Paper* **272**, 56 (1955); L. B. Leopold and J. P. Miller, *U.S. Geol. Surv. Prof. Paper* **282-A**, 1956); J. J. Flint, *Water Resours. Res.* **10**, 969 (1974).

[13] T. Sun, P. Meakin and T. J$\phi$ssang, *Water Resour. Res.* **29**, 1219 (1993); *ibid* **30**, 2599 (1994); P. Meakin, J. Feder and T. J$\phi$ssang, *Physica A* **176**, 409 (1991).

[14] F. Colaiori, A. Flammini and A. Maritan, (unpublished).

[15] A. E. Scheidegger, *Bull. Ass. Sci. Hydrol.* **12**, 15 (1967).

[16] A. Rinaldo and I. Rodriquez-Iturbe, (unpublished).

[17] R. Chandler, J. Koplik, K. Lerman and J. F. Willemsen, *J. Fluid Mech.* **119**, 249 (1982); D. Wilkinson and J. f. Willemsen, *J. Phys. A* **16**, 3365 (1983).

[18] R. Leheny and S. R. Nagel, *Phys. Rev. Lett.* **71**, 1470 (1993).

[19] S. F. Edwards and D. Wilkinson, *Proc. Roy. Soc. London, Ser. A* **381**, 17 (1982).

[20] G. I. Barenblatt, *Similarity, Self-Similarity and Intermediate Asymptotics*, (Consultants Bureau, New York 1979); see also Chapter 10 of N. Goldenfeld, *Lectures on Phase Transitions and the Renormalization Group*, (Addison-Wesley, New York, 1993).

[21] W. S. Glock, *Geographical Review* **21**, 475 (1931).

# MULTI-INTERACTIONS IN FEED FORWARD NEURAL NETWORKS

R.M.C. de Almeida, E. Botelho, and J. R. Iglesias
*Instituto de Física, Universidade Federal do Rio Grande do Sul*
*Caixa Postal 15051 91501-970 Porto Alegre, RS, Brazil*
*E-mail:rita@if.ufrgs.br*

We consider learning and generalization in a multi-interacting feed-forward network model. In the binary output model, with an *a priori* definition of the net architecture, based on the symmetries presented by the function to be learnt, we define a generalized Hebb rule, extend the maximum stability algorithm to multi-interactions, and obtain training and generalization curves. For rules where different orders of synapses are not correlated, the results obtained for the simple perceptron concerning the Hebb rule and through replica calculations in the space of couplings may be straightforwardly adapted to multi-interactions through a simple renormalization of the total number of independent couplings. Analytical and numerical simulation results are compared and show excellent agreement. When continuous outputs are considered, the expansion in different multi-interacting orders suggests the learning rules, that reduce to the Hebb-learning rule only for the first order, linear perceptron. The student oversophistication problem is straightforwardly overcome by a natural cutoff in the multi-interacting synapses: the student is able to learn the architecture of the target rule, that is, the simpler a rule is, the faster the multi-interacting perceptron may learn. Again, simulation results are in excellent agreement with analytical calculations.

## 1 Introduction

Feed-forward neural networks are made of neurons disposed in ordered layers that can only act on next layers, with a well defined direction (for a recent review see Seung, Sompolinsky and Tishby[1] and Watkin, Rau and Biehl[2]) and the interest in such systems lays on their ability to learn rules or tasks from examples. The most studied model of such systems is the simple perceptron[3,4].

Roughly speaking, a feed-forward neural network may be conceived as a device with one output and $N$ input units that are somehow connected. Usually the input units are considered to be binary variables represented by a vector $\vec{S} = (S_1, S_2, \ldots, S_N)$, with $S_i = \pm 1$, while the output unit may be taken as a binary or continuous variable $S_o$. The way these variables are connected defines the different models for feed-forward neural networks. The most studied model, the simple perceptron, for example, considers that each input $S_i$ is coupled to the output through a synaptic weight $J_i$, such that the output state is determined by the sign of the field acting over it, that is $S_o = \text{sign} \sum_i J_i S_i$.

However, only linearly separable rules may be exactly solved by the simple perceptron. To bypass these limitations, new architectures are proposed, mainly by introducing new layers of hidden units ( for a review on this subject see Watkin, Rau and Biehl[2]). The multi-layered neural networks have been successful in solving many problems, but they are also difficult to treat analytically and each new problem requires a certain amount of intuition to devise an adequate architecture.

We have recently proposed an alternative way to increment the ability of solving rules of neural nets: we consider multi-interactions among input and output units, that is, connections involving more than one input unit are also allowed. The advantage of this model lays on its analytical simplicity, besides yielding a natural classification for the rules that goes beyond the presence or absence of linear separability.

The paper is organized as follows. In the next section, we define a multi-interacting network and discuss the target rule architecture. In section 3 we present error curves for the binary multi- interacting network using an extended version of the Hebb rule and the maximum stability algorithm. In section 4 we present the results for the continuous output model that may be devised to learn not only the target rule but also its architecture, overcoming the over-sophisticated student problem. Finally, in section 5 we conclude.

## 2 Multi-Interactions

The multi-interacting perceptron consists of two layers of neurons, the input one, containing $N$ neurons $S_i = \pm 1$ $i = 1, 2, \ldots, N$ interacting on a single output neuron $S_o$ through the synaptic intensities $J$, whose multi-interacting character will be soon put in evidence. A function - or rule - performed by the perceptron may be written as

$$\sigma_o = f\left(F(\vec{S})\right) \tag{1}$$

where $F(\vec{S})$ is a real function of $\vec{S}$. When $\sigma$ is also taken as a binary variable, $f$ is a boolean function and may be chosen to be the sign function, while when $\sigma$ is a real variable, other perceptron models are realized depending on the choice of $f(x)$; $f(x) = x$ defines the linear perceptron model [5] while $f(x) = \tanh(\beta x)$, with $\beta$ being a parameter, defines the analog (stochastic units) perceptron [5], for instance.

Now, consider a delta function defined as

$$\delta_{\vec{\xi},\vec{S}} = \frac{1}{2^N}\prod_{i=1}^{N}(1 + S_i\xi_i) \tag{2}$$

that returns the value 1 if $\vec{S} = \vec{\xi}$ and zero otherwise. Any real function $F$ of $\vec{S}$ may be written as

$$F(\vec{S}) = \sum_{\{\vec{\xi}\}} F(\vec{\xi})\delta_{\vec{S},\vec{\xi}} \tag{3}$$

where the sum over $\{\vec{\xi}\}$ stands for the sum over all $2^N$ input configurations. Expanding the product in eq.(3) to be used in eq.(1)

we show the multi-interacting character of a typical rule:

$$
\begin{aligned}
F(\vec{S}) &= \frac{1}{2^N}\sum_{\{\vec{\xi}\}} F(\vec{\xi}) + \frac{1}{2^N}\sum_{i=1}^{N}\left[\sum_{\{\vec{\xi}\}} F(\vec{\xi})\xi_i\right] S_i \;+ \\
&+ \frac{1}{2^N}\sum_{i<j}^{N}\left[\sum_{\{\vec{\xi}\}} F(\vec{\xi})\xi_i\xi_j\right] S_iS_j + \dots \\
&+ \frac{1}{2^N}\left[\sum_{\{\vec{\xi}\}} F(\vec{\xi})\xi_1\xi_2\dots\xi_N\right] S_1S_2\dots S_N
\end{aligned}
\tag{4}
$$

The different multilinear forms may be taken as different orders of multi-interactions, the zeroth order one representing a threshold and the other ones involve $1, 2, \dots, N$ input units. Should the sum be restricted to a smaller set of input states, eq.(4) would return zero for every state outside the summing set. As remarked by Carmesin, who obtained an equivalent equation [6], this expansion does not compress information, that is, after summing over $P$ states, the function cannot "guess" the right value for a $(P+1)^{th}$ state, regardless how large $P$ is.

Anyway, we can define a rule to be learnt by a perceptron by the set of all multi-interacting weights present in its multilinear expansion given by eq.(4). Depending on the symmetries of the rule, some of the multilinear forms are zero: the constant function has only the zeroth order term, a linearly separable rule presents only first order couplings, odd (even) functions have only odd (even) terms present in the multilinear expansions, for example. A natural classification for the rules is then given by the set of non vanishing orders present in their multilinear expansions. A second classification step could be the kind of correlations existing between different coupling orders. We shall write the threshold and the multi-interaction connections of a target rule as

$$B_0 = \frac{1}{2^N} \sum_{\{\vec{\xi}\}} F(\vec{\xi}) \tag{5}$$

and

$$B^{(\lambda)}_{i_1 i_2 \ldots i_\lambda} = \frac{1}{2^N} \sum_{\{\vec{\xi}\}} F(\vec{\xi}) \xi_{i_1} \xi_{i_2} \ldots \xi_{i_\lambda} \tag{6}$$

for $\lambda = 1, 2, \ldots, N$.

A way to devise models for multi-interacting perceptrons is to prescribe how to build from examples the perceptron connections to emulate the different multilinear orders present in the target rule. And that means first of all to define the architecture of the student net. We shall first define $\Lambda_T$ and $\Lambda_S$ as the set of non-vanishing multilinear forms respectively in the teacher and student net.

Each order $\lambda$ contain a number $M_\lambda$ of different $\lambda$-interactions, given by

$$M_\lambda = \frac{N!}{\lambda!(N-\lambda)!} \tag{7}$$

such that the total number $M$ of couplings in a net is given by the sum

$$M_{S,T} = \sum_{\lambda \in \Lambda_{S,T}} M_\lambda \tag{8}$$

When all orders are present, then $M = 2^N$.

Considering a student multi-interacting perceptron $J$, the couplings may be displayed as

$$|J\rangle_{\Lambda_S} = (J^{(0)}, J^{(1)}_1, J^{(1)}_2, \ldots, J^{(1)}_N, J^{(2)}_{12}, J^{(2)}_{13}, \ldots, J^{(N)}_{12\ldots N}) \tag{9}$$

and a $\Lambda_S$-scalar product may be written as

$$\langle J|S\rangle_{\Lambda_S} = \sum_{\lambda \in \Lambda_S} \sum_{i_1 < i_2 < \ldots < i_\lambda}^{N} J^{(\lambda)}_{i_1 i_2 \ldots i_\lambda} S_{i_1} S_{i_2} \ldots S_{i_\lambda} \tag{10}$$

where only the orders in $\Lambda_S$ are considered. The student output is then $S_o = f(\langle J|S\rangle_{\Lambda_S})$. Analogously, a scalar product with the teacher net may be defined and the target rule may be written as

$$\sigma_o = f\left(\langle B|S\rangle_{\Lambda_T}\right) \tag{11}$$

The amount of information presented by each example question-answer determines the learning rate and, in what follows, we shall consider separately the binary and continuous output models.

## 3 The Binary Model

The binary model considers the sign function, that is not invertible, as the transmission function $f$. As a consequence, the value of $F(\vec{S})$ may not be inferred from an example pair $(\vec{S}, \sigma_o(\vec{S}))$, and eqs.(5) and (6) may only be taken as an indicative of the adequate architecture for the perceptron. For the binary model we shall always take $\Lambda_S = \Lambda_T = \Lambda$, that is, the symmetries of the target rule define the architecture of the student perceptron. Moreover, we shall also consider $B^{(0)} = 0$.

We present here the results obtained for the binary multi- interacting perceptron in a previous paper[11], considering the optimal perceptron and learning by an extension to multi- interactions of the Hebb rule.

First we obtain the generalization rate as proposed by Opper *et al.* [7] as a function of the average $\Lambda$-scalar product of the net $|J\rangle_\Lambda$ and the teacher $|B\rangle_\Lambda$. We define

$$\begin{aligned} x &= \langle J|S\rangle & (12)\\ y &= \langle B|S\rangle & (13)\end{aligned}$$

where $\langle J|J\rangle = \langle B|B\rangle = 1$ and we have dropped the subindex $\Lambda$ for simplicity. As in the simple perceptron case, for random inputs $\vec{S}$, $x$ and $y$ are correlated gaussian variables with zero mean, unit variance and covariance $\langle xy \rangle = R$, where $R$ is the $\Lambda$-scalar product between the net and the teacher:

$$R = \langle J|B \rangle \tag{14}$$

We remark at this point that the absence of correlation between different orders of the teacher $\Lambda$-vector is a necessary condition to obtain $x$ and $y$ as gaussian correlated variables. The generalization error $\epsilon_g$ is then given by the probability that $xy < 0$; the calculations are the same as for the simple perceptron [7], that is

$$\epsilon_g = \frac{1}{\pi} \cos^{-1} R \tag{15}$$

Now, to obtain $R$ we must first choose the learning strategy. For the optimal perceptron, the calculations are performed on the space of couplings, as proposed by Gardner [8]. The multi- interactions were treated in the same way as done by Kohring[9], what is possible when the different orders of couplings in the target rule are not correlated. The problem results to be equivalent to the simple perceptron, but for the renormalization of $\alpha$, here defined as

$$\alpha_S = \frac{P}{M_S} \tag{16}$$

In the case where only the first coupling order is present in the multilinear expansion of the target rule, that is, the rule is linearly separable, $M = N$ and the simple perceptron result is recovered.

Figure(1) shows the results for the optimal perceptron together with data from numerical simulations where the maximum stability algorithm has been applied.

We also considered an extension to the Hebb learning rule, that is for every order of interaction $\lambda$ present in $\Lambda$, we consider the following:

$$J^{(\lambda)}_{j_1,\dots,j_\lambda} = \frac{1}{P} \sum_{\mu} \text{sign}\left(\langle B|S^\mu \rangle\right) S^\mu_{j_1} \dots S^\mu_{j_\lambda} \tag{17}$$

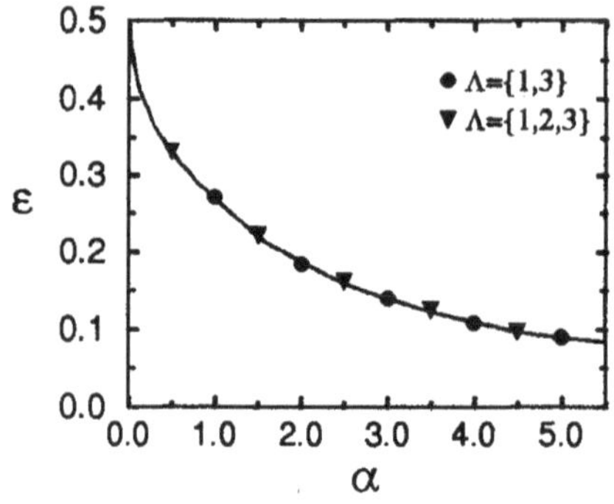


Figure 1: Generalization error curve for the optimal perceptron (solid line) and numerical simulation results for $N$=16 and $\Lambda = \{1,3\}$ (circles) and $\Lambda = \{1,2,3\}$ (triangles)

Observe that here we consider averages over samples of $P$ items of $\text{sign}(F(\vec{S}))$ instead of averages of $F(\vec{S})$ as it appears in the multilinear expansions of the target rule, eq.(6). This is so because the sign function is non- invertible. Also, for the first order couplings the above equation reduces to the Hebb rule. To obtain the generalization and training error curves we follow the prescription by Vallet [10] and calculate the probability that the teacher and student field have the same sign for a given input $\vec{S}$ [11]. Again, when the different coupling orders in the target rule are not correlated, the results are the same as for the simple perceptron with $\alpha$ given by eq.(16): the generalization error reads

$$\epsilon_g = \frac{1}{\pi}\tan^{-1}\left(\sqrt{\frac{\pi}{2\alpha}}\right) \tag{18}$$

and the training error rate, that is, the relative number of wrong answers to questions drawn from the set of examples may also be obtained from the simple perceptron results by Vallet [10]:

$$\epsilon_t = \frac{1}{2} - \int_0^\infty \frac{du}{\sqrt{\pi}} \exp\left(-u^2\right) \text{erfc}\left(-u\sqrt{\frac{2\alpha}{\pi}} - \frac{1}{\sqrt{2\alpha}}\right) \tag{19}$$

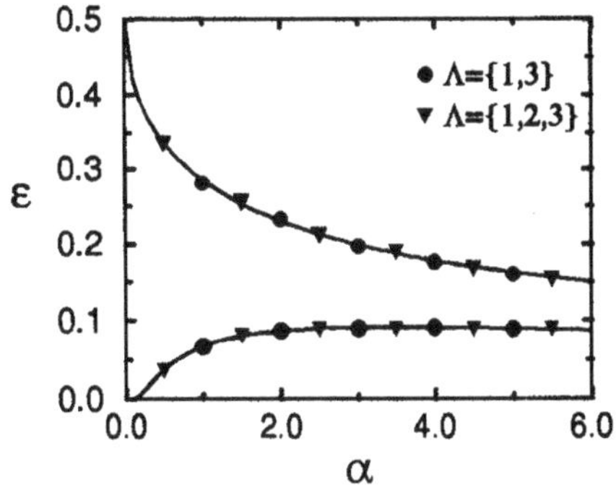


Figure 2: Generalization (upper) and training (lower) error curves for the extended Hebb learning rule (solid line) and numerical simulation results for $N$=20 and $\Lambda = \{1, 3\}$ (circles) and $\Lambda = \{1, 2, 3\}$ (triangles)

Fig.(2) shows the theoretical generalization and training error curves for the generalized Hebb rule together with the results for numerical simulations, presenting an excellent agreement.

From the results for the optimal perceptron and the Hebb learning rule, we can observe that the oversophisticated student problem is present: we could have considered students with higher coupling orders than those presented by the target rule. In this case, the student must learn that these higher orders are zero and the learning still scales with $\alpha_S = P/M_S$ defined with the number of couplings in the student net.

We performed numerical simulations for the Hebb rule and with an extension of the maximum stability algorithm [12] to multi-interactions. Actually, the extension to multi-interactions is quite straightforward. We considered perceptrons with multi-interactions of first and third orders ($\Lambda = \{1, 3\}$) and for first, second and third orders ($\Lambda = \{1, 2, 3\}$). Averages were taken over 10 to 20 different sets of examples and 1000 random questions for each load parameter $\alpha = P/M$. The simulations were performed on a CRAY Y-MP2E of the Centro Nacional de Supercomputação of Universidade Federal do Rio Grande do Sul.

## 4 Continuous Output Model

Now we consider continuous output $\sigma$ and transmission functions that are invertible. In this case, the amount of information contained in each example is strongly increased in comparison with the binary model.

We first consider $\Lambda_S = \Lambda_T = \Lambda$ and the threshold term may be present. When an appropriate learning rule is considered, in principle, a network student defined with the same transmission function $f$ as the teacher should be able to reproduce the rule when its architecture contains all orders present in the teacher. Eqs.(5) and (6) suggest these learning rules; the synaptic strengths are built from a randomly chosen set of $P$ correct examples using the following:[13]

$$
\begin{aligned}
J^{(0)} &= \frac{1}{P}\sum_{\mu=1} f^{-1}(\sigma_o(\vec{\xi^\mu})) \\
&= \frac{1}{P}\sum_{\mu=1}^{P} F(\vec{\xi^\mu})
\end{aligned}
\tag{20}
$$

and

$$
\begin{aligned}
J^{(\lambda)}_{i_1\ldots i_\lambda} &= \frac{1}{P}\sum_{\mu=1}^{P} f^{-1}(\sigma_o(\vec{\xi^\mu}))\xi^\mu_{i_1}\cdots\xi^\mu_{i_\lambda} \\
&= \frac{1}{P}\sum_{\mu=1}^{P} F(\vec{\xi^\mu})\xi^\mu_{i_1}\cdots\xi^\mu_{i_\lambda},
\end{aligned}
\tag{21}
$$

for the orders $\lambda$ present in the student net. Here $(\vec{\xi^\mu}, \sigma_o(\vec{\xi^\mu}))$, for $\mu = 1, \ldots, P$, are the $P$ training example pairs.

The generalization and training errors are respectively defined as

$$
\epsilon_g = \frac{1}{2^N}\sum_{\{\vec{S}\}}\frac{1}{2}[\sigma_o(\vec{S}) - S_o(\vec{S})]^2 \tag{22}
$$

$$\epsilon_t = \frac{1}{P}\sum_{\mu}\frac{1}{2}[\sigma_o(\vec{S^{\mu}}) - S_o(\vec{S^{\mu}})]^2 \tag{23}$$

and may be calculated following what has been done for the simple perceptron with continuous outputs [14] that is, writing the convenient the joint probability distribution $P(x,y)$, which gives the probability density of simultaneously occurring $x = \langle B|S\rangle$ and $y = \langle J|S\rangle$ for $|S\rangle$ being any input vector (generalization error) or drawn from the set of training examples (training error). For continuous outputs the modules of the student net $J$ is relevant and besides $R = \langle B|J\rangle$, a second order parameter $Q = \langle J|J\rangle$ is required. As we are considering the possibility of non-vanishing threshold, a dependence on $B^{(0)}$ is also expected. Taking the teacher modules as unity, that is $\langle B|B\rangle = 1$, and after averaging over the training sets we get

$$\begin{aligned}\epsilon_g &= \int_{-\infty}^{\infty} Du \int_{-\infty}^{\infty} Dz \\ &\times \frac{1}{2}\left[f\left(\sqrt{1-B^{(0)2}}u + B^{(0)}\right)\right. \\ &- \left. f\left(\frac{z}{\sqrt{\alpha}} + \sqrt{1-B^{(0)2}}u + B^{(0)}\right)\right]^2\end{aligned} \tag{24}$$

and

$$\begin{aligned}\epsilon_t &= \int_{-\infty}^{\infty} Du \int_{-\infty}^{\infty} Dz \\ &\times \frac{1}{2}\left[f\left(\sqrt{1-B^{(0)2}}u + B^{(0)}\right)\right. \\ &- \left. f\left(\frac{z}{\sqrt{\alpha}} + (1+\frac{1}{\alpha})(\sqrt{1-B^{(0)2}}u + B^{(0)})\right)\right]^2\end{aligned} \tag{25}$$

where $Dx = dx\exp(-x^2/2)/\sqrt{2\pi}$. Eqs.(24) and (25) reproduce the simple perceptron results with continuous outputs [14] but here $\alpha =$

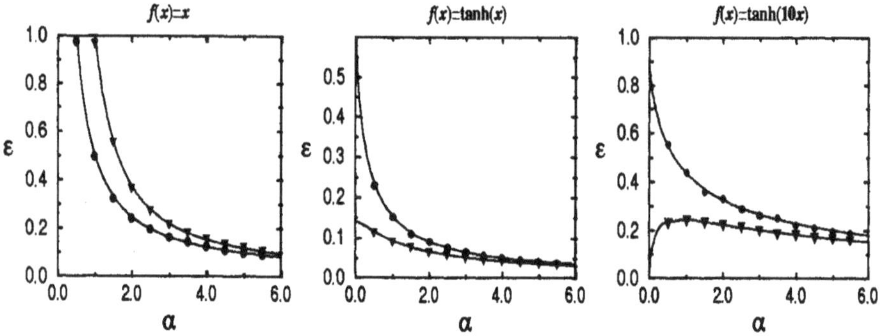


Figure 3: Generalization (circles) and training (triangles) error curves for $f(x) = x$, $f(x) = \tanh(x)$ and $f(x) = \tanh(10x)$, for $N = 20$.

$P/M$. Observe that the error curves depend on the choice of the transmission function $f(x)$; in particular for the linear perceptron, $f(x) = x$, eq.(24) reduces to $\epsilon_g = 1/2\alpha$ and eq.(25) to $\epsilon_t = 1/2\alpha + 1/2\alpha^2$.

In figure (3) we present generalization and training error curves together with numerical simulations results, for different transmission functions. For these transmission functions the dependence of the error curves on the threshold is only noticeable if $B^{(0)} \approx 1$. However that means the relative intensity of the threshold is dominating and the target rule is tending to a constant function. The analysis of the behavior of the error curves with different relative intensities of the different orders is left for further investigation. In what follows we considered small $B^{(0)}$.

The oversophisticated student problem is still present and may be inferred from both generalization and training error curves: the learning rate is ruled by $P/M$, where $M$ is determined by the student architecture. However it is desirable that the perceptron could learn simpler tasks faster; higher order interactions that are approaching zero should be neglected in early stages of the training sessions and a dependence on the number $M_T$ of independent synapses in the teacher net should appear in the error curves. On the other hand, diversely from other models, the learning rules eqs.(20) and (21) may be regarded as the measurement of population averages from averages over $P$- items samples. In this case the dispersion around the target values may be estimated as $1/\sqrt{P}$.

Consider now a student containing all possible coupling orders, learning target rules whose architectures are defined by $\Lambda_T$. After a learning $P$ examples, for large $P$, the absent orders in the teacher net ($B^{(\lambda)}_{i_1,i_2,\ldots,i_\lambda} = 0$) lead to couplings in the student net with norms $\langle J^{(\lambda)}|J^{(\lambda)}\rangle$ typically less than $M_\lambda/P$. A lower cut off may then be introduced in the learning rules:

$$J^{(\lambda)}_{i_1,i_2,\ldots,i_\lambda} = \frac{c_\lambda}{P}\sum_{\mu=1}^{P}\langle B|S^\mu\rangle S^\mu_{i_1}S^\mu_{i_2},\ldots,S^\mu_{i_\lambda} \tag{26}$$

with

$$c_\lambda = \Theta(\langle J^{(\lambda)}|J^{(\lambda)}\rangle - M_\lambda/P) \tag{27}$$

where $\Theta(x)$ is the step function. Hence, the role played by $c_\lambda$ is to test whether the population average approached in a given coupling order is zero and, in the affirmative case, neglect it. Naturally, this cut-off will only work satisfactorily for randomly chosen examples. In the large $P$ limit we use the approximations

$$\langle\langle c_\lambda\langle J^{(\lambda)}|J^{(\lambda)}\rangle\rangle\rangle \approx \Theta\left(\langle\langle\langle J^{(\lambda)}|J^{(\lambda)}\rangle\rangle\rangle - \frac{M_\lambda}{P}\right)\langle\langle\langle J^{(\lambda)}|J^{(\lambda)}\rangle\rangle\rangle \tag{28}$$

$$\langle\langle c_\lambda\langle J^{(\lambda)}|B^{(\lambda)}\rangle\rangle\rangle \approx \Theta\left(\langle\langle\langle J^{(\lambda)}|J^{(\lambda)}\rangle\rangle\rangle - \frac{M_\lambda}{P}\right)\langle\langle\langle J^{(\lambda)}|B^{(\lambda)}\rangle\rangle\rangle \tag{29}$$

where $\langle\langle\ldots\rangle\rangle$ stands for the average over all possible training sets. With this approximation, the cut-off does not introduce any new dependence on the fields $x$ and $y$, the calculation performed in the previous section are still valid, changing $\alpha$ for $\alpha_T$. Thus, generalization and training error curves scale with the number of couplings in the teacher net, that is, the student net is able to learn also the architecture of the teacher and may now learn simpler rules faster. The student oversophistication is then overcome.

The simulations of the continuous output model were performed on the CRAY Y-MP2E of the Centro de Supercomputação of the Universidade Federal do Rio Grande do Sul. We used networks with $N = 20$ input units and synaptic couplings of zeroth, first, second and third orders ($\Lambda = \{0, 1, 2, 3\}$). The averages were taken over 20 samples for 1000 questions each run.

## 5 Conclusions

The multilinear expansion of real functions $F$ of $N$-dimensional binary vectors $\vec{S}$ are especially suited to classify target functions to be learnt by feed-forward neural networks: the different order terms may be mapped on different orders of multi-couplings that may involve one or more input units. And the non-vanishing terms define the adequate architecture of a multi-interacting perceptron devised to solve the rule.

In the case of a binary output, the transmission function is the sign function that is not invertible. We considered the case where the architecture of the student net is equal to the teacher one and obtained the error curves for the optimal case and for extended Hebb learning rules. We found that for rules where the different multi-interaction order couplings are not correlated, the results obtained for the simple perceptron still apply, but for $\alpha = M/P$, where $M$ is the total number of synapses in the student net. The choice of the student architecture as being equal to the target rule can also be argued as being the adequate way to compress information, since a given architecture is equivalent to a definite way of compressing the relevant information contained in the supplied data (examples) . As an example, the information contained in the training set of a linearly separable rule are conveniently compressed by the simple perceptron architecture because every other order in the multilinear expansion of such rules is zero.

When a continuous output is considered, the amount of information contained in each example is strongly increased and the inversibility of the transmission function led to a network model able not only to learn the rules, but also to adequate the student architecture in order to speed up the learning of simpler rules. The overcome of the student oversophistication problem was possible by using a cut-off in the unnecessary coupling orders. As here the learning rules are in fact sample averages that approach a population average, the cut-off definition from a random walk dispersion is a natural choice. However, due to the learning rules defined in eqs.(20) and (21), each order in the student net approaches the corresponding order in the target rule, that is, absent orders in the student are not in any way compensated by other orders. The analytical calculations do not involve greater efforts than for the simple perceptron with continuous outputs and the simulations are in excellent agreement with analytical results.

However many points are still left for further investigation: correlated target rules, where different orders may not be independent as the many teacher problem, or other learning strategies, where there may be compensation of absent orders, are two examples. Also, an investigation of the best transmission function $f$ could be interesting: although the learning rules may not depend on the choice of $f$, the error curves do. However, these points are beyond the scope of this paper and shall be addressed elsewhere.

Finally, we remark that multineuron interactions have been extensively considered showing some success in enhancing the performance of attractor neural network models [15,16] and a recent review may be found in ref.17. For feed-forward neural networks, previous multineuron interaction models, besides the recent work by Carmesin [6], are the parity machine [18,19] and the sigma-pi units proposed by Rumelhart *et al* [20].

## Acknowledgement

This work has been partially supported by brazilian agencies Conselho Nacional de Desenvolvimento Científico e Tecnológico (CNPq), Financiadora de Estudos e Projetos (FINEP) e Fundação de Ampara à Pesquisa do Estado do Rio Grande do Sul (FAPERGS). The simulations were performed at the Cray supercomputer in Centro Nacional de Supercomputação, Porto Alegre, RS.


## References


1. H.S. Seung, H. Sompolinsky, and N. Tishby, *Phys. Rev. A* **45**, 6057 (1992).
2. T. L.H. Watkin, A. Rau, and M. Biehl, *Rev. Mod. Phys.* **65**, 499 (1993).
3. F. Rosenblatt, *Principles of neurodynamics*, Spartan, New York (1962).
4. M. L. Minsky and S. A. Papert, *Perceptrons*, MIT, Cambridge, MA (1969).
5. J.A. Herz, A. Krogh and G. I. Thorbergsson, *J. Phys A* **22**, 2133 (1989).
6. H.-O. Carmesin, *Phys. Lett. A* **188**, 27 (1994).
7. M. Opper, W. Kinzel, J. Kleinz, and R. Nehl, *J.Phys A*, **23** L581 (1990).
8. E. Gardner, *J. Phys A*, **21** 257 (1988).
9. G.A. Kohring, *J. Phys (France)*, **51** 145 (1990).
10. F. Vallet, *Europhys. Lett.*, **8** 747 (1989).
11. E. Botelho, R.M.C. de Almeida, and J.R. Iglesias, *J. Phys A* **28**, 1879 (1995).
12. W. Krauth and M. Mezard, *J. Phys. A*, **20** L745 (1987).
13. R.M.C. de Almeida and E. Botelho, *A multi- interacting perceptron model with continuous outputs*, preprint (1995).

14. S. Bös, W. Kinzel, and M. Opper, *Phys. Rev. E* **47**, 1384 (1993).
15. E. Gardner, *J. Phys A* **20**, 3453 (1987).
16. R.M.C. de Almeida e J.R. Iglesias, *Phys. Lett. A* **146**, 239 (1990).
17. R.M.C. de Almeida, T.J.P. Penna, and P.M.C. de Oliveira, "Multineuron interaction effects", in *Annual Reviews of Computational Physics I*, World Scientific, Singapore (1994), 193.
18. G.J. Mitchison and R.M. Durbin, *Biol. Cybern.* **60**, 345 (1989).
19. D. Hansel, G. Mato, and C. Meunier, *Europhys. Lett.* **20**, 471 (1992).
20. D.E. Rumelhart, J.L. Mclelland and the PDP research group, *Parallell Distributed Processing* vol. 1, (Cambridge:MIT) 1986.

# VORTEX TRANSPORT AND FLUCTUATIONS IN HIGH-TEMPERATURE SUPERCONDUCTORS

DAVID R. NELSON
*Lyman Laboratory of Physics, Harvard University*
*Cambridge, MA 02138, USA*
E-mail: nelson@cmt.harvard.edu

Concepts from quantum mechanics can be used to understand pinning, transport and thermal fluctuations of vortices in high temperature superconductors. The Abrikosov flux lattice in clean materials melts like ice, i.e., the vortices become denser at a first order melting transition to an entangled flux liquid. The theory can be extended to treat effects due to correlated disorder in the form of twin boundaries, grain boundaries and columnar pins induced by heavy ion irradiation. Mapping vortex trajectories onto the world lines of quantum mechanical particles allows easy evaluation of the thermal renormalizaiton of critical currents, the decoupling field and the flux lattice melting temperature. This approach applied to many flux lines leads to a "superfluid" (i.e.,entangled) phase, as well as "Bose glass" and "Mott-insulator" phases in which the vortices are localized. Double-sided flux decoration experiments [1] have now allowed direct observation of the phonon-roton spectrum in the entangled "superfluid" state. The dynamics in the Bose glass proceeds in close analogy with variable range hopping in semiconductors, and leads to highly usual current-voltage characteristics. Both flux creep and critical currents should improve dramatically in materials with splayed columnar defects. A complete understanding of this last problem requires a nontrivial extension of conventional many particle quantum mechanics.

Extensive reviews on this subject are available. A particularly comprehensive treatment is the the recent Reviews of Modern Physics article by G. Blatter et al. [2] The material described above is discussed in greater detail in Refs. [3-5].

1. Z. Yao, S. Yoon, H. Dai, S. Fan and C.M. Lieber, Nature **371** (1994) 771; S. Yoon, Z. Yao, H. Dai and C. Lieber, Harvard Chemistry Department preprint.
2. G. Blatter, M.V. Feigel'man, V.B. Geshkenbein, A.I. Larkin, and V. Vinokur, Rev. Mod. Phys. **66** (1994) 1125.
3. D.R. Nelson, in *Phenomenology and Applications of High-Temperature Superconductors*, edited by K. Bedell et al. (Addison-Wesley, New York, 1991).
4. D.R. Nelson, in *Phase Transitions and Relaxation in Systems with Competing Energy Scales*, edited by T. Riste and D. Sherrington (Kluwer, Dodrecht, 1993).
5. D.R. Nelson, Nature **375** (1995) 356.

# QUANTUM PHASE TRANSITIONS IN SPIN SYSTEMS AND THE HIGH TEMPERATURE LIMIT OF CONTINUUM QUANTUM FIELD THEORIES

SUBIR SACHDEV
*Department of Physics, Yale University,*
*P.O. Box 208120, New Haven CT 06520-8120, U.S.A.*

We study the finite temperature crossovers in the vicinity of a zero temperature quantum phase transition. The universal crossover functions are observables of a continuum quantum field theory. Particular attention is focussed on the high temperature limit of the continuum field theory, the so-called "quantum-critical" region. Basic features of crossovers are illustrated by a simple solvable model of dilute spinless fermions, and a partially solvable model of dilute bosons. The low frequency relaxational behavior of the quantum-critical region is displayed in the solution of the transverse-field Ising model. The insights from these simple models lead to a fairly complete understanding of the system of primary interest: the two-dimensional quantum rotor model, whose phase transition is expected to be in the same universality class as those in antiferromagnetic Heisenberg spin models. Recent work on the experimental implications of these results for the cuprate compounds is briefly reviewed.

## 1 Introduction

Consider a quantum system on an infinite lattice described by the Hamiltonian $\mathcal{H}(g)$, with $g$ a dimensionless coupling constant. For any reasonable $g$, all observable properties of the *ground state* of $\mathcal{H}$ will vary smoothly as $g$ is varied. However, there may be special points, like $g = g_c$, where there is a non-analyticity in some property of the ground state: we identify $g_c$ as the position of a quantum phase transition. In finite lattices, non-analyticities can only occur at level crossings; the possibilities in infinite systems are richer as avoided level crossings can become sharp in the thermodynamic limit. In this paper, I will restrict my discussion to second order quantum transitions, or transitions in which the correlation length and correlation time diverge as $g$ approaches $g_c$. As I will review below, any such quantum transition can be used to define a continuum quantum field theory (CQFT): the CQFT has no intrinsic short-distance (or ultraviolet) cutoff. The main purpose of this paper is to review some recent work[1,2,3,4,5] on the properties of $\mathcal{H}(g)$ at *finite temperatures* ($T$) in the vicinity of $g = g_c$. This is equivalent to a study of the finite $T$ crossovers of the associated CQFT. We shall focus especially on the dynamic properties of a ubiquitous finite $T$ region, usually called "quantum critical" [6] (as we shall see below, there are reasons why this name is misleading and not quite appropriate; nevertheless, I will use it here). The quantum-critical region appears as the high $T$ limit of the CQFT; unlike the statics of classical lattice models, the high $T$ limit of a CQFT is usually highly non-trivial. All of this discussion will take place in the context of some simple examples drawn from quantum spin systems.

I set the stage by reviewing the Wilsonian approach to critical phenomena and field theories [7], using the perspective of quantum critical phenomena. By the usual Trotter product decomposition, we can set up the partition function of $\mathcal{H}(g)$ as a functional integral over degrees of freedom which fluctuate as a function of the spatial co-ordinate $x$ and imaginary time $\tau$. Let us now examine the behavior of this functional integral under the rescaling transformation [8,9]

$$x \rightarrow e^{-\ell} x \qquad\qquad \tau \rightarrow e^{-z\ell} \tau \tag{1}$$

The dynamic exponent $z$ determines the relative scaling dimensions of space and time co-ordinates. The critical point at $g = g_c$ is a fixed point of (1), and $g - g_c$ is a relevant perturbation away from this point. We have therefore the flow equation

$$\frac{dg}{d\ell} = \frac{1}{\nu}(g - g_c) \tag{2}$$

which defines the critical exponent $\nu$. (For simplicity I do not discuss the case of fixed points with more than one relevant perturbation, as they can be treated in a similar manner). In the long-distance, long-time limit, this deviation from the critical point will be characterized by some *renormalized* energy scale, $G$. I emphasize that $G$ is a dimensionful parameter, expressed in the laboratory units of energy, and directly measurable in an experiment; a typical example would be an energy gap. In the vicinity of the critical point, the renormalized energy scale $G$ will be related to the bare coupling $g$ by

$$G \sim \Lambda |g - g_c|^{z\nu} \tag{3}$$

where $\Lambda$ is an ultraviolet cutoff, measured for convenience in the units of energy too. From the perspective of a field theorist, the CQFT associated with the quantum critical point is now defined by taking the limit $\Lambda \rightarrow \infty$ at fixed $G$; from (3) we see that, because $z\nu > 0$, it is possible to take this limit by tuning the bare coupling $g$ closer and closer to the critical point as $\Lambda$ increases. (A condensed matter physicist would take the complementary, but equivalent, perspective of keeping $\Lambda$ fixed but moving closer to criticality by lowering his probe frequency $\omega \sim G$). The resulting CQFT then contains only the energy scale $G$. At finite temperatures, there is a second energy scale $T$ (using units in which $k_B = 1$); its thermodynamic properties will then be a universal function of the only dimensionless ratio available—$G/T$. It is the purpose of this paper to review recent work on the crossovers as a function of $G/T$ in a number of systems, and to highlight the unusual properties of the heretofore unexplored high-temperature, "quantum critical", limit of the CQFT, $T \gg G$.

It is now easy to see why the high $T$ limit of the CQFT can be non-trivial. A conventional high $T$ expansions of the lattice model $\mathcal{H}$ proceeds with the series

$$\mathrm{Tr}e^{-\mathcal{H}/T} = \mathrm{Tr}1 - \frac{1}{T}\mathrm{Tr}\mathcal{H} + \frac{1}{2T^2}\mathrm{Tr}\mathcal{H}^2 + \ldots \tag{4}$$

The successive terms in this series are well-defined and finite because of the ultraviolet cutoffs provided by the lattice. Further, the series is well-behaved provided $T$ is larger than all other energy scales; in particular we need $T \gg \Lambda$. In contrast, the CQFT was defined by the limit $\Lambda \to \infty$ at fixed $T$, $G$, so the high $T$ limit of the CQFT corresponds to the intermediate temperature range $G \ll T \ll \Lambda$ of the lattice model. It is not possible to access this temperature range by an expansion as simple as (4), and more sophisticated techniques, to be discussed here, are necessary. (One could also, of course, determine a large number of terms in (4) and then use some Padé extrapolation methods to access the $T \ll \Lambda$ region: this method has been used by Sokol *et. al.*[10] and I will not discuss it here).

In contrast to the static properties, the dynamic properties of $\mathcal{H}$ are already non-trivial in the high $T \gg \Lambda$ limit of the lattice model. Although one expects some sort of incoherent, dissipative dynamics, the damping co-efficients cannot be determined directly—all the approaches used so far are essentially variants of the methods discussed by Moriya [11] and Forster [12], and use a short-time expansion, coupled with an ansatz for the spectral function, to extrapolate to the long-time limit. In this paper, we will discuss the dynamics in the high $T$ limit of the CQFT, $G \ll T \ll \Lambda$, or the "quantum-critical dynamics". The dynamics continues to be dissipative and relaxational, but is not amenable to a description either by a classical Boltzmann equation for a dilute gas of quasiparticle excitations, or by a classical Langevin-like models of the types discussed in the classic review of Hohenberg and Halperin [13]. However, as we shall show, the scaling structure of the CQFT does permit some progress to be made; indeed we will discuss below the complete solution of the quantum-critical dynamics of a simple spin model, including the exact determination of a damping co-efficient.

We will begin our discussion with a simple solvable model of spinless fermions in Section 2: this will allow introduction of the main concepts in a very simple setting. Section 3 will extend these results to a related but more complex model of dilute bosons in spatial dimensions $d < 2$. An explicit solution of the relaxational dynamics of the quantum-critical region will be obtained in the discussion in Section 4 on the Ising model in a transverse field. The expository analyses of these toy models will lead in Section 5 to the main system of interest—the $O(3)$ quantum rotor model in $d = 2$. We will also review applications of these results to numerical simulations and experiments. Finally Section 6 will conclude by highlighting the main results and noting recent work on related subjects.

## 2 Dilute gas of spinless fermions

Much of the physics I wish to discuss is displayed in a surprisingly simple model of a dilute gas of spinless fermions at finite temperature: its scaling forms have a structure identical to those of much more complicated models. The main shortcoming of the model is that the associated CQFT has no interactions, and there is therefore no relaxational behavior in the scaling functions.

Consider the following Hamiltonian:

$$\mathcal{H}_F = -t \sum_{<ij>} \left( c_i^\dagger c_j + c_j^\dagger c_i - c_i^\dagger c_i - c_j^\dagger c_j \right) - \mu \sum_i c_i^\dagger c_i \tag{5}$$

where $c_i$ is a spinless fermion annihilation operator at the site $i$ of a $d$-dimensional hypercubic lattice, and $< ij >$ are nearest neighbors. There are no interactions in $\mathcal{H}_F$, so it is trivially solvable. Consider the ground state of $\mathcal{H}_F$ as a function of the dimensionless coupling constant $g = \mu/t$. For $g < g_c = 0$, the ground state has no particles. There is a non-analytic onset in the density of particles at $g = g_c$, signaling a quantum phase transition. The bandwidth $\sim t$ of the fermions plays the role of the upper cutoff in energy ($\Lambda$) for this transition, and the critical region defining the applicability of a CQFT is roughly $T, \mu \ll t$. In fact, it is not difficult to determine the exact effective action of the CQFT (in units with $\hbar = 1$):

$$\mathcal{L}_F = \int d^d x \int_0^{1/T} d\tau \Psi_F^\dagger(x,\tau) \left( \frac{\partial}{\partial \tau} - \frac{\nabla^2}{2m} - \mu \right) \Psi_F(x,\tau) \tag{6}$$

where $\Psi_F(x_i) = a^{-d} c_i$, $m \sim 1/(ta^2)$, and $a$ is the lattice spacing. For this action we can identify by the usual methods the exponents $\nu = 1/2$ and $z = 2$ at the $\mu = 0$ critical point; taking the density $n = \langle \Psi_F^\dagger \Psi_F \rangle$ as the order parameter, we get the exponents $\beta = d/2$ and $\eta = d$. Further, it is also easy to see that all interactions are irrelevant at this critical point in all dimensions $d > 0$ (the least irrelevant interaction term, $|\Psi_F \nabla \Psi_F|^2$, becomes relevant only

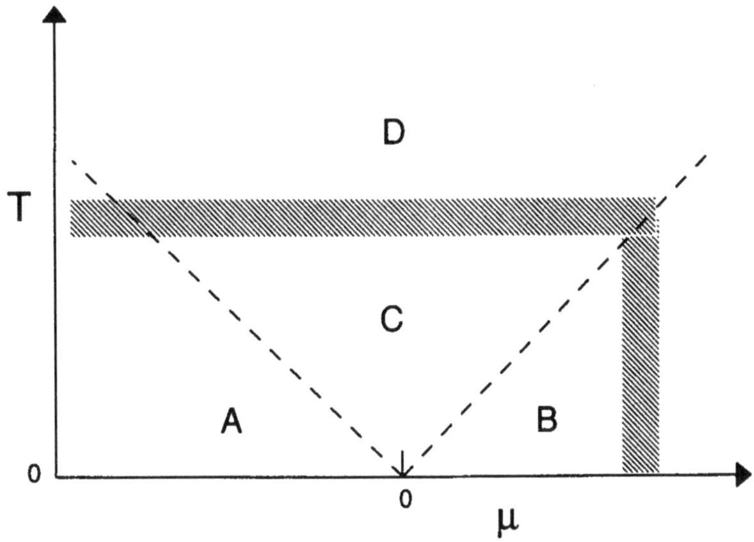


Figure 1: Phase diagram of the dilute Fermi gas $\mathcal{H}_F$ (Eqn (5)) as a function of the chemical potential $\mu$ and the temperature $T$. The regions A, B, C, are separated by crossovers; all observables in these regions are described universal crossover functions of the CQFT $\mathcal{L}_F$ (Eqn (6)). Region A has an exponentially activated fermion density, region B is a Fermi liquid (in $d = 1$, a Luttinger liquid), and region C is quantum critical. The hatched region marks the boundary of applicability of the CQFT and occurs at $\mu, T \sim \Lambda = t$. The same phase diagram also applies to the dilute Bose gas of Sec. 3, but the interpretation of the regions is different.

for $d < 0$). Finally, for the renormalized energy scale measuring the deviation from the critical point, $G$, we may take $G = \mu$, the bare chemical potential in $\mathcal{H}_F$. Note that there is no non-universal scale-factor in the relationship between $G$ and $\mu$—this is a consequence of the triviality of the critical exponents. More typical models with anomalous exponents will have non-universal scale-factors.

We show in Fig 1 the phase diagram of $\mathcal{H}_F$ as a function of $G = \mu$ and $T$. Our interest is primarily in the regions within the hatched lines where the CQFT (6) applies. Within this region there are three physically distinct types of behavior (A, B and C): as $\mathcal{L}_F$ is trivially solvable, the universal properties of A, B, and C and the crossovers between them can be exactly determined. Let us describe the regions in turn:
(A) *Activated* $\mu \ll -T$: The fermions are very dilute, with a density $\sim e^{\mu/T}$. Quantum effects are suppressed and the particles behave classically.
(B) *Fermi or Luttinger liquid* $\mu \gg T$: Now quantum degeneracy effects are paramount. At $T = 0$, the ground state is a Fermi liquid (in $d = 1$, a Luttinger liquid); at finite $T$ thermal effects lead to a small number of particle and hole

excitations near the Fermi surface.

(C) *Quantum Critical* $|\mu| \ll T$: Unlike A and B, the temperature $T$ is the most important energy scale in this region. We can set $\mu = 0$ here without much damage (all corrections involve positive powers of $\mu/T$). The energy of a typical excitation in this region is of order $T$ and as a result, quantum and thermal fluctuations are equally important.

The relationships between the regions becomes clearer upon considering an explicit example of a crossover function. The density of particles $n = \langle \Psi_F^\dagger \Psi_F \rangle$ obeys the scaling form

$$n = (2mT)^{d/2} \Phi_n \left( \frac{\mu}{T} \right) \tag{7}$$

where the universal scaling function $\Phi_n(\overline{\mu})$ is given by

$$\Phi_n(\overline{\mu}) = \int \frac{d^d k}{(2\pi)^d} \frac{1}{e^{k^2 - \overline{\mu}} + 1}. \tag{8}$$

Notice that there are no arbitrary scale-factors in (7). In the activated region A ($\overline{\mu} \ll -1$), we have $\Phi_n \approx e^{\overline{\mu}}/(4\pi)^{d/2}$, which is exactly the result we would have obtain from the classical Maxwell-Boltzmann statistics. In the Fermi liquid region B ($\overline{\mu} \gg 1$), $\Phi_n \approx \overline{\mu}^{d/2} V_d/(2\pi)^d$, where $V_d$ is the volume of the unit sphere in $d$ dimensions; this is the fully quantum result obtained by filling up the Fermi sphere. Most interesting is the quantum critical region C ($|\overline{\mu}| \ll 1$) where

$$\Phi_n(\overline{\mu}) = \zeta \left( \frac{d}{2} \right) \left( \frac{\pi}{2} \right)^d \left( 1 - 2^{1-d/2} \right) + \mathcal{O}(\overline{\mu}) \tag{9}$$

The value of $\Phi_n(0)$ depends upon the details of the Fermi distribution function, and not just its forms in the classical and quantum limits: this illustrates our assertion that quantum and classical effects are equally important in region C.

It is also interesting to compare the behavior of the density in the universal region C with the true high temperature limit of the lattice model - region D. In other words, we are going vertically upwards in $T$ at $\mu = 0$ in Fig 1. It is easy to compute:

$$n = \begin{cases} (2mT)^{d/2} \Phi_n(0) & |G| \ll T \ll \Lambda \quad \text{(region C)} \\ 1/2a^d - c_1/T & T \gg \Lambda \quad \text{(region D)} \end{cases} \tag{10}$$

where $c_1$ is a non-universal constant (recall that in this model $G = \mu$ and $\Lambda = t$). Notice the difference between the universal high $T$ limit of the CQFT (the first result) and the lattice high $T$ limit. I hope that this example has illustrated the general principle, and in the remainder of the paper I will make no further reference to non-universal regions like D. It will be implicitly assumed that we are working with the universal continuum theory, and we will describe only regions like A, B and C.

Before closing our discussion on this deceptively simple model, we discuss the scaling form of observables as a function of momentum $k$ and frequency $\omega$. Because of the absence of interactions, the single-particle Green's function is trivial; so we discuss the density-density correlator $\chi_n(k,\omega)$ which has a slightly more interesting structure. As the particles are free, $\chi_n$ is of course given simply by the Lindhard function, which can be manipulated into the scaling form

$$\chi_n(k,\omega) = \frac{(2mT)^{d/2}}{T}\Phi_{\chi_n}\left(\frac{k}{\sqrt{2mT}}, \frac{\omega}{T}, \frac{\mu}{T}\right) \tag{11}$$

where $\Phi_{\chi_n}$ is a universal complex-valued function related to the Lindhard function. Notice that, like the scaling form (7), there are again no arbitrary scale factors. Further, $\Phi_{\chi_n}$ is well-defined at $\mu/T = 0$, where it yields the dynamic susceptibility of the quantum-critical region C. We will see several other examples of scaling forms like (7) and (11) in this paper, but the scaling functions will not be as simple as they are here.

## 3 Dilute Bose gas

Now we consider the same density onset quantum transition considered in Sec 2, but for the case of bosons. The discussion here is drawn from that of Sachdev, Senthil and Shankar [4] to which the reader is referred for further details.

Unlike the case for spinless fermions, it is no longer possible to ignore the interactions between the particles. We consider the properties of the following continuum model

$$\begin{aligned}\mathcal{L}_B &= \int d^dx \int_0^{1/T} d\tau \Psi_B^\dagger(x,\tau)\left(\frac{\partial}{\partial\tau} - \frac{\nabla^2}{2m} - \mu\right)\Psi_B(x,\tau) \\ &+ \frac{1}{2}\int d^dx \int d^dx' \int_0^{1/T} d\tau |\Psi_B(x,\tau)|^2 v(x-x')|\Psi_B(x',\tau)|^2 \end{aligned} \tag{12}$$

where $\Psi_B$ is a boson annihilation operator, and $v(x)$ is a repulsive interaction of range $\sim a$. Like $\mathcal{L}_F$, $\mathcal{L}_B$ has a quantum phase transition at $\mu = T = 0$, and we will discuss its universal properties here. Because of the finite range of $v$, the universality only sets in at distances larger than $a$. A straightforward RG analysis [14] of the vicinity of the quantum critical point shows that $v$ flows into the $v = 0$ fixed point for $d \geq 2$. It turns out that $v$ is actually dangerously irrelevant for $d \geq 2$: we do not wish to enter into a discussion of such effects here, and so most of our remaining discussion will be restricted to $d < 2$.

For $d < 2$, $v(x)$ flows into a universal fixed point interaction $v(x) = u^* \delta(x)$ for some $d$ dependent constant $u^*$. The scaling structure of this fixed point turns out to be very closely related to the non-interacting spinless fermion model of Sec 2 for the same value of $d$. All exponents and scaling forms of the boson and fermion models are identical, but the scaling functions themselves are different. The crossover phase diagram of the $d < 2$ dilute Bose gas is essentially identical to the fermion phase diagram in Fig 1, but the physical interpretation of the phases is somewhat different:
(A) *Activated* $\mu \ll -T$: This is essentially identical to the fermion case as the particles are dilute and their quantum statistics plays a negligible role.
(B) *Incipient Superfluid* $\mu \gg T$: The ground state is now a superfluid (in $d = 1$ a Luttinger liquid), but classical thermal fluctuations destroy the long range order at any non-zero temperature. Nevertheless, the phase coherence length is large and system behaves like a superfluid at short scales.
(C) *Quantum Critical* $|\mu| \ll T$: This is similar to the fermion case in that $T$ is the most important energy scale. However, there are now strong interactions among the particles, leading to an incoherent excitation spectrum. The system does not display characteristics of a superfluid ground state at any length scale, but instead crosses over directly from free particle behavior at short time scales, to dissipative, relaxational dynamics at long time scales. The single particle Green's function $G(x, \tau) = \langle \Psi_B(x, \tau) \Psi_B^\dagger(0, 0) \rangle$ has a non-trivial scaling function $\Phi_G$ (this form actually holds in all three regions A, B, and C)

$$G^R(k, \omega) = \frac{1}{T} \Phi_G \left( \frac{k}{\sqrt{2mT}}, \frac{\omega}{T}, \frac{\mu}{T} \right). \tag{13}$$

We have Fourier transformed and analytically continued to the retarded Green's function at real frequencies. This scaling form also holds for the spinless fermions of Section 2, but the scaling function then is simply the free fermion form $\Phi_G(\overline{k}, \overline{\omega}, \overline{\mu}) = 1/(\overline{\omega} - \overline{k}^2 - \overline{\mu} + i\eta)$ where $\eta$ is a positive infinitesimal. Computing $\Phi_G$, and other scaling functions, for bosonic quantum critical dynamics is not as easy. One approach [4] is to expand in powers of the fixed-point interaction $u^*$ which becomes small as $d$ approaches 2 from below: this becomes an expansion in $\varepsilon = 2 - d$.

In $d = 1$, it is possible to make more explicit progress. It has been argued [4] that now $u^* = \infty$. The bosons thence become impenetrable, and their quantum mechanics becomes identical to those of free fermions. Hence, in $d = 1$, the quantum critical dynamics of dilute gases of spinless fermions and boson are described by the same CQFT, $\mathcal{L}_F$ of Eqn (6). For the bosonic system we have to supplement $\mathcal{L}_F$ with the following non-local relationship between the boson and fermion operators (essentially a continuum Jordan-Wigner transformation)

$$\Psi_B(x) = \exp\left(i\pi \int_{-\infty}^{x} dx' \Psi_F^\dagger(x')\Psi_F(x')\right) \Psi_F(x) \tag{14}$$

We are not home yet, as evaluating correlators of (14) under (6) is not easy. Korepin and Slavnov [15,16] have succeeded in showing how this problem may be reduced to determining the solution and Fredholm determinant of a linear Fredholm integral equation. At this stage, numerical analysis is required, and some scaling functions have been determined to essentially arbitrary accuracy[4].

Finally, an additional comment about the $d \geq 2$ case. There is now true superfluidity and a finite temperature phase transition to a normal state, all within region B. The remainder of the phase diagram remains the same as in the $d \leq 2$ case.

## 4 Ising Model in a transverse field

Unlike the dilute Fermi and Bose gases, the Ising model possesses anomalous exponents. Yet it is simple enough in $d = 1$ to allow exact computation of a quantum-critical dynamic correlation function. We will also review, in this section, the very useful and general mapping between the quantum model and an equivalent classical statistical mechanics model; we will then discuss the crossovers in the phase diagram like Fig 1 in the context of the classical model. Some of of the following discussion is a review of well-known properties of the Ising model [17,18,19,20]; our main purpose here is to use the explicit solution in $d = 1$ to present a physical interpretation which generalizes to other quantum phase transitions.

We will explicitly discuss the following Hamiltonian, describing the Ising model in a transverse field in $d = 1$ (we will remark briefly on the generalization to higher $d$):

$$\mathcal{H}_I = -J \sum_i (g\sigma_{x,i} + \sigma_{z,i}\sigma_{z,i+1}) \tag{15}$$

where $J > 0$ in an overall energy scale, $g > 0$ is a dimensionless coupling constant, and $\sigma_{x,i}, \sigma_{z,i}$ are Pauli matrices on a chain of sites, $i$. Consider the ground states of $\mathcal{H}_I$ for small and large $g$ in turn. For small $g$, the second term in (15) dominates and the spins all align themselves either in the $+z$ or $-z$ directions: there is a spontaneous magnetization and spin-reversal symmetry is broken. On the other hand, for large $g$, the first term in (15) prefers a state which is in a superposition of $\sigma_z$ eigenstates, with different sites uncorrelated:

the wavefunction looks like $\prod_i(|+\rangle_i + |-\rangle_i)$. These two limits are separated by a phase transition at $g = g_c$ (in fact, $g_c = 1$, exactly, because of a self-duality property of (15)). In this section, we shall discuss the finite $T$ dynamic properties in the vicinity of $g = g_c$.

To begin, we recall the well-known fact [17,18] that the $T = 0$ correlators of $\mathcal{H}_I$ are similar to those in the classical, two-dimensional, Ising model given by the partition function Tr $e^{-F}$ with

$$F = -K \sum_{<ij>} \sigma_{z,i}\sigma_{z,j} \tag{16}$$

where the sites $i, j$ now lie on a square lattice (say). This classical model has a phase transition at $K = K_c$. At the level of the critical continuum theories, the mapping between the classical and quantum models become exact [18]. There is a simple relationship between the two-dimensional field theory, with classical degrees of freedom, describing $F$ in the vicinity of $K = K_c$ and the one-dimensional quantum field theory describing $\mathcal{H}_I$ near $g = g_c$: one simply identifies one of the spatial directions of the classical theory as an imaginary time, and then analytically continues the correlators to real time, to obtain observables of the quantum theory. This mapping leads immediately to some useful information. As the classical theory is spatially isotropic, the quantum theory has dynamic exponent $z = 1$. Further, in the classical model the $\sigma_z$ $\sigma_z$ correlator behaves like $\sim p^{-7/4}$ in momentum space [19] ($p$ is the two-dimensional momentum of the classical model); analytically continuing this to real time, we obtain for the dynamic susceptibility $\chi(k, \omega)$ (this is the $\sigma_z$ $\sigma_z$ correlator of the quantum $\mathcal{H}_I$ and $k$ is now a one-dimensional spatial momentum) as a function of $d = 1$ momentum $k$ and frequency $\omega$:

$$\chi(k, \omega) = \frac{Z}{(c^2k^2 - \omega^2)^{7/8}} \qquad T = 0, g = g_c. \tag{17}$$

Here $Z$ and $c$ (an excitation velocity) are non-universal constants. We plot $\text{Im}\chi(k, \omega)/\omega$ in Fig 2. Notice that there are no delta functions in the spectral density, indicating the absence of any well-defined quasiparticles. Instead, we have a critical continuum of excitations. We can also compute the dynamic local susceptibility,

$$\chi_L''(\omega) = \int \frac{dk}{2\pi} \text{Im}\chi(k, \omega), \tag{18}$$

a quantity often measured in neutron scattering experiments:

$$\chi_L''(\omega) = \text{sgn}(\omega) \frac{\sqrt{\pi} Z}{2\Gamma(7/8)\Gamma(5/8)c} \left(\frac{1}{|\omega|}\right)^{3/4} \qquad T = 0, g = g_c. \tag{19}$$

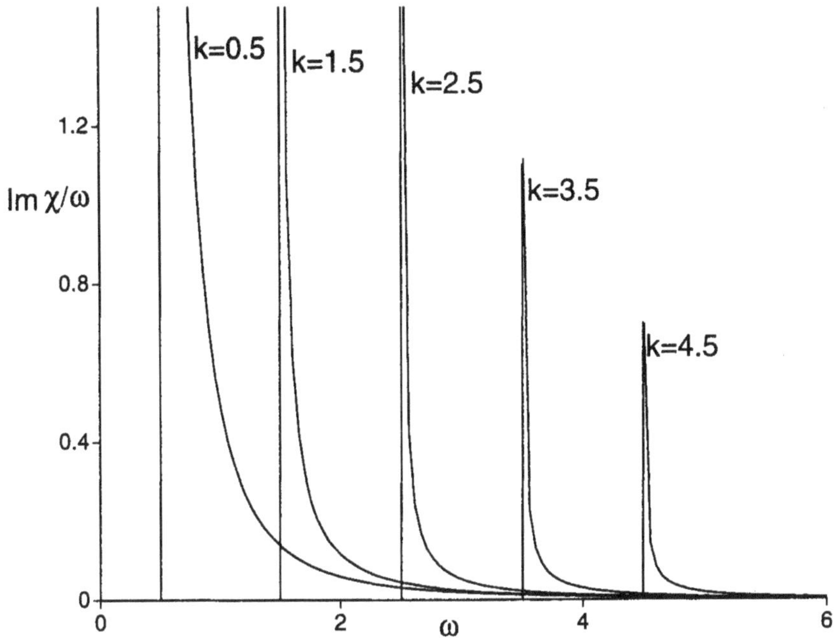


Figure 2: Spectral density of the transverse field Ising model (15) at its critical point $g = g_c$ at $T = 0$. We chose $Z = c = 1$ in Eqn (17)

This quantity is the density of states of local spin-flip excitations, and has a divergence as $\omega \to 0$.

Our discussion so far has been at $T = 0$, and let us turn now to non-zero $T$. A finite $T$ translates into a *finite size* $L_\tau = 1/T$ for the classical model along imaginary time direction. Periodic boundary conditions are imposed along the finite direction, and so the classical model has the geometry of a cylinder with circumference $L_\tau = 1/T$. We now discuss the crossovers at finite $T$ in the context of the classical model. At $L_\tau = \infty$ ($T = 0$) we can characterize the deviations from criticality by the correlation length $\xi \sim |g - g_c|^{-1}$ (for the quantum model we may choose our renormalized energy scale $G = 1/\xi \sim |g - g_c|$). At distances shorter than $\xi$, the spins display critical correlations characteristic of the point $K = K_c$; it is only at distances larger than $\xi$ that they become sensitive to the value of $K - K_c$ and display characteristics of the ordered (for $K > K_c$) or the paramagnetic (for $K < K_c$) phase. Now consider the effect of a finite $L_\tau$; there are two distinct possibilities:

(*i*) $\xi < L_\tau$ (for the quantum model, $T < G$): Moving from the shortest to largest length scales, the crossover from critical to non-critical (either ordered or paramagnetic) behavior still occurs at a scale $\sim \xi$; the length scale $L_\tau$ has little effect at this point. The effects of $L_\tau$ only become apparent at larger scales, at which point it is permissible to use an effective model which characterizes the non-critical ground state.
(*ii*) $L_\tau < \xi$ (for the quantum model, $T > G$): Now the short-distance critical fluctuations see a finite size $L_\tau$ *before* they have had a chance to become sensitive to $K - K_c$. These critical fluctuations are quenched by finite size effects in a universal way. The resulting non-critical theory then responds only weakly at the scale $\xi$. Note that the system does not display characteristics of the ordered or the paramagnetic state, of the $L_\tau = \infty$ system, at *any* length scale.

The above arguments are summarized in Figs 3 and 4; notice that Fig 3 is quite similar to Fig 1. In all three regions of Fig 3, at the largest frequencies, $\omega$, $\mathcal{H}_I$ displays the critical correlations of the $g = g_c$ point as depicted in Fig 4 (we have returned now to the language of the quantum model $\mathcal{H}_I$).

In regions A and B there is a crossover from these critical fluctuations, at an energy scale $G$, to the behavior of the ordered ($g < g_c$) or paramagnetic ($g > g_c$) ground state of $\mathcal{H}_I$ (see Fig 4). Both ground states have a gap, and thermal fluctuations will lead to dilute gas of quasiparticle excitations. We expect that an effective classical model (like Glauber [21] dynamics, which is similar in spirit to the Langevin models of Hohenberg and Halperin [13]) will provide a suitable description of these thermal fluctuations. In $d = 1$, on the ordered side ($g < g_c$), these quasiparticle excitations are the 'kink' and 'anti-kink' solitons; even an infinitesimal concentration of these is sufficient to destroy long-range order at any finite temperature. However, for $d > 1$, long-range order is not immediately destroyed: as a result there is finite temperature phase transition within region A where the magnetic moment disappears; this transition will be in the universality class of the $d$-dimensional classical Ising model [17].

In the quantum-critical region C, the critical fluctuations are quenched by thermal effects at the energy scale $T$ (see Fig 4; an early analysis of finite $T$ crossovers in the transverse-field Ising model by Suzuki [17] failed to identify region C). This quenching is completely universal and will be described explicitly below. The system has had no chance to display any characteristic of either non-critical ground state at any frequency scale. At low frequencies, the system realizes a new quantum relaxational regime. It is this regime which is really characteristic of the region C, and not the high frequency critical behavior which is present in all three regions. It is in this sense that the name "quantum-critical" of region C is a misnomer.

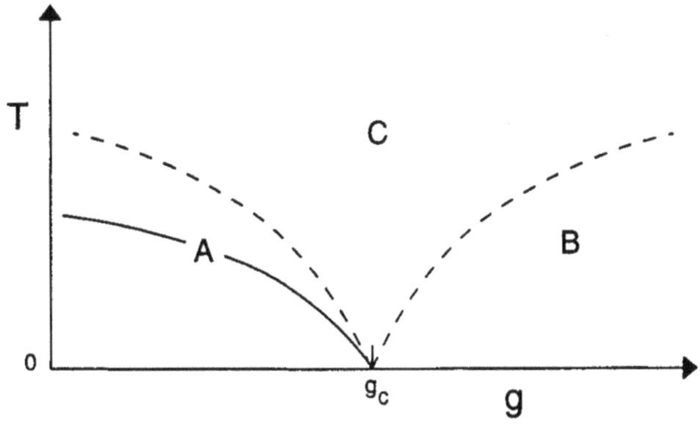


Figure 3: Phase diagram of the transverse field Ising model in $d \geq 1$ dimensions; dashed lines denote crossover boundaries, while the solid line is a phase transition. In region A the system displays characteristics of the ordered ground state; thermal fluctuations about this state can destroy long-range order at any non-zero $T$ in $d = 1$, but for $d > 1$ there is a phase transition (in the universality class of the $d$-dimensional classical Ising model). In region B the system is a gapped quantum paramagnet, and region C is quantum-critical. The crossover boundaries obey $T \sim |g - g_c|^{z\nu}$ with $z = 1$ and $\nu$ the correlation length exponent of the $d+1$ dimensional classical Ising model ($\nu = 1$ in $d = 1$). The same phase diagram also applies to the quantum rotor model of Section 5 in $d \geq 2$, with the difference that the phase transition in region A is present only for $d > 2$ and is then in the universality class of the $d+1$ dimensional classical Heisenberg ferromagnet. Chakravarty *et. al.*[6] obtained this phase diagram for the $d = 2$ quantum rotor model.

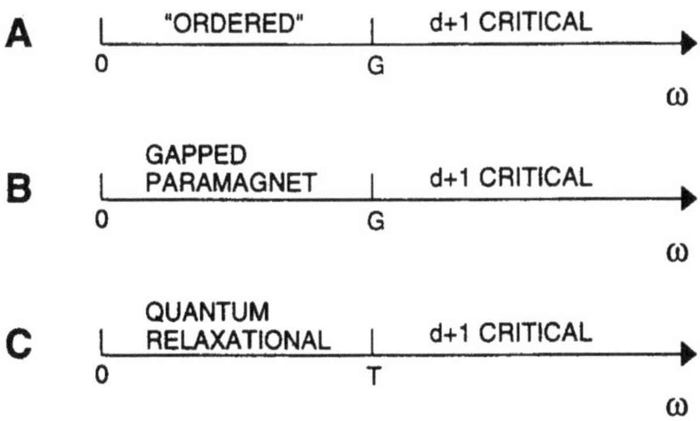


Figure 4: Crossovers as a function of probe frequency in the regions of Fig. 3 for the transverse-field Ising model in $d$ dimensions. The $d+1$ critical regime is described by the CQFT at $g = g_c$ and $T = 0$. The "ordered" regime is where a classical description in terms of a gas of well-separated domain walls in an otherwise ordered state is appropriate. In $d = 1$ such "kinks" destroy long-range-order at any non-zero $T$. The same crossovers also apply to the quantum rotor model of Section 5: the "ordered" regime is now one where spin-wave fluctuations about an ordered ground state dominate, and these fluctuations destroy long-range-order at any non-zero $T$ for $d = 2$.

As in the case of the dilute Fermi and Bose gases considered earlier, the dynamic susceptibility $\chi(k,\omega)$ will satisfy a universal scaling form over the regions of Fig 3:

$$\chi(k,\omega) = \frac{Z}{T^{7/4}}\Phi_{\chi}\left(\frac{ck}{T},\frac{\omega}{T},\frac{G}{T}\right) \tag{20}$$

In the following we will determine the leading term in $\Phi_\chi$ in the quantum-critical region C *i.e.* we will present an exact expression for $\Phi_\chi(\overline{k},\overline{\omega},0)$.

In the classical model $F$ at $K = K_c$ and $L_\tau = \infty$, we know from (17) that the Green's function $G(x,\tau) = \langle\sigma_z(0,0)\sigma_z(x,\tau)\rangle \sim (x^2+\tau^2c^2)^{-1/8}$; we are using the label $\tau$ for spatial direction corresponding to imaginary time. We can now use a remarkable result of Cardy [22], which relies on the conformal invariance of this critical theory, to obtain an exact result for $G$ in a system with a finite $L_\tau$:

$$G(x,\tau) = \frac{\Gamma(1/8)Z}{2^{13/8}\pi^{3/4}\Gamma(7/8)c}\left(\frac{1}{L_\tau}\right)^{1/4}\left(\frac{1}{\cosh(2\pi x/L_\tau c) - \cos(2\pi\tau/L_\tau)}\right)^{1/8} \tag{21}$$

This result has been asserted earlier by an inspection of the partial differential equation satisfied by $G$ [23]. Although results like (21) have been known to conformal field theorists for some time, they usually interchange the roles of $x$ and $\tau$ *i.e.* they consider systems of finite spatial length $L_x$, and infinite temporal length, so that the system is in its ground state. For us, the spatial extent is infinite, and there is a finite length $L_\tau$ along the $\tau$ direction, with the correlator (21) periodic in $\tau$ with period $L_\tau$ (such a perspective has also been discussed by Shankar [24] and by Korepin *et. al.* [16]). Note that it doesn't really make sense to talk about the long imaginary time limit, $\tau \gg L_\tau$. However, after analytic continuation to real time, the long time limit of the quantum problem, $t \gg 1/T$ or $\omega \ll T$, is eminently sensible, and is precisely the new quantum relaxational regime that we wish to access. The analytic continuation is a little more convenient in Fourier space: we Fourier transform (21) to obtain $G(k,\omega_n)$ at the Matsubara frequencies $\omega_n$ and then analytically continue to real frequencies (there are some interesting subtleties in the Fourier transform to $G(k,\omega_n)$ and its analytic structure in the complex $\omega$ plane, which are discussed elsewhere [4]). This gives us the universal function $\Phi_\chi$ in the quantum-critical region:

$$\Phi_\chi(\overline{k},\overline{\omega},0) = \frac{1}{(4\pi)^{7/4}}\frac{\Gamma\left(\frac{1}{16}+i\frac{\overline{\omega}+\overline{k}}{4\pi}\right)\Gamma\left(\frac{1}{16}+i\frac{\overline{\omega}-\overline{k}}{4\pi}\right)}{\Gamma\left(\frac{15}{16}+i\frac{\overline{\omega}+\overline{k}}{4\pi}\right)\Gamma\left(\frac{15}{16}+i\frac{\overline{\omega}-\overline{k}}{4\pi}\right)}. \tag{22}$$

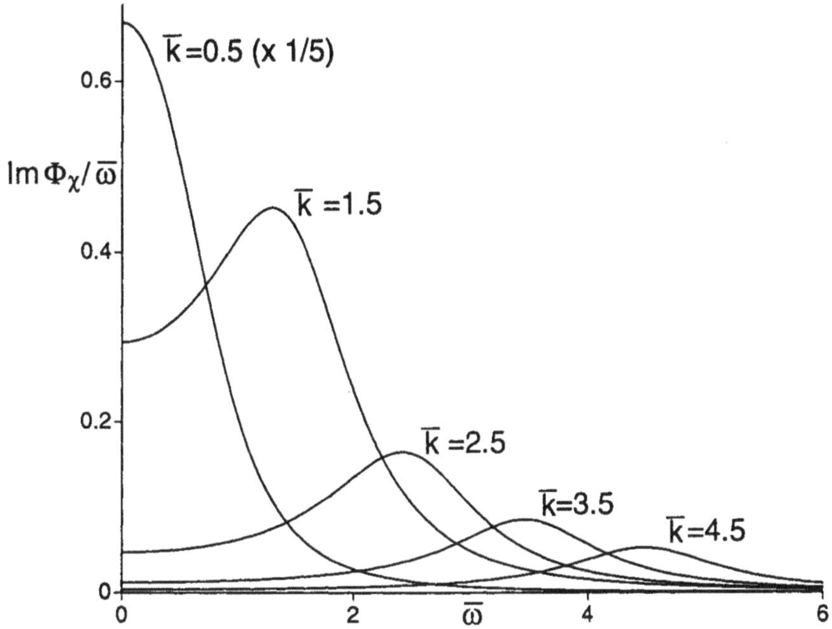


Figure 5: Scaling function for the imaginary part of the dynamic susceptibility in the quantum critical region C of the transverse field Ising model in $d = 1$. The susceptibility $\chi = Z\Phi_\chi/T^{7/4}$ and $\overline{k} = ck/T$, $\overline{\omega} = \omega/T$.

We show a plot of $\text{Im}\Phi_\chi/\overline{\omega}$ in Fig 5. This result is the finite $T$ version of Fig 2. Notice that the sharp features of Fig 2 have been smoothed out on the scale $T$, and there is non-zero absorption at all frequencies. We can also observe the crossover as a function of frequency claimed earlier in Fig 4. Notice that for $\overline{\omega}, \overline{k} \gg 1$ there is a well-defined peak in $\text{Im}\Phi_\chi/\overline{\omega}$ (Fig 5) rather like the $T = 0$ critical behavior of Fig 2. However, for $\overline{\omega}, \overline{k} \ll 1$ we cross-over to the quantum relaxational regime and the spectral density $\text{Im}\Phi_\chi/\overline{\omega}$ is similar to a Lorentzian around $\overline{\omega} = 0$. This relaxational behavior can be characterized by a relaxation rate $\Gamma_R$ defined as [13]

$$\Gamma_R^{-1} = -i \left. \frac{\partial \ln \chi(0,\omega)}{\partial \omega} \right|_{\omega=0} ; \tag{23}$$

(this is motivated by the phenomenological relaxational form $\chi(0,\omega) = \chi_0/(1-$

$i\omega/\Gamma_R + \mathcal{O}(\omega^2)))$. From (20) and (22) we determine:

$$\Gamma_R = \left(2\tan\frac{\pi}{16}\right)\frac{k_B T}{\hbar}, \tag{24}$$

where we have returned to physical units. The ease with which this result was obtained belies (I claim) its remarkable nature. Notice that we are working in a closed Hamiltonian system, evolving unitarily in time with the operator $e^{-i\mathcal{H}_I t}$, from an initial density matrix given by the Gibbs ensemble at a temperature $T$. Yet, we have obtained relaxational behavior at low frequencies, and determined an exact value for a dissipation constant. Such behavior is more typically obtained in phenomenological models which couple the system to an external heat bath and postulate an equation of motion of the Langevin type. Notice also that $\mathcal{H}_I$ in (15) is known to be integrable with an infinite number of conservation laws [18]. However, the conservation laws are associated with a mapping to a free fermion model and are highly non-local in our $\sigma_z$ degrees of freedom; they play essentially no role in our considerations, and do not preclude relaxational behavior in the $\sigma_z$ variables.

For completeness we also present results on a related observable which shows the crossover from critical to quantum relaxation behavior. We consider the local susceptiblity $\chi_L''$, and obtain the finite $T$ form of (19) by integrating (22) over momenta:

$$\chi_L''(\omega) = \frac{Z}{cT^{3/4}}\Phi_L\left(\frac{\omega}{T},\frac{G}{T}\right) \tag{25}$$

$$\Phi_L(\overline{\omega},0) = \frac{1}{2^{7/4}\pi^{5/4}\Gamma(5/8)\Gamma(7/8)}\sinh\left(\frac{\overline{\omega}}{2}\right)\left|\Gamma\left(\frac{1}{8}-i\frac{\overline{\omega}}{2\pi}\right)\right|^2 \tag{26}$$

A plot of the scaling function $\Phi_L$ is shown in Fig 6. The function has the asymptotic limits:

$$\Phi_L(\overline{\omega},0) = \begin{cases} \dfrac{\Gamma^2(1/8)}{2^{11/4}\pi^{5/4}\Gamma(5/8)\Gamma(7/8)}\overline{\omega} & |\overline{\omega}| \ll 1 \\ \mathrm{sgn}(\overline{\omega})\dfrac{\sqrt{\pi}}{2\Gamma(7/8)\Gamma(5/8)}\left(\dfrac{1}{|\overline{\omega}|}\right)^{3/4} & |\overline{\omega}| \gg 1 \end{cases} \tag{27}$$

The small $\overline{\omega}$ behavior is relaxational as $\Phi_L$ is linear in frequency, and the critical behavior at large frequencies agrees with (19).

## 5 Quantum rotors in two dimensions

The study of this model is of direct experimental interest, as it is believed[6] to be a reasonable model of the spin fluctuations in antiferromagnetic compounds

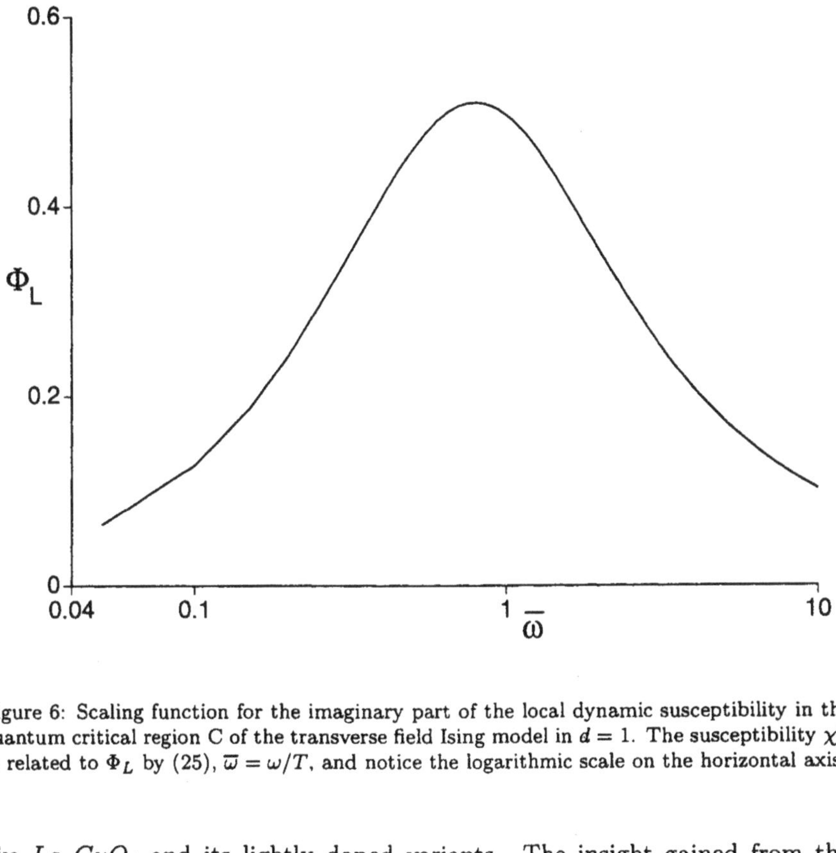


Figure 6: Scaling function for the imaginary part of the local dynamic susceptibility in the quantum critical region C of the transverse field Ising model in $d = 1$. The susceptibility $\chi_L''$ is related to $\Phi_L$ by (25), $\overline{\omega} = \omega/T$, and notice the logarithmic scale on the horizontal axis.

like $La_2CuO_4$ and its lightly doped variants. The insight gained from the simple models studied in the previous sections will now be of great use, and we will rapidly be able to present a scaling analysis of its quantum phase transition.

The Hamiltonian of the quantum rotor model is

$$\mathcal{H}_R = J\sum_i \frac{g}{2}\vec{L}_i^2 - J\sum_{<ij>} \vec{n}_i.\vec{n}_j \tag{28}$$

where $J > 0$ is an overall energy scale, $g > 0$ is a dimensionless coupling constant, and $i$,$j$ are the sites of a two dimensional lattice ($< ij >$ denotes nearest neighbors). Notice the similarity between the forms of $\mathcal{H}_R$ and $\mathcal{H}_I$ in (15): it will turn out that the corresponding terms play a similar role. On each site $i$ of the lattice we have the 3-component vector operators $\vec{L}$, $\vec{n}$ (dropping the site index), which obey the commutation relations:

$$[n_a, n_b] = 0 \quad , \quad [L_a, n_b] = i\epsilon_{abc}n_c \quad , \quad [L_a, L_b] = i\epsilon_{abc}L_c. \tag{29}$$

The vector $\vec{n}$ is of unit length $\vec{n}^2 = 1$, and its orientation identifies direction of the local magnetic order; the quantum rotor model is usually considered as an effective model for an underlying system of Heisenberg spins—in this case the magnetic order can be any ordering which is specified by a single vector and has no spatially averaged magnetic moment. The simplest example of this is the two sublattice Néel ordering, and we will therefore refer to $\langle\vec{n}\rangle$ as the Néel order parameter. The $\vec{L}$ operator measures the angular momentum, and as all phases have no net magnetic moment, we will always have $\langle\vec{L}\rangle = 0$.

For further insight into the meaning of $\mathcal{H}_R$, consider the eigenstates of a single site Hamiltonian $Jg\vec{L}^2/2$. This describes a particle moving on a unit sphere with angular co-ordinate $\vec{n}$ and kinetic energy $Jg\vec{L}^2/2$. Its eigenenergies are $Jg\ell(\ell+1)/2$ with degeneracy $2\ell+1$ where $\ell = 0, 1, 2, 3 \ldots$. The ground state is a non-degenerate singlet ($\ell = 0$) and has maximum uncertainty in the orientation of $\vec{n}$. For large $g$, the ground state of $\mathcal{H}_R$ can be approximated by the tensor product of $\ell = 0$ states on each site. This state is clearly a quantum paramagnet and has a gap, $\Delta$, to all excitations. Notice the similarity between this state and the large $g$ quantum paramagnet of the Ising model $\mathcal{H}_I$; in both cases the order parameters $\sigma_z$, $\vec{n}$ are in a state of maximum uncertainty. The small $g$ limit of $\mathcal{H}_R$ is also similar to the small $g$ limit of $\mathcal{H}_I$: now the exchange interactions between the sites prefer a state in which $\vec{n}$ has the same definite orientation on each site. Therefore, we expect long-range Néel order in the small $g$ ground state. These two limiting states will be separated by a quantum phase transition at $g = g_c$, which is, of course, the main subject of interest in this section.

Before discussing the critical properties, we pause to remark on the relationship between the rotor model and Heisenberg antiferromagnets. Consider a pair of antiferromagnetically coupled spin-$S$ Heisenberg spins: the eigenstates of this pair will have energies $\propto \ell(\ell+1) - \text{const}$ for $\ell = 0, 1, \ldots 2S$. Notice the similarity between these states and those of a single quantum rotor; the only difference is that there is no upper limit on the maximum value of $\ell$ in the rotor case. However, it is reasonable to expect that these extra high energy states will not modify the low energy properties of lattice models. Therefore, there is little reason to doubt that the critical properties of Heisenberg antiferromagnets with a natural pairing of spins (or more generally, a natural clustering into an even number of spins) will be same as those of the rotor model. Antiferromagnets with no such pairing contain net Berry phase terms in their imaginary time path integral, beyond those present for the rotor model. However these

Berry phases cancel between the sites, except for "hedgehog"-like spacetime singularities [25], and it has been argued [1,3] that these remnant Berry phases have no effect of the leading critical singularities. The reader is referred to the original papers for further discussion on these subtle issues [26,3]: we will restrict our discussion here to the much simpler rotor model $\mathcal{H}_R$.

The critical properties and phase diagram of the $d = 2$ rotor model $\mathcal{H}_R$ turn out to be remarkably similar to those of the transverse field Ising model $\mathcal{H}_I$ of (15) in $d = 1$. Like the $d = 1$ Ising model, the $d = 2$ rotor model has no phase transition at any finite $T$: so the phase diagram of Fig 3 applies to $\mathcal{H}_R$, with no phase boundary in region A (the phase diagram for $\mathcal{H}_R$ in $d = 2$ was obtained first by Chakravarty *et. al.* [6]). In both systems, the $T = 0$, $g = g_c$ critical point has $z = 1$. In the case of $\mathcal{H}_I$ this critical point was described by a CQFT which upon analytic continuation to imaginary time was the field theory of the two-dimensional classical Ising model. The analogous mapping for $\mathcal{H}_R$ yields the CQFT associated with the field theory for the three-dimensional, classical, Heisenberg ferromagnet. The universal scaling functions describing the crossovers in Fig 3 and 4 have an identical form in both theories, although, because the critical field theories are different, the critical exponents, universal amplitude ratios, and scaling functions will have different numerical values. The explicit results presented in Sec 4 for quantum-critical region C of the Ising model, all apply, unchanged in form, to the region C of the $d = 2$ rotor model: the qualitative features of the spectral functions in Figs 2, 5, and 6 remain the same, and the relaxation rate $\Gamma_R$ (defined in (23)) satisfies (24) but with a different universal numerical prefactor. However, unlike the $d = 1$ Ising model, we cannot now get exact numerical results for the scaling functions of $\mathcal{H}_R$: this is because the three-dimensional classical Hiesenberg ferromagnet is not exactly solvable (unlike the two-dimensional classical Ising model). Instead we have to be satisfied by approximate methods; reasonably accurate numerical estimates can be obtained in the $1/N$ expansion which has been discussed at length by Chubukov, Sachdev and Ye [3].

There is a small, but significant, difference between the Ising model in $d = 1$ and the rotor model in $d = 2$ which cannot go unmentioned. This difference applies mainly to "ordered" regime in region A (See Figs 3 and 4). The $T = 0$ ground state of $\mathcal{H}_I$ for $g < g_c$ has a gap, associate with the finite energy cost of creating a kink or anti-kink soliton. In contrast, the $g < g_c$ ground state of $\mathcal{H}_R$ has gapless spin-wave excitations because of the broken continuous $O(3)$ symmetry of the ordered state. So we can no longer use the gap, $\Delta$, as the energy scale, $G = \Delta \sim (g_c - g)^{z\nu}$ for measuring deviations from $g = g_c$ for $g < g_c$. A convenient substitute turns out to be the spin stiffness $G = \rho_s$ which has the physical dimensions of energy in $d = 2$, and which also vanishes as $\rho_s \sim (g_c - g)^{z\nu}$. At finite $T$ in region A, it is known [27,6] that the spin correlation length $\sim \exp(2\pi\rho_s/T)$ as $T \to 0$. It is interesting to note that the behavior of $\mathcal{H}_I$ in $d = 1$ is very similar: in this case the correlation length is determined by the mean spacing between kinks, and therefore behaves as $\sim \exp(\Delta/T)$ for low $T$ in region A. (Related to the exponential divergence of the correlation length, there is a further sub-division of the "ordered" regime of region A (Figs 3 and 4) at the energy scale $c$/(correlation length); this complication occurs for both the $d = 2$ rotor and the $d = 1$ Ising models, and has been discussed elsewhere [3,6].)

Associated with the continuous $O(3)$ symmetry of $\mathcal{H}_R$, there is an important observable whose properties cannot be deduced by an analogy with the the Ising model. This is the uniform susceptibility $\chi_H$, the response to a field, $H$ which couples to the global conserved charge associated with the continuous symmetry:

$$\mathcal{H}_R \to -H \sum_i L_{zi} \tag{30}$$

The scaling dimension of $\chi_H$ can be determined exactly using symmetry arguments and the assumption of hyperscaling [2,5]: this yields the scaling form

$$\chi_H = \frac{T}{c^2} \Phi_H \left( \frac{\rho_s}{T} \right) \tag{31}$$

Here $c$, is the same velocity that appears in a $T = 0$ correlator like (17), and $\Phi_H$ is a fully universal function. This function was computed exactly [2,3] in a $O(N = \infty)$ rotor model, along with $1/N$ corrections in some limits; it has the limiting behavior

$$\Phi_H(r) = \begin{cases} \dfrac{\sqrt{5}}{\pi} \ln \left( \dfrac{\sqrt{5}+1}{2} \right) \left[ 1 - \dfrac{0.6189}{N} + \ldots \right] + \ldots & r \to 0 \\ \dfrac{2r}{N} + \dfrac{N-2}{N} + \ldots & r \to \infty \end{cases} \tag{32}$$

The two terms in the second result ($r \to \infty$) are expected to be exact to all orders in $1/N$; the same two terms were also obtained by Hasenfratz and Niedermayer [28]. Notice from (31) and (32) that $\chi_H$ has a linear dependence on $T$ both for $T \ll \rho_s$ and $T \gg \rho_s$; the slopes however differ by a factor of about 3 (for $N = 3$) and this will be important for experimental comparisons [2,3].

### 5.1 *Comparison with simulations and experiments on Heisenberg antiferromagnets*

The most straightforward comparison is with the double-layer, spin-1/2 Heisenberg antiferromagnet [29]. This model consists of spin-1/2 Heisenberg spins on two adjacent square lattices, with an intralayer antiferromagnetic exchange $J$ and an interlayer antiferromagnetic exchange $K$. The ratio $K/J$ acts much like the dimensionless coupling $g$, with the large $K/J$ a gapped quantum paramagnet of singlet pairs of spins in opposite layers, and the small $K/J$ magnetically ordered. Extensive numerical simulations have been carried out on this model by Sandvik and collaborators [30], and the critical point $K = K_c$ identified rather precisely. It is then possible to study the quantum-critical region C quite carefully as it extends over the maximum $T$ range. A number of universal amplitude ratios, including those associated with $\chi_H$, and all results are now in good agreement with the $1/N$ expansion on the $O(N)$ quantum rotor model. Results from high temperature series expansions on the double-layer model also support this conclusion [31].

Secondly, comparisons have been made with the single-layer, square lattice spin-1/2 Heisenberg antiferromagnet, both via simulations and by experimental measurements on $La_2CuO_4$. This model has long-range order at $T = 0$, so must map onto the non-linear sigma model with $g < g_c$. The low $T$ region A was studied in the paper of Chakravarty, Halperin and Nelson [6], with good experimental agreement. Here we focus on the issue of whether this lattice model exhibits the CQFT high $T$ behavior of region C, or it goes directly from region A to a non-universal, lattice dominated high $T$ region like D of Fig 3. It was first argued by Chubukov and Sachdev [2] that this model does indeed possess a significant intermediate temperature regime of region C: this was based on comparisons with the $T$ dependence of $\chi_H$, with the factor of 3 alluded to above playing an important role. They also noted that it would be difficult to identify this region in the correlation length, an observation that was subsequently re-iterated by Greven *et. al.* [32]. A rather convincing demonstration of the presence of region C was given recently by Elstner *et. al.* [31,10] who examined a large number of observables in a high $T$ expansion, and found good consistency with the universal rotor model results.

Also significant in this context have been nuclear magnetic resonance experiments of Imai *et. al.* [33] on $La_2CuO_4$. They have measured the $T$ dependence of the $1/T_1$ and $1/T_2$ relaxation rates at intermediate temperatures. Their observations are in reasonable agreement with the predictions [2,3,34,35] that can be obtained from the universal scaling result for $\chi(k,\omega)$ in the quantum-critical region C.

## 6 Conclusions

This paper has presented a discussion of the vicinity of a second-order quantum phase transition in the context of a number of simple models. The overall

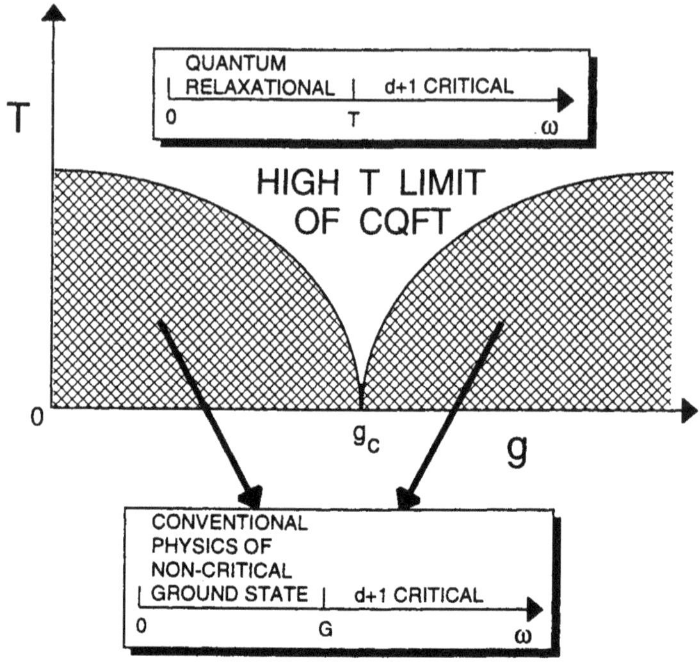


Figure 7: Generic phase diagram of a $d$-dimensional system displaying a second-order quantum transition at $g = g_c$ and $T = 0$. The boundary of the shaded region represents a crossover and not a phase transition. See the text for more information.

picture that emerges is summarized in Fig 7, which shows a generic phase diagram in the plane of a coupling constant $g$, and the temperature $T$ of a $d$-dimensional system; this phase diagram is a generalization of a diagram obtained first by Chakravarty *et. al.*[6] for the $d = 2$ quantum rotor model. The quantum phase transition occurs at the point $g = g_c$, $T = 0$. Associated with this critical point, we can define a continuum quantum field theory (CQFT) over spacetime. In general, space ($x$) and time ($\tau$) do *not* play a similar role in the CQFT and have different scaling dimensions: $x \to x/s$, $\tau \to \tau/s^z$ under a spatial rescaling by $s$, with $z$ the dynamic exponent. Further, even in imaginary time, the action for the CQFT can be complex due to the presence of Berry phases, and therefore corresponds to a statistical mechanics model with complex weights. Correlators of the CQFT are the universal functions describing crossovers in the vicinity of $g = g_c$, $T = 0$. The CQFT has no ultraviolet cutoff, but is characterized solely by two energy scales: the temperature $T$ and an energy scale $G \sim |g - g_c|^{z\nu}$ characterizing the deviation of the ground state from the critical point (here $\nu$ is the correlation length exponent). The value of the ratio $G/T$ determines two distinct regions of the CQFT shown in Fig 7. In both regions there is a high frequency regime ($\omega > \max(T, G)$) which is dominated by excitations of the critical CQFT of the $g = g_c$, $T = 0$ point. The regions are distinguished only by their low frequency behavior, which we discuss in turn:

(*i*) The low $T$ region ($T \ll G$, shown shaded in Fig 7) is a region of "conventional" physics for frequency scales $\omega < G$. This frequency regime can be understood by beginning with the non-critical ground state and examining the particle-like excitations above it. A simple classical model (a Boltzmann equation for a gas of quasiparticle excitations, or a Langevin model of the types discussed by Hohenberg and Halperin [13]) is usually adequate for describing the long-distance, long-time dynamics of these excitations. The shaded region can also contain thermally driven, phase transitions; these transitions will be described by a classical field theory.

(*ii*) The high $T$ region of the CQFT has a novel quantum relaxational regime. This regime is not described by an effective classical model, and displays intrinsic quantum-mechanical effects at the longest time and distance scales. A scaling analysis for this regime was reviewed in this paper. In some cases, as in the model of Section 4, it is possible to obtain an exact value for the relaxation constant.

This paper has reviewed only a small portion of what is a rapidly developing subject. We list below a number of recent (and not so recent) developments in related areas:

- Quantum transitions between Fermi liquids and states with various types of spin or charge density wave orderings were discussed in important early work by Hertz [8]. He focussed on the immediate vicinity of the finite termperature transition, like that within region A in Fig 3. In particular, Hertz missed the existence of the "quantum-critical" region (as was pointed out recently by Millis [36]—this oversight is similar to Suzuki's [17] for the Ising model), which has been the main focus of this paper. More detailed studies of quantum transitions involving Fermi liquids have appeared recently [36,38,39,40,41,42]. A related, but different, perspective is provided by studies of critical phenomena in rotor models with doped electrons [43,39].

- Related ideas on scaling in the quantum critical region have been presented by Tsvelik and collaborators [44].
- A great deal of work has been done recently on "quantum impurity" models [45,46] like the multi-channel Kondo effect. These models also display quantum phase transitions, with crossovers bearing some similarity to those discussed here. However the transitions do not modify the bulk properties, and are related instead to boundary critical phenomena.
- As we indicated briefly in the discussion on the Bose gas in Section 3, dangerously irrelevant operators sometimes need to be considered, as they do in the dilute Bose gas for $d > 2$. In fact such effects arise somewhat more frequently than they do in classical critical phenomena, as the upper critical dimension of the quantum transition is often quite low. An early analysis of the dilute Bose gas in $d = 3$ by Weichmann *et. al.* [47] was dominated by such effects, although they did not present their results in the general context of quantum phase transitions. Such a perspective can be found in more recent work [36,37].
- An important subject, on which much is not understood, is the effect of quenched randomness on quantum phase transitions. An early analysis for the random quantum rotor model was given by Boyanovsky and Cardy [9]. A recent exact solution by Fisher [48] of the random transverse-field Ising model in $d = 1$ represents significant progress. The nature of spin glass ordering at $T = 0$ and its destruction by quantum fluctuations has also been studied recently [49,37].
- A number of experiments [50,51,52] have reported scaling of the type in Eqn (25) in the local dynamic susceptibility. However the universality classes controlling these systems are not understood and quenched randomness appears to play a significant role. Nevertheless, it is interesting to note the qualitative similarity between the experimental measurements in Fig. 4 of Aronson *et. al.*[52] and our result for the local susceptibility in Fig 6. The latter measures spin correlations on a single Ising spin induced via its coupling to its environment of other Ising spins. However, there is a fundamental equivalence between the spin being measured and its environment; this is an important difference between our bulk approach and alternative descriptions of the experiments using "quantum impurity" models [45,46,52] which clearly distinguish between the impurity and environment degrees of freedom.
- A recent study [53] has examined the CQFT of a quantum ferromagnet. This system does not display a quantum phase transition of the type

discussed here. Nevertheless, its phase diagram has regions similar to those in Fig 7 and aspects of its scaling properties are related to those of the dilute Bose gas of Section 3.

## Acknowledgments

I thank N. Read, R. Shankar, A. Sokol, T. Senthil, J. Ye, and especially A.V. Chubukov for collaborations on the topics reviewed in this paper. I am grateful to A.V. Chubukov and T. Senthil for valuable comments on the manuscript and R.E. Shrock for helpful discussions. This research was supported by National Science Foundation Grant DMR-9224290.

## References

1. S. Sachdev and J. Ye, Phys. Rev. Lett. **69**, 2411 (1992).
2. A.V. Chubukov and S. Sachdev, Phys. Rev. Lett. **71**, 169 (1993).
3. A.V. Chubukov, S. Sachdev, and J. Ye, Phys. Rev. B **49**, 11919 (1994).
4. S. Sachdev, T. Senthil, and R. Shankar, Phys. Rev. B **50**, 258 (1994).
5. S. Sachdev, Z. Phys. B **94**, 469 (1994).
6. S. Chakravarty, B.I. Halperin, and D.R. Nelson, Phys. Rev. B **39**, 2344 (1989).
7. E. Brezin, J.C. Le Guillou and J. Zinn-Justin, in *Phase Transitions and Critical Phenomena*, edited by C. Domb and M.S. Green, (Academic Press, New York, 1975) Vol. 6.
8. J.A. Hertz, Phys. Rev. B **14**, 525 (1976).
9. D. Boyanovsky and J.L. Cardy, Phys. Rev. B **26**, 154 (1982)
10. A. Sokol, R.L. Glenister, and R.R.P. Singh, Phys. Rev. Lett. **72**, 1549 (1994).
11. T. Moriya, Prog. Theor. Phys. **16**, 23 (1956).
12. *Hydrodynamic Fluctuations, Broken Symmetry, and Correlation Functions*, by D. Forster, Benjamin Cummings, Reading (1975).
13. P.C. Hohenberg and B.I. Halperin, Rev. Mod. Phys. **49**, 435 (1977).
14. M.P.A. Fisher, P.B. Weichmann, G. Grinstein, and D.S. Fisher, Phys. Rev. B **40**, 546 (1989).
15. V.E. Korepin and N.A. Slavnov, Commun. Math. Phys. **129**, 103 (1990).
16. *Quantum Inverse Scattering Method and Correlation Functions* by V.E. Korepin, N.M. Bogoliubov, and A.G. Izergin, Cambridge University Press, Cambridge (1993).
17. M. Suzuki, Prog. Theor. Phys. **56**, 1454 (1976).

18. J.B. Kogut, Rev. Mod. Phys. **51**, 659 (1979).
19. *Statistical Field Theory* by C. Itzykson and J.-M. Drouffe, Cambridge University Press, Cambridge (1989).
20. E. Lieb, T. Schultz, and D. Mattis, Ann. of Phys. **16**, 406 (1961); Th. Niemeijer, Physica **36**, 377 (1967), **39**, 313 (1968); P. Pfeuty, Ann. of Phys. **57**, 79 (1970); E. Barouch and B.M. McCoy, Phys. Rev. B **3**, 786 (1971); B.M. McCoy, J.H.H. Perk, and R.E. Shrock, Nucl. Phys. B**220**, 35 (1983).
21. R.J. Glauber, J. Math. Phys. **4**, 294 (1963).
22. J.L. Cardy, J. Phys. A **17**, L385 (1984).
23. J.H.H. Perk, H.W. Capel, G.R.W. Quispel, and F.W. Nijhoff, Physica **123A**, 1 (1984); A. Luther and I. Peschel, Phys. Rev. B **12**, 3908 (1975).
24. R. Shankar, Int. J. Mod. Phys. B **4**, 2371 (1990).
25. F.D.M. Haldane, Phys. Rev. Lett. **61**, 1029 (1988); N. Read and S. Sachdev, Nucl. Phys. B**316**, 609 (1989).
26. N. Read and S. Sachdev, Phys. Rev. B **42**, 4568 (1990).
27. E. Brezin and J. Zinn-Justin, Phys. Rev. B **14**, 3110 (1976).
28. P. Hasenfratz and F. Niedermayer, Phys. Lett. B **268**, 231 (1991); Z. Phys. B **92**, 91 (1993).
29. J. M. Tranquada, G. Shirane, B. Keimer, S. Shamoto, and M. Sato, Phys. Rev. B **40**, 4503 (1989); A. J. Millis and H. Monien, Phys. Rev. Lett. **70**, 2810 (1993); Phys. Rev. B **50**, 16606 (1994).
30. A.W. Sandvik and D.J. Scalapino, Phys. Rev. Lett. **72**, 2777 (1994); A. W. Sandvik, A.V. Chubukov and S. Sachdev, Phys. Rev. B, **51**, 16483 (1995).
31. N. Elstner, R.L. Glenister, R.R.P. Singh, and A. Sokol: Phys. Rev. B **51**, 8984 (1995).
32. M. Greven *et. al.*, Phys. Rev. Lett. **72**, 1096 (1994); Z. Phys. B **96**, 465 (1995).
33. T. Imai *et. al.* Phys. Rev. Lett. **70**, 1002 (1993); *ibid* **71**, 1254 (1993).
34. A. Sokol and D. Pines, Phys. Rev. Lett. **71**, 2813 (1993).
35. A.V. Chubukov, S. Sachdev, and A. Sokol, Phys. Rev. B **49**, 9052 (1994).
36. A.J. Millis, Phys. Rev. B **48**, 7183 (1993).
37. S. Sachdev, N. Read and R. Oppermann, Phys. Rev. B, **52**, Oct. 1 (1995); cond-mat/9504036.
38. L.B. Ioffe and A.J. Millis, cond-mat/9411021; U. Zulicke and A.J. Millis, Phys. Rev. B **51**, 8996 (1995); B.L. Altshuler, L.B. Ioffe, and A.J. Millis, cond-mat/9504024.

39. S. Sachdev, A.V. Chubukov and A. Sokol, Phys. Rev. B, **51**, 14874 (1995).
40. S. Sachdev and A. Georges, Phys. Rev. B, **52**, Oct. 1 (1995); cond-mat/9503158.
41. A.V. Chubukov, cond-mat/9504009.
42. G. Murthy and R. Shankar, cond-mat/9505106.
43. S. Sachdev Phys. Rev. B **49**, 6770 (1994); A.V. Chubukov and S. Sachdev, Phys. Rev. Lett. **71**, 3615 (1993).
44. B. Andraka and A.M. Tsvelik, Phys. Rev. Lett. **67**, 2886 (1991); A.M. Tsvelik and M. Reizer, Phys. Rev. B **48**, 9887 (1993).
45. I. Affleck and A.W.W. Ludwig, Nucl. Phys. B**360**, 641 (1991).
46. I.E. Perakis, C.M. Varma, and A.E. Ruckenstein, Phys. Rev. Lett. **70**, 3467 (1993); C. Sire, C.M. Varma, A.E. Ruckenstein, and T. Giamarchi, Phys. Rev. Lett. **72**, 2478 (1994).
47. P.B. Weichmann, M. Rasolt, M.E. Fisher, and M.J. Stephen, Phys. Rev. B **33**, 4632 (1986).
48. D.S. Fisher, Phys. Rev. Lett. **69**, 534 (1992); Phys. Rev. B **50**, 3799 (1994).
49. N. Read, S. Sachdev and J. Ye, Phys. Rev. B, **52**, 384 (1995) and references therein.
50. G. Aeppli, private communication.
51. B. Keimer *et.al.*, Phys. Rev. B **46**, 14034 (1992).
52. M.C. Aronson *et. al.*, Phys. Rev. Lett. **75**, 725 (1995).
53. N. Read and S. Sachdev, Phys. Rev. Lett. in press, cond-mat/9507103.

# Replica Symmetry Breaking and the Renormalization Group Theory of the Weakly Disordered Ferromagnet

Viktor Dotsenko and D.E.Feldman

*Landau Institute for Theoretical Physics, Moscow, Russia*

**Abstract**

We study the critical properties of the weakly disordered $p$-component ferromagnet in terms of the renormalization group (RG) theory generalized to take into account the replica symmetry breaking (RSB) effects coming from the multiple local minima solutions of the mean-field equations. Recently it has been shown that for $p < 4$ the traditional RG flows at dimensions $D = 4 - \epsilon$, which are usually considered as describing the disorder-induced universal critical behavior, are unstable with respect to the RSB potentials as found in spin glasses, and a new type of the stable 1-step RSB fixed points has been discovered[1]. Here it is demonstrated that for a general type of the Parisi RSB structures there exists no stable fixed points, and the RG flows lead to the ***strong coupling regime*** at the finite scale $R_* \sim \exp(1/u)$, where $u$ is the small parameter describing the disorder. The physical consequences of the obtained RG solutions are discussed. In particular, we argue, that discovered RSB strong coupling phenomena indicate on the onset of a new spin glass type critical behaviour in the temperature interval $\tau < \tau_* \sim \exp(-\frac{1}{u})$ near $T_c$. Possible relevance of the considered RSB effects for the Griffith phase is also discussed.

# 1 Introduction

In this paper we study the effects produced by weak quenched disorder on the critical phenomena in the ferromagnetic spin systems near the phase transition point. According to the usual fluctuation theory of the second order phase transitions the leading singularities of the thermodynamical functions near the critical temperature $T_c$ are fully described in terms of the set of the universal critical exponents [2]. The only relevant spatial scale in the critical region is the correlation length $R_c$ which scales as $\sim \tau^{-\nu}$, where $\tau \equiv (T/T_c - 1) << 1$ is the reduced temperature parameter and $\nu$ is the correlation length critical exponent.

According to the traditional point of view worked out many years ago [3]-[5] the effect produced by weak disorder on the critical behavior remains negligible so long as the correlation length $R_c$ is not too large, i.e. for temperatures $T$ not too close to $T_c$. However, in the close vicinity of the critical point, where the effective strength of the disorder measured with respect to the correlation length, becomes non-small, one has no grounds, in general, for believing that the effect of the disorder will be small. According to the so-called Harris criterion [3], weak quenched disorder does strongly effects the critical behaviour only if $\alpha$, the specific heat exponent of the pure system, is positive (i.e. the specific heat of the pure system is divergent at the critical point). In this case a new universal critical behavior, with new critical exponents, is established sufficiently close to the phase transition point for $\tau \ll \tau_u \equiv u^{1/\alpha}$ [4,5]. In contrast, when $\alpha < 0$ (the specific heat is finite), the disorder appears to be irrelevant for the critical behavior.

Let us consider the usual $D$-dimensional scalar field Ginsburg-Landau Hamiltonian with a double-well potential:

$$H = \int d^D x \left[ \frac{1}{2}(\nabla\phi(x))^2 + \frac{1}{2}[\tau - \delta\tau(x)]\phi^2(x) + \frac{1}{4}g\phi^4(x) \right] . \tag{1.1}$$

Here $\tau = (T/T_c - 1) \ll 1$, and quenched disorder is introduced by random fluctuations of the effective transition temperature $\delta\tau(x)$. The probability distribution for $\delta\tau(x)$ is taken to be symmetric and Gaussian:

$$P[\delta\tau] = p_0 \exp\left( -\frac{1}{4u} \int d^D x (\delta\tau(x))^2 \right) , \tag{1.2}$$

where $u \ll 1$ is the small parameter which describes strength of the disorder, and $p_0$ is the normalization constant.

In general terms, to derive the critical properties of such system, one has to integrate over all local field configurations up to the scale of the correlation length. This type of calculation is usually performed using a Renormalization Group (RG) scheme, which self-consistently takes into account all the fluctuations of the field on length scales up to $R_c$. In actual calculations the traditional results for the critical properties of the system under consideration are obtained in terms of the standard RG procedure developed for dimensions $D = 4 - \epsilon$, where $\epsilon \ll 1$. Then one finds that in the presence of the quenched disorder the pure system fixed point becomes unstable, and the RG rescaling trajectories are arriving to another (universal) fixed point $g_* \neq 0$; $u_* \neq 0$, which yields the new critical exponents describing the critical properties of the system with disorder.

However, there exists an important point which missing in the traditional approach. Consider the ground state properties of the system described by the Hamiltonian (1.1). Configurations of the fields $\phi(x)$ which correspond to local minima in $H$ satisfy the saddle-point equation:

$$-\Delta\phi(x) + (\tau - \delta\tau(x))\phi(x) + g\phi^3(x) = 0 . \tag{1.3}$$

Clearly, the solutions of this equations depend on a particular configuration of the function $\delta\tau(x)$ being inhomogeneous. In such solutions the non-zero values of $\phi(x)$

exist in regions of space where $(\tau - \delta\tau(x)) < 0$. Moreover, one finds a *macroscopic* number of local minima solutions of the saddle-point equation (1.3). Indeed, for a given realization of the random function $\delta\tau(x)$ there exists a macroscopic number of spatial "islands" where $(\tau - \delta\tau(x))$ is negative (so that the local effective temperature is below $T_c$), and in each of these "islands" one finds two local minimum configurations of the field: one which is "up", and another which is "down". These local minimal energy configurations are separated by finite energy barriers, whose heights become larger as the size of the "islands" are increased.

The problem is that the traditional RG approach is only a perturbative theory in which one integrates over the deviations of the field around the ground state configuration, and it can not take into account other local minimum configurations which are "beyond barriers". This problem does not arise in the pure systems, where the solution of the saddle-point equation is unique. However, in the presence of quenched disorder, when one gets numerous local minimum configurations separated by finite barriers, the direct application of the traditional RG scheme may be questioned. In this situation a systematic approach must consist of both integration (in an RG way) over fluctuations around the local minima configurations, and summation over all these local minima. In view of the fact that the local minima configurations are defined by the random quenched function $\delta\tau(x)$ in an essentially non-local way, the possibility to implement successfully such a systematic approach seems rather hopeless.

On the other hand there exists another technique which has been developed specifically for dealing with systems which exhibit numerous local minima states. It is the Parisi Replica Symmetry Breaking (RSB) scheme which has proved to be crucial in the mean-field theory of spin-glasses (see e.g. [6]). Recent studies show that in certain cases the RSB approach can also be generalized for situations where one has to deal with fluctuations as well [7]-[9]. Moreover, recently it has been shown that the RSB technique can be successfully applied for the RG studies of the critical phenomena in the Sine-Gordon model where remarkable instability of the RG flows with respect to the RSB modes has been discovered [10].

In the paper [1] qualitative arguments were presented showing that the summation over multiple local minima configurations in the $p$-component spin system could provide additional non-trivial RSB interaction potentials for the fluctuating fields [1]. The idea was that hopefully after summation over these discrete degrees of freedom the critical phenomena could be studied then in term of the usual RG scheme for the fluctuating fields with modified effective interaction potentials. It is believed that this type of generalized RG scheme self-consistently takes into account relevant degrees of freedom coming from the numerous local minima. In particular, the instability of the traditional Replica Symmetric (RS) fixed points with respect to the RSB would indicate that the multiplicity of the local minima can be relevant for the critical properties in the fluctuation region. In the paper [1] due to several simplifying assumptions, the effective $\phi^4$ interaction potentials have been calculated explicitly, and it was demonstrated that the structure of replica interactions belongs to the so-called, 1-step RSB class. It was also shown that, whenever the disorder is relevant for the critical behavior (for the number of spin components $p < 4$), the usual RS fixed points (which were usually considered as describing the disorder-induced universal critical behavior) are unstable with respect to "turning on" the RSB potential, and a new type of the stable 1-step RSB fixed points has been discovered. These results made it possible to calculate the corresponding critical exponents, which (unlike the traditional RS situation) appeared to be *non-universal* being dependent on the concrete statistical properties of the disorder.

The important point, however, is that the non-trivial fixed points obtained in [1] can be shown to be stable only within the 1-step RSB subspace, being unstable with respect to continuous RSB modes. Until it is certain that the starting $\phi^4$ potentials have strictly 1-step RSB structure, this type of instability is not important (just like the instability with respect to the RSB is irrelevant until the starting $\phi^4$ interactions are strictly replica symmetric). However, at the present stage one can not be sure that considered systems are indeed adequately described by the 1-step RSB ansatz. Moreover, according to [1], in the most interesting Ising case ($p = 1$) there exist no stable fixed points within 1-step RSB subspace at all.

Here we consider the critical properties of weakly disordered $p$-component systems (including $p = 1$ case) along the same lines of the generalized RG approach taking into account the possibility of a *general type* of the RSB potentials for the fluctuating fields. It can be shown [11] that in this case the RG flows does not arrive to any fixed point, and instead they actually lead to the *strong coupling regime* at the finite spatial scale $R_* \sim \exp(1/u)$ (which corresponds to the temperature scale $\tau_* \sim \exp(-\frac{1}{u})$). At this scale the structure of the renormalized interactions develops strong RSB, and the values of the corresponding interaction parameters are getting non-small. It can also be shown that due to the absence of fixed points at the disorder dominated scales $R >> u^{-\nu/\alpha}$ (or at the corresponding temperature scales $\tau << u^{1/\alpha}$) there must be no simple scaling behavior of the physical quantities. Moreover, the specific heat is getting *non-singular* at the critical point being "smoothed out" in the temperature interval $\sim \tau$ around $T_c$. Besides, we demonstrate, that the replica structure of the SG type two-points correlation functions is characterized by the strong RSB. We also briefly discuss possible relevance of the considered RSB phenomena for the Griffith phase which is known to exist in a finite temperature interval near $T_c$ [13].

# 2 Replica Symmetry Breaking in the Renormalization Group Theory

We consider the $p$-component ferromagnet with quenched random effective temperature fluctuations, which near the transition point can be described by the usual Ginzburg-Landau Hamiltonian:

$$H = \int d^D x \left[ \frac{1}{2} \sum_{i=1}^{p} (\nabla \phi_i(x))^2 + \frac{1}{2}(\tau - \delta\tau(x)) \sum_{i=1}^{p} \phi_i^2(x) + \frac{1}{4} g \sum_{i,j=1}^{p} \phi_i^2(x) \phi_j^2(x) \right] , \quad (2.1)$$

where the quenched random temperature $\delta\tau(x)$ is described by the Gaussian distribution (1.2). To carry out the appropriate average over quenched disorder we can use the standard replica approach, in which one averages the $n$-th ($n \to 0$) power of the partition function, $Z_n \equiv \overline{Z^n[\delta\tau]}$, where $\overline{(...)}$ denotes the averaging over $\delta\tau(x)$ with the probability distribution (1.2). Simple integration yields:

$$Z_n \equiv \overline{Z^n(\delta\tau)} = \int D\phi_i^a(x) \exp\Bigg[ - \int d^D x \Bigg( \tfrac{1}{2} \sum_{i=1}^{p} \sum_{a=1}^{n} [\nabla \phi_i^a(x)]^2 +$$
$$+ \tfrac{1}{2} \tau \sum_{i=1}^{p} \sum_{a=1}^{n} [\phi_i^a(x)]^2 + \tfrac{1}{4} \sum_{i,j=1}^{p} \sum_{a,b=1}^{n} g_{ab} [\phi_i^a(x)]^2 [\phi_j^b(x)]^2 \Bigg) \Bigg] , \quad (2.2)$$

where

$$g_{ab} = g\delta_{ab} - u . \quad (2.3)$$

To study the critical properties of this system we use the standard RG procedure developed for dimensions $D = 4 - \epsilon$, where $\epsilon \ll 1$. Along the lines of the usual rescaling scheme (see e.g. [2]) one easily gets the following (one-loop) RG equations for the interaction parameters $g_{ab}$:

$$\frac{dg_{ab}}{d\xi} = \epsilon g_{ab} - \frac{1}{8\pi^2} (4g_{ab}^2 + 2(g_{aa} + g_{bb}) g_{ab} + p \sum_{c=1}^{n} g_{ac} g_{cb}) , \quad (2.4)$$

where $\xi$ is the standard rescaling parameter. Note that the above equations are valid for *arbitrary* structure of the interaction matrix $g_{ab}$, and not only for the replica-symmetric type (2.3).

If one takes the matrix $g_{ab}$ to be replica symmetric, as in the starting form of Eq. (2.3), then one would recover the usual RG equations for the parameters $g$ and $u$, and eventually one would obtain the well known results for the fixed points and the critical exponents [4,5]. Here we leave apart the question as to how perturbations out of the RS subspace could arise (see discussion in [1]) and formally consider the RG eqs.(2.4) assuming that the matrix $g_{ab}$ has a general Parisi RSB structure.

According to the standard technique of the Parisi RSB algebra (see e.g. [6],[14]), in the limit $n \to 0$ the matrix $g_{ab}$ is parametrized in terms of its diagonal elements $\tilde{g}$

and the off-diagonal *function* $g(x)$ defined in the interval $0 < x < 1$. Changing $g_{ab} \to 8\pi^2 g_{ab}$, and $g_{a\neq b} \to -g_{a\neq b}$ (so that the off-diagonal elements would be positively defined), instead of the matrix RG eqs.(2.4) in terms of the Parisi representation one gets the following RG equations (for details see [1],[11]):

$$\frac{d}{d\xi}g(x) = (\epsilon-(4+2p)\tilde{g})g(x)+4g^2(x)-2pg(x)\int_0^1 dyg(y)-p\int_0^x dy(g(x)-g(y))^2 \quad (2.5)$$

$$\frac{d}{d\xi}\tilde{g} = \epsilon\tilde{g} - (8+p)\tilde{g}^2 + p\overline{g^2} \quad (2.6)$$

where $\overline{g^2} \equiv \int_0^1 dxg^2(x)$.

Usually in the studies of the critical behavior one is looking for the stable fixed-points solutions of the RG equations. From the eq.(2.5) one can easily find out what should be the structure of the function $g(x)$ at the fixed point, $\frac{d}{d\xi}g(x) = 0$, $\frac{d}{d\xi}\tilde{g} = 0$. Taking the derivative over $x$ twice, one gets, from Eq.(2.5): $g'(x) = 0$. This means that either the function $g(x)$ is constant (which is the RS situation), or it has the step-like structure. It is interesting to note that the structure of fixed-point equations is similar to those for the Parisi function $q(x)$ near $T_c$ in the Potts spin-glasses [15], and it is the term $g^2(x)$ in Eq.(2.5) which is known to produce 1step RSB solution there. The numerical solution of the above RG equations convincingly demonstrates that whenever the trial function $g(x)$ has the many-step RSB structure, it quickly develops into the 1-step one with the coordinate of the step being the most right step of the original many-step function.

In the paper [1] the following 1-step RSB ansatz for the function $g(x)$ has been considered:

$$g(x) = \begin{cases} g_0 & \text{for } 0 \le x < x_0 \\ g_1 & \text{for } x_0 < x \le 1 \end{cases} \quad (2.7)$$

where $0 \le x_0 \le 1$ is the coordinate of the step.

In terms of this ansatz the fixed-point equations have several non-trivial solutions:

1) The RS fixed-point which corresponds to the pure system: $g_0 = g_1 = 0$; $\tilde{g} = \epsilon/(8+p)$. This fixed point (in accordance with the Harris criterion) is stable for the number of spin components $p > 4$, and it is getting unstable for $p < 4$.

2) The disorder-induced RS fixed point: $g_0 = g_1 = \epsilon\frac{4-p}{16(p-1)}$; $\tilde{g} = \epsilon\frac{p}{16(p-1)}$ [4,5]. It was usually considered to be the one which describes the new universal critical behavior in systems with impurities. This fixed point has been shown to be stable (with respect to the RS deviations!) for $p < 4$, which is consistent with the Harris criterion. (For $p = 1$ the corresponding fixed point is of the order of $(\epsilon)^{1/2}$ and it is is only revealed within a two-loop approximation). However, the stability analysis with respect to the RSB deviations shows that this fixed point is *always unstable*

[1]. Therefore, whenever the disorder is relevant for the critical behavior, the RSB perturbations must be getting the dominant factor in the asymptotic large scale limit.

3) The 1-step RSB fixed point [1]:

$$g_0 = 0; \quad g_1 = \epsilon \frac{4-p}{16(p-1)-px_0(8+p)}, \qquad \tilde{g} = \epsilon \frac{p(1-x_0)}{16(p-1)-px_0(8+p)}. \tag{2.8}$$

This fixed point can be shown to be stable (within 1-step RSB subspace!) for:

$$1 < p < 4, \qquad 0 < x_0 < x_c(p) \equiv \frac{16(p-1)}{p(8+p)}. \tag{2.9}$$

Using the result (2.8) one can easily obtain the corresponding critical exponents which are now getting to be *non-universal* being dependent on the starting parameter $x_0$. In particular, for the critical exponents of the correlation length and the specific heat one finds [1]:

$$\nu(x_0) = \frac{1}{2} + \frac{1}{2}\epsilon \frac{3p(1-x_0)}{16(p-1)-px_0(p+8)} \tag{2.10}$$

$$\alpha(x_0) = -\frac{1}{2}\epsilon \frac{(4-p)(4-px_0)}{16(p-1)-px_0(p+8)} \tag{2.11}$$

Unfortunately, to obtain the first correction to the critical exponent $\eta$ of the two-point correlation function one has to study the RG fixed points in the next order ($\sim \epsilon^2$) approximation. Technically such type of calculations are much more cumbersome and the result for the critical exponent $\eta$ still remains to be found.

However, if the parameter $x_0$ of the starting function $g(x; \xi = 0)$ (or, more generally, the coordinate of the most right step of the many-steps starting function) is beyond the "stability interval", eq.(2.9), such that $x_c(p) < x_0 < 1$, then there exist *no stable fixed points* of the RG eqs.(2.5),(2.6). One faces the same situation, of course, in the case of a general continuous starting function $g(x; \xi = 0)$. Moreover, according to eq.(2.9) there exist no stable fixed points out of the RS subspace in the most interesting Ising case, $p = 1$.

It should be stressed that unlike the RS situation for $p = 1$, where one finds the stable $\sim \sqrt{\epsilon}$ fixed point in the two-loop RG equations [5], here adding next order terms in the RG equations does not cure the problem. In the considered RSB case one finds that in the two-loops RG equations the values of the parameters in the fixed point are formally getting of the order of one, and it signals that we are entering the strong coupling regime where all the orders of the RG are getting relevant.

Nevertheless, to get at least some information about the physics behind this instability phenomena, one can proceed analyzing the actual evolution of the above one-loop RG equations. The scale evolution of the parameters of the Hamiltonian would still adequately describe the properties of the system until we reach a critical scale $\xi_*$, at which the strong coupling regime begins.

The evolution of the renormalized function $g(x;\xi)$ can be analyzed both numerically and analytically. It can be shown[11] that in the case $p < 4$ for a general continuous starting function $g(x;\xi = 0) \equiv g_0(x)$ the renormalized function $g(x;\xi)$ tends to zero everywhere in the interval $0 \leq x < (1 - \Delta(\xi))$, while in the narrow (scale dependent) interval $\Delta(\xi)$ near $x = 1$ the values of the function $g(x;\xi)$ grow:

$$g(x;\xi) \sim \begin{cases} a\frac{u}{1-u\xi}; & \text{at } (1-x) << \Delta(\xi) \\ \\ 0; & \text{at } (1-x) >> \Delta(\xi) \end{cases} \tag{2.12}$$

$$\tilde{g}(\xi) \sim u \ln \frac{1}{1 - u\xi} \tag{2.13}$$

where $\Delta(\xi) \simeq (1 - u\xi)$. Here $a$ is a positive non-universal constant, and the critical scale $\xi_*$ is defined by the condition that the values of the renormalized parameters are getting of the order of one: $(1 - u\xi_*) \sim u$, or $\xi_* \sim 1/u$. Correspondingly, the spatial scale at which the system is entering the strong coupling regime is:

$$R_* \sim \exp(\frac{1}{u}) \tag{2.14}$$

Note that the value of this scale is much bigger than the usual crossover scale $\sim u^{-\alpha/\nu}$, at which the disorder is getting relevant for the critical behavior. The corresponding crossover temperature is:

$$\tau_* \sim \exp(-\frac{const}{u}) \tag{2.15}$$

According to the asymptotic solution (2.12),(2.13), in the disorder dominated region $\tau_* << \tau_0 << u^{\nu/\alpha}$ (which corresponds to scales $\xi_u << \xi << \xi_*$) the RG trajectories for the interaction parameters $\tilde{g}(\xi)$ and $\overline{g}(\xi)$ does not arrive to any fixed point. Thus, in the disorder dominated critical region the temperature and scale dependences of the physical quantities such as specific heat, correlation length etc. should not have the traditional scaling form [11].

In the close vicinity of $T_c$, at $\tau << \tau_*$, we are facing the situation that at large scales the interaction parameters of the asymptotic (zero-mass) Hamiltonian are getting non-small, and the properties of the system can not be analyzed in terms of simple one-loop RG approach.

Qualitatively similar asymptotic behavior for $g(x;\xi)$ is obtained for the case when the starting function $g_0(x)$ has the 1-step RSB structure (2.7), and the coordinate of the step $x_0$ is in the "instability region" (or for any $x_0$ in the Ising case

$p = 1$). In this case again, the system arrives into the strong coupling regime at scales $\xi \sim 1/u$.

# 3 Correlation Functions

The calculations of the critical behavior of the usual two-point correlation function involves next-order (two-loop) approximation of the RG analysis. In the framework of the present RSB scheme such type of calculations appear to be extremely cumbersome, and they are not done yet. On the other hand, the scaling properties of the so-called spin-glass type connected correlation function:

$$K(R) = \overline{(\langle\phi(0)\phi(R)\rangle - \langle\phi(0)\rangle\langle\phi(R)\rangle)^2} \equiv \overline{\langle\langle\phi(0)\phi(R)\rangle\rangle^2} \tag{3.1}$$

can be calculated within the usual one-loop approximation, and it is the properties such type of the correlation function, which is of the main interest for the considered spin-glass effects in the critical phenomena (see Section 4).

It is well known [6] that in terms of the replica formalism the correlation function (3.1) can be represented as follows:

$$K(R) = \lim_{n\to 0} \frac{1}{n(n-1)} \sum_{a\neq b}^{n} K_{ab}(R) \tag{3.2}$$

where

$$K_{ab}(R) = \langle\langle\phi_a(0)\phi_b(0)\phi_a(R)\phi_b(R)\rangle\rangle \tag{3.3}$$

It is also well known that in terms of the standard RG formalism for the scaling behavior of the above replica correlation function one finds:

$$K_{ab}(R) \sim (G_0(R))^2 (Z_{ab}(R))^2 \tag{3.4}$$

where $G_0(R) = R^{-(D-2)}$ is the free-field correlation function, and in the one-loop approximation the scaling of the mass-like object $Z_{ab}(R)$ (with $a \neq b$) is defined by the following RG equation:

$$\frac{d}{d\xi} \ln Z_{ab}(\xi) = 2g_{ab}(\xi) \tag{3.5}$$

where $\xi = \ln R$, and $Z_{ab}(0) \equiv 1$. Thus, in the Parisi representation for the correlation function (3.4) one finds:

$$K(x;R) \sim (G_0(R))^2 \exp\{4\int_0^{\ln R} d\xi g(x;\xi)\} \tag{3.6}$$

In the traditional RS case the interaction parameter $g_{a\neq b}(\xi) \equiv u(\xi)$ arrives to the fixed point $u_* = \epsilon\frac{4-p}{16(p-1)}$, (for $1 < p < 4$), and according to eq.(3.6) one obtains simple scaling:

$$K_{rs}(R) \sim R^{-2(D-2)+\theta_{RS}} \tag{3.7}$$

with the universal disorder induced critical exponent $\theta_{RS} = \epsilon\frac{4-p}{4(p-1)}$.

In the case of the 1-step RSB fixed point, eq.(2.8), the situation is getting somewhat more complicated. Here one finds that the correlation function $K(x; R)$ also have 1RSB structure:

$$K(x;R) \sim \begin{cases} K_0(R) = R^{-2(D-2)}; & \text{for } 0 \le x < x_0 \\ K_1(R) = R^{-2(D-2)+\theta_{1RSB}}; & \text{for } x_0 < x \le 1 \end{cases} \tag{3.8}$$

with *non-universal* critical exponent $\theta_{1RSB}$ explicitly depending on the coordinate of the step $x_0$:

$$\theta_{1RSB} = \epsilon\frac{4(4-p)}{16(p-1) - px_0(8+p)} \quad (> \theta_{RS}) \tag{3.9}$$

Since the critical exponent $\theta_{1RSB}$ is positive, the leading contribution to the asymptotic behavior of the "observable" quantity $K(R) = \overline{\langle\langle\phi(0)\phi(R)\rangle\rangle^2}$, eq.(3.2), is defined only by the function $K_1(R)$:

$$K(R) = \int_0^1 dx K(x; R) \sim (1-x_0)K_1(R) + x_0 K_0(R) \sim R^{-2(D-2)+\theta_{1RSB}} \tag{3.10}$$

In the case of a general type of the RSB, according to the qualitative solution (2.12)-(2.13), at scales $\xi >> \xi_u \sim \frac{\nu}{\alpha}\ln\frac{1}{u}$ the function $g(x;\xi)$ does not arrive to any fixed point. Therefore, at the disorder dominated scales $R >> R_u \sim u^{-\nu/\alpha} >> 1$ there must be no scaling behavior of the correlation function $K(R)$. According to eq.(3.6), near the critical scale $R_* \sim \exp(1/u)$ for the correlation function $K(x; R)$ one obtains "quasi-1step" structure:

$$K(x;R) \sim \begin{cases} R^{-2(D-2)}(1-u\ln R)^{-4a} \equiv K_1(R); & \text{for } (1-x) << \Delta(R) \\ \\ R^{-2(D-2)} \equiv K_0(R); & \text{for } (1-x) >> \Delta(R) \end{cases} \tag{3.11}$$

where $\Delta(R) = (1 - u\ln R) \to u << 1$ as $R \to R_*$.

# 4 Discussion

According to the results obtained in the papers [1],[11] spontaneous replica symmetry breaking in the effective interaction potential for the fluctuating fields has a dramatic effect on the renormalization group flows and on the critical properties.

The problem now is to understand, how all those formal results should be interpreted in qualitative terms for the observable physics. The first qualitative result which comes out from the calculations of Section 3 is that the physical quantities exhibit no scaling behavior in the critical region. Actually, this is just the general consequence of the absence of the fixed point in the RG. Although it is the RSB phenomena which provide the absence of the fixed point in the considered problem, the non-scaling behavior itself does not give insight into specific effects of the RSB which is of the main interest of the present study.

The key question which remains unanswered, is whether or not the obtained RSB strong coupling phenomena in the RG flows could be interpreted as the onset of a kind of the spin-glass phase near $T_c$. Since it is the RSB interaction parameter describing the disorder, $g(x;\xi)$, which is the most divergent, it is tempting to argue that in the temperature interval $\tau << \tau_* \sim \exp(-1/u)$ near $T_c$ the properties of the system should be essentially SG-like.

It should be stressed, however, that in the present study we observe only the crossover temperature $\tau_*$, at which the change of the critical regime may occur, and it is hardly possible to associate this temperature with any kind of phase transition. Actually, if the RSB effects could indeed provide any kind of true thermodynamic order parameter, then this must be true in a whole temperature interval where the RSB potentials exist.

The true spin-glass order (in the traditional sense) arises from the onset of non-zero order parameter $Q_{ab}(x) =< \phi_a(x)\phi_b(x) >; a \neq b$, and, at least for the infinite-range spin-glasses, $Q_{ab}$ develops the hierarchical dependence on replica indices obtained by Parisi [16]. In the present problem we only find that the coupling matrix $g_{ab}$ for the fluctuating fields develops strong RSB structure and its elements are getting non-small at the finite scale. Therefore, it seems more realistic to interpret discovered RSB strong coupling phenomena in the RG just as a new type of the critical behavior characterized by strong SG-effects in the scaling properties rather then in the ground state.

In spin-glasses it is generally believed that RSB phenomenon can be interpreted as a factorization of the phase space into (ultrametric) hierarchy of "valleys", or local minima pure states, separated by macroscopic (infinite) barriers [6]. Although in the systems considered here the local minima configurations responsible for the RSB are not likely to be separated by infinite barriers (otherwise it would mean true SG freezing), it would be natural to interpret obtained phenomenon as effective factorization of the phase space into a hierarchy of valleys separated by *finite* barriers. In this situation one could expect that besides the usual critical slowing down (corresponding to the relaxation inside one valley) qualitatively much bigger (exponentially large) relaxation times would be required for overcoming barriers separating different valleys. Therefore, the traditional measurements (made at finite equilibration times) can actually correspond to the equilibration within one valley only, and not to the true thermal equilibrium. Then in a close vicinity of the critical point different measurements of the critical properties of e.g. spatial correlation functions (in the same sample) would exhibit different results as if the state of the system is getting effectively "trapped" in different valleys, and thus the traditional spin-glass situation will be observed.

In this respect the *replica structure* of the two-point correlation function $K_{ab}(R) = \langle\phi_a(0)\phi_b(0)\phi_a(R)\phi_b(R)\rangle$ (studied in Section 3.3) is of fundamental importance. Of course, the "observable" quantity (averaged over the disorder):

$$K(R) \equiv \overline{\langle\phi(0)\phi(R)\rangle^2} = \lim_{n\to 0} \frac{1}{n(n-1)} \sum_{a\neq b}^{n} K_{ab}(R) \tag{4.1}$$

has no replica structure, and correspondingly, besides its scaling properties, it can not give any direct information about the RSB. Nevertheless, according to the traditional SG philosophy [6],[14] the result that the behavior of the replica quantity $K_{ab}(R)$ does depend on the replica indices $(a, b)$, indicates that in different measurements of the correlation functions for a *a given realization* of the disorder one has to be obtaining *different* results. To describe this phenomenon in more concrete terms, like in spin-glasses, it appears to be more convenient to deal with the so-called "overlap" quantities. In our case the *spatially averaged* overlap quantity, which corresponds to the *replica* correlation function $K_{ab}(R)$ could be defined as follows:

$$K_{ij}(R) \equiv \frac{1}{V} \int d^D r \langle\phi(r)\phi(r+R)\rangle_i \langle\phi(r)\phi(r+R)\rangle_j \tag{4.2}$$

Here $i$ and $j$ label two different realizations of the disorder, and it is assumed (like in spin-glasses) that the measurable thermal average corresponds to a particular valley in the space of states for a given realization of the disorder, and not to the true thermal average.

Apparently, if the usual RS situation takes place (so that only one global valley exists), then for the correlation function $K_{ij}(R)$ one will be obtaining the same result $K(R)$, eq.(4.1), in all measurements for any pair of realizations of the disorder. However, in the case of the RSB solution, the situation is getting qualitatively different. The point is that if the phase space is factorized into a multiple valley structure (and each thermal average corresponds to one valley only), then the correlation function (4.2), although being spatially averaged, becomes a *probabilistic* quantity, and according to the SG theory of the RSB, the fluctuations of $K_{ij}(R)$ are described by the distribution function, which is defined by the RSB structure of the replica quantity $K_{ab}(R)$.

In particular, if the 1-step RSB result (3.8) takes place, then for the correlation function $K_{ij}(R)$ one has to be obtaining the values $K_0(R)$ and $K_1(R)$, with the probabilities $x_0$ and $(1 - x_0)$ correspondingly. Similarly, in the case of the non-scaling result (3.11), which also exhibits quasi 1-step RSB asymptotic structure, the probabilities to obtain the values $K_0(R)$ and $K_1(R)$ must be $(1 - \Delta(R))$ and $\Delta(R)$ correspondingly.

The other unsolved problem is about possible relevance of the RSB phenomena considered here for the so-called Griffith phase [13].

Apparently, if the summation over multiple local minima configurations yields the effective Hamiltonian with the RSB in the fourth-order potential, then formally one is dealing with a phase exhibiting a different symmetry than the conventional RS paramagnetic phase. Thus there would have to be a temperature $T_{RSB}$ at which this change in symmetry occurs, since for large enough $\tau$ there must be no RSB. Although in the present study we were concentrated at the effects of the RSB for the fluctuating fields and for the critical phenomena, it should be stressed that it is the statistics of the saddle-point solutions only, which is responsible for the appearance of the RSB. Therefore, one can consider separate problem of summing over saddle-point solutions without fluctuating fields, keeping arbitrary parameter $\tau$, and aiming to find finite value of $\tau_{RSB}$ at which the RSB solution for this problem disappears.

Of course, in general this problem is very difficult to solve, but one can easily obtain an estimate for the value of $\tau_{RSB}$ (assuming that at $\tau = 0$ the RSB situation takes place). According to the qualitative study of this problem in the paper [1], the RSB solution can occur only when the effective interactions between the "islands", where the system is effectively below $T_c$, are getting non-small. The islands are the regions where $\delta\tau(x) > \tau$. According to the Gaussian distribution for $\delta\tau(x)$, the average distance between them must be of the order of $\exp[-\tau^2/u]$, so that the islands are getting distant at $\tau > \sqrt{u}$. The interaction between the islands is exponentially small in their separation. Therefore at $\tau > \sqrt{u}$ they must be getting weakly interacting, and there must be no RSB.

Note now that the shift of $T_c$ with respect to the corresponding pure system is also of the order of $\sqrt{u}$. On the other hand, it is the presence of multiple local minima configurations which is believed to be the fundamental reason for the existence of the Griffith phase [13] which is claimed to be observed in the temperature interval between $T_c$ of the disordered system and $T_c$ of the corresponding pure system. On these grounds it is tempting to associate the (hypothetical) RSB transition in the statistics of the saddle-point solutions with the Griffith transition. Correspondingly, it would also be natural to suggest that discovered RSB phenomena in the scaling properties of weakly disordered systems could be associated with the Griffith effects.

# References

[1] Vik.S.Dotsenko, B.Harris, D.Sherrington and R.Stinchcombe, J.Phys. **A 28**, 3093 (1995)

[2] A.Z.Patashinskii and V.L.Pokrovskii "Fluctuation Theory of Phase Transitions", Pergamon Press, 1979;
G.Parisi "Statistical Field Theory", Addison-Wesley, 1988.

[3] A.B.Harris, J.Phys. **C 7**, 1671 (1974).

[4] T. C. Lubensky and A. B. Harris, AIP Conf. Proc. **24** 311 (1974);
G. Grinstein, AIP Conf. Proc. **24** 313 (1974);
A. B. Harris and T. C. Lubensky, Phys.Rev.Lett., **33**, 1540 (1974);
G. Grinstein and A. Luther, Phys.Rev. **B 13**, 1329 (1976).

[5] D. E. Khmelnitskii, ZhETF (Soviet Phys. JETP) **68**, 1960 (1975).

[6] M.Mezard, G.Parisi and M.Virasoro "Spin-Glass Theory and Beyond", (World Scientific, Singapore, 1987).

[7] M.Mezard and G.Parisi, J.Phys. I **1**, 809 (1991).

[8] M.Mezard and A.P.Young, Europhys.Lett., **18**, 653 (1992).

[9] S. Korshunov, Phys.Rev. **B48**, 3969 (1993)

[10] P. Le Doussal and T.Giamarchi, Phys.Rev.Lett., **74**, 606 (1995)

[11] Vik.S.Dotsenko and D.E.Feldman, Preprint cond-mat/9502061, (to be published in J.Phys.A:Math.Gen).

[12] G.Parisi, Proceedings of the Les Houches 1982, Session XXXIX, edited by J.B.Zuber and R.Stora (North Holland, Amsterdam) 1984.

[13] R.Griffiths, Phys.Rev.Lett., **23**, 17 (1969);
A.T.Ogielski, Phys.Rev., **B32**, 7384 (1985);
A.J.Bray, Phys.Rev.Lett., **59**, 586 (1987).

[14] Vik.S.Dotsenko "Introduction to the Theory of Spin-Glasses and Neural Networks", World Scientific, 1994.

[15] D.Gross, I.Kanter, and H.Sompolinsky, Phys.Rev.Lett. **55**, 304 (1985)

[16] G.Parisi, J.Phys. **A13**, L-115 (1980)

# WAVES IN DISORDERED MEDIA: THE INTERMEDIATE FREQUENCY REGIME*

PING SHENG

*Department of Physics*
*Hong Kong University of Science and Technology*
*Clear Water Bay*
*Kowloon, Hong Kong*
E-mail: phsheng@usthk.ust.hk

ABSTRACT

The intermediate wave frequency regime, where the wavelength is comparable to the scale of the scatterers, is rich in wave phenomena but traditionally the most difficult regime to treat theoretically. This talk will present a simple theoretical approach for the quantitative prediction of wave phenomena beyond the long wavelength regime, when the coherent potential approximation fails. The effectiveness of the approach is illustrated by comparing theoretical predictions with experimental measurements of acoustic excitations in colloidal suspensions, where two distinct acoustic modes were observed.

## 1. Introduction

Wave propagation in disordered media can be qualitatively divided into three regimes: the high-frequency limit, where wave propagation can be reduced to geometric optics; the low-frequency limit, where the wave cannot resolve the inhomogeneities and the effective medium concept prevails; and the intermediate frequency regime, defined as the one in which the wavelength is comparable to the scale of the inhomogeneities. The intermediate frequency regime is traditionally the most difficult one to treat theoretically, due to the lack of a small parameter for perturbation calculations. However, the regime is rich in physical phenomena, such as resonant scattering, diffusive transport, localization, etc. Recently, experiment on colloidal suspensions has added to this list the surprising result that showed there can be two distinct acoustic modes in the intermediate frequency regime, in direct contradiction to the usual expectation

* Work done in collaboration with X. Jing and M. Zhou. Experimental results on colloids are due to Ling Ye, Dave Weitz, and Jing Liu.

that there can be only one acoustic mode in a medium that has no shear modulus.

In this talk, I present a simple theoretical approach for the quantitative calculation of wave behavior in the intermediate frequency regime, where the coherent potential approximation fails. The effectiveness of the approach is illustrated by comparing theoretical predictions with measured dispersion curves of colloidal suspensions. Excellent agreement was obtained.

## 2. **Formulation of the Approach**

Consider the scalar wave equation

$$(\nabla^2 + \frac{\omega^2}{v^2(\vec{r})})\phi = 0 \ , \qquad (1)$$

where $v(\vec{r})$ is the wave speed at location $\vec{r}$. By adding and subtracting a constant term, Eq. (1) can be written as

$$[(\nabla^2 + \frac{\omega^2}{v_o^2}) - \frac{\omega^2}{v_o^2}(1-\varepsilon(\vec{r}))]\phi = 0 \ , \qquad (2)$$

where $\varepsilon(\vec{r}) = v_o^2 / v^2(\vec{r})$. Equation (2) is seen to consist two parts: a part associated with a uniform medium with wave speed $v_o$ and another part associated with deviation from the uniform medium. The second part is responsible for all the scatterings. In order to obtain the coherent part of the wave behavior [1], it is desirable to calculate the configurationally-averaged Green's function $\langle G\rangle_c$:

$$\langle G\rangle_c(\omega,q) = \frac{1}{\dfrac{\omega^2}{v_o^2} - q^2 - \Sigma_{v_o}(\omega,q)} \ , \qquad (3)$$

where $\Sigma_{v_o}(\omega,q)$ is the self energy term *evaluated relative to* $v_o$. One well-known property of the Green's function is that its poles give the excitations of the system. In the present case, we wish to use the poles of $\langle G\rangle_c$ to delineate the wave excitations in a disordered medium. This can be accomplished by defining a spectral function,

$$S(\omega,q) = -\mathrm{Im}\langle G\rangle_c \ , \qquad (4)$$

where the peaks of $S(\omega,q)$ directly give the locations of the excitations in the $(\omega,q)$ plane, i.e., the dispersion relations of the modes.

The self energy is a quantity that is usually difficult to calculate accurately. What we propose to do first is using the freedom in choosing $v_o$ to define a $v_e$ as

$$v_e = \frac{\omega}{q} \quad . \tag{5}$$

Then the spectral function is given by

$$S(\omega,q) = \frac{-\operatorname{Im}\Sigma_{\omega/q}}{(\operatorname{Re}\Sigma_{\omega/q})^2 + (\operatorname{Im}\Sigma_{\omega/q})^2} \quad , \tag{6}$$

where it is explicitly noted that the self-energy should be evaluated at $v_e = \omega / q$. The calculation of $\Sigma_{\omega/q}$ is done to the leading order in the weak scattering limit, i.e.,

$$\Sigma_{\omega/q} \cong n t_{\omega/q}(0) \ , \tag{7}$$

where $n$ is the density of the scatterers, $t_{\omega/q}(0) = f_{\omega/q}(0)/4\pi$, $f_{\omega/q}(0)$ being the forward-scattering amplitude calculated as follows. In a colloidal suspension, the complex medium may be viewed as composed of basic units of coated spheres, shown in Fig. 1. The scattering problem is therefore that of a coated sphere embedded in a homogeneous effective medium whose wave speed is given by $v_e = \omega / q$. By thus calculating the value of the spectral function at every point in the $(\omega,q)$ plane, the peak positions of $S(\omega,q)$ can be unambiguously determined. The error made in using the weak scattering limit to evaluate the self-energy is noted to be smallest at the peak positions, and largest at places where one does not care anyway.

The present approach can be seen to be exactly equivalent to the coherent potential approximation (CPA) at the long wavelength limit, where the CPA condition of $\operatorname{Re}\Sigma_{v_e} \cong n\operatorname{Re}t_{v_e}(0) = 0$ for the determination of $v_e$ is exactly equivalent to the condition of maximun in $S(\omega,q)$, as $\operatorname{Im}t_{v_e}(0)$ is always small in the low frequency limit

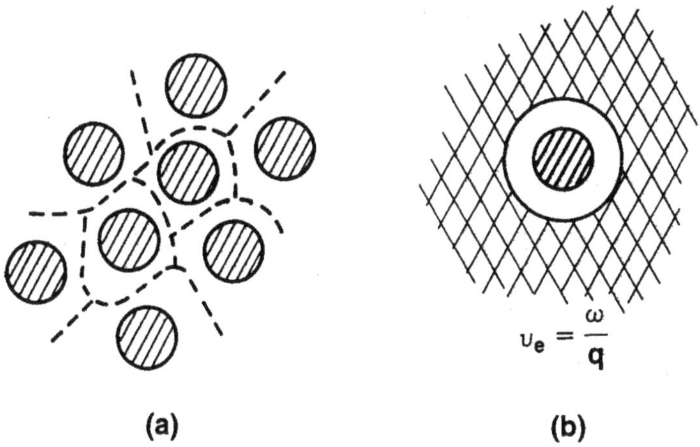


Fig. 1. (a) In a random medium composed of a dispersion of spheres, the basic structural unit may be regarded as a coated sphere, as delineated by the dashed lines. (b) To calculate the scattering amplitude, a coated sphere that represents that structural unit of (a) is embedded in a uniform medium of speed $v_e = \omega / q$. The coating thickness is given by Eq. (8).

because it is determined by Rayleigh scattering. However, when the wavelength becomes comparable to the size of the scatterers as in the intermediate frequency regime, the CPA condition generally fails to yield a solution for $v_e$. Yet even in those cases the present approach can still give peaks in $S(\omega,q)$, since it is obvious that the vanishing of $\mathrm{Re}\,\Sigma_{\omega/q}$ is not a necessary condition for having a peak in the spectral function. One may denote those modes given by the maxima in the spectral function in the intermediate frequency regime as "quasi modes," which can have non-vanishing real part of the self-energy, in contrast to the low-frequency CPA modes, where the real part of the self-energy is always zero.

## 3. Application to the Description of Acoustic Modes in Colloidal Suspensions

In applying the above-described approach to the explanation of the experimental results on acoustic excitations in colloidal suspensions [2,4], one needs to (1) set the coating thickness $s$ of the structural unit by the relation

$$\phi = \text{volume fraction of suspended solid particles}$$

$$= (1 + 2s/d)^{-3}, \tag{8}$$

where $d$ =370 nm is the particle diameter, and (2) take into account the fact that the solid particles are elastic, so that the elastic wave equation has to be used to describe the wave excitation inside the particle. In order to simplify the model, the liquid is assumed to be inviscid, so that the wave equation

$$\frac{\partial^2 \vec{u}}{\partial t^2} = v_t^2 \nabla^2 \vec{u} + (v_p^2 - v_t^2)\nabla(\nabla \cdot \vec{u}) \tag{9}$$

is applicable to both the solid and the liquid, where $v_t = 0$. Here $\vec{u}$ is the displacement,

$$v_p = \begin{cases} 2.7\times10^5 \text{ cm/s} & \text{inside solid particle (PMMA)} \\ c_\ell = 1.2\times10^5 \text{ cm/s} & \text{for liquid coating} \\ \omega / q & \text{for the effective medium} \end{cases}$$

$$v_t = \begin{cases} 1.1\times10^5 \text{ cm/s} & \text{inside solid particle (PMMA)} \\ 0 & \text{otherwise.} \end{cases}$$

The density of the solid and liquid is about the same, both about equal to 1 g/cm$^3$. The forward scattering amplitude for the coated sphere embedded in the effective medium (as in Fig. 1(b)) may be solved straightforwardly by matching the boundary conditions at the two interfaces. At the outer interface, the condition is simply the continuity of both the normal displacement and normal stress. At the inner solid-liquid interface, the conditions are the continuity of normal displacement and stress, and the vanishing of the tangential stress. Once $t_{\omega/q}(0)$ is evaluated, the spectral function is given by

$$S(\omega,q) = -\text{Im}[n t_{\omega/q}(0)]^{-1} \ , \tag{10}$$

where $n$ is $[4\pi(s+d/2)^3/3]^{-1}$.

The results for $\phi=0.38$ are plotted in grey scale as a function of $\omega$ and $q$ in Fig. 2. In the same figure the experimental data are also plotted, together with the bars indicating the width of the measured frequency peaks. The readers are referred to Ref. [4] for experimental details. The calculated results clearly indicate two ridges, where the spectral function has maxima. The wavelength range whee this happens is at $qd \geq \pi$, i.e., beyond the low frequency regime where the traditional CPA is valid. Indeed, at low frequencies our approach yields a nearly delta-function-like peak which coincides exactly with the CPA solution. With no adjustable parameter, the agreement with the experimental data is seen to be

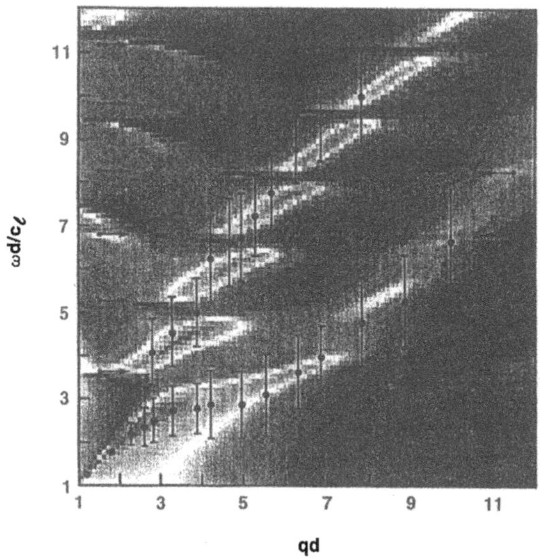


Fig. 2. The spectral function calculated as a function of $\omega d/c_\ell$ and $qd$, where $c_\ell$ is the speed of sound in the liquid. The magnitude of the spectral function is delineated by grey scale, where high is indicated by the dark patches inside the light halos and low is indicated by dark. Light is intermediate. Experimental data (from [4]) are shown as solid circles. The bars indicate their frequency widths. The arrows on the sides indicate the frequency positions of the scattering cross-section peaks for a single solid sphere in liquid. $\phi=0.38$.

excellent, including the excitation gaps seen both in the calculated and the measured results. The experimental results therefore convincingly demonstrate that the quasi modes are observable, and their dispersion relations may be quantitatively calculated by the approach presented.

The physics of the two acoustic modes can best be delineated by noting the arrows denoting the frequency positions of the resonant peaks for a single solid sphere immersed in liquid. For the high-frequency mode, the positions of the arrows correspond directly with the gaps in the calculated dispersion relations. The first gap is clearly seen in the experimental data, and the second gap and the hint of a third are also visible. This indicates that the high frequency mode may be regarded as resulting from the the *antiresonance* condition for the single-sphere scattering.

For the low frequency mode, on the other hand, the arrows are noted to coincide with the peaks of the spectral function. The physics of this mode is best described as the coherent transfer of resonant energy from one sphere to the next. This is possible because the amplitude of the sphere resonance decays away from the sphere surface with a tail in the liquid, which can overlap with that of the neighboring sphere resonance. Due to this coupling, the frequency of the resonance is split into two, with a minimum in between. The *antiresonance of two coupled spheres* is therefore what gives rise to the low frequency mode. While the low frequency mode is a result of coupling between two neighboring spheres, it can be picked out by the single-sphere calculation because from the point of view of an individual sphere, the interaction with neighboring spheres may be represented by some particular value of wave impedance at the fluid coating-effective medium interface. Since the effective medium $q$ is treated as a variable, we are thus able to pick up the antiresonance condition by scanning its value.

The nature of the elastic sphere resonances is responsible for the slow wave speed of the low frequency mode. It turns out that the sphere resonances in this case are nothing more than the Stoneley wave[2], a fluid-solid interfacial mode, specialized to the spherical geometry. Since the Stoneley mode has a wave speed that is lower than both the fluid wave speed and the solid shear wave speed, it follows that as the volume fraction of suspension is increased, the interfacial area is correspondingly increased, and the low frequency mode speed thus moves further away from that of the fluid and closer to that of the Stoneley wave speed.

## 4. Concluding Remarks

We have shown that by calculating the spectral function, one can monitor not only those modes predicted by the CPA, but also the quasi modes that are beyond the predictive ability of the CPA. In applying the approach to the acoustic excitations in colloidal suspensions, it is shown that the present scheme can give a quantitative account of the experimental data in the intermediate frequency regime. In particular, it reproduces the surprising multiple quasi modes behavior.

## References

[1] P. Sheng, *Introduction to Wave Scattering, Localization, and Mesoscopic Phenomena* (Academic Press, Boston, 1995).

[2] X. Jing, P. Sheng and M. Zhou, Phys. Rev. Lett. **66** (1991) 1240.

[3] M. Lax, Rev. Mod. Phys. **23** (1951) 287.

[4] J. Liu, L. Ye, D. A. Weitz and P. Sheng, Phys. Rev. Lett. **65** (1990) 2602.

# Clustering, Hidden Coherence, Chaotic Itinerancy, and Information Cascade in Globally Coupled Dynamical Systems

Kunihiko KANEKO

*Department of Pure and Applied Sciences*

*University of Tokyo, Komaba, Meguro-ku, Tokyo 153, JAPAN*

**Abstract**

General features of globally coupled dynamical systems are briefly reviewed. According to the nonlinearity in the system, the degree of synchronization changes, and there appears a sequence of coherent, ordered, partially-ordered, and turbulent phases. In the turbulent phase, there remains hidden coherence among elements, although collective dynamics is not clearly discernible, while the system splits into few synchronized clusters in the ordered phase. Partition to clusters is complex and diverse in the partially ordered phase, where the orbits show "chaotic itinerancy" over ordered states. Cascade process of clustering is also found, leading to marginal stability of mutual coherence. Relevance of theses studies to biological systems are briefly discussed, with the emphasis on homeochaotic stability in ecological networks and a dynamic mechanism of cell differentiations.

## 1 Introduction

Globally coupled dynamical systems [1, 2] are of importance in physical, chemical, and biological problems. Examples are seen in systems with some constraints, such as the parallelly coupled Josephson junction arrays. Globally couplings here appear through the competition for total currents by all elements. A similar example is seen in optical systems, where many modes compete for energy.

In biological systems, nonlinear dynamics with global couplings is rather common. In neural systems, roles of chaotic dynamics have been recognized by Freeman[3] and Tsuda [4], while global couplings among neurons or neuronal groups are well known. Partial synchronization of oscillations has been discovered in the visual cortex of a cat [5, 6, 7]. Vaadia and Aertsen [7, 8] have found that the degree of synchronization among pairs of neurons changes both temporally and by the choice of pairs.

In cellular biology, metabolic reactions often show nonlinear oscillations through autocatalytic processes, while global interaction among cells is important in developmental process and cell differentiation. In an immune network, oscillations of antibodies, coupled with each other, are studied theoretically, while globally coupled population dynamics is rather common in an ecological system.

In these systems, we need to reveal some universal mechanisms for the origin, maintenance, and evolution of diversity and complexity. It is requested to clarify complex dynamics of many-to-many relationships. As will be seen, studies of globally coupled dynamical systems provide a new conceptual framework for these issues. Here we survey some universal aspects of globally coupled dynamical systems, with the help of coupled map systems, and discuss their relevance to physical and biological systems.

# 2 Clustering in Globally Coupled Maps

The simplest case of global interaction is studied as the "globally coupled map" (GCM) of chaotic elements [1]. A standard example is given by

$$x_{n+1}(i) = (1-\epsilon)f(x_n(i)) + \frac{\epsilon}{N}\sum_{j=1}^{N} f(x_n(j)) \tag{1}$$

where $n$ is a discrete time step and $i$ is the index of an element ($i = 1, 2, \cdots, N$ = system size), and $f(x) = 1 - ax^2$. The model is just a mean-field-theory-type extension of coupled map lattices (CML)[9]. The above dynamics consists of parallel nonlinear transformation and a feedback from the "mean-field". It is equivalent to $y_{n+1}(i) = f[(1-\epsilon)y_n(i) + \frac{\epsilon}{N}\sum_{j=1}^{N} y_n(j)]$, with the aid of transformation $y_n(i) = f(x_n(i))$. In this form, one can see clear correspondence with spinglass-type dynamics: If one chooses a sigmoid function (e.g.,tanh($\beta x$)) as $f(x)$ and a random or coded coupling $\epsilon_{i,j}$, a typical spinglass-type dynamics is obtained[10].

Through the interaction, elements are tended to oscillate synchronously, while chaotic instability leads to destruction of the coherence. When the former tendency wins all elements oscillate coherently, while elements are completely desynchronized in the limit of strong chaotic instability. Between these cases, elements split into clusters in which they oscillate coherently. Here a cluster is defined as a set of elements in which $x(i) = x(j)$ [1]. Attractors in GCM are classified by the number of synchronized clusters $k$ and the number of elements for each cluster $N_k$. Each attractor is coded by the clustering condition $[k, (N_1, N_2, \cdots, N_k)]$.

It is useful to introduce the split exponent for the stability of a clustered state. For this, we note that the distance between the values of two elements $x_n(j)$ and $x_n(j) = x_n(i) + \delta_n$ is amplified or reduced by

$$\delta_{n+1} = x_{n+1}(j) - x_{n+1}(i) = (1-\epsilon)f'(x_n(i))\delta_n, \tag{2}$$

for small $\delta_n$. Thus the criterion of splitting between an element $i$ and another element (taking a close value) is given by the sign of the exponent

$$\lambda_{spl}(i) = \lim_{T\to\infty}\frac{1}{T}\sum_{n}^{T} log|(1-\epsilon)f'(x_n(i))|. \tag{3}$$

We note that this split exponent $\lambda_{spl}(i)$ is different from the Lyapunov exponent, which gives the amplification rate of a tiny difference of two orbits in phase space, while our exponent gives the rate of the amplification of a tiny difference between two elements for a single orbit. If two elements belong to a same synchronized cluster, the exponent $\lambda_{spl}(i)$ should be negative; otherwise the clusters would split at some time.

As a simple consequence of the above condition, a coherent state is stable if the Lyapunov exponent of the single logistic map is less than $-\log(1-\epsilon)$. In a globally coupled system, coherent chaos can be stable, in contrast with a locally coupled map lattice system.

Generally speaking, there are three types in clusterings; phase, amplitude, and frequency of oscillations. So far the clustering we have studied does not purely consist of only one of the above three types. Phase, amplitude, and frequency clusterings are mixed, although the phase difference is most relevant to clusterings. In neural systems, it should be noted that synchronizations seem to split into clusters, as discussed as "gravitational clustering" by Vaadia and Aertsen [8]. These clusterings are rather complicated, although the phase differences seem to be important again.

An interesting possibility in the clustering is that it provides a source for diversity. In clustering it should be noted that identical chaotic elements differentiate spontaneously into different groups: Even if a system consists of identical elements, they split into groups with different phases of oscillations. Hence a network of chaotic elements gives a theoretical basis for isologous diversification and provides a mechanism for the origin of diversity and complexity in biological networks [11].

In a globally coupled chaotic system in general, the following phases appear successively with the increase of nonlinearity in the system ($a$ in the above logistic map case) [1]:

(i) **Coherent phase**: Only a coherent attractor ($k = 1$) exists.

(ii) **Ordered phase**: All attractors consist of few ($k = o(N)$) clusters.

(iii) **Partially ordered phase**: Attractors with a variety of clusterings coexist, while most of them have many clusters ($k = O(N)$).

(iv) **Turbulent phase**: Elements are completely desynchronized, and all attractors have $N$ clusters.

The above clustering behaviors have universally been confirmed in a variety of mappings [13], as well as coupled ordinary differential equations for Josephson junction [14]. Similar clusterings are also found in globally coupled Ginzburg-Landau equations [15].

# 3 Hidden Coherence in Turbulent State

In the turbulent phase, although $x(i)$ takes almost random values almost independently, there remains some coherence among elements. Indeed the distribution of the mean field $h \equiv (1/N)\sum_j f(x(j))$ does not obey the law of large numbers. If the law were obeyed, the mean square deviation (MSD) of the mean-field $h$ would decrease with $N$ as $N^{-1}$. It is found numerically that with the increase of the size $N$, the variance of the mean field decreases up to a crossover size $N_c$ but then stays constant. The distribution of the mean field approaches a Gaussian form up to $N_c$, but keeps some deviation for $N > N_c$ [16]. As the origin of the violation of the law of large numbers, it might be possible to think of the following two mechanisms: (i) clear collective behaviors (with low dimensional dynamics), or (ii) the frozen relationships among elements. Neither of the mechanisms, however, can explain the mean-field dynamics of the above turbulent state. The mean-field dynamics is aperiodic, and shows high-dimensional chaos (i.e., at least the dimension is larger than 10). To check the possibility of the frozen relationship, we have measured the ratio of pairs of two elements remaining close. It decays exponentially to zero with time [1].

It still remains an important problem to clarify the nature of this mean-field dynamics, although the instability of the fixed point state of the mean field dynamics is discussed in terms of Perron-Frobenius (PF) operator [16, 18, 21, 22, 23]. Let us first define the snapshot distribution $\rho_n(y)$ in the limit of $N \to \infty$, as the probability that $x(i)$ assumes the value $y$ ($\rho_n(y) \equiv (1/N)\sum_{j=1}^{N}\delta(y - x_n(j))$). The dynamics of this probability distribution is written as

$$\rho_n(x) = \frac{1}{1-\epsilon}\sum_{pre=+,-}\frac{\rho_{n-1}(y_{pre})}{|f'(y_{pre})|} \tag{4}$$

where the two preimages are given by

$$y_{pre} = f^{-1}\left(\frac{x - \epsilon\int f(z)\rho_{n-1}(z)dz}{1-\epsilon}\right), \tag{5}$$

and $f^{-1}(y) = \pm\sqrt{\frac{1-y}{a}}$ for our logistic map. If the law of large numbers is valid, the mean-field is constant; the probability function remains time-invariant whose form is given by the invariant measure $\rho^*(y)$ of the logistic map $x_{n+1}(i) = (1-\epsilon)f(x_n(i)) + \epsilon h^*$,, where $h^*$ is determined by the condition $h^* = \int f(x)\rho^*(x)dx$. We have computed the above self-consistent solution $\rho^*(y)$, and studied its stability in the original Perron-Frobenius equation. Due to the fine structure in the bifurcation of the logistic map, fixed point solution $\rho^*(x)$ turns out to be unstable [21] ( see also [23]). Thus the remnance of the mean-field fluctuation in the infinite size limit is explained, although the nature of the mean-field dynamics itself is not yet clarified.

The emergence of hidden coherence is a general property in a globally coupled chaotic system [13, 19, 20], and has gathered much attention as a new type of subtle collective behavior [17].

# 4 Partition Complexity and Chaotic Itinerancy in Partially Ordered State

In the partially ordered (PO) phase, there are a variety of attractors with a different number of clusters, and a different way of partitions $[N_1, N_2, \cdots, N_k]$. The clustering here is typically inhomogeneous: The partition $[N_1, N_2, \cdots, N_k]$ is far from equal partition. An idealized partition is formed as follows: First split the system into two equal clusters. Take one of them and split it again into two equal clusters, while leave the other without split. By repeating this process, the partition is given by $[N/2, N/4, N/8, \cdots]$. In this case, the difference of phase of oscillations is also hierarchical. The difference is roughly $\pi$ for the first split, and then $\pi/2$ for the second, and so forth. Although this partition is a vast simplification, such hierarchical structure in partition and in the phase space is typically observed in the PO phase. Thus the partition

[1] It should be noted that collective behaviors in the globally coupled tent map are explained by the above mechanisms. Collective low-dimensional dynamics or frozen relationships between elements are observed [16, 21]. In the tent map, a chaotic orbit is always expanding, which is essentially different from an orbit in the logistic map, where contraction occurs for some time steps.

is organized as an inhomogeneous tree structure, as in the spin glass model [10].

We have also measured the fluctuation of the partitions, using the probability $Y$ that two elements fall on the same cluster. In the PO phase, this $Y$ value fluctuates by initial conditions, and the fluctuation remains finite even if the size goes to infinity [24]. It is noted that such remnance of fluctuation of partitions is also seen in spin glass models [10].

Besides the above *static* complexity, *dynamic* complexity is more interesting at the PO phase. Here orbits make itinerance over ordered states with partial synchronization of elements, via highly chaotic states. This dynamics, called chaotic itinerancy (CI), is a novel universal class in high-dimensional dynamical systems. Our CI consists of a quasi-stationary high-dimensional state, exits to "attractor-ruins" with low effective degrees of freedom, residence therein, and chaotic exits from them. In the CI, an orbit successively itinerates over such "attractor-ruins", ordered motion with some coherence among elements. The motion at "attractor-ruins" is quasistationary. For example, if the effective degrees of freedom is 2, the system splits into two groups, in each of which elements oscillate almost coherently. The system is in the vicinity of a two-clustered state, which is not a stable attractor, but whose stable manifolds are rather high-dimensional. After staying at an attractor-ruin, an orbit exits from it due to chaotic instability, and shows a high-dimensional chaotic motion without clear coherence. This high-dimensional state is again quasistationary, although there are some holes connecting to the attractor-ruins from it. Once the orbit is trapped at a hole, there appears a sudden collapse of the high-dimensional chaos, and ordered motion replaces.

This CI dynamics has been found in a model of neural dynamics by Tsuda [4], optical turbulence [12], and in GCM. It should be noted that it provides an example of successive changes of relationships among elements. Vaadia and Aertsen [7, 8] have found that the effective coupling among neurons varies temporally in a rather short time scale, by showing that the degree of synchronization among pairs of neurons change both temporally and by the choice of pairs.

# 5 Marginal Stability and Information Cascade

In the PO phase, a cascade process to successively smaller clusters is found [25]. It is shown that the splitting of clusters and the synchronization of elements are balanced, where marginal stability is sustained over an interval of the bifurcation parameters. This marginal stability is seen over a wide parameter region in the partially ordered phase ( between the ordered and coherent phases). Its mechanism is explained in terms of the avalanche of partition information creation (PIC) in bit space.

The PIC is represented as the mutual information of partitions between a given bit and its next neighbor. The flow of mutual information in bit space is intermittent here. There is an intermittent propagation of the partition information in the bit space, corresponding to the amplification of desynchronization. Similar intermittent flow of mutual information was also noted by Matsumoto and Tsuda for a class of one-dimensional maps [26]. In contrast with their study in low-dimensional chaos, the information here measures a relation between elements rather than between two time steps. Changes in mutual relationship of elements are measured by PIC.

The avalanche of the PIC demonstrates a cascade process of clusters. In analogy with the cascade process of fluid turbulence, power-law-like behavior of power spectra and accumulation of Lyapunov exponents at the null value are found. Conversely, it will also be interesting to study the cascade process of fluid turbulence in terms of the mutual information of vortices (instead of clusters here), along the line sketched here.

# 6 Relevance to Biological Networks

Here we have discussed some aspects of globally coupled dynamical systems. As mentioned[27] these results have some relevance to to biological networks. Since the relevance to neural systems has been often mentioned, we briefly mention two other examples.

Ikegami and the author have studied an ecological model with mutation of types of hosts and parasites, and mutation of mutation rates. Population dynamics of each type is nonlinear, and the mutation rate corresponds to the coupling parameter $\epsilon$. Thus the change of mutation rates leads to the change of the clustering of population dynamics. We have found that the system evolves to a weak and high-dimensional chaotic state, called *homeochaos*, which are sustained by chaotic itinerancy of clusterings [28]. The homeochaos allows for coexistence of a variety of types, and is robust against external perturbation. Thus it provides a mechanism of maintenance of diversity in an ecological system.

The clustering leads to a mechanism of differentiation of identical units. Based on this idea, Yomo and the author have studied cell differentiation [29], where each cell has a few chemicals with metabolic reaction process. This metabolic reaction leads to nonlinear oscillations, while competition for chemical resources among cells leads to a global coupling. Thus the model belongs to a class of globally coupled dynamical systems. One novel feature in the cell differentiation process lies in that cell divides, and thus the number of elements increases with cell divisions. The model [29] with this cell division shows the phase clustering initially, and then the amplitude clustering. The amplitude clustering is fixed by cells, and is inherited to daughter cells by the cell division process. Cells are differentiated, determined, and lose the totipotent ability successively. A unified scenario for the cell differentiation and development is constructed in this manner, which is so far consistent with experiments in cellular biology.

The work is partially supported by Grant-in-Aids for Scientific Research from the Ministry of Education, Science, and Culture of Japan.

# References

[1] K. Kaneko, Phys. Rev. Lett. **63** (1989) 219; Physica **41 D** (1990) 137; Physica **54 D** (1991) 5

[2] K. Wiesenfeld and P.Hadley, Phys. Rev. Lett. **62** (1989) 1335

[3] W. Freeman and C. A. Skarda, Brain Res. Rev. **10** (1985) 147;
W. Freeman, Physica **75D** (1994) 151

[4] I. Tsuda, in *Neurocomputers and Attention*, (eds. A.V. Holden and V. I. Kryukov, Manchester Univ. Press, 1990); I. Tsuda, Physica **75 D** (1994) 163

[5] R. Eckhorn et al., Biol. Cybernetics **60** (1988) 121

[6] C.M. Gray, P Koenig, A.K. Engel, and W. Singer, Nature **338** (1989) 334

[7] Kruger, J., eds. 1991, Neuronal Cooperativity, New York,Springer-Verlag

[8] A. Aertsen and E.Vaadia (1992) in *Information Processing in the Cortex: Experiments and Theory*, (eds. A. Aertsen A, V. Braitenberg . Berlin, Heidelberg, New York, Tokyo: Springer.): Ad Aertsen, M. Erb and G. Palm, Physica **75D** (1994) 103

[9] K. Kaneko, Prog. Theor. Phys. **72** (1984) 480; **74** (1985) 1033; Physica **34D** (1989) 1;
in *Formation, Dynamics, and Statistics of Patterns* (ed. K. Kawasaki, A. Onuki, and M. Suzuki, World. Sci. 1990)
*Theory and Applications of Coupled Map Lattices* (K. Kaneko ed., Wiley 1993 ).

[10] M. Mezard, G. Parisi, and M.A. Virasoro eds., *Spin Glass Theory and Beyond* (World Sci. Pub., Singapore, 1988)

[11] K. Kaneko, Artificial Life, **1** (1994) 163

[12] K.Ikeda, K. Matsumoto, and K. Ohtsuka, Prog. Theor. Phys. Suppl. **99** (1989) 295

[13] G. Perez et al., Phys.Rev. **A 45** (1992) 5469; S.Sinha et al., Phys Rev. **A 46** (1992) 3193

[14] D.Dominguez and H.A. Cerdeira, Phys. Rev. Lett. **71** (1993) 3359

[15] N. Nakagawa and Y. Kuramoto, Prog. Theor. Phys. **89** (1993) 313, and Physica **75 D** (1994) 74; Physica D (1994), in press; see also V. Hakim and W.J. Rappel, Phys. Rev. **A 46** (1992) 7347

[16] K. Kaneko, Phys. Rev. Lett. **65** (1990) 1391; Physica **55D** (1992) 368

[17] Collective behavior of locally coupled dynamical systems does not show such hidden coherence, but there are some interesting behaviors such as collective quasiperiodic motion; see e.g., H. Chaté and P. Manneville, Prog. Theo. Phys. 87 (1992) 1

[18] A.S.Pikovsky and J. Kurths, Phys. Rev. Lett. **72** (1994) 1644

[19] G. Perez, H.A. Cerdeira, Phys. Rev. **A 46** (1992) 7492; G. Perez et al., Physica **63 D** (1993) 341

[20] S.Sinha et al., Phys Rev. **A 46** (1992) 6242

[21] K. Kaneko, PhysicaD, in press

[22] W. Just, Physica D, in press

[23] S. V. Ershov and A. B. Potapov, Physica D, in press

[24] K. Kaneko, J. Phys. A, **24** (1991) 2107

[25] K. Kaneko, Physica **77 D** (1994) 456

[26] K. Matsumoto and I. Tsuda, J. Phys. A 21 (1988) 1405

[27] K. Kaneko, Physica **75 D** (1994) 55

[28] K. Kaneko and T. Ikegami, Physica **56 D** (1992), 406; T. Ikegami and K. Kaneko, Chaos **2** (1992) 397

[29] K. Kaneko and T. Yomo, Physica **75D** (1994) 89; in *Advances in Artificial Life*, Springer (1995 E. Moran et al eds.); preprint (1995) submitted to J. Theor. Biol.

# TRANSIENT CHAOS: A TYPE OF METASTABLE STATE

TAMÁS TÉL

*Institute for Theoretical Physics,*
*Eötvös University. H-1088 Budapest, Puskin u. 5-7, Hungary*

Chaotic dynamics can often be observed for finite times only. This transient chaos represents a kind of nonequilibrium state different from the asymptotic steady state. The underlying nonattracting chaotic set in phase space is the fractal analog of a saddle point and is often called a chaotic saddle. Basic concepts and fundamental recent applications are reviewed. Besides discussing the analogy with nonequilibrium processes, we summarize how a chaotic saddle can be reconstructed in experiments. The phenomenon of chaotic transport in open hydrodynamical flows is discussed in some detail.

## 1. Introduction: What is Transient Chaos?

The appearance of chaos with finite lifetime is known as *transient chaos* (for a review see [1]), and provides an example of a kind of "nonequilibrium state" that cannot be understood as an asymptotic behavior. In such cases one observes a moving around of the system in an apparently chaotic manner and then, rather suddenly, a settling down to a steady state which is either a periodic or a chaotic motion (but of different type than the transients). Studying only the asymptotic behavior of such systems would mean loosing the interesting, chaotic part contained in the transients.

### 1.1. *Every-day-life examples*

1. *Chaos machines.* We all know the simple double-rotor-like battery-driven devices taylored for demonstrating permanent chaotic motion. In practice, however, nearly all of them settle down to a periodic motion after a chaotic swinging of no longer than one hour. This clearly shows the robustness of transient chaos that - as we shall see in the paper - is a more general phenomenon than stationary or permanent chaos.

2. *Shimmying wheels.* Trolleys (like the ones used in supermarkets or at airports) have wheels which can choose the direction of their rollings. Even at a constant speed of towing one can observe a periodic motion of the wheels' orientation ('shimmy'). At higher speeds this motion might become chaotic which is often of transient type: after a while the irregular oscillation of the wheel disappears and the motion becomes regular and stationary. A detailed mechanical model of this phenomenon has been investigated in the engineering literature [2]. Chaotic shimmying was

shown to be the consequence of the non-rigid suspension of the wheel's axis and the possibility of both rolling and slipping. The orientational angle of the wheel observed at a given position of its axis is governed by an approximate one-dimensional map whose form was determined numerically. In the case of transient chaos, there is an interval around an unstable fixed point of the map which is mapped partially outside of the interval. The motion in this interval is chaotic and might be very long, but outside of it the effect of a coexisting attracting point (the origin) becomes essential and oscillations gradually die out [2].

### 1.2. *Experiments*

The experimental investigation of transient chaos has received, in spite of its relevance, relatively little attention (for a few examples see [3] –[12]) and, with the exception of very recent efforts [11,12], has mainly concentrated on determining the most striking characteristics, the average chaotic lifetime only. Here we briefly discuss some of these experiments.

1. *Convection loop experiment* [4,9]. The apparatus consists of a loop-shaped reservoir filled with water that is heated from below along the lower semicircle and cooled from above. The system is the one-dimensional analog of the Rayleigh-Bénard convection problem. After convection sets in at sufficiently high heat flux, the velocity of the fluid along the loop changes its sign and modulus randomly. There is a broad range of system parameters where these chaotic oscillations do not last forever, but sooner or later (depending on the initial conditions) the system reaches a stage where oscillations cease to exist and a uniform rotation in one direction sets in.

2. *Chemical reactions preceeding thermal equilibrium.* It is well known that stirred chemical reactions in closed containers cannot be chaotic in a stationary manner since the system approaches thermal equilibrium corresponding to a fixed point attractor in the space of concentrations. It has been conjectured, however, that the *approach to equilibrium*, i.e., the kinetics before reaching the final attractor can be chaotic. This has to be a kind of transient chaos [13]. Recently, the conjecture has been verified with the example of the Belousov-Zhabotinsky reaction, and chaotic oscillations of the light transmission have been observed over several hours before a stage very close to thermal equilibrium was reached [10].

3. *NMR laser and pendulum experiments.* These are the first experiments in which the chaotic set responsible for the transient behavior could be

reconstructed. This has been achieved by considering only those parts of several thousands transient signals that did *not* belong to the attractor.
*NMR laser:* The low noise and drift free data of an NMR laser made this system appropriate for the experimental investigation of a variety of chaotic phenomena. The laser output is a time-dependent voltage, and the control parameter can be considered to be the modulation amplitude. By gluing several truncated signals (that contained points far away from the attractor only) together at a level corresponding to a background voltage very close to zero, a long artificial time series was created. It could then be investigated by standard methods of time series analysis worked out originally for permanent chaotic signals. This way not only the fractal chaotic set was reconstructed, but also its dynamical properties, like the average Lyapunov exponent or the correlation decay could be determined [11].
*Pendulum:* The parametrically driven damped pendulum is another example where high precision experiments can be carried out. The angle and the angular velocity of the pendulum were measured, and the damping constant played the role of the control parameter [12]. From several distinct truncated signals the chaotic set was reconstructed on a stroboscopic map. This experiment clearly demonstrates the fractality along the unstable foliation (cf. Section 2). The generalized dimensions of the chaotic set were also determined.

## 2. Basic Properties of Transient Chaos

### 2.1. *The chaotic set*

In transiently chaotic systems one finds a *nonattracting chaotic set* in the phase space. In physical cases characterized by *invertible* dynamics this set is a *chaotic saddle* [14,15,17,1]. Similarly to a chaotic attractor, a chaotic saddle also contains an *infinity* of unstable orbits, but is globally *not* attracting. In this paper we shall mainly consider systems with three-dimensional phase spaces, or equivalently, two-dimensional discrete-time dynamics. As an example, Fig. 1 shows the chaotic saddle of the Hénon map (at parameter values different from the usual ones $a = 1.4, b = 0.3$ belonging to the Hénon attractor).

This observation can be used to define transient chaos. We say that a dynamical process is chaotic in a transient sense if *there exists an underlying nonattracting chaotic set* in the phase space. As a consequence of the fractal structure and the infinity of unstable orbits, the chaotic saddle's *topological entropy has to be positive*, in contrast to the attractor's topological entropy.

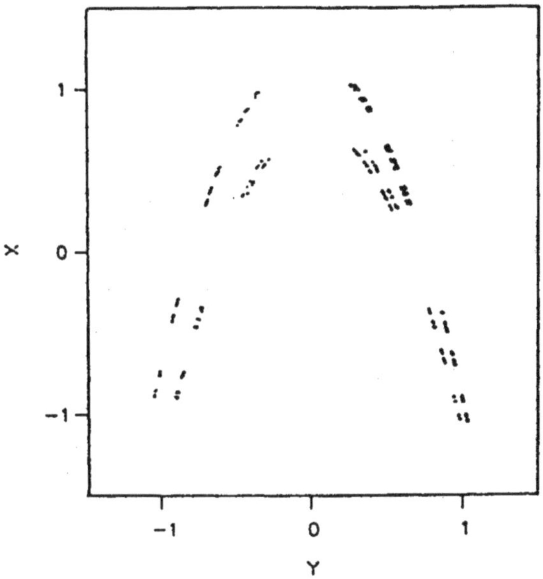


Fig. 1. The chaotic saddle of the Hénon map $x' = 1 - ax^2 + by, y' = x$ at parameters $a = 2, b = 0.3$ where no other invariant sets of the map exist. Trajectories escaping from this set go to minus infinity. Note the pronounced fractal structure resembling the direct product of two Cantor sets.

### 2.2. *Invariant manifolds*

A chaotic saddle is the chaotic analog of a saddle point that repells trajectories from its neighborhood, but has nevertheless an invariant subspace of volume zero along which it is attractive. This subspace is called the *stable manifold* of the chaotic saddle. The stable manifold provides a fractal foliation of the phase space (see Fig. 2). Similarly, there is an invariant subspace along which the chaotic set repells out those exceptional trajectories that have hit it infinitely long time ago, the set's *unstable* manifold. More precisely, this can also be defined as the stable manifold of the time reversed dynamics. Sometimes it is worth considering the chaotic saddle as the intersection of its invariant manifolds.

### 2.3. *Fractal properties*

A chaotic saddle has *more pronounced fractal properties* than a chaotic attractor. Due to the escape, the chaotic saddle has holes on all scales *along its unstable manifold*, too. The saddle of the Hénon map appears to be the direct

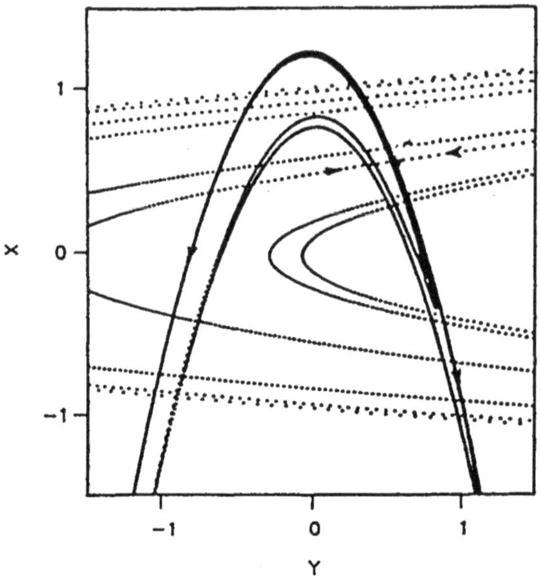


Fig. 2. The stable (dotted lines) and unstable (full lines) manifolds of the Hénon map chaotic saddle at $a = 2, b = 0.3$.

product of two Cantor sets. The full fractal dimension $D_0$ of the saddle can thus be written as the sum of the Cantor sets' dimensions: $D_0 = D_0^{(s)} + D_0^{(u)}$, where $D_0^{(s)}$ ($D_0^{(u)}$) is called the saddle's *partial* fractal dimension along the stable (unstable) direction [15]. Because the invariant manifolds provide a foliation of the phase space with smooth curves, their fractal dimension is obtained as the corresponding partial dimension plus unity. Thus, for the fractal dimension of the stable (unstable) manifold we find [17] $d^{(s)} = 1 + D_0^{(u)}$ ($d^{(u)} = 1 + D_0^{(s)}$). For a chaotic attractor the partial dimension along the unstable manifold is always unity (no escape), thus its stable manifold is space filling in certain regions at least. This is why chaotic attractors have finite basins of attractions. A chaotic saddle's stable manifold, and consequently its basin of attraction, is, however, a fractal of *zero volume*. Chaotic saddles typically coexist in dissipative cases with an attractor, and in Hamiltonian systems with an exit channel leading to an asymptotic motion. Trajectories starting from randomly chosen initial points then approach the attractor or the asymptotic state with probability one. Because of the saddle's stable foliation, however, they might come close to the saddle and stay in its vicinity for a longer time. What is observable numerically or experimentally is not the chaotic saddle but rather a *small neighborhood* of it. This results in the appearance of a chaotic motion for finite times. The actual time spent around the saddle depends sensitively on initial conditions but the average transient lifetime is well defined.

### 2.4. *The escape rate*

When distributing a large number $N_0$ of initial points in a region overlapping with the chaotic saddle (but not overlapping with the attractor(s)), the trajectories emanating from these points will leave the region with certainty. Those with initial points close enough to a branch of the stable manifold have a long time before escaping. Thus the number of points $N(t)$ staying still in the preselected neighborhood after a time $t$ ($t$ can also denote discrete time) is a function monotonously descreasing to zero. How rapidly it decreases is an important characteristics of the saddle. Often one finds that the decay is exponential for long times [15], i.e.,

$$N(t) \sim N_0 e^{-\kappa t} \tag{1}$$

for $t \to \infty$. In what follows we shall deal with such hyperbolic cases. The positive number $\kappa$ is called the *escape rate* of the saddle and turns out to be independent of the shape of the region containing the initial points. The escape rate measures the saddle's strength of repulsion: a large escape rate marks a rather unstable saddle. Conversely, $\tau = 1/\kappa$ is considered to be the average lifetime of transient chaos.

### 2.5. *The natural distribution and Lyapunov exponents*

Just like on chaotic attractors, there exist invariant distributions on chaotic saddles, too. The so-called *natural distribution* [15,1] on a saddle is obtained by distributing an *ensemble* of points around it and following those with long lifetimes. The frequency of visiting different regions of the saddle by these trajectories defines the natural probability measure. One can then speak about characteristic numbers taken with respect to this measure. The largest average Lyapunov exponent $\lambda$ on the saddle is positive, and is of basic importance since it characterizes the typical divergence of nearby trajectories staying for long times around this set. The Lyapunov exponent describes the local instability of the saddle, while the escape rate is a global measure of instability. It is natural to believe that these quantities are intimately related.

### 2.6. *Fundamental relations*

A basic relation [15] connecting the escape rate with metric properties says that

$$\kappa = (1 - D_1^{(u)})\lambda = \lambda - K_1 \tag{2}$$

where $D_1^{(u)}$ is the *information* dimension (which is typically very close to the fractal dimension) along the unstable direction, and $K_1$ represents the metric or Kolmogorov-Sinai entropy of the saddle. (For a chaotic attractor $\kappa = 0$ and we recover the well-known relations $D_1^{(u)} = 1$ and $K_1 = \lambda$.) As a consequence, for a saddle

$$D_1^{(u)} = 1 - \frac{\kappa}{\lambda}, \tag{3}$$

that is, the deviation of the partial information dimension along the unstable manifold from unity is given by the ratio of the global and local instabilities' measures. This deviation is the stronger the larger is the escape rate measured on the Lyapunov time scale. The partial information dimension $D_1^{(s)}$ along the stable direction turns out to be $\lambda/\lambda'$ times $D_1^{(u)}$, where $\lambda'$ is the modulus of the negative average Lyapunov exponent [15]. This leads to a generalization of the Kaplan-Yorke formula valid for transient chaos. The saddle's full information dimension can be written as the sum of the partial dimensions [16,15]

$$D_1 = (1 - \frac{\kappa}{\lambda})(1 + \frac{\lambda}{\lambda'}) = K_1(\frac{1}{\lambda} + \frac{1}{\lambda'}). \tag{4}$$

These relations connect fractal and dynamical properties in a unique and elegant fashion.

## 3. Types of Transient Chaos

### 3.1. *Type I — in coexistence with simple attractors*

The cleanest form of transient chaos occurs when the system has a periodic attractor coexisting with a chaotic saddle [14,15]. Then the asymptotic behaviour is simple, the maximal Lyapunov exponent on the attractor is negative. The transients, however, exhibit chaotic properties, and depending on the parameter can be rather persistent. Cases when the attractor is at infinity (see e.g., the Hénon example) or open Hamiltonian systems when the role of the attractor(s) is taken over by different exit possibilities are also of this type. In such cases the dynamical complexity (and the positive Lyapunov exponent and metric entropy around the saddle) would be completely missed when studying the asymptotic properties only.

### 3.2. *Type II — in coexistence with chaotic attractors*

The situation when a chaotic attractor coexists with a chaotic saddle is a more complicated and more interesting case. Then two different kinds of chaos are present. The motion around the saddle is, however, typically *more* chaotic than on the attractor (both the Lyapunov exponent and the metric entropy are larger), as, e.g. in the NMR laser experiment [11]. Thus, studying asymptotic properties only would lead to neglecting the dominant chaotic part of the full complex dynamics.

## 4. Occurrence of Transient Chaos

Here we give a list of phenomena from the class of deterministic dynamical systems that are inevitably related to transient chaos.

### 4.1. *Periodic windows*

Periodic windows are ubiquitous in the chaotic regime of dynamical systems [18]. In such windows chaos is present in the sense that there exists an infinity of periodic orbits but their union is not neccessarily attractive. Transient chaos of Type I (II) occurs inside (outside) the period doubling regime of these windows where the attractor is a cycle of length $2^n$ with an integer $n$. Since the total measure of windows is finite in the parameter space, just like that of strictly chaotic cases, the probability to find transient chaos is on the *same* order as that of permanent chaos.

### 4.2. *Crises*

Transient chaos can also be a sign of permanent chaos to be born, hence the name "preturbulence" or "metastable chaos" [19]. More generally, all types of crisis configurations: attractor destructions, explosions or mergings [14] are accompanied with long lived transient chaos. Large attractors born at crises incorporate into themselves the chaotic saddles existing before. Consequently, the dynamical properties of the saddle are partially inherited by the large attractor. This is the origin for the scaling behaviour found in such systems [20]. The average time $\tau_1$ trajectories of the large attractor spend in the region where the saddle existed (and continuos to exist) is practically the same as the average lifetime of transient chaos in the pre-crisis regime (cf. NMR laser experiment [11]). The average time $\tau_2$ of visiting the pre-crisis attractor, however, changes as a negative power of the control parameter difference $a - a_c$. This is nothing but the transient chaotic lifetime on a newborn chaotic saddle into which the pre-crisis chaotic attractor turned at crisis. Transient chaos can thus provide a backbone of the motion on composed attractors [21,22]. More generally, one can select any finite subregion of the phase space and determine the set of orbits never escaping this region. If the subregion overlaps with an attractor (or the chaotic sea in Hamiltonian cases), these orbits form a fractal chaotic saddle responsible for the transient chaotic dynamics exhibited by trajectories not bounded to this subregion forever [23].

### 4.3. *Fractal boundaries*

Fractal basin boundaries [24] are another common properties of dynamical systems. If two or more simple or chaotic attractors coexist, trajectories may hesitate for a long time to which of the attractors they will asymptote to. On fractal basin boundaries such trajectories exhibit transient chaos. In fact, fractal basin boundaries turned out to be *stable manifolds of chaotic saddles* [25].

Uncertainties in initial conditions on fractal boundaries lead to enhanced uncertainties in the final state. This can be characterized by an uncertainty exponent $\alpha$ [24]: the fraction $f(\epsilon)$ of uncertain initial conditions typically decreases with the box size $\epsilon$ according to a power law: $f(\epsilon) \sim \epsilon^{\alpha}$. The fractal dimension of the intersection of the saddle's stable manifold with a one-dimensional line is given by $d = 1 - \alpha$. In two-dimensional maps $\alpha = 1 - D_0^{(u)}$. It has recently been shown that in higher dimensional systems the boundaries can be widely interwoven [26]. Such *riddled basins* are characterized by near-zero uncertainty exponents. This means that an increase in the resolution hardly improves the prediction of the system's final outcome. Simultaneously, the fractal dimension of the set of intersecting points of the stable manifold with a one-dimensional line is a rather dense fractal with dimension close to unity.

### 4.4. *Controlling chaos*

The OGY method of controlling permanent chaos [27] is based on the requirement that control sets in if the trajectory visits a preselected target region. The set of points never reaching this target region form a fractal subset whose escape rate determines the time needed to achieve control. Thus, the OGY method converts the motion on a chaotic attractor into a kind of transient chaos before control sets in. The OGY method has been extended to controlling transient chaos, too, both in dissipative and in Hamiltonian cases [28,29].

## 5. Applications of Transient Chaos

The concept of transient chaos has proved to be rather useful outside the realm of dynamical systems, as well. Below we give a few examples belonging to entirely different fields of physics. We can just mention here that interesting applications can be found in other disciplines, too, among which epidemiology is a notable example [30].

### 5.1. *Chaotic scattering*

For scattering processes in open Hamiltonian systems the only way chaos can appear is in the form of transients, because of the incoming and outgoing motion's asymptotic freedom. Trajectories are then trapped in a given region of the configuration space for a while. Because of the Hamiltonian character of the dynamics the chaotic saddle has a special symmetry that implies the coincidence of the stable and unstable manifolds' dimensions. In systems with two degrees of freedom or for two dimensional area preserving maps

$$D_0^{(u)} = D_0^{(s)} \qquad D_1 = 2(1 - \frac{\kappa}{\lambda}). \tag{5}$$

A detailed characterization of the trapping process is based on the so-called time delay function. It tells us how the time $\tau(b)$ spent around the chaotic saddle depends on the impact parameters $b$ of initial conditions taken along a line. A unique sign of chaotic scattering is the rather *irregular* appearance of the time delay function. The function $\tau(b)$ is singular in points that lie on branches of the saddle's stable manifold. Consequently, the time delay function has singularities on a fractal whose dimension is the partial fractal dimension of the saddle. The phenomenon of chaotic scattering [31] has attracted great interest recently: the field of its application ranges from chemical reactions, via acustics and elasticity to celestial mechanics.

### 5.2. *Chaotic mixing in open hydrodynamical flows*

The passive advection of tracer particles (e.g., small dye droplets) in open hydrodynamical flows with uniform inflow and outflow velocities is one of the most appealing applications of chaotic scattering [32] - [36]. The reason is that in two-dimensional incompressible fluids the tracer dynamics turns out to be Hamiltonian with the plane of the flow as its phase space. Thus, the fractal structures appearing in the phase space become now observable *by naked eye.* This phenomenon is thus a good candidate for being the first experimental investigation of chaotic scattering. The chaotic saddle consists now of an infinity of unstable tracer orbits, restricted to a finite domain of the flow. Tracers injected into the flow before this domain are either transported away quickly or, if being sufficiently close to the stable manifold, stay a long time around the saddle. These particles finally escape very close to the unstable manifold. Thus the *unstable manifold of the saddle becomes a direct physical observable*: the object traced out by particles with long life times. Its fractal dimension is $d^{(u)} = 1 + D_0^{(s)} = 1 + D_0^{(u)}$ and is definitely less than two. Transient dynamics is thus accompanied with pronounced and persistent fractality. The latter is reflected by the singularity distribution of the particles' time delay function, too. These general features do not depend on whether the fluid is viscid or not. The key ingredients are the time dependence and incompressibility of the flow.

As an example we consider a model of the so-called 'leapfrogging' motion of smoke rings. If the rings have the same sense of rotation and move along the same axis, the rear vortex ring attempts to pass through the front one. The leading ring then widens due to the mutual interaction and travels more slowly. Simultaneously, the other ring shrinks, travels faster and penetrates the first one. This process is then repeated continuously, with some period. In [35] we examined the two-dimensional analog of this process: the advection in the field of two coaxial pairs of ideal point vortices of the same strength, which also exhibit a strictly periodic motion of some period $T$. It is worth then investigating the particle dynamics on a stroboscopic map taken at integer multiples of $T$. Figure 3 shows the chaotic saddle of the map, as well as its unstable manifold, in a frame comoving with the vortex system. The manifold has been obtained by injecting a droplet of tracers in front of the system and plotting its shape after a long time. It qualitatively agrees with a photograph of smoke rings taken in a laboratory experiment [37].

### 5.3. *Random systems*

Renormalization group transformations or, more generally, length scaling or decimation procedures when applied to models of disordered systems often lead to unbounded iterations. A classical example is provided by the ***phonon spectrum of fractals***. Imagine a two-dimensional Sierpinski gasket lattice of point masses linked by identical strings and allowed to move perpendicularly to the plane of the lattice. A decimation procedure applied to frequency $\omega$ of the collective motion yields [38] a new frequency $\omega' = \omega(5 - \omega)$. This is the logistic map at a parameter where it generates transient chaos. A quantity of central interest is the spectrum of eigenfrequencies. The invariant set of the frequency iteration is obviously part of the spectrum. Thus, each point of the

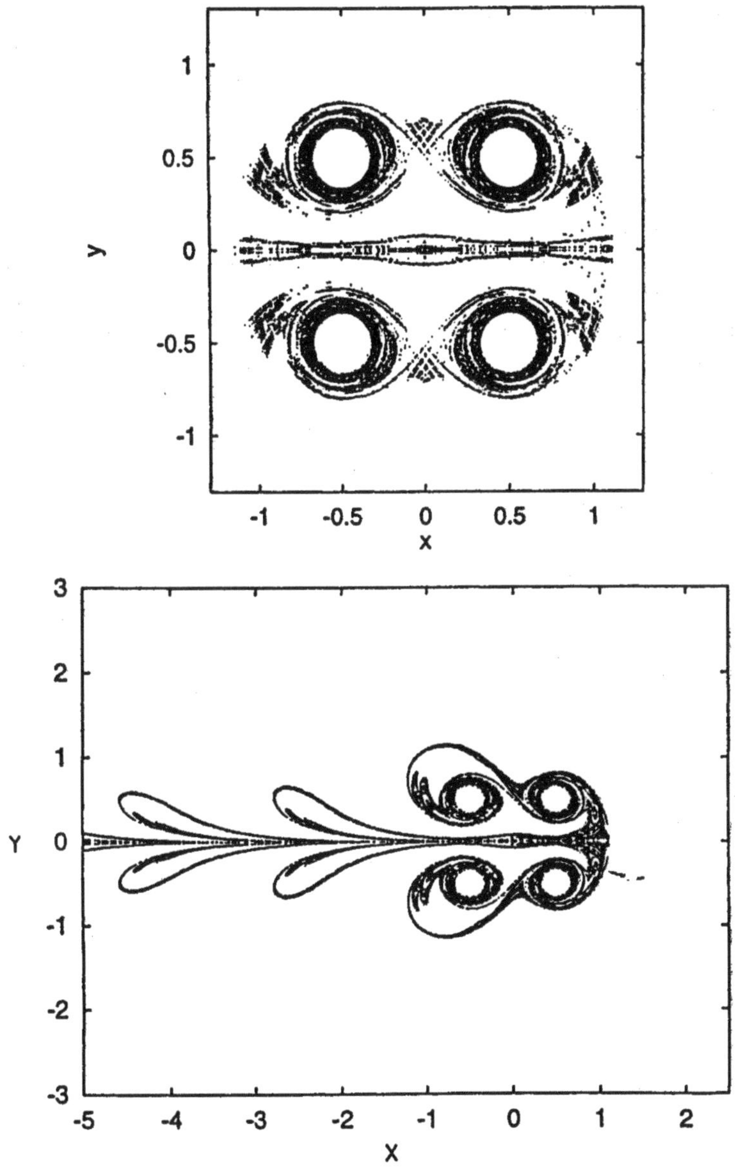


Fig. 3. Top: chaotic saddle around four leapfrogging ideal vortices as appearing in the plane of the flow. Bottom: the unstable manifold of the same saddle. The snapshots are taken in both cases when the two upper vortices fall on the same horizontal line. The white regions around the vortex centers are never reachable by tracers coming form outside. They can be considered as the vortex cores from the point of view of the tracer dynamics.

chaotic set represents an element of the vibration spectrum which exhibits, therefore, a *Cantor set structure*. More recent examples are provided by the theory of random walks in random environments [39], and the learning process of neural networks with random input data [40].

### 5.4. *Noise induced chaos*

In systems subjected to external random forces the form of the attractor observed might depend on the noise intensity. The phenomenon when a system with simple periodic attractors turns to be chaotic at sufficiently strong noise is called noise induced chaos [41]. It has recently be interpreted as a consequence of transient chaos [42]. In such systems there is always a chaotic saddle coexisting with the simple attractors. At increasing noise intensity the saddle becomes embedded, together with the simple attractors, into a noisy chaotic attractor. Another interesting effect of weak noise is that it decreases the escape rate from a saddle, i.e., it stabilizes transient chaos [43].

### 5.5. *Transport phenomena*

Diffusion and other transport phenomenona can be interpreted as consequences of chaotic scattering and transient chaos [44]. This new deterministic way of describing transport phenomena in a single particle picture is based on the idea of considering an open Hamiltonian system that is of finite but large extent $L$ along a given direction. The phase space is low-dimensional but of large linear size. One can calculate the escape rate from the chaotic saddle and show that it can be connected with transport coefficients [44,45,46]. This escape rate formalism has previously been applied to cases corresponding to a symmetric random walk only. When the system is also subjected to a bias causing a nonzero average velocity, it has recently been shown [47] that the escape rate can be written as

$$\kappa = \frac{1}{2}\frac{v^2}{D} + \frac{1}{2}\frac{D\pi^2}{L^2}. \tag{6}$$

Here $v$ and $D$ are the (constant) drift and diffusion coefficients of the system on a macroscopic level that also appear in the Fokker-Planck equation governing the macro dynamics. This formula says that escape is responsible for two different ways of transport: drift and diffusion, and that the corresponding coefficients can be extracted via finite size scaling. It is worth mentioning that other characteristics of the chaotic scattering process cannot be expressed solely by means of macro parameters. It is the escape rate alone that has a well defined thermodynamical limit.

### 5.6. *Supertransients*

Transient chaos also occurs in spatiotemporal dynamical systems having high-dimensional phase spaces. These transients differ from their low-dimensional counterparts in that the average lifetime can be extremely long before settling down into a final attractor which is usually nonchaotic. Often the lifetime grows exponentially with the system size $L$, i.e,

$$\kappa \sim \exp\left(-aL^{\sigma}\right) \tag{7}$$

where $a$ and $\sigma$ are positive constant, the latter on the order of unity. The orgin of such *supertransients* [48,49] is not yet fully understood. Their presence implies that the observation of the systems's actual attractor is practically impossible. It has recently been shown [50] that supertransients are accompanied with very dense chaotic saddles: their stable manifolds have fractal dimensions arbitrary close to the phase space dimension. In some other cases the escape rate decreases slower; as a negative power of the system size [48,51].

## 6. Conceptual Novelties

- Transient chaos plays a similar role in the realm of chaotic processes as an unstable equilibrium point in mechanical systems. Starting an ensemble of trajectories around such a hyperbolic fixed point, particles escape from any neighborhood of it with a rate governed by the positive Lyapunov exponent. A chaotic saddle is a fractal set of such hyperbolic points. It is, however, globally *less* repelling than the component periodic orbits because the escape rate is less than the average Lyapunov exponent by $D_1^{(u)}\lambda$. The fractal structure thus stabilizes the saddle, and a dense set implies rather small escape rates.

- Transient chaos is an example for a phenomenon where long time single particle and short time ensemble averages are different. An ensemble of trajectories staying around the chaotic saddle yields averages characterizing the saddle, different from the long time asymptotics.

- Transient chaos might also shed new light on previous observations. For instance, the famous "period-3 implies chaos" statement [52] means in modern terminology: period-3 implies transient chaos.

In conclusion, we believe that transient chaos is an essential part of nonlinear dynamics accessible in experiments, too, and is needed to the correct interpretation of many interesting phenomena. It is especially well suited to characterize nonequilibrium processes preceding the approach to steady states.

## Acknowledgments

I would like to thank for the enjoyable and fruitful collaboration W. Breymann, L. Fepp, C. Grebogi, R. Graham, A. Hamm, I. Jánosi, C. Jung, Z. Kovács, Y.-C. Lai, E. Ott, Á. Péntek, K. G. Szabó, Z. Toroczkai, C. Van den Broeck, J. Vollmer, J. Yorke, and E. Ziemniak. I am greatful for sharing their views and experiences with me during our joint works. Useful discussions with P. Reimann and G. Stépán are acknowledged. This work has been supported by the Hungarian National Science Foundation (OTKA) grants T4439 and T17493.

## References

[1] T. Tél, Transient chaos in *Directions in Chaos*, Vol.3, ed.: Bai-lin Hao (World Scientific, Singapore, 1990) pp. 149-221.
[2] G. Stépán, Vehicle System Dynamics **20**, 341 (1991); G. Stépán, in: *Future Direction of Nonlinear Dynamics in Physical and Biological Systems* (Plenum, New York, 1992) p.169; L. Palkovics, G. Stépán, and P. Michelberger, Machine Vibration **2**, 47 (1993).
[3] R. W. Leven et al, Physica (Amsterdam) **D 16**, 371 (1985).
[4] P. J. Widmann, M. Gorman and K. A. Robbins, Physica (Amsterdam) **D 36**, 157.
[5] T. L. Carroll, L. M. Pecora and F. J. Rachford, Phys. Rev. Lett. **59**, 2891 (1987).
[6] Z. J. Kowalik, M. Franaszek and P. Pieranski, Phys. Rev. **A37** (1988).
[7] W. L. Ditto et al., Phys. Rev. Lett. **63**, 923 (1989).
[8] R. Stoop and J. Parisi, Phys. Rev. **A 43**, 1802 (1991).
[9] Y. Wang, J. Singer, and H. H. Bau, J. Fluid. Mech. **237**, 479 (1992).
[10] J. Wang, P. G. Sorensen, and F. Hynne, J. Chem. Phys. **98**, 725 (1994).
[11] I. M. Jánosi, L. Flepp, and T. Tél, Phys. Rev. Lett. **73**, 529 (1994).
[12] R. W. Leven and M. Selent, Chaos Sol. Frac. **4**, 2217 (1994).
[13] S. K. Scott et al, J. Chem. Phys. **94**, 1134 (1991).
[14] C. Grebogi, E. Ott and J. Yorke, Phys. Rev. Lett. **48**, 1507 (1982); Physica (Amsterdam) **D 7**, 181 (1983).
[15] H. Kantz and P. Grassberger, Physica **D 17**, 75 (1985).
[16] L. S. Young, Ergod. Th. Dynam. Syst. **2**, 109 (1982).
[17] G. H. Hsu, E. Ott, and C. Grebogi, Phys. Lett. **127A**, 199 (1988).
[18] J. A. Yorke et al., Phys. Rev. Lett. **54**, 1095 (1985).
[19] J. A. Yorke and E. D. Yorke, J. Stat. Phys. **21**, 263 (1979).
[20] C. Grebogi, E. Ott, and J. Yorke, Phys. Rev. Lett. **57**, 1284 (1986).
[21] M. Franaszek, Phys. Rev. **A46**, 6340 (1992); M. Franaszek and A. Nabaglo, Phys. Lett. **A178**, 85 (1993); **A182**, 99 (1993).
[22] K. G. Szabó and T. Tél, Phys. Lett. **A196**, 173 (1994).

[23] H. E. Nusse and J. Yorke, Physica **36D**, 137 (1989).
[24] C. Grebogi et al, Phys. Rev. Lett. **50**, 935 (1983); E. Ott et al, Phys. Rev. Lett. **71**, 4134 (1993).
[25] C. Grebogi et al, Lect. Notes in Math. **1342**, 220 (1988).
[26] E. Ott et al., Phys. Rev. Lett. **71**, 4134 (1993).
[27] E. Ott, C. Grebogi, and J. Yorke, Phys. Rev. Lett **64**, 1196 (1990).
[28] T. Tél, J. Phys. **A24**, L1359 (1991).
[29] Y. Lai, T. Tél, and C. Grebogi, Phys. Rev. **E48** 709 (1993); Z. Kovács, K. G. Szabó, and T. Tél, in: *Nonlinearity and Chaos in Engineering Dynamics*, ed.: J. M. T. Thompson and S. R. Bishop (Wiley and Sons, Chichester, 1994) p. 155; Y. C. Lai, C. Grebogi and T. Tél, in: *Towards the Harnessing of Chaos* ed.: M. Yamaguti (Elsevier, Amsterdam, 1994) p. 153.
[30] D. Rand and H. B. Wilson, Proc. Roy. Soc. **246**, 179 (1991); R. Engbert and F. R. Drepper, Chaos Sol. Frac. **4**, 1147 (1994).
[31] E. Ott and T. Tél, Chaos **3**, 417 (1993), and the entire special issue of Chaos **3**, no.4 (1993).
[32] C. Jung and E. Ziemniak, J. Phys A **25** (1992), 3929.
[33] C. Jung, T. Tél, and E. Ziemniak, Chaos **3** (1993), 555.
[34] E. Ziemniak, C. Jung, and T. Tél, Physica D **76** (1994), 123; C. Jung and E. Ziemniak, in: *Fractals in the Natural and Applied Sciences*, M. M. Novak, ed. (North Holland, Amsterdam, 1994).
[35] Á. Péntek, T. Tél, and Z. Toroczkai, J. Phys. A **28** (1995), 2191; Fractals **3**, 33 (1995).
[36] Á. Péntek et al, Phys. Rev. **E51**, 4074 (1995).
[37] M. Van Dyke, *An Album of Fluid Motion* (Parabolic Press, Stanford, 1988).
[38] R. Rammal and G. Toulouse, J. Physique Lett. **44**, L13 (1983).
[39] C. Van den Broeck and T. Tél, in *From Phase Transitions to Chaos*, eds.: G. Györgyi et al., (World Scientific, Singapore, 1992) p. 227.
[40] G. Radons, H. G. Schuster, and D. Werner, Phys. Lett. **A174**, 293 (1993); G. Radons, J. Stat. Phys. **72**, 227 (1993); U. Behn et al, Physica **D68**, 401 (1993).
[41] M. Iansiti et al, Phys. Rev. Lett. **55**, 746 (1985); H. Herzel et al., Z. Naturforsch. **42a**, 136 (1987); A. R. Bulsara et al., Phys. Rev. **A42**, 4614 (1992).
[42] A. Hamm, T. Tél, and R. Graham, Phys. Lett. A **185**, 313 (1994); P. Reimann, R. Müller, and P. Talkner, Phys. Rev. **E49**, 3670 (1994).
[43] M. Franaszek, Phys. Rev. **A44**, 4065 (1991); P. Reimann, Z. Naturforsch **49a**, 1248 (1994).
[44] P. Gaspard and G. Nicolis, Phys. Rev. Lett. **65**, 1693 (1990)).
[45] J. R. Dorfman and P. Gaspard, Phys. Rev. **E51**, 28 (1995).
[46] H. van Beijeren and J. R. Dorfman, Phys. Rev. Lett. **74**, 4412 (1995); M. H. Ernst et al., Phys. Rev. Lett. **74**, 4416 (1995).
[47] T. Tél, J. Vollmer, and W. Breymann, preprint, 1995.
[48] J. P. Chrutchfield and K. Kaneko, in *Directions in Chaos*, ed.: B. L. Hao, Vol.I (World Scientific, Singapore, 1987); K. Kaneko, Phys. Lett. **149A**, 105 (1990).
[49] A. Politi et al., Europhys. Lett. **22**, 571 (1993); A. S. Pikovsky and J. Kurths, Phys. Rev. **E49**, 898 (1994); R. Kapral et al., Phys. Rev. **E49**, 2009 (1995); Y.-C. Lai, Phys. Lett. **A200**, 418 (1995).

[50] Y.-C. Lai and R. L. Winslow, Phys. Rev. Lett. **74**, 5208 (1995).
[51] H. Fujisaka, K. Egami, and T. Yamada, Phys. Lett. **174A**, 103 (1993).
[52] T. Y. Li and J. A. Yorke, Am. Math. Monthly **82**, 985 (1975).

# ORDERED PHASES OF COMPLEX FLUIDS

MICHAEL SCHICK
*Department of Physics Box 351560, University of Washington*
*Seattle WA 98195-1560, U.S.A.*
*E-mail schick@dirac.phys.washington.edu*

Amphiphilic systems assemble themselves into interesting, and sometimes unusual, ordered phases. The different theoretical methods which are used to describe and predict them are outlined, and it is shown that the appearance of these phases depends only on general, not specific, properties of the amphiphile.

## 1 Introduction

My subject is the description of self-assembly in systems of oil, water, and amphiphile[1,2]. The latter is a molecule with a polarizable head group, which prefers a water environment, and a tail consisting of one or two carbon chains, which prefers the oil environment. As a consequence, the amphiphile tries to organize the oil and water by creating internal interfaces between them at which energetics favors it to be. At very low concentrations, the amphiphiles are molecularly dispersed. Beyond a certain concentration, called the critical micelle concentration, spherical micelles are formed in water-rich regions. The micelle is an aggregation of amphiphiles with the polar head group towards the water, and the carbon chains shielded from it within. If there is a small amount of oil, it will be found within the micelle. In water-poor regions, inverted micelles are formed in which the head groups are within and the tails point outward. At larger concentrations, ordered phases (denoted "lyotropic") are formed. It is these phases which I will emphasize in this paper. One of the simplest of these is a body-centerd cubic packing of spherical micelles. A second phase consists of cylindrical micelles which are packed in a hexagonal array. The simplest phase is a lamellar one consisting of sheets of amphiphile which separate oil and water. These phases also appear in the two-component system of water and amphiphile. A single bilayer sheet of lipids is, of course, the absolutely simplest model of a biological membrane. As one increases the concentration of amphiphile in the system, the phases often, but not always, appear in the order body-centered cubic, hexagonal, lamellar, inverted hexagonal, inverted body-centered cubic. Some phases may not appear in a given system. Between the hexagonal and lamellar phases, another phase of cubic symmetry[3] often appears; a bicontinuous phase of symmetry $Ia\bar{3}d$. It is highly unusual, to say the least, consisting of two three-fold coordinated networks of the minority component, each of which span the system. Each network has a chirality, the one being a mirror image of the other.

To describe these phases, and other properties of ternary amphiphilic systems, various theoretical models have been employed which are usually of one of the following three types; a microscopic lattice model, a Ginzburg-Landau model, or a mesoscopic membrane model. They are outlined briefly below.

## 2 Microscopic Lattice Models

In these models, space is divided into cells in each of which is found a molecule which can be of three types, water, oil, or amphiphile. The basic idea is that a configuration in which an amphiphile sits between a water and an oil, $W - A - O$, competes with the other energetically favorable configurations of pure water $W - W$, or pure oil $O - O$. This is just like a magnetic system in which an antiferromagnetic interaction between second-neighbors competes with a ferromagnetic interaction between nearest neighbors. It is well known[4] that models of this type, such as the "Anisotropic Next Nearest Neighbor Ising Model" exhibit lamellar phases of long periods. Furthermore, when fluctuations are taken into account, one expects a roughening transition to occur which will render irrelevant the fact that an underlying lattice was used. A nice property of the lattice formulation is that the only length scale put into the problem is a microscopic one, the lattice parameter, which is of molecular size. The long periods characteristic of the lyotropic phases emerge from the model.

A simple example of such a lattice model is the following[5]. On every site, $i$, of a face-centered cubic lattice there is a variable $\sigma_i$ which takes the values $1, 0, -1$ representing the presence at that site of a water, amphiphile, or oil molecule. There is a second variable, $s_i$ at each site which is only non-zero if $\sigma_i = 0$; in the latter case, $s_i$ is a unit vector representing the orientation of the amphiphile. It points to any one of the twelve nearest neighbors of the site $i$. The Hamiltonian of the system is

$$H = -J_1 \sum_{<ij>} \sigma_i \sigma_j - J_2 \sum_{<ij>} (\sigma_i s_j \cdot \mathbf{r}_{ji} + \sigma_j s_i \cdot \mathbf{r}_{ij}) + \mu_s \sum_i \sigma_i^2, \qquad (1)$$

where $< ij >$ denotes summation over all distinct pairs of nearest-neighbor sites $i$ and $j$, and $\mathbf{r}_{ij}$ specifies the direction from site $i$ to site $j$. The parameter $J_1 > 0$ favors the separation of oil and water, while the parameter $J_2 > 0$ gives the amphiphile head group its hydrophilic behavior and the tail group its hydrophobic behavior. The chemical potential of the amphiphile is $\mu_s$. Lamellar phases appear both when the model is solved within mean-field approximation and when simulated. A bicontinuous phase has been observed in a lattice model by Ciach *et al.*[6], and a hexagonal phase by Larson[7]. The lattice model used by the latter is notable in that it is one of the few in which the architecture of the molecule, *i.e.* the sizes of the head and tail groups, can be incorporated.

## 3 Landau-Ginsburg Theories

The basic length of a Landau-Ginsberg theory is somewhat larger than a microscopic one. At the outset, one must define the order parameters in the theory. While an incompressible ternary mixture is characterized by two independent densities, all structures of interest are characterized by the relative locations of oil and water. Once they are known, the location of the amphiphile is as well, because it will be at the interfaces. Thus we can define a single order parameter $\psi(\mathbf{r})$ which is proportional to the difference in water and oil concentrations at the point $\mathbf{r}$. From the observation[8] that the water-water structure functions of scattering experiments are well fit by the form $S^{-1}(q) \propto cq^4 + gq^2 + a$, it follows that the Landau Ginsburg free energy functional is

$$F\{\psi\} = \int d\mathbf{r}[c(\nabla^2\psi)^2 + g(\nabla\psi)^2 + f\{\psi\}], \tag{2}$$

where $f\{\psi\}$ is the bulk free energy density, and $f'' = 2a$ in the bulk phase. This functional differs from those which describe ordinary fluids by the presence of the Laplacian squared term. This is needed because the coefficient $g$ can be negative, which reflects the physical fact that the amphiphile likes large gradients in the oil, water density at which it is favorable for it to sit. The implications of this functional are reviewed in[1]. An obvious consequence is the appearance of a new length scale, $(c/a)^{1/4}$, which characterizes the scale of the lyotropic phases. The architecture of the amphiphile enters the theory in principle via the coefficients $c$ and $g$ and the bulk free energy $f\{\psi\}$. In practice however, they have not been derived from underlying microscopic models except in the simplest possible case[9].

Gompper and Zschocke[10] looked for lyotropic phases using the above model and found lamellar, hexagonal, and cubic micellar crystals, and their inverses. They also find a bicontinuous phase, but one which is only metastable.

It should be noted that the Landau-Ginsburg free energy functional of Eq.(2) is commonly employed in the theory of diblock copolymers which also self-assemble into the same lyotropic phases as described above[11].

## 4 Membrane Theories

Membrane theories assume that the amphiphile is essentially insoluble, and that the amphiphilic sheets are thin and well-defined. One ignores the environment in which the sheets are embedded, and describes the system in terms of the sheets alone. At each point on the sheet, or membrane, one can define the two principle curvatures, $C_1$ and $C_2$, and the invariant combinations $H \equiv (C_1+C_2)/2$, the mean curvature, and $K \equiv C_1C_2$, the Gaussian curvature. The elastic energy of the membrane is[12], to second order in the curvatures,

$$H = \int dS[\mu + \lambda_s H + 2\kappa H^2 + \bar{\kappa}K], \tag{3}$$

where the integral is over the entire surface and $\mu$ is a chemical potential for the creation of surface. There are three material constants. Because $\lambda_s$ couples linearly to the mean curvature, it favors a particular value, $c_0 = -\lambda_s/4\kappa$ of the mean curvature. Its presence reflects the physical fact that it may be energetically favorable for the membrane to have a curvature due to different sizes of the head and tail groups. The bending modulus, $\kappa$, and saddle-splay modulus $\bar{\kappa}$ determine the energetic cost of non-zero curvatures. These material parameters are determined, in principle, by the architecture of the molecules.

The great advantage of this approach is its simplicity. For example, let us consider the free energy of a system of spherical micelles. Assume that oil is the minority component, with density $n_o$, and that the density of amphiphile is $n_a$. Because the oil is found within the spheres of volume $4\pi R^3/3$ and the amphiphile on the surface of volume $4\pi R^2\delta$, with $\delta$ a molecular thickness, it follows that the curvatures $C_1 = C_2 = 1/R = 2q/3$, where $q \equiv n_a/n_o\delta$ and is varied simply by changing the relative concentrations of the components. The energy per unit area of the system is then

$$\frac{H_{sphere}}{A} = \mu + \frac{2\lambda_s q}{3} + \frac{(2\kappa + \bar{\kappa})q^2}{9}. \tag{4}$$

Similarly

$$\frac{H_{cylinder}}{A} = \mu + \frac{\lambda_s q}{2} + \frac{\kappa q^2}{8}, \tag{5}$$

$$\frac{H_{lamellae}}{A} = \mu. \tag{6}$$

It is now simple to determine the regions of stability of the various configurations[13]. We find that as $q$ decreases, one goes from lamellar to cylindrical to spheres in analogy with the lamellar to hexagonal to body-centered cube of spheres encountered earlier. The calculation can also be carried out for bicontinuous phases[14].

The effects of fluctuations are extremely important in membrane theories. In particular they cause the bending modulus $\kappa$ to become scale dependent[15]

$$\kappa(L) = \kappa(a) - (3T/4\pi)\log(L/a), \tag{7}$$

where $T$ is the temperature. This equation states that at a length scale $L$, the bending modulus is reduced from its value on microscopic length scales $a$. It also defines a new length scale via $\kappa(\xi) = 0$, the persistence length[16]

$$\xi = a\exp[4\pi\kappa(a)/3T]. \tag{8}$$

Other lengths are generated by the renormalization of $\bar{\kappa}$ and combinations of the two moduli[17], and it is the shortest of these in a given region of the phase diagram which sets the scale of the lyotropic phase found there.

## 5 Summary

The fact that amphiphilic systems assemble themselves into an array of lyotropic phases is clearly captured by the three common forms of theory which I have discussed. In a sense, this is a surprise because all of the models are extremely simplified; the architecture of the amphiphile at best is folded into a very few parameters. On the one hand, the appearance of these phases after so little detailed input to the theory assures us that they are robust; that they occur for general reasons of competition between the energy of the configuration and the number of ways in which they can be formed. On the other hand, the very generality provides little guidance as to the results to be expected for specific systems. Although there have been notable attempts to relate bending modulii to the architecture of amphiphiles[18], the ability to make specific predicitions as to the location of lyotropic phases in amphiphilic systems is not at all near the level which exists in self-assembling diblock copolymers[19]. More efforts along these lines would be most challenging, and most welcome.

## Acknowledgments

This work was supported in part by grant number DMR9220733 of the National Science Foundation.

## References


1. G. Gompper and M. Schick, "Self -Assembling Amphiphilic Systems", volume 16 of "Phase Transitions and Critical Phenomena", C. Domb and J.L. Lebowitz eds., (Academic Press, London, 1994).
2. G.J.T. Tiddy, Physics Reports **57** (1980) 1.
3. V. Luzzatti, A. Tardieu, T. Gulik-Krzywicki, E. Rivas, and F. Reiss-Husson, Science **220** (1968) 485.
4. W. Selke in vol 15 "Phase Transitions and Critical Phenomena", C. Domb and J.L. Lebowitz eds., (Academic Press, London, 199?), p1.
5. M.W. Matsen and D.E. Sullivan, Phys. Rev. **E51** (1995) 548.
6. A. Ciach, J.S. Høye, and G. Stell, J. Chem. Phys. **95** (1991) 5300.
7. R.G. Larson, J. Chem. Phys. **96** (1992) 7904.
8. M. Teubner and R. Strey, J. Chem. Phys. **87** (1987) 3195.
9. G. Gompper and M. Schick, Phys. Rev. **B41** (1990) 9148.
10. G. Gompper and S. Zschocke, Phys. Rev. **A46** (1992) 4836.
11. L. Leibler, Macromolecules**13** (1980) 1602.
12. P.B. Canham, J. Theor. Biol. **26** (1970) 61; W. Helfrich, Z. Naturforsch. **28c** (1973) 693; E. Evans, Biophys. J. **14** (1974) 923.
13. S.A. Safran, L.A. Turkevich, and P.A. Pincus, J. Phys. (Paris) Lett. **45** (1984) L69.
14. Z.G. Wang and S.A. Safran, Europhys. Lett. **11** (1990) 425.
15. L. Peliti and S. Leibler, Phys. Rev. Lett. **54** (1985) 1690.
16. P.G. de Gennes and C. Taupin, J. Phys. Chem. **86** (1982) 2294.
17. L. Golubović, Phys. Rev. **E50** (1994) R2419; D.C. Morse, Phys. Rev. **E50** (1994) R2423.
18. I. Szleifer, D. Kramer, A. Ben-Shaul, D. Roux, and W.M. Gelbart, Phys. Rev. Lett. **60** (1988) 1966.
19. M. Matsen and M. Schick, Macromolecules**27** (1994) 6761.

# Generalized Maxwell Model Equations for the Static and Dynamic Nonlinear Flow Behavior

Ortwin Hess[1,2] and Siegfried Hess[1]

[1]Institut für Theoretische Physik, Technische Universität Berlin
Hardenbergstr. 36, D - 10623 Berlin, Germany

[2] Institut für Technische Physik, DLR,
Pfaffenwaldring 38 - 40, D - 70569 Stuttgart, Germany

**Abstract**

A nonlinear relaxation equation for the friction pressure tensor is fomulated as a generalization of the Maxwell model equation. It involves the derivative of a Landau type potential with respect to the symmetric traceless part of the pressure tensor. Various forms of this potential are stated. Results are presented for one specific form chosen in analogy to the Landau-de Gennes potential of nematic liquid crystals. Some remarks on the occurrence of a yield stress are made. The special case of a plane Couette flow is studied. Consequences of the model for the shear rate dependence of the viscosity, of the shear stress and of the normal stress differences are presented for a stationary flow. Both shear thinning and shear thickening are found. The response of the components of the stress tensor to a ramp like increasing and decreasing time dependent shear rate is computed numerically. A hysteresis like behavior is observed in the shear thickening regime. The response to a shear rate switched on in a step like manner and reversed after some time is analyzed and also displayed graphically.

# 1 Introduction

Many complex fluids, such as polymers, dense colloidal dispersions and micellar solutions show a dependence of the viscosity on the shear rate and the occurence of normal pressure differences. These phenomena, termed "nonlinear fluid behavior" or also "non-newtonian behavior" are observed for shear rates $\gamma$ with $\gamma\tau \gtrsim 1$, where $\tau$ is the relevant relaxation time. The kinetic theory of the nonlinear flow behavior has been developed for several types of fluids ranging from gases [1, 2] and simple fluids [3, 4] to colloidal solutions and polymeric liquids [4, 5]. Non-equilibrium molecular dynamics (NEMD) computer simulations have provided insight into the shear flow induced structural changes and anisotropies of various distribution functions [6]–[8]. These microscopic theories take the specific properties of the fluids into consideration.

A phenomenological description, analogous to the time honored Maxwell model equation, can be formulated for a wider class of fluids. It is useful for a qualitative understanding and also for engineering applications. Quite a number of phenomenological models suitable for shear thinning, have been proposed and applied to various flow situations [5, 9]. Recently, a relatively simple model has been introduced [10] which can be used to describe both shear thinning and shear thickening behavior, i.e. the decrease and the increase of the the viscosity with increasing shear rate. The motivation for these considerations comes from experiments with dense colloidal dispersions where a dramatic shear thickening occurs [11, 12].

Here, the generic form of a nonlinear Maxwell model equation is formulated in section 2. It involves a damping term proportional to the derivative of a free energy functional with respect to the symmetric traceless part of the pressure or stress tensor. General properties and specific expressions for this potential are discussed. Particular attention is paid to the special case where the potenial is analogous to the Landau-de Gennes potential used for a description of the phase tansition isotropic-nematic in liquid crystals. A plane Couette flow geometry is considered in section 3. With a symmetry adapted ansatz for the pressure tensor, the tensorial relaxation equation reduces to three coupled differential equations for the relevant components of the pressure tensor. The principal axes and the biaxiality are discussed. The definitions of viscosity coefficients and the relevant components of the stress are stated. Section 4 is devoted to an analysis of the stationary solutions of the relaxation equations. The time dependent response of the shear stress and the normal stress difference to imposed ramp like and step like time dependent shear rate is presented in section 5. A hysteresis like behavior is found in the shear thickening regime. Some pecularities are observed in response to a shear rate step and a subsequent reversal of the shear rate.

# 2 The Model

## 2.1 Nonlinear Relaxation Equation

Let $p_{\mu\nu} = -\sigma_{\mu\nu}$ be the pressure tenor of a fluid, $\sigma_{\mu\nu}$ is the stress tensor. Greek subsripts indicate cartesian components, the summation convention will be used for them. In the following, the symmetric traceless part of the pressure tensor (negative stress tensor) is characterized by the dimensionless "friction pressure tensor" $\pi_{\mu\nu}$ introduced by:

$$\overleftrightarrow{p_{\mu\nu}} = \sqrt{2}G\,\pi_{\mu\nu}, \tag{1}$$

where $G$ is a (reference value of) the (high frequency) shear modulus. The numerical factor $\sqrt{2}$ has been inserted for convenience. The double arrow $\overleftrightarrow{\cdots}$ indicates the symmetric traceless part of a tensor, e.g. $\overleftrightarrow{p_{\mu\nu}} = \frac{1}{2}(p_{\mu\nu} + p_{\nu\mu}) - \frac{1}{3}p_{\lambda\lambda}\delta_{\mu\nu}$, where $\delta_{\mu\nu}$ is the unit tensor.

In terms of $\pi_{\mu\nu}$, the generic form of the generalized Maxwell model equation is

$$\frac{\partial}{\partial t}\pi_{\mu\nu} - 2\,\overleftrightarrow{\epsilon_{\mu\lambda\kappa}\omega_\lambda\pi_{\kappa\nu}} + 2\kappa\,\overleftrightarrow{\gamma_{\mu\lambda}\pi_{\lambda\nu}} + \tau_0^{-1}\Phi_{\mu\nu} = -\sqrt{2}\gamma_{\mu\nu}. \tag{2}$$

The quantities $\underline{\omega}$ and $\gamma_{\mu\nu}$ are the vorticity and the (symmetric traceless) deformation rate tensor, viz.

$$\underline{\omega} = \frac{1}{2}\underline{\nabla} \times \underline{v}, \quad \gamma_{\mu\nu} = \overleftrightarrow{\nabla_\mu v_\nu}. \tag{3}$$

Here, $\underline{v} = \underline{v}(t, \underline{r})$ is the flow velocity. The quantity $\epsilon_{\mu\nu\lambda}$ is the totally antisymetric isotropic tensor of rank 3 which can e.g. be used to express the $\mu$-component of the cross product of two vectors $\underline{a}$ and $\underline{b}$ as $\epsilon_{\mu\nu\lambda}a_\nu b_\lambda$.

The relaxation time coefficient $\tau_0$ and the quantity $\kappa$ are model parameters. In (2), $\Phi_{\mu\nu} = \partial\Phi/\partial\pi_{\mu\nu}$ is the derivative of a "potential" function $\Phi$ with respect to the tensor $\pi_{\mu\nu}$. In general, $\Phi$ is a function of the second and third order scalar invariants $I_2$ and $I_3$ which are essentially the square of the magnitude and the determinant $\det(\pi)$ of the tensor, viz.: $I_2 = \pi_{\mu\nu}\pi_{\mu\nu}$, $I_3 = \sqrt{6}\,\pi_{\mu\nu}\pi_{\nu\lambda}\pi_{\lambda\mu} = 3\sqrt{6}\,\det(\pi)$. The first order invariant, the trace $I_1 = \pi_{\mu\mu}$ vanishes. The factor $\sqrt{6}$ has been inserted for convenience.

The simplest case is the generalized linear Maxwell model. It follows from a potential where only terms of second order in the pressure tensor are considered: $\Phi = \frac{1}{2}AI_2$ , which implies: $\Phi_{\mu\nu} = A\pi_{\mu\nu}$. Here $A > 0$ is assumed. For $\kappa = 0$, Eq. (2), in this case, reduces to the *Jaumann-Maxwell* model [5] with the Maxwell relaxation time $\tau$ and the newtonian viscosity $\eta = \eta_{New}$ given by:

$$\tau = \tau_0 A^{-1}, \quad \eta_{New} := G\tau = G\tau_0 A^{-1}. \tag{4}$$

The cases $\kappa = 1$ or $\kappa = -1$ correspond to "codeformational" models [5].

The stationary solution of (2) for the case of small velocity gradients where also the terms involving the vorticity and the deformation rate can be disregarded on the l.h.s. of (2) is $\Phi_{\mu\nu} = -\sqrt{2}\tau_0\gamma_{\mu\nu}$. It corresponds to $\tau_0\gamma_{\mu\nu} = -H_{\mu\nu}$ where $H_{\mu\nu}$ is the derivative of the enthalpy $\mathcal{H} = G\,\Phi$ with respect to the symmetric traceless part $\overleftrightarrow{p_{\mu\nu}}$ of the pressure tensor. When $\tau_0\gamma_{\mu\nu}$ is identified with the symmetric traceless part $\overleftrightarrow{u_{\mu\nu}}$ of the deformation tensor $u_{\mu\nu}$, this equation corresponds to a standard relation of elasticity theory. When the free energy $\mathcal{F}$ is known as function of $\overleftrightarrow{u_{\mu\nu}}$, one can compute $\Phi$ according to $G\,\Phi = \mathcal{F} + \overleftrightarrow{p_{\mu\nu}}\,\overleftrightarrow{u_{\mu\nu}}$ with $\overleftrightarrow{p_{\mu\nu}}$ given by the derivative of $\mathcal{F}$ with respect to $\overleftrightarrow{u_{\mu\nu}}$.

In the absense of a flow, the stationary solution of (2) is $\Phi_{\mu\nu} = 0$. The system under consideration is in a fluid state when $\pi_{\mu\nu} = 0$, which is always a solution of $\Phi_{\mu\nu} = 0$, is actually a stable solution corresponding to the absolute minimum of the potential function $\Phi$. When a nonzero $\pi_{\mu\nu}$ is a stable solution, the sytem possesses a yield stress.

With terms up to sixth order in the pressure tensor taken into account, the potential function can be written as

$$\Phi = \frac{1}{2}AI_2 - \frac{1}{3}BI_3 + \frac{1}{4}CI_2^2 + \frac{1}{5}DI_2\,I_3 + \frac{1}{6}EI_2^3 + \frac{1}{6}FI_3^2. \tag{5}$$

The dimensionless coefficients $A$, $B$, $C$, $D$, $E$, $F$ are model parameters. Notice that the derivatives of $I_2$ and $I_3$ with respect to $\pi_{\mu\nu}$ are $2\pi_{\mu\nu}$ and $3\sqrt{6}\;\overleftrightarrow{\pi_{\mu\lambda}\pi_{\lambda\nu}}$ .

For $D = E = F = 0$ and $C > 0$ (5) reduces to the above mentioned Landau-de Gennes type of potential which is of forth order in the pressure tensor. The quantity $\Phi_{\mu\nu}$ then is given by:

$$\Phi_{\mu\nu} = A\,\pi_{\mu\nu} - \sqrt{6}B\;\overleftrightarrow{\pi_{\mu\lambda}\pi_{\lambda\nu}} + C\,\pi_{\mu\nu}\,\pi_{\lambda\kappa}\pi_{\lambda\kappa}. \tag{6}$$

The results to be presented later are for this specific case.

For gases, a relaxation equation of the type (2) with (6) has been derived from the nonlinear Boltzmann equation [2]. The Maxwell model equation which was originally obtained from the linearized Boltzmann equation has been applied successfully to various types of fluids whose microscopic dynamics differ drastically from that of a simple gas. In the same spirit, we consider Eq.(2) with (6) as a phenomenological model equation for complex fluids. The choice of these equations was also motivated by a formally eqivalent nonlinear relaxation equation for the alignment tensor of molecular fluids and nematic liquid crystals which has been derived within the framework of irreversible thermodynamics [13] and from a Fokker-Planck-equation [14]. The flow alignment both in the isotropic and nematic phases of liquid crystals as well as the influence of a shear flow on the phase transition isotropic-nematic were treated starting from that equation [15]. A close similarity in the behavior of the pressure or stress and the alignment tensor is found in many fluids and referred to as stress-optical law [5].

In the following, the flow velocity field $\underline{v}$ and consequently $\underline{\omega}$ and $\gamma_{\mu\nu}$ are considered as given, "external" variables. The analysis of the full hydrodynamical equations, viz. the insertion of $\pi_{\mu\nu}$ as obtained from (2) into the momentum balance equation and its solution for a specific geometry are not considered here.

In general, the symmetric traceless friction pressure tensor has 5 independent components. The same applies to the dimensionless tensor $\pi_{\mu\nu}$. In the following, the special case of a simple shear flow, i.e. a plane Couette (or plane Poiseuille) flow is studied, where only 3 relevant components have to be considered.

# 3 Plane Couette Flow

## 3.1 Symmetry Adapted Ansatz for the Pressure Tensor

For a flow velocity $\underline{v}$ in $x$-direction between flat walls perperdicular to the $y$-direction one has $\underline{\nabla v} = \gamma \underline{e}^y \underline{e}^x$ with the shear rate

$$\gamma = \frac{\partial v_x}{\partial y}. \tag{7}$$

Furthermore, $\underline{\omega} = \frac{1}{2}\underline{\nabla} \times \underline{v} = -\frac{1}{2}\gamma \underline{e}^z$. Here $\underline{e}^x, \underline{e}^y, \underline{e}^z$ are unit vectors parallel to the coordinate axes. For this geometry, the symmetry adapted ansatz [2, 15]

$$\pi_{\mu\nu} = \pi_+ T^+_{\mu\nu} + \pi_- T^-_{\mu\nu} + \pi_0 T^0_{\mu\nu} \tag{8}$$

is made with the basis tensors

$$T^+_{\mu\nu} = \sqrt{2}\, \overleftrightarrow{e^x_\mu e^y_\nu}, \quad T^-_{\mu\nu} = \frac{1}{2}\sqrt{2}\left(e^x_\mu e^x_\nu - e^y_\mu e^y_\nu\right), \quad T^0_{\mu\nu} = \sqrt{\frac{3}{2}}\, \overleftrightarrow{e^z_\mu e^z_\nu}\,. \tag{9}$$

The shear stress $\sigma_{yx} = -p_{yx}$ ist proportional to $\pi_+$; $\pi_-$ and $\pi_0$ are proportional to the normal pressure differences $p_{xx} - p_{yy}$ and $p_{zz} - \frac{1}{2}(p_{xx} + p_{yy})$, respectively. Furthermore, the quantities $\pi_+, \pi_-$ are proportional to the imaginary and real parts of the spherical component with $m = 2$ of the symmetric traceless part of the pressure tensor, $\pi_0$ is proportional to the $m = 0$ component (with the z-axis chosen as reference axis).

The variables $\pi_+, \pi_-$ can be expressed in terms of their magnitude $p = \sqrt{\pi_+^2 + \pi_-^2}$ and an angle $\varphi$, for convenience $\pi_0$ is replaced by $q$: $\pi_+ = -p\sin 2\varphi$, $\pi_- = -p\cos 2\varphi$, $p^2 = \pi_+^2 + \pi_-^2$, $\pi_0 = q$. The angle $\varphi$ determines the direction of the two principal axes of the tensor $\pi_{\mu\nu}$ which are in the shear plane. More specifically, these principal axes enclose the angles $\varphi$ and $\varphi + 90^o$ with the flow direction (x-axis). The third principal axis, of course, is parallel to the z-axis. Notice that positivity of the viscosity implies $\pi_+ < 0$, and $\varphi \leq 45^o$ is expected. In the newtonian limit one has $\varphi = 45^o$. The variable $q$ may have either sign.

The variables $p$ and $q$ introduced above can further be expressed in terms of $r = \sqrt{p^2 + q^2} = \sqrt{\pi_+^2 + \pi_-^2 + \pi_0^2}$, the magnitude of the total tensor $\pi_{\mu\nu}$, and of an angle $\alpha$ which determines the degree of biaxiality of this tensor: $p = r\sin\alpha$, $q = r\cos\alpha$. More specifically, the biaxiality is chacterized by $b$ defined by [16] $b^2 = (I_2^3 - I_3^2)/I_2^3 = \sin^2 3\alpha$. Here, $I_2 = p^2 + q^2 = r^2$, $I_3 = q^3 - 3qp^2 = r^3 \cos 3\alpha$, has been used. The tensor $\pi_{\mu\nu}$ is uniaxial, i.e. $b = 0$, for $\alpha = 0^o,\ 60^o,\ 120^o$. The two latter cases correspond to uniaxial states with the symmetry direction parallel to the x- or the y-axis rather than the z-axis. The biaxality assumes its extremal values $b = \pm 1$ in a planar biaxial state for $\alpha = 90^o$ and $\alpha = 30^o,\ 150^o$.

Notice that $I_2$ and $I_3$ do not depend on $\varphi$. It seems worth mentioning that the potential $\Phi$ as given by (5) essentially reduces to the potential of the *Henon-Heiles* model [17] for $C = D = E = F = 0$.

## 3.2 Relaxation Equations for the Components

Use of the symmetry adapted ansatz presented above in (2) with (6) leads to a specific form of nonlinear coupled equations for the components $\pi_+$, $\pi_-$, $\pi_0$ of the friction pressure tensor. In the following, these variables are expressed in units of

$$\pi_c = \frac{2}{3}\frac{B}{C}\,. \tag{10}$$

With the substitutions

$$A = A^* A_c, \quad A_c := 2B^2\,(9C)^{-1}\,,\ \tau_c = \tau_0 A_c^{-1} = \tau_0\left(\frac{1}{2}C\pi_C^2\right)^{-1}, \quad C > 0; \tag{11}$$

$$t = \tau_c t^*, \quad \Gamma = \tau_c \gamma = -\Omega\,, \tag{12}$$

the model coefficients $B$ and $C$ no longer appear explicitely in the relaxation equations. Due to $\sqrt{2}\tau_c\gamma_{\mu\nu} = \Gamma\, T^+_{\mu\nu}$ these equations read:

$$\begin{aligned}
\dot{\pi_+} - \Omega\pi_- + A^*\pi_+ + 6\pi_0\pi_+ + 2\pi_+\left(\underline{\underline{\pi}}\right)^2 &= -\pi_c^{-1}\Gamma + \kappa^*\Gamma\pi_0,\\
\dot{\pi_-} + \Omega\pi_+ + A^*\pi_- + 6\pi_0\pi_- + 2\pi_-\left(\underline{\underline{\pi}}\right)^2 &= 0,\\
\dot{\pi_0} + A^*\pi_0 + 3\left(\pi_+^2 + \pi_-^2 - \pi_0^2\right) + 2\pi_0\left(\underline{\underline{\pi}}\right)^2 &= \kappa^*\Gamma\pi_+.
\end{aligned} \tag{13}$$

Here $\left(\underline{\underline{\pi}}\right)^2$ stands for $\pi_+^2 + \pi_-^2 + \pi_0^2$, furthermore $\kappa^* = \kappa/\sqrt{3}$, and the dot denotes the differentiation with respect to $t^*$.

In (13) the coefficient $A^*$ determines whether terms of higher order in in the components of the pressure tensor are of relevance ($A^* \approx 1$) or not ($A^* \gg 1$). It is assumed that $A^*$ strongly depends on the temperature $T$ or the number density $n$ of a dispersion in a way typical for Landau type theory of a phase transition. In particular, it is expected that $A^*$ will decrease with decreasing $T$ or increasing $n$. This implies that the newtonian viscosity, cf.(4) increases with decreasing $T$ or increasing $n$ due to the increase of the relaxation time $\tau$. In general, the shear modulus $G$ will also increase but this effect is not considered explicitly here. In the absense of a flow, the fluid state coexists with a state possessing a yield stress at $A^* = 1$. For $A^* > 1.125$ no stationary solution with finite values of the tensor $\pi_{\mu\nu}$ exists in equilibrium.

Equations equivalent to (13) can be given for the variables $r$, $\varphi$, $\alpha$ but are not presented here.

### 3.3 Viscosity and Stress Components

The non-newtonian viscosity $\eta$ and a pertaining dimensionless coefficient $H$ are defined by:

$$p_{xy} = -\eta_+ \gamma = -\eta_c H \Gamma, \quad \eta_c = G\tau_c. \tag{14}$$

The dimensionless variables

$$\sigma = -p_{xy}/G, \quad \nu = (p_{yy} - p_{xx})/G, \tag{15}$$

which are proportional to $\pi_+$ and $\pi_-$, respectively, are used to characterize the shear stress and the first normal stress difference. The second normal stress difference $p_{zz} - p_{yy}$ is proportional to a linear combination of $\pi_0$ and $\pi_-$.

Positive entropy production requires $\eta > 0$ and hence $H > 0$. For $A^* > 1.125$ and for very small shear rates, the stationary value of the dimensionless viscosity is its newtonian limit $H = H_{New} := {A^*}^{-1}$.

Next, results of an analytic solution for the stationary values of $H, \sigma, \nu$ as functions of $\Gamma$ and of numerical solutions for the dynamic behavior are displayed in graphical form. In all cases presented here, the parameter $\kappa$ is put equal to zero.

## 4 Stationary Solution

### 4.1 Non-Newtonian Viscosity

In Fig.1, $\log H$ is plotted versus $\log \Gamma$ for $\pi_c = 0.1$ (thick curves), $\pi_c = 1$ (thinn curves), and the values 1.3, 1.5, 1.7, 2.0, 2.5, 3.0, 4.0, 6.0, and 10.0 for the parameter $A^*$ in both cases. The lowest curve on the right, pertaining to $\pi_c = 1$ and $A^* = 10.0$, shows the typical shear thinning behavior. For decreasing $A^*$, the newtonian viscosity (small $\Gamma$) increases and, in a limited range of $\Gamma$-values, the viscosity increases with increasing shear rate: shear thickening. Qualitatively, these curves are very similar to those observed in experiments with dense colloidal dispersions of spherical particles [11, 12] where smaller values of $A^*$ correspond to larger densities $n$ and where one can assume $A^* \sim 1 - \frac{n}{n_0}$ with a reference density $n_0$.

### 4.2 Stress Components

In Fig.2 linear scales are used to display the dimensionless shear stress $\sigma$ and the first normal stress difference $\nu$ as functions of the shear rate $\Gamma$ for $\pi_c = 0.1$ and $A^* = 1.125$ (dashed curves), 1.2(from left to right), $1.3, 1.4, 1.45, 1.5$ (thick curve), 1.55, and 1.6. For $A^* < 1.5$, these functions become multivalued. This will lead to a hysteresis-like behavior (to be discussed in section 5) which is also found experimentally [11, 12].

# 5 Dynamics

## 5.1 Response to a Shear Ramp, Stress Hysteresis

The dimensionless shear stress $\sigma$, the dimensionless viscosity $H$ and the first normal stress difference $\nu$ as functions of the shear rate $\Gamma$ can be obtained from a numerical solution of eqs.(13) with a ramp-like time dependent shear rate which increases linearly from 0 to a value $\Gamma_{max}$ over the time $t_{rise}$ and then linearly falls to 0 over the time $t_{fall}$. Both a Runge-Kutta method with self adaptive time steps and *NDSolve* of *Mathematica* were employed. In the limit of long rise and fall times, one follows a sequence of stationary states of the flow curve.

A hysteresis-like behavior is found for the range of parameter values where $\sigma, \nu$ and $H$ are multivalued functions of $\Gamma$. The curves shown in Fig.3 were computed for $A^* = 1.3, \kappa = 0$, $\pi_c = 1$, $\Gamma_{max} = 0.4$, and $t_{rise} = t_{fall} = 400$. The (initially) lower branches of the curves correspond to rising shear rates. The "stress overshoot" seen here is typical for the dynamic hysteresis behavior. The jumps of the $\sigma$ vs. $\Gamma$ curves from the lower to the upper branch and back occur at the values and are of the size as expected from the corresponding static curve. The unstable region in the static $\sigma$ vs. $\Gamma$ curves, where the differential viscosity (derivative of $\sigma$ with respect to $\Gamma$) is negative, are not accessible in the dynamic case.

## 5.2 Shear Reversal

The response of the three components $\sigma$, $\nu$, $q$ of the stress tensor (for $q$ see section 3.1) to a shear step at $t = 0$ where a shear rate $\Gamma_0$ is suddenly switched on and where subsequently, after the time $t_{rev}$, the shear rate is reversed to $-\Gamma_0$, is shown in Fig.4 for the case $A^* = 1.5$, $\kappa = 0$, $\pi_c = 1$, $\Gamma_0 = 10$ and $t_{rev} = 10$. Notice the strong stress overshoot after switching the shear rate. The shear stress $\sigma$ changes sign when the shear rate is reversed. The coefficients $\nu$, $q$ associated with the first and second normal stress difference do not change sign. These features explain the intriguing phase portrait Fig.5 where $\nu$ is plotted versus $\sigma$ for this shear rate reversal.

# 6 Concluding Remarks

The relatively simple model introduced here allows to describe qualitatively a transition from shear thinning to a flow behavior where also shear thickening occurs by varying one model parameter, viz. $A^*$ in (13). The study of the case where $\kappa^*$ is nonzero and a quantitative comparison with experimental data for colloidal dispersions [11, 12] and also for surfactant solutions [18] seems to be worthwhile. This means one has to determine the relevant model parameters $G, \tau_0, \kappa, A, B, C$. The analysis of other cases for the Landau potential function (5) are of interest, e.g. $B = 0, C > 0$ or $C < 0$, where terms up to sixth order in $\pi_{\mu\nu}$, at least $E > 0$, have to be taken into account. The coefficient $F > 0$ is needed if one wants to describe a system which has biaxial rather than uniaxial symmetry in the absence of a flow when a yield stress exists. The consequences of the present model for a range of parameters where a yield stress occurs and to situations where no stationary solution exists have not been exploited here. Furthermore, flow fields with other symmetries such as a four roller flow or an extensional flow, as well as flow fields with spatially inhomogeneous velocity gradients and spatially inhomogeneous pressure or stress tensors should be studied. In spatially inhomogeneous situations, the basic equation (2) has to be supplemented by (at least) a term $\tau_0^{-1}\xi^2\Delta\pi_{\mu\nu}$ where $\Delta$ is the Laplacian and $\xi$ is a characteristic length. Furthermore, the velocity gradient can no longer be treated as a prescribed "external" variable but has to be inferred from the simultaneous solution of the momentum conservation equation and of the extended version of eq.(2).

## 6.1 Acknowledgements

Many thanks are due to H.M. Laun and R. Bung (BASF, Ludwigshafen) for helpful discussions.

# References

[1] M.H. Ernst, B. Cichocki, J.R. Dorfman, J. Sharma, and H. van Beijeren, J.Stat.Phys. **18** (1978) 237;
R. Zwanzig, J.Chem.Phys. **71** (1979)

[2] N. Herdegen and S. Hess, Physica A **115** (1982) 281; W. Loose and S. Hess, Phys. Rev. Lett. **58** (1988) 2443; Phys. Rev. A **37** (1988) 2099

[3] S. Hess, in J. Casas-Vázques and D. Jou (eds.) *Rheological Modelling: Thermodynamical and statistical approaches*, Lecture Notes in Physics **381**, Springer, Berlin 1991, p. 51

[4] S. Hess, Physica A **118** (1983) 79; J. C. Rainwater and S. Hess, Physica A **118** (1983) 371

[5] R. B. Bird, R. C. Armstrong, and O. Hassager, *Dynamics of polymeric liquids,vol1: Fluid Mechanics*, J. Wiley, New York 1977; R. B. Bird, O. Hassager, R. C. Armstrong, and C. F. Curtiss, *Dynamics of polymeric liquids, vol 2:Kinetic Theory*, J. Wiley, New York 1977;
H. Janeschitz-Kriegel, *Polymer melt rheology and flow birefrengence*, Springer, Berlin 1983

[6] W. G. Hoover, Annu. Rev. Phys. Chem. **34** (1983) 103 ; B. D. Holian and D. J. Evans, J. Chem. Phys. **78** (1983) 5157 ; D. J. Evans and G. P. Morriss, Comp. Phys. Rep. **1** (1984) 287 ; D. J. Evans and W. G. Hoover, Ann. Rev. Fluid Mech. **18** (1986) 243; M. P. Allen and D. J. Tildesley, *Computer Simulation of Liquids*, Clarendon, Oxford (1987); W. G. Hoover *Molecular Dynamics*, Springer, Berlin 1986

[7] O. Hess, W. Loose, T. Weider, and S. Hess, Physica B **156/157** (1989) 505; W. Loose and S. Hess, Rheol. Acta **28** (1989) 91; in *Microscopic simulation of complex flows*, M. Mareschal (ed.), ASI series, Plenum Press (1990); S. Hess and W. Loose, in D. Axelrad and W. Muschik (eds.) *Constituive laws and microstructure*, Springer, Berlin 1988, p. 92; Physica A **162** (1989) 138; Ber. Bunsenges., Phys. Chem. **94** (1990) 216; S. Hess, D. Baalss, O. Hess, W. Loose, J. F. Schwarzl, U. Stottut and T. Weider in G. A. Maugin (ed) *Continuum models and discrete systems*, Longman, Essex, 1990, vol. 1, p. 18 - 30;
T. Weider, U. Stottut, W. Loose, and S. Hess, Physica A **174** (1991) 1

[8] M. Kröger and S. Hess, Physica **A 195** (1993) 336;
M. Kröger, W. Loose and S. Hess, J. Rheol. **37** (1993) 1057

[9] Encyclopidia of Fluid Mechanics, vol. 7, *Rheology and Non-Newtonian flows*, ed. N. P. Cheremisinoff, Gulf, Houston 1988

[10] O. Hess and S. Hess, Physica **A 207** (1994) 517;

[11] H.A. Barnes, J. Rheol. **33** (1989) 329; H. M. Laun, 10th International Congress on Rheology, ed. P.H.T. Uhlherr, Australian Society of Rheology, Sidney, 1988, vol.1; H. M. Laun, R. Bung, and F. Schmidt, J. Rheol. **35** (1991) 999;
W.H. Boersma, J. Laven, and H.N. Stein, J.Colloid Interface Sci. **149** (1992) 10

[12] H. M. Laun, R. Bung, S. Hess, W. Loose, O. Hess, K. Hahn, E. Hädicke, R. Hingmann, F. Schmidt, and P. Lindner, J. Rheol. **36** (1992) 743

[13] S. Hess, Z.Naturforsch.**30a** (1975) 728, 1224 ; J. Non-Equilibr. Thermodyn. **11** (1986) 176

[14] S.Hess, Z.Naturforsch. **31a** (1976) 1034

[15] S.Hess, Z.Naturforsch. **31a** (1976) 1507;
P. M. Goldbart and P. D. Olmsted, in *Complex Fluids*, ed. L.Garrido, Lecture Notes in Physics, Springer, Berlin 1993, vol. 415, p. 381

[16] P. Kaiser, W. Wiese and S. Hess, J. Non-Equilibr. Thermodyn. **17** (1992) 153

[17] M. Henon and C. Heiles, Astron.J. **69** (1964) 73;
H.G.Schuster, Deterministic Chaos, VCH Weinheim, 2nd rev. ed. 1988

[18] H. Thurn, M. Löbl, and H. Hoffmann, J. Phys. Chem. **89**, (1985) 517;
H.Rehage and H.Hoffmann, Mol.Phys. **74** (1991) 933

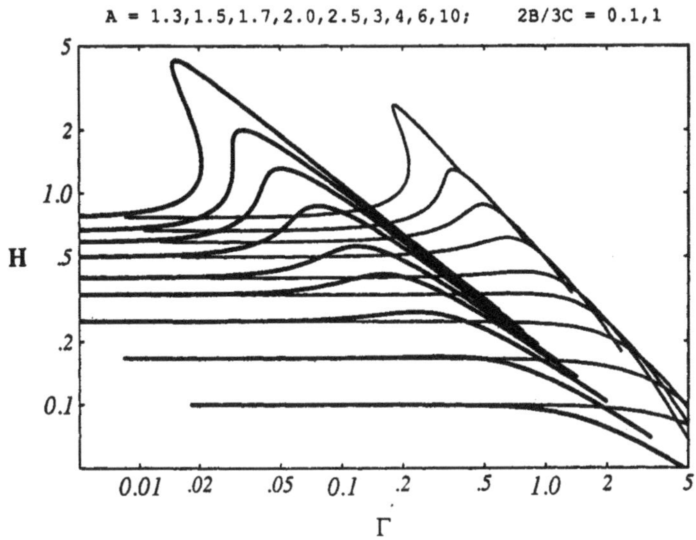


Fig.1 From shear thinning to shear thickening: $\log H$ vs. $\log \Gamma$ for $\kappa = 0$, and $\pi_c = 2B/(3C) = 0.1$ (thick curves), $\pi_c = 1$ (thinn curves) for $A^* = 10$ (lowest curves) 6.0, 4.0, 3.0, 2.5, 2.0, 1.7, 1.5, 1.3(highest curves)

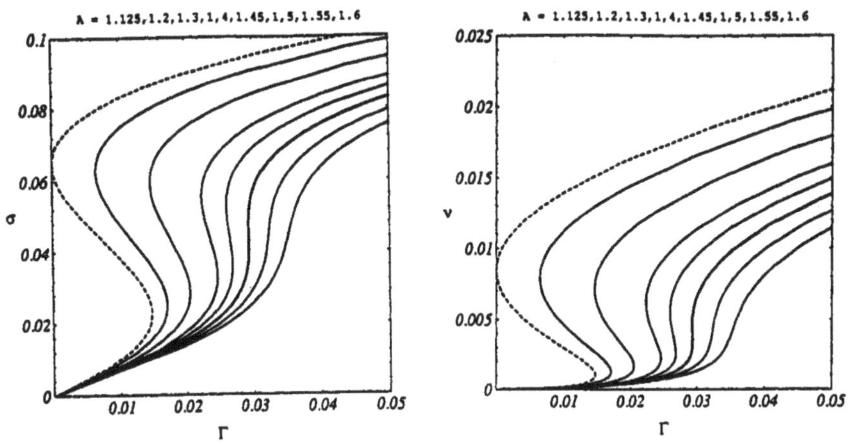


Fig.2 The dimensionless shear stress $\sigma$ and the first normal stress diference $\nu$ versus the dimensionless shear rate $\Gamma$ for $\kappa = 0$, $\pi_c = 0.1$ and the values of $A^*$ indicated in the graph.

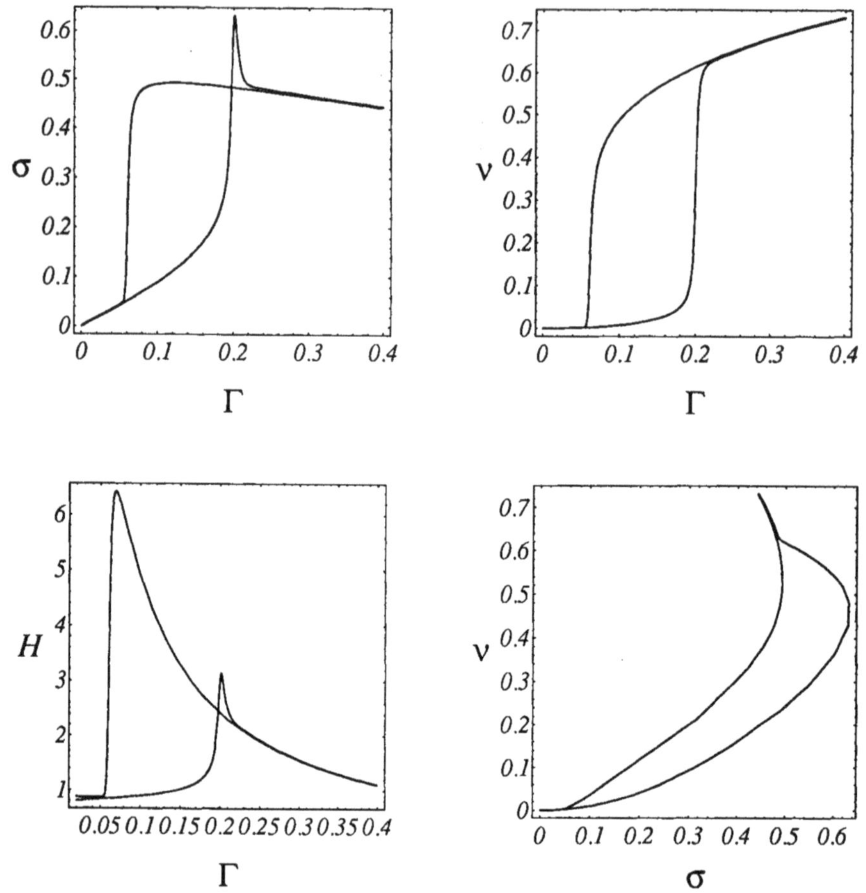


Fig.3 The stress components $\sigma$, $\nu$ and the dimensionless viscosity $H$ versus the dimensionless shear rate $\Gamma$, as well as $\nu$ versus $\sigma$ for $A^* = 1.3$, $\kappa = 0$, $\pi_c = 1$ obtained from a shear ramp with $\Gamma_{max} = 0.4$ and the rise and fall times $t_{rise} = t_{fall} = 400$.

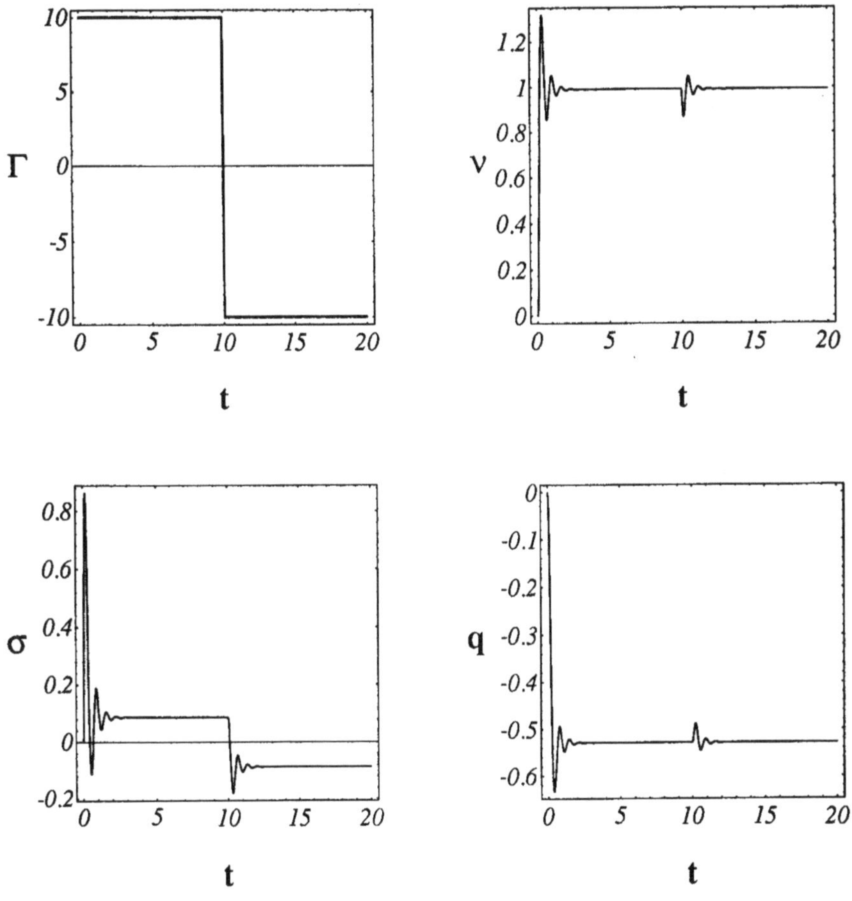


Fig.4 Shear reversal: The shear rate Γ and the components $\sigma$, $\nu$, $q$ of the stress tensor as functions of the dimensionless time $t$ for $A^* = 1.5$, $\kappa = 0$, $\pi_c = 1$.

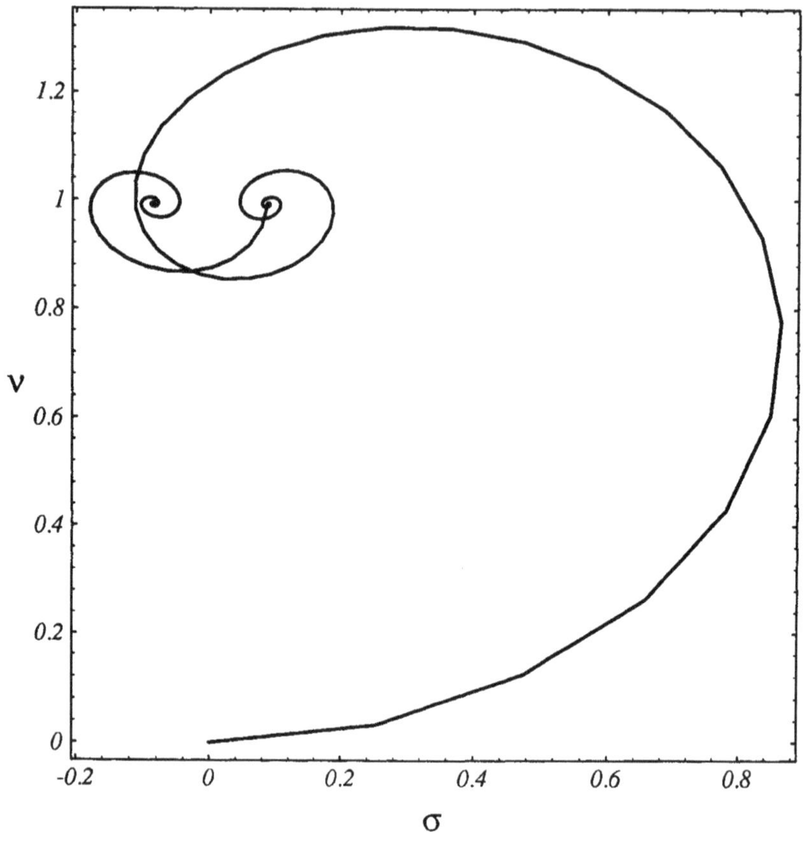


Fig.5 Shear reversal phase portrait: The first normal stress difference $\nu$ versus the shear stress $\sigma$ for $A^* = 1.5$, $\kappa = 0$, $\pi_c = 1$, $\Gamma_0 = 10$, $t_{rev} = 10$.

# WELL-TEMPERED SEQUENCES OF RANDOM NUMBERS FOR RELIABLE LARGE-SCALE MONTE CARLO CALCULATIONS

A. COMPAGNER

*Laboratory of Applied Physics, Technological University,*
*P.O.Box 5046, 2600 GA Delft, the Netherlands*
E-mail: ac@dutncp8.tudelft.nl

ABSTRACT

Based on emsemble theory and complex correlation coefficients of any range and order, a general description is given of arbitrary sequences of integers modulo $m$. Several definitions of a random sequence are compared. For a given value of $m$, the total amount of correlation is the same for all sequences of the same length (without internal periodicities), but the correlations that are present may vary widely in range and order. A few qualitative conditions are given for efficient linear production rules that generate well-tempered sequences, in which low-order correlations of short range are absent.

## 1. Introduction

The two desirable properties of random-number generators for large-scale Monte Carlo calculations, efficiency and reliability, are difficult to combine: the deterministic algorithms on which efficient generators are based disagree with any notion of randomness. This conflict has dominated the history of the subject, leading to the proliferation of generators and numerical tests reviewed, for instance, by Knuth[1]. The present paper describes a qualitative but operational method to find a suitable compromise between efficiency and reliability for arbitrary Monte Carlo applications. It is an abridged version of a paper appearing elsewhere[2], in which details, proofs and many references can be found.

The starting point is the observation that notions like 'randomness' and 'a truly random sequence' have never been defined properly. This preliminary problem is solved by the use of ensemble theory and a complete set of complex correlation coefficients, which are the Fourier transform of the ensemble weights. These coefficients are closely related to the main quantity that occurs in the Fourier analysis carried out, many years ago, by Coveyou and MacPherson[3] for the special case of linear-congruence sequences, which lead to the spectral test for these sequences. However, even the most time-consuming numerical tests are inconclusive: the total amount of correlation for any given sequence is too large and the possible correlations are too many for a complete check.

To avoid constructive interferences between unknown correlations in the random numbers and the equally unknown correlations in the problem to which the Monte Carlo method is applied, the best strategy is to use a sufficiently complicated generator. The regions in correlation space where the unavoidable correlations between the random numbers are hiding should be so specific, that any interference is unlikely *a priori*. So, what is a sufficiently complicated generator? A surprisingly simple answer can be given.

## 2. Ensemble Weights and Correlation Coefficients

Consider an arbitrary sequence $r_1, r_2, \ldots, r_N$ of $N$ integers modulo $m$. The elements of this object sequence are the components of a vector $\boldsymbol{r} \in \mathbf{Z}_m^N$. A general ensemble is introduced by assigning a probability, the ensemble weight $p(\boldsymbol{r}) \geq 0$, to each sequence $\boldsymbol{r}$. Together, the ensemble weights add up to 1. A natural measure for the amount of randomness in the ensemble is the entropy, defined by

$$S(p(\boldsymbol{r})) \equiv - \sum_{\text{all } \boldsymbol{r}} p(\boldsymbol{r}) \log_2 p(\boldsymbol{r}) . \tag{1}$$

When all weights are equal, $p(\boldsymbol{r}) = m^{-N}$, the ensemble is called the gambling ensemble, for which the entropy $S(p(\boldsymbol{r})) = N \log_2 m$ is maximal. For the singular ensemble, in which only one sequence is present (with weight 1), the entropy is zero.

Next, the test sequences $\boldsymbol{k} \equiv (k_1, k_2, \ldots, k_N) \in \mathbf{Z}_m^N$ of $N$ integers modulo $m$ are introduced, where the hypercubic lattice $\mathbf{Z}_m^N$ serves the dual purpose of being a test or correlation space as well as an object or probability space. The correlation coefficient belonging to $\boldsymbol{k}$,

$$C(\boldsymbol{k}) \equiv \sum_{\text{all } \boldsymbol{r}} p(\boldsymbol{r}) \exp [ (2\pi i/m) \boldsymbol{k} \cdot \boldsymbol{r} ] , \tag{2}$$

is the Fourier transform of the ensemble weights. It is the characteristic function of the multivariate distribution $p(\boldsymbol{r})$, which upon expansion of the exponential generates the moments of that distribution. For the present purpose, $C(\boldsymbol{k})$ itself is the most suitable quantity. Its values are located on or in the unit circle: $|C(\boldsymbol{k})| \leq 1$.

A few trivial Fourier properties follow. Inversion of the Fourier transform gives

$$p(\boldsymbol{r}) = m^{-N} \sum_{\text{all } \boldsymbol{k}} C(\boldsymbol{k}) \exp [ -(2\pi i/m) \boldsymbol{k} \cdot \boldsymbol{r} ] . \tag{3}$$

For $\boldsymbol{r} = 0$ one finds the special value

$$\langle C(\boldsymbol{k}) \rangle \equiv m^{-N} \sum_{\text{all } \boldsymbol{k}} C(\boldsymbol{k}) = p(0) . \tag{4}$$

The Fourier transformation is norm-conserving (Parseval's theorem):

$$\langle \mid C(\boldsymbol{k}) \mid^2 \rangle \equiv m^{-N} \sum_{\text{all } \boldsymbol{k}} \mid C(\boldsymbol{k}) \mid^2 = \sum_{\text{all } \boldsymbol{r}} [p(\boldsymbol{r})]^2 . \tag{5}$$

For the gambling ensemble, with $p(\boldsymbol{r}) = m^{-N}$, one finds $C(\boldsymbol{k}) = 0$ for $\boldsymbol{k} \neq 0$, whereas for $\boldsymbol{k} = 0$, which does not measure a true correlation, one finds $C(0) = 1$. If randomness is defined as the absence of correlations, it can only be achieved by means of the gambling ensemble (which is useless in Monte Carlo). For the opposite case of the singular ensemble, with $p(\boldsymbol{r}') = 1$ for the single sequence $\boldsymbol{r}'$, one finds $|C(\boldsymbol{k})| = 1$ for all $\boldsymbol{k}$, showing that in a single sequence everything is completely correlated. These properties agree with the values obtained for the entropy in the two ensembles.

## 3. The Scanning Ensemble

Clearly, something else is needed to discuss the stochastic properties of a single sequence. This is the scanning ensemble, in which a particular sequence $\boldsymbol{r}$ and all its translated versions have the same weight:

$$p(T^j \boldsymbol{r}) = 1/N , \qquad j = 0, \ldots, N-1 . \tag{6}$$

Here, $T$ is the translation matrix which shifts the elements of $\boldsymbol{r}$ one position to the left. To avoid boundary problems, the sequence is taken to be periodic, with $r_{i+N} = r_i$ (for simplicity, $N$ is assumed to be the smallest period).

For the entropy one finds $S = \log_2 N$, much less than for the gambling ensemble, but at least not zero as in the singular ensemble. The correlation coefficient is

$$C(\boldsymbol{k}) = \sum_{\text{all } \boldsymbol{r}} \exp [ (2\pi i/m) \boldsymbol{k} \cdot T^j \boldsymbol{r} ] . \tag{7}$$

Adding the limit $N \to \infty$, Coveyou and MacPherson[3] used the same quantity to analyze linear-congruence sequences. Eqs. (4) and (5) now read:

$$\langle C(\boldsymbol{k}) \rangle = 0 , \qquad \langle \mid C(\boldsymbol{k}) \mid^2 \rangle = 1/N . \tag{8}$$

The last quantity behaves properly in the limit $N \to \infty$, but multiplying it with $m^N$, the number of different test sequences $\boldsymbol{k}$, one finds that the total amount of correlation $\Gamma$ is given by $\Gamma = m^N/N$. This large value, which is the same for all sequences of length $N$ of which all translated versions are different, explains why tests for randomness are inconclusive; by minimizing the correlations that a test happens to be sensitive for, one may even select sequences in which more harmful correlations are larger.

The scanning ensemble can be used to study the stochastic properties of arbitrary sequences, such as $N$ outcomes of an $m$-valued roulette or $N$ decimals of $\pi$ (to which periodicity is added). In these examples, the distribution of the values $C(k)$ within the unit circle, while obeying strictly to Eqs. (8), will be approximately Gaussian (as one may check numerically). This Gaussian behaviour of the correlation coefficients corresponds closely to the commonly expected 'normal behaviour' of the digits of $\pi$ described, for instance, by Wagon[4].

## 4. Linear Production Rules

A Gaussian behaviour of the correlation coefficients is the best result achievable for a random sequence. However, the non-linear character of the generators involved makes them more difficult to analyse and less efficient, and may cause even worse problems. Therefore, the attention is restricted to sequences produced by a linear algorithm,

$$r_{j+n} = \sum_{i=1}^{n} K_i \, r_{i+j-1} + c \mod m \,, \tag{9}$$

where the constant $c$ and the coefficients $K_i$ (with at least $K_1 \neq 0$) are integers mod $m$. From an initial seed of $n$ integers $r_1, \ldots, r_n$ all other elements of the sequence can be generated. The period $N$ now is a direct consequence of the algorithm.

For $n = 1$, Eq. (9) is the linear-congruence generator, and for $m = 2$ the shiftregister algorithm results. Other well-known recipes, such as the lagged-Fibonacci generator, are also special cases of Eq. (9). For many different values of $m$ and $n$, values of $K_i$ and $c$ are known for which the period of the resulting sequence $r$ is close to or equal to the maximum period $N = m^n$. Then, all possible seeds of $n$ integers mod $m$ occur just once somewhere along the sequence, which is equidistributed with respect to them: the scanning ensemble for $N$ integers becomes equivalent to the gambling ensemble for $n$ integers. This is an important property, causing all test sequences of at most $n$ subsequent non-zero elements to be uncorrelated. Henceforth, all sequences considered will be assumed to be of maximum-length, at least to a very good approximation.

The constant $c$ in Eq. (9) does not play an important role: in the Fourier transform it ends up in a phase factor which does not change the correlation properties. Complications due to 'carry bits' or to combining linear generators with different moduli can be ignored for similar reasons. Furthermore, all elements of the sequences and all coefficients $K_i$ in Eq. (9) are thought to be replaced by their binary representations, allowing easy comparison between cases with different $m$. The difficulty that an equidistributed sequence at the level of $m$ is not uniform at binary level when $m$ is not a power of 2, is irrelevant in practice, when $m$ is either 2 or very large.

Defining the vector $\boldsymbol{K} \equiv (K_1, \ldots, K_n, K_{n+1} = m-1)$, of which all elements beyond $K_{n+1}$ are zero, one may rewrite Eq. (9) in the following form, where the constant is independent of $j$:

$$\boldsymbol{K} \cdot T^j \boldsymbol{r} \bmod m \quad = \quad \text{constant} . \tag{10}$$

Hence, $|C(\boldsymbol{K})| = 1$ holds. Indeed, $\boldsymbol{K}$ is the first (non-zero) completely correlated test sequence, and the linear production rule identified by it may be called $\boldsymbol{K}$ too. Further complete correlations, with $|C(\boldsymbol{K}')| = 1$, are due to linear combinations

$$\boldsymbol{K}' \quad \equiv \quad \sum_{i=0}^{N-1} \quad b_i \, T^{-i} \boldsymbol{K} \bmod m \tag{11}$$

of $\boldsymbol{K}$ and its translated versions $T^{-i}\,\boldsymbol{K}$, where $T^{-1}$ is a shift to the right over one position. Counting the possiblilities for the coefficients $b_i$ one finds[2] that these linear combinations of $\boldsymbol{K}$ and its translations generate the total amount of correlation $\Gamma = m^N/N$ present in the period of a maximum-length sequence. All correlation coefficients due to these linear combinations are located on the unit circle, and all other correlation coefficients are zero. Eqs. (8) remain valid, but the contrast with the Gaussian behaviour of the correlations in the digits of $\pi$ could not be greater. The resulting sequences can still be used in reliable Monte Carlo calculations, if a few conditions are obeyed.

## 5. Range and Order of Correlations

Consider a test sequence $\boldsymbol{k}$ of which $k_i$ and $k_j$ are the first and the last non-zero element. Its range $\Lambda(\boldsymbol{k})$ is the total number of bits involved,

$$\Lambda(\boldsymbol{k}) \quad \equiv \quad (\, j-i+1 \,) \ \log_2 m \ . \tag{12}$$

The ambiguity that arises for $\Lambda > N/2$ when $\boldsymbol{k}$ contains many zeros is easily dealt with. Various choices can be made[2] for the order of a test sequence. The simplest one is

$$q(\boldsymbol{k}) \quad \equiv \quad \sum_{i=1}^{N} \ H(k_i) \ , \tag{13}$$

where $H(k_i)$, the Hamming weight of $k_i$, is the number of bits equal to 1 in the binary form of $k_i$. In particular, the range and order of the production rule $\boldsymbol{K}$, consisting of $n+1$ elements of which at least the first and the last one are not zero, are given by

$$\Lambda(\boldsymbol{K}) \quad = \quad (n+1) \log_2 m \ , \qquad\qquad q(\boldsymbol{K}) \quad = \quad \sum_{i=1}^{n+1} \quad H(K_i) \ . \tag{14}$$

Together, the range and order of $K$ serve as an operational measure for its amount of complication. A certain analogy exists between this measure and the complexity of a sequence, defined by Kolmogorov[5] and Chaitin[6] as the number of bits in the shortest algorithm generating the sequence, which is a more subtle quantity than the range of $K$. In the 'complexity definition' of randomness, a sequence is called random if its complexity, in an asymptotic sense, is equal to its binary length, but this definition is as unsuitable for Monte Carlo calculations as the one based on the gambling ensemble.

Consider the cumulative distribution $d(\Lambda,q)$, defined as the number of completely correlated test sequences of range $\Lambda$ or smaller and of order $q \leq \Lambda$. The following general result for $\Lambda \geq \Lambda(K) = (n+1)\log_2 m$ is easily proven[2]:

$$D(\Lambda) \equiv \sum_{q=0}^{\Lambda} d(\Lambda,q) = 2^{\Lambda} / N , \tag{15}$$

where $D(\Lambda)$ is the total number of complete correlations of maximum range $\Lambda$ and of any order $q$, generated by linear combinations of $K$ and its translations. For $0 \leq \Lambda < \Lambda(K)$ one finds $D(\Lambda) = 1$, due to $K' = 0$; there are no true correlations of range below $\Lambda(K)$. The first non-trivial value, $D(\Lambda(K)) = m$, is due to the $m$ multiples of $K$ (including, again, $K' = 0$). For larger $\Lambda$, the linear combinations of Eq. (11) start to contribute.

Since in correlation space every correlated test sequence is surrounded by many uncorrelated ones of (almost) the same range and order, only one or two bits being different, the distribution $d(\Lambda,q)$ of the correlated test sequences will be similar, at least asymptotically, to the much more massive distribution $f(\Lambda,q)$ obeyed by all test sequences, whether completely correlated or uncorrelated. The latter distribution is given by the binomial coefficient:

$$f(\Lambda,q) = \binom{\Lambda}{q} \equiv \Lambda! \,/\, q!\,(\Lambda-q)! \,. \tag{16}$$

Therefore, $d(\Lambda,q)$ is split into an asymptotic 'stochastic' and a 'deterministic' part,

$$d(\Lambda,q) \equiv d_s(\Lambda,q) + d_d(\Lambda,q) , \tag{17}$$

where the stochastic part, obeying Eq. (15) all by itself, is defined by

$$d_s(\Lambda,q) \equiv \binom{\Lambda}{q} / N . \tag{18}$$

This part is the same for all maximum-length sequences of period $N$ generated by linear production rules. In simple cases, when $\Lambda(K)$ or $\Lambda - \Lambda(K)$ is small, it is easy to verify numerically[2] that the deterministic part $d_d(\Lambda,q)$, which summed over $q$ is zero and which depends strongly on the production rule $K$, for most values of $q$ is only a minor correction to the stochastic part given by Eq. (18). However, its low-order contributions may completely spoil the reliability of a linear random-number generator by constructive interference with the inherent correlations in the problem to which the Monte Carlo method is applied.

## 6. Well-tempered Sequences

A few conditions suffice to ensure that the unavoidable correlations in random numbers are moved to regions in correlation space where they are unlikely to effect the results of a Monte Carlo calculation. These conditions are not sharp, and the phrase 'unlikely' is admittedly vague. Further experience may seem to be needed, but testing random numbers is an ever receding research goal. The evasive character of the deterministic part $d_d(\Lambda,q)$, which in practical cases, due to the large value $\Gamma = m^N/N$ of the total amount of correlation, is as hard to find as a needle in a haystack, constitutes an intrinsic difficulty of the subject. Moreover, the desirable quality of a generator depends strongly on the Monte Carlo problem in hand. The conditions below provide at least a qualitative guideline.

A standard requirement is maximum length, such that the generated sequence is equidistributed. To ensure that a linear maximum-length sequence is a reliable source of random numbers, its production rule $\boldsymbol{K}$ should obey the following conditions:

A) the range $\Lambda(\boldsymbol{K})$ is large enough ($\Lambda(\boldsymbol{K}) > 200$),
B) the order $q(\boldsymbol{K})$ is close to the optimal value $\Lambda(\boldsymbol{K})/2$,
C) the bit pattern of $\boldsymbol{K}$ is irregular.

When these conditions are met, correlations of short range and low order are absent, and the resulting sequences will be called 'well-tempered'.

Condition A is needed to ensure that the stochastic part $d_s(\Lambda,q)$ of Eq. (18) of the cumulative distribution for all correlations is harmless. This unavoidable stochastic part is concentrated in a narrow region around the maximum at $q = \Lambda/2 \geq \Lambda(\boldsymbol{K})/2$. Clearly, $\Lambda(\boldsymbol{K})$ needs to be sufficiently large. There is no strict lower bound, but a crude estimate, $\Lambda(\boldsymbol{K}) > 200$, can be based[2] on the requirement that the tail of Eq. (18) below the arbitrary value $q = 10$ remains negligible over a subsequence equivalent to at least a few hundred subsequent random numbers of 32 bits.

Condition B tries to control the deterministic part of the distribution. When $q(\boldsymbol{K})$ is small, low-order correlations are generated right from the start and new ones will continue to be generated: most coefficients $b_i$ in Eq. (11) may be zero while the remaining ones are of low order. To avoid low-order correlations, bit mixing should not depend on the coefficients $b_i$ but should start already in Eq. (10), due to the number of non-zero elements of sufficiently high order in the production rule $\boldsymbol{K}$ itself. Preferably, the distribution $d(\Lambda,q)$ should be fed as close to the maximum of the stochastic part as is possible, but the optimal value $q(\boldsymbol{K}) = \Lambda(\boldsymbol{K})/2$ does not need to be obeyed strictly.

Condition C takes into account that the bit pattern of $\boldsymbol{K}$ should be sufficiently irregular to prevent that self-interference in Eq. (11) could still lead to early low-order correlations: random numbers need random production rules. The more Conditions A and B are obeyed, the less likely it is that such a self-interference of the production rule occurs; also, the early 'overcorrelation' caused by self-interference is in conflict with the assumed maximum length. In practical cases, Condition C is almost automatically obeyed. The circularity entailed in Condition C could be removed by using a frequency-based entropy as a more fundamental measure for the amount of complication of the bit pattern of $\boldsymbol{K}$ than the combination of range and order. Conditions A and B would be effected also. However, the numerical aspects of such an approach are not attractive.

To see how the conditions can be applied, consider standard linear-congruence generators, with $n = 1$ and $m = 2^{31}$ or $2^{32}$ and hence $\Lambda(\boldsymbol{K}) \approx 60$. Studying the bit patterns of the various multipliers involved, one finds that all three conditions are disobeyed. Indeed, they have fallen in disrepute for large-scale Monte Carlo calculations; for modest applications they may still suffice. Further examples are given elsewhere[2].

## 7. Combinations of Shiftregister Sequences

To conclude, a practical recipe for random-number generation is given, fit for all occasions. It is based on combinations of the two-bit-feedback shiftregisters described, for instance, by Golomb[7]. The characteristic function $R(n, i)$ of such a shiftregister,

$$R(n, i) \equiv 1 + x^i + x^n , \tag{19}$$

is equivalent to a production rule $\boldsymbol{K}$ of range $\Lambda(\boldsymbol{K}) = n+1$ and of order $q(\boldsymbol{K}) = 3$, with $K_1 = K_{i+1} = K_{n+1} = 1$ (all other elements are zero). When the trinomial in Eq. (19) is primitive, the generated sequence is of maximum length, the period being $N = 2^n-1$. All primitive trinomials with a Mersenne exponent as degree $n$ (with a Mersenne prime as period) up to $n = 9689$ were listed by Zierler[8]. Recently, the list was extended[9] up to the Mersenne exponent $n = 132049$, but for most purposes Zierler's list will suffice.

When the elements of $M$ of these two-bit-feedback shiftregister sequences, each with a different Mersenne prime as period, are added mod 2 (which can be realized in a cascade of EXOR-operations), the period of the resulting sequence is the product of the periods of the participating sequences. The difference with maximum length can be shown to be negligible [10]. The corresponding characteristic polynomial is the product of the primitive trinomials that belong to the participating sequences. The degree of this polynomial is the sum of the degrees of the trinomials, and its number of terms is $3^M$

(apart from occasional cancellations due to the mod-2 arithmetic, which seldom happen when the degree is large in comparison with $3^M$). This is equivalent to a production rule of order $q(K) = 3^M$, its range being equal to the degree of the characteristic polynomial.

In an earlier paper[11], it was suggested to combine $M = 8$ trinomials of Zierler's list of varying degrees, including the one of degree $n = 9689$. This would be an ultimate random-number generator, of range $\Lambda(K) > 10000$ and of order $q(K) \approx 3^M \approx 6000$; its bit pattern would be very irregular, and the resulting flow in probability space would never be parallel to one of the orthogonal subspaces that belong to the participating trinomials. However, a combination of $M = 4$ trinomials of rather modest degrees, with $\Lambda(K) \approx 500$ and $q(K) \approx 3^M \approx 80$, will have similar properties and will very probably be sufficient for even the most-demanding Monte Carlo calculations, without slowing them down considerably when efficient programming techniques are used. Often, for less-demanding tasks, the case $M = 2$ will do, while $M = 1$ should be avoided under all circumstances. A suitable choice for $M$ should depend on the expected correlation properties inherent to an application.

Combinations of shiftregister generators can be used to obtain reliable Monte Carlo results. Of course, combinations of other types of linear generators that obey the conditions may also be used.

**Acknowledgements**

Special thanks are due to Dakai Wang in Xian and Jouke Heringa in Delft, who contributed to the early stages of this research.

**References**

1. D.E. Knuth, *The Art of Computer Programming* **2** (Addison-Wesley, Reading, 1981).
2. A. Compagner, *Phys.Rev.E.*, to be published.
3. R.R. Coveyou and R.D. MacPherson, *J.Ass.Comp.Mach.* **14** (1967) 100.
4. S. Wagon, *The Mathematical Intelligencer* **7**, No. 3 (1985) 65.
5. A.N. Kolmogorov, *Russian Math. Surveys* **38**, No.4 (1983) 29.
6. G.J. Chaitin, *Information, Randomness and Incompleteness* (World Scientific, Singapore, 1987).
7. S.W. Golomb, *Shift Register Sequences* (Holden-Day San Fransisco, 1967).
8. N. Zierler, *Inform. and Control* **15** (1969) 67.
9. J. Heringa *et al.*, *Int. J. of Mod. Phys.* **C 3** (1992) 561.
10. D. Wang and A. Compagner, *Math. of Comput.* **60** (1993) 363.
11. A. Compagner, *Am. J. of Phys.* **59** (1991) 700.

# GLASSY DYNAMICS

M. MÉZARD

*Laboratoire de Physique Théorique de l'Ecole Normale Supérieure*
*24 rue Lhomond*
*75231 Paris Cedex 05 France*

We describe some recent developments on the mean field dynamics of spin glasses. This dynamics is necessarily out of equilibrium, but can be described in a general framework involving for instance a modification of the usual fluctuation dissipation theorem. We discuss several indications coming from recent work which point to the relevance of these spin glass ideas for the description of the glass transition.

## 1 Introduction

Many systems exhibit at low temperature a dynamical behaviour which does not reach equilibrium within experimental time scales. This is a constitutive property of the glass phase in magnetic systems (spin glasses), systems of interacting particles (window glasses), or extended objects (vortex lines or interfaces) pinned by impurities. It clearly relates to some important practical problems (like the physical aging of polymeric glasses, or the critical current in a type II superconductor where the flux line lattice is pinned by impurities). It is also a major challenge to the theory.

Recent analytical progress on the dynamics of mean field spin glasses indicates that:

i) a theoretical description of glassy dynamics is within reach.

ii) it requires to abandon such well accepted basic concepts as the time translation invariance (TTI), and the fluctuation dissipation theorem (FDT). They are substituted by new behaviours that we shall describe. One might hope that these new behaviours will be as basic for glassy dynamics as their famous counterparts (TTI and FDT) were for equilibrium dynamics.

iii) This new theoretical description, developed for systems with quenched disorder, is not restricted to them. It provides for instance a framework for extending the mode coupling theory of glasses to temperatures below the glass transition temperature. More generally this approach seems to open a route to a mean field description of the glass phase.

## 2 Spin glass mean field dynamics

### *2.1 General framework*

Consider a spin glass which is a set of $N$ real classical spins $\sigma_i$ with a Hamiltonian

$$H(\vec{\sigma}) = \sum_i V_0(\sigma_i) + V_d(\sigma) \; , \tag{1}$$

where $V_0(\sigma_i)$ is the local natural measure on the spins (for instance one can take $V_0(\sigma) = \lambda(\sigma^2 - 1)^2$, which reduces to Ising spins in the large $\lambda$ limit), and $V_d(\sigma)$ is an interaction term between spins, involving some quenched randomness. A class of problems which has received a lot of attention in recent years involve p-spin interactions:

$$V_d(\sigma) = -\sum_{i_1,\ldots,i_p} J_{i_1\ldots i_p} \sigma_{i_1} \ldots \sigma_{i_p} \tag{2}$$

(the usual exchange couplings correspond to $p = 2$). It is also interesting to study quenched interaction potential $V_d$ with a gaussian distribution characterized by:

$$\overline{V_d(\vec{\sigma})} = 0 \quad , \quad \overline{V_d(\vec{\sigma}) V_d(\vec{\sigma}')} = f\left(\frac{(\vec{\sigma} - \vec{\sigma}')^2}{N}\right) \; , \tag{3}$$

with an arbitrary correlation function $f$ (the overline denotes an average with respect to the distribution of quenched disorder). This last formulation is also useful in that it describes the diffusion of a particle in a random potential in a $N$ dimensional space. (A more elaborate problem consists in replacing the particle by a $D$-dimensional manifold).

Our aim is to study the dynamics of the spin glass, described by a Langevin process:

$$\frac{\partial \vec{\sigma}}{\partial t} = -\frac{\delta H}{\delta \vec{\sigma}} + \vec{\eta} \; , \tag{4}$$

where $\vec{\eta}$ is a thermal white noise with $< \eta_i(t) >= 0$ and $< \eta_i(t)\eta_j(t') >= 2T\delta_{ij}\delta(t - t')$ (the brackets denote an average with respect to the distribution of thermal noise). There is by now a well established technique to work out this problem. It involves introducing a field theoretic description of the dynamics which provides a generating function for the various correlation and response functions. As noted long ago [1], in this generating function one can average explicitely over the quenched disorder. Therefore this dynamical approach also provides a no replica route towards the study of spin glasses. In the mean field case, where each spin is coupled to all other ones, one can write for all above models, in the thermodynamic limit $N \to \infty$, a closed set of coupled equations involving the (two point) correlation and response:

$$C(t,t') \equiv \overline{<\sigma_i(t)\sigma_i(t')>} \quad , \quad r(t,t') \equiv \overline{<\frac{\partial \sigma_i(t)}{\partial \eta_i(t')}>} \,. \tag{5}$$

These equations take the form of first order in time differential equations:

$$\frac{\partial C(t,t')}{\partial t} = \mathcal{F}_c[C,r] \quad , \quad \frac{\partial r(t,t')}{\partial t} = \mathcal{F}_r[C,r] \,, \tag{6}$$

(we always suppose that $t > t'$). Here the origin of time is taken at $t = t_0$, and the initial condition $\sigma_i(t_0)$ is either fixed or random, but in any case not correlated with the quenched disorder. The functionals $\mathcal{F}_c$ and $\mathcal{F}_r$ are non linear functionals of $C$ and $r$ which have been computed in the literature for the various cases. An a priori obvious, but important, point is that these equations are causal: $\mathcal{F}_c(t,t')$ (and $\mathcal{F}_r$) involves the correlation $C(s,s')$ and response $r(s,s')$ only at times $s, s'$ which are smaller than $t$. At the mean field level, the whole complexity of glassy dynamics is contained in the dynamical equation (6). An alternative, less formal derivation of these equations can be obtained with the cavity method [2].

### *2.2 Time translation invariant approach*

The dynamical study of the spin glass phase has been pioneered by Sompolinsky and Zippelius [3] for the case of the infinite range p=2 model. They study the long time behaviour by sending the initial time $t_0$ to $-\infty$ and assume TTI, namely $C(t,t') = C(t-t')$ and similarly for the response $r$. The spin glass phase is characterized by a long time persistence of the correlations, $\lim_{t-t'\to\infty} C(t-t') = q_{EA}$, as originally found by Edwards and Anderson[4]. One also finds an anomalous contribution to the response, in the sense that the Fourier transform $r(\omega)$ of $r(t-t')$ acquires a singular contribution at $\omega = 0$, of the form $\Delta\delta(\omega)$. The meaning of this anomaly is not transparent. As we shall see, the problem lies in the $t_0 \to -\infty$ limit and the assumption of TTI. As shown in details by the static approach, in the spin glass phase ergodicity is broken. There exist a time scale $\tau_\infty$ which diverges in the thermodynamic limit $N \to \infty$. To be consistent, this approach needs a regularisation in order to have a finite thermalisation time. Several possibilities have been explored, either using a finite system with a finite (but large) $\tau_\infty$ [3,5], or letting the external parameters like the temperature $T$, evolve slowly with time [6]. In this case the rate of change provides a regularisation time equal to $T/(dT/dt)$. For an infinite system at fixed temperature one needs to give up TTI, as shown in the following.

*2.3 Dynamics of infinite systems: two times correlations and response*

An important step forward was taken recently by Cugliandolo and Kurchan [7] in their study of the p spin model with a spherical constraint. They realized the importance of studying the dynamics with a fixed initial time $t_0 = 0$. The basic equations (6) apply to an infinite system ($N = \infty$) studied at finite (large) times. The problem is to find the solution to these equations. So far there is no full solution of the equations for any model in all the time regimes (Here the word 'regime' refers to the way one takes the limit $t, t' \to \infty$; in any case let us underline once again that the $N \to \infty$ limit has to be taken first). Beside the original p-spin spherical model, and the SK model close to its critical temperature [8], a careful numerical integration of the dynamical equations has been performed in the case of the particle in a random potential [9]. A detailed overview of all these works goes much beyond this presentation, but let us point out the following salient features.

In all cases there exists an asymptotic regime. It corresponds to taking $t \to \infty$, $t' \to \infty$,with a fixed value of $\tau = t - t'$. Then the correlation and response go to their asymptotic forms $C_{as}(\tau)$ and $r_{as}(\tau)$. The solution of these asymptotic equations gives three main results: The FDT is satisfied: $dC_{as}/d\tau = T r_{as}(\tau)$. It agrees with the statics as far as the Edwards Anderson order parameter is concerned: $\lim_{\tau\to\infty} C_{as}(\tau) = q_{EA}$ (Notice that this is a non trivial result in the sense that the static result referred to here needs a breaking of replica symmetry). Furthermore the equations for the asymptotic forms involve the non asymptotic regime only through the value of a long term memory of the response, $R$, defined by:

$$R = \lim_{t_w \to \infty} \int_0^{t_w} ds\ r(t_w, s) - \int_0^{\infty} d\tau\ r_{as}(\tau)\ . \tag{7}$$

In this two times description, $R$ has replaced the 'anomaly of the response' which was found when assuming TTI. Its physical meaning is clear: A nonzero $R$ means that the system keeps a memory of what has happened at all times since it was prepared at $t_0 = 0$ (Fortunately this memory is blurred: the memory of the past history is kept only through one -or actually a few- numbers; this is the important point which allows for a solution of the problem.) We shall discuss the experimental relevance of this memory in the next section.

Beside the asymptotic regime there may exist one or several other regimes. The simplest case is that of a single non asymptotic regime. This is found so far in systems for which the static solution involves a single step of replica symmetry breaking. For instance in the $p > 2$-spin spherical model it has been argued convincingly (but not yet proven) that this regime is found in the limit

$t \to \infty$, $t' \to \infty$,with a fixed value of $\lambda = t'/t$ (or may be $\lambda = h(t')/h(t)$ with $h$ a non polynomial function, the theory is not yet able to give the value of $h$). In this limit one gets correlations and response $C(t,t') = \hat{C}(t'/t)$, $r(t,t') = \hat{r}(t'/t)/t$. Therefore TTI is violated. The FDT is also violated, it is substituted by:

$$Tr(t,t') = x\frac{\partial C(t,t')}{\partial t'} \quad (8)$$

where the fluctuation dissipation (FD) ratio $x$ is smaller than one. In more complicated cases there exist several non asymptotic regime. For instance one could take the limits $t \to \infty$, $t' \to \infty$, with a fixed value of $\lambda = (t-t')/t^u$. In each such regime there will be a modified FDT as in (8), with a FD ratio which depends on the regime.(For some non understood reason the set of values of the FD ratio is related to the Parisi order parameter function in the replica solution of the statics[7,9,8]).

### *2.4 A side remark on dynamics and barriers in large dimensions*

As already hinted at by the last comment above, the results of this nice computation are not necessarily very intuitive. One important property which is found is the 'weak ergodicity breaking'[16]: For an infinite system which has aged for an arbitrary large (but finite) waiting time $t_w$, the correlation eventually decays to zero: $\lim_{\tau\to\infty} C(t_w+\tau, t_w) = 0$, while the opposite order of limits (concerning a hypothetical system which started at equilibrium) would give: $\lim_{\tau\to\infty} lim_{t_w\to\infty} C(t_w+\tau, t_w) = q_{EA}$. It is not obvious how to build up an intuitive picture of this kind of behaviour (valid at any temperature below $T_C$) in terms of barriers and Arrhenius like arguments. In fact one must be careful of the fact that the system is infinite dimensional, and there may well be purely entropic barriers playing a crucial role. This name has been given[10] to systems which get trapped into some regions of phase space without energy barriers: there exist some paths to get out of these regions without increasing the energy, but they are rare. Purely entropic barriers can lead to a glassy dynamics.

## 3 Aging

The previous section has shown us that a spin glass can keep some memory of its history in a remote past. In fact this is well known in experiments where an aging effect was observed more than ten years ago[11]. The experiments on the decay of the thermoremanent magnetization (TRM) in spin glasses tell us a few facts of fundamental importance about the spin glass phase. Below the freezing temperature, the relaxation takes place on very long time scales (days or more). There is an aging effect, which means that the susceptibility of the system (its ability to respond to a change in the magnetic field) decreases with its age. For instance the susceptibility is different for a spin glass which has been 'at rest' in its spin glass phase for two hours or for ten hours (even if one does not see any noticeable change of its properties (e.g. the magnetization) during this waiting time). This agrees with the existence of non asymptotic regimes found in the mean field dynamics. For instance in the simplest scenario described above where the response goes as $r(t,t') = \hat{r}(t'/t)/t$, the TRM at time $t_w + t$ for a system which has been aging for a time $t_w$ before one switches off the magnetic field, given by $\int_0^{t_w} dt'\ r(t,t') = \int_0^{t_w/t} d\lambda\ \hat{r}(\lambda)$ , depends on $t/t_w$ as observed (roughly) in experiments. It is important to notice that similar aging effects have been observed also in some polymeric glasses like polyvinilchloride[12]. Let us also mention some very nice and subtle effects which have been observed in temperature cycling experiments: when the temperature is changed during the waiting process, there is an asymmetry with respect to the temperature change. If one lowers the temperature from $T$ to $T - \Delta T$ at time $t1$ the dynamics is restarted (there is a new value of the susceptibility), but when one gets back to the original temperature at time $t2$ the system has kept the memory of its state before the temerature change. Instead if one raises the temperature the dynamics is restarted, but there is no memory effect. So far these effects have not been explained starting from a microscopic dynamics.

Aging is not a very precise notion. Experimentally it means that the thermalisation time is larger than the time of the experiment. The mean field microscopic dynamics provide us an example of a system where the thermalisation time is infinite (in the thermodynamic limit). It also allows for one possible rigorous definition of aging as the existence of a non zero long term memory of the response -see Eq. (7). A simpler and weaker definition has been used also, especially in numerical simulations[13], namely the existence of a $t'/t$ dependance of the correlation function $C(t,t')$. Such a dependance is found in many instances of systems[14,13] which are not necessarily frustrated - nor disordered. A typical example is the growth of domains in a ferromagnet quenched below its critical temperature. This system will age because the size of the domain grows with time (as $L(t) \simeq \sqrt{t}$). This idea of domain growth (with a much slower growth law due to disorder pinning) has been used for a phenomenological description of the aging effect in spin glasses[15]. Another successful phenomenological description is in terms of phase space traps with a broad distribution of trapping times[16]. The relationship of these approaches to the mean field microscopic dynamics is not yet understood, but it has been argued recently that the aging effects found in the microscopic dynamics and in the domain growth might be fundamentally different[17].

## 4 Towards glasses

A priori, there seems to be some fundamental differences between the spin glass systems on which our attention has focused, and the real glasses. In the latter there is no quenched disorder, rather the disorder is 'self induced'. Furthermore in a glass there exists generally a crystalline state with a lower free energy; the quench must be such that this state will not be reached. In experimental spin glasses the nonlinear susceptibility diverges, while the glass transition looks like a purely dynamical effect.

It was suggested already some time ago by Kirkpatrick and Thirumalai that these differences might not be as crucial as one might think [19]. One first important step is to understand that there exist two different classes of spin glasses. In the usual exchange interaction ($p = 2$) case, relevant to experimental spin glasses, the dynamical transition appears at the same temperature as the static transition. However for the (theorist's) $p > 2$ case (described statically by a one step replica symmetry breaking [20]), it happens that the dynamical transition temperature $T_D$ is higher than the static transition temperature $T_S$ [19,21]. Therefore there will not be any singularity of static quantities at the dynamical transition: these are the best candidates for the description of glasses. Note that this second class of spin glasses is also the one for which there is a single non asymptotic regime in the dynamics. We would like to mention some recent developments which also point towards the similarity between spin glasses and glasses, and make us think that the techniques used in spin glass studies can also be useful for glasses.

First we mention the study of some frustrated spin systems without disorder, like the problem of low binary autocorrelation sequences [18]. In some cases it has been possible to study the statics with the replica method, and a low temperature phase with replica symmetry breaking has been found. It may look surprising to use replicas in systems without any disorder. The trick was to introduce a family of disordered systems, to which the ordered system belongs. Of course one needs that the ordered system be a typical instance of this family, which is very far from obvious (for instance a ferromagnet is one possible instance of a spin glass, but it is not a typical one!). An alternative approach would be to use the dynamics as above. In any case these non disordered spin systems have a low temperature phase which is a spin glass phase. In some cases there also exists a cristalline phase which is not reached by cooling the system at any computer accessible rate. These systems are good candidates to be spin analogs of glasses, without quenched disorder.

The mode coupling theory [22], which is successful in describing the approach to the glass transition, can be understood as an exact description for a frustrated system with deterministic [23] or disordered [24] interactions. In fact in the simplest case the mode coupling equations are extremely similar to the dynamical equations for a p-spin spin glass in the high temperature (TTI) phase. This analogy allows to derive the generalized mode coupling equations to describe the glass phase. From what we have seen above, these equations lack the properties of TTI and FDT.

All these are very encouraging facts. In order to elaborate a full theory of glasses, one needs to understand what will be the modification of these mean field theories for finite dimensional systems. Crucial to the existence of the class of spin glasses with $T_D > T_S$ is the fact that they are mean field systems: they possess many metastable states with an energy *density* which is larger than the one of the ground state. Clearly this is not possible in a finite dimensional system, because of the possibility of nucleating a bubble of ground state with an energy barrier which is non extensive. Therefore the tunneling effects should round up the dynamical transition. Some challenging recent speculations[25] suggest that one could derive in this way a Vogel Fulcher like divergence of the relaxation times of the type $\tau \simeq \exp\left(\frac{A}{(T-T_S)^\gamma}\right)$. It will be very interesting to pursue this approach further. It would also be interesting to have more detailed aging experiments in glasses.

## Acknowledgments

It is a pleasure to thank my collaborators on these topics, A. Barrat, J.P. Bouchaud, R. Burioni, L. Cugliandolo, S. Franz, J. Kurchan, G. Parisi. Laboratoire de Physique Théorique de l'Ecole Normale Supérieure is a "laboratoire propre du CNRS, associé à l'Ecole Normale Supérieure et à l'Université de Paris Sud".

## References

1. C. DeDominicis, *Phys. Rev.* B **18**, 4913 (1978).
2. M. Mézard, G. Parisi and M.A. Virasoro, *Spin glass theory and beyond* (World Scientific, Singapore, 1987).
3. H. Sompolinsky and A. Zippelius, *Phys. Rev. Lett.* **47**, 359 (1981); *Phys. Rev.* A **25**, 6860 (1982).
4. S. Edwards and P.W. Anderson, *J. Phys.* F **5**, 965 (1975).
5. H. Sompolinsky , *Phys. Rev. Lett.* **47**, 935 (1981).

6. H. Horner, *Z. Phys.* B **66**, 175 (1987); L.B. Ioffe,*Phys. Rev.* B **38**, 5181 (1988).
7. L.F. Cugliandolo and J. Kurchan *Phys. Rev. Lett.* **71**, 173 (1993).
8. L.F. Cugliandolo and J. Kurchan *J. Phys.* A **27**, 5749 (1994).
9. S. Franz and M. Mézard, *Europhys. Lett.* **26**, 209 (1994); *Physica* A **210**, 48 (1994).
10. F. Ritort, preprint cond-mat/9504081; A. Barrat and M. Mézard, preprint cond-mat/9504088.
11. L. Lundgren, P. Svedlindh, P. Nordblad and O. Beckman, *Phys. Rev. Lett.* **51**, 911 (1983); references to recent work can be found in E. Vincent, J. Hammann and M. Ocio, in *Recent progress in random magnets*, ed. D.H. Ryan (World Scientific, Singapore, 1992) and references therein.
12. L.C.E. Struik,*Physical aging in amorphous polymers and other materials* (Elsevier, Amsterdam, 1978).
13. Because of lack of space we cannot mention at all the numerous and interesting numerical simulations on aging. We refer the reader to the references in the recent works by H. Rieger, preprint cond-mat/9507047, and E. Marinari, G. Parisi and F.Ritort, *J. Phys.* A **28**, 327 (1995).
14. L.F. Cugliandolo, J. Kurchan and G. Parisi, *J. Phys. I (France)* **4**, 1641 (1994).
15. G. J. M. Koper and H. J. Hilhorst, *J. Phys. France* **49**, 429 (1988); D. S. Fisher and D.A. Huse, *Phys. Rev.* B **38**, 373 (1988).
16. J.P. Bouchaud, *J. Phys. I* **2**, 1705 (1992); J.P. Bouchaud and D. Dean, *J. Phys. I* **5**, 265 (1995).
17. A. Barrat, R. Burioni, M. Mézard, in preparation.
18. J.P. Bouchaud and M. Mézard, *J. Phys. I France* **4**, 1109 (1994); E. Marinari, G. Parisi and F. Ritort, *J. Phys.* A **27**, 7615 (1994), *J. Phys.* A **27**, 7647 (1994).
19. T.R. Kirkpatrick and D. Thirumalai, *Phys. Rev. Lett.* **58**, 2091 (1987), *J. Phys.* A **22**, L149 (1989) and references therein.
20. D.G. Gross and M. Mézard, *Nucl. Phys.* B **240**, 431 (1984).
21. A. Crisanti, H. Horner and H.J. Sommers, *Z. Phys.* B **92**, 257 (1992).
22. For a review see W. Götze, in *Liquids, freezing and glass transition*, J.P. Hansen et al Ed. (North Holland, 1989).
23. S. Franz and J. Hertz, *Phys. Rev. Lett.* **74**, 2114 (1995).
24. J.P. Bouchaud, L.F. Cugliandolo, J. Kurchan and M. Mézard, in preparation.
25. G. Parisi, preprint cond-mat/9412034.

# MACRODYNAMICS OF DISORDERED AND FRUSTRATED SYSTEMS

D. SHERRINGTON, A.C.C. COOLEN AND S.N. LAUGHTON
*Theoretical Physics, University of Oxford, 1 Keble Road,*
*Oxford, OX1 3NP, England*
E-mail: sherr@thphys.ox.ac.uk, coolen@thphys.ox.ac.uk, stephen@thphys.ox.ac.uk

ABSTRACT

It is shown how the macroscopic non-equilibrium dynamics of systems whose microscopic stochastic dynamics involves disordered and frustrated interactions can be well described by closed deterministic flow equations; this requires an appropriate choice of order parameter/function and ansätze.

## 1. Introduction

One would often wish to describe the macroscopic non-equilibrium dynamics of systems of many stochastically interacting microscopic units in terms of a closed set of deterministic flow equations. This is a non-trivial exercise for systems which are highly disordered and frustrated, but below we describe a procedure which works very well for two such problems, the Sherrington-Kirkpatrick spin glass[1] and the Hopfield neural network[2], and, we believe, provides a framework for more general application.

It is based on an appropriate choice of macroscopic order parameters together with two ansätze. Here we outline the philosophy and the essentials of the technique and present a few results to illustrate the degree of success. Further details can be found elsewhere[3,4,5].

## 2. The problem

In general terms, we are concerned with systems whose microscopic state is described by a set of $N$ variables $\boldsymbol{S}_i$ at 'sites' $i = 1...N$ and which obey random stochastic microdynamics leading to known master equations for the time-dependence of the microstate probability distributions $p_t(\{\boldsymbol{S}\})$, involving only the instantaneous time $t$ but with disorder and frustration in the local and intersite controlling elements. Our objective is to devise a description of the macrostate dynamics in terms of closed equations for few-parameter sets of macrovariables $\Omega_\mu(\{\boldsymbol{S}\}); \mu = 1...n$.

More specifically, we concentrate on systems in which the variables $\boldsymbol{S}$ are Ising spins $\{\sigma = \pm 1\}$ and obey random sequential Glauber stochastic dynamics via local effective fields determined through pairwise exchange interactions with other spins and external stimuli. The evolution of the microstate distribution satisfies the master equation

$$\frac{d}{dt}p_t(\boldsymbol{\sigma}) = \sum_{k=1}^{N}[p_t(F_k\boldsymbol{\sigma})W_k(F_k\boldsymbol{\sigma}) - p_t(\boldsymbol{\sigma})W_k(\boldsymbol{\sigma})] \tag{1}$$

where $F_k$ is the spin-flip operator $F_k\Phi(\boldsymbol{\sigma}) = \Phi(\sigma_i, ..., -\sigma_k, ..., \sigma_N)$, the transition rates and local fields are

$$W_k(\boldsymbol{\sigma}) = \frac{1}{2}[1 - \sigma_k \tanh(\beta h_k(\boldsymbol{\sigma}))] \qquad h_k(\boldsymbol{\sigma}) = \sum_{\ell \neq k} J_{kl}\sigma_\ell + \theta_k, \tag{2}$$

$\beta$ is the inverse temperature and we are now using the vector notation $\boldsymbol{\sigma} = (\sigma_1, ...\sigma_N)$. From (1) we may derive an equation for the evolution of the macrovariable probability distribution

$$P_t[\boldsymbol{\Omega}] = \sum_{\boldsymbol{\sigma}} p_t(\boldsymbol{\sigma})\delta[\boldsymbol{\Omega} - \boldsymbol{\Omega}(\boldsymbol{\sigma})]; \quad \boldsymbol{\Omega} \equiv (\Omega_1, ...\Omega_n) \tag{3}$$

in the form

$$\frac{d}{dt}P_t[\boldsymbol{\Omega}] = \sum_{\ell \geq 1} \frac{(-1)^\ell}{\ell!} \sum_{k_1=1}^{n} .. \sum_{k_\ell=1}^{n} \frac{\partial^\ell}{\partial\Omega_{k_1}...\partial\Omega_{k_\ell}} \{P_t[\boldsymbol{\Omega}]F^{(\ell)}_{k_1..k_\ell}[\boldsymbol{\Omega};t]\} \tag{4}$$

where

$$F^{(\ell)}_{k_1...k_\ell}[\boldsymbol{\Omega};t] = \langle \sum_{j=1}^{N} W_j(\boldsymbol{\sigma})\Delta_{jk_1}(\boldsymbol{\sigma})...\Delta_{jk_\ell}(\boldsymbol{\sigma})\rangle_{\boldsymbol{\Omega};t} \quad \Delta_{jk}(\boldsymbol{\sigma}) \equiv \Omega_k(F_j\boldsymbol{\sigma}) - \Omega_k(\boldsymbol{\sigma}) \tag{5}$$

and the notation $\langle\rangle_{\boldsymbol{\Omega};t}$ refers to a sub-shell average

$$\langle f(\boldsymbol{\sigma})\rangle_{\boldsymbol{\Omega};t} \equiv \frac{\sum_{\boldsymbol{\sigma}} p_t(\boldsymbol{\sigma})\delta[\boldsymbol{\Omega} - \boldsymbol{\Omega}(\boldsymbol{\sigma})]f(\boldsymbol{\sigma})}{\sum_{\boldsymbol{\sigma}} p_t(\boldsymbol{\sigma})\delta[\boldsymbol{\Omega} - \boldsymbol{\Omega}(\boldsymbol{\sigma})]} \quad . \tag{6}$$

In several cases of interest and for finite times only the first term on the right hand side of (4) survives in the limit $N \to \infty$, yielding the deterministic flow

$$\frac{d}{dt}\boldsymbol{\Omega}_t = \langle \sum_i W_i(\boldsymbol{\sigma})[\boldsymbol{\Omega}(F_i\boldsymbol{\sigma}) - \boldsymbol{\Omega}(\boldsymbol{\sigma})]\rangle_{\boldsymbol{\Omega};t} \quad . \tag{7}$$

In general this does not yet constitute a closed set of equations due to the appearance of $p_t(\boldsymbol{\sigma})$ in the sub-shell average. This requires an appropriate choice of $\boldsymbol{\Omega}$ and possibly further ansätze.

## 3. The specific physical systems

Here we concentrate on the two specific model systems, the Sherrington-Kirkpatrick (SK) spin glass[1] and the Hopfield neural network[2]. In the SK model the $\{\sigma\}$ represent true magnetic spins and the $\{J_{ij}\}$ are chosen randomly from a Gaussian distribution

$$J_{ij} = J_o/N + Jz_{ij}/\sqrt{N} \qquad z_{ij} = z_{ji} \quad \langle z_{ij}\rangle = 0 \quad \langle {z_{ij}}^2\rangle = 1 \quad . \tag{8}$$

In the Hopfield model the $\{\sigma\}$ represent states of McCulloch-Pitts neurons, $\sigma = \pm 1$ corresponding to firing/non-firing, and the $\{J_{ij}\}$ provide for the storage and retrieval of random patterns $\{\xi_i^\mu = \pm 1\}$; $\mu = 1...p = \alpha N$, via the Hebb rule $J_{ij} = N^{-1}\sum_{\mu=1}^{p}\xi_i^\mu\xi_j^\mu$. Concentrating for simplicity on the region of phase space within the basin of attraction of one pattern, $\mu = 1$, it is convenient to apply the gauge transformation $\sigma_i \to \sigma_i\xi_i^1$, $J_{ij} \to \xi_i^1\xi_j^1 J_{ij}$, $\theta_i \to \xi_i^1\theta_i$ to re-write this in the form of (8) with

$$z_{ij} = \frac{1}{\sqrt{p}}\sum_{\mu>1}^{p}\xi_i^1\xi_i^\mu\xi_j^1\xi_j^\mu \qquad J_o = 1 \qquad J = \sqrt{\alpha} \qquad \langle\xi_i\rangle = 0 \ . \tag{9}$$

The symmetry $z_{ij} = z_{ji}$ provides for the steady state distribution $p_\infty(\boldsymbol{\sigma})$ to be expressible in the Boltzmann form $p_\infty(\sigma) \sim \exp(-\beta H)$ with the Hamiltonian

$$H = -\sum_{i<j} J_{ij}\sigma_i\sigma_j - \sum_i \theta_i\sigma_i \ . \tag{10}$$

Further specializing to $\theta_i = \theta$, all $i$, the Hamiltonian is expressible in terms of the two macroscopic parameters

$$m(\boldsymbol{\sigma}) = N^{-1}\sum_i \sigma_i \qquad r(\boldsymbol{\sigma}) = N^{-3/2}\sum_{i<j}\sigma_i z_{ij}\sigma_j \tag{11}$$

$$H(\boldsymbol{\sigma})/N = [-\frac{1}{2}J_o m^2(\boldsymbol{\sigma}) + \theta m(\boldsymbol{\sigma})] - Jr(\boldsymbol{\sigma}) + 0(N^{-1}) \tag{12}$$

where the term in [ ] is disorder-independent with all the disorder effects in the $Jr(\boldsymbol{\sigma})$ term. Hence $m, r$ suffice as order parameters to describe the equilibrium probability distribution. Coupled with the need for $m$ to study magnetization or overlap with the condensed pattern they represent a minimum choice for the set $\boldsymbol{\Omega}$. Thus we shall first discuss an attempt to find a non-equilibrium macrodynamics in terms of $m, r$, alone, and show that it provides a reasonable but imperfect description. We shall then go on to a more sophisticated theory in terms of a generalized order function which provides a very good fit to the results of microscopic simulation.

## 4. The simple version of the theory

In this section we choose the minimal form

$$\boldsymbol{\Omega}^s(\boldsymbol{\sigma}) \equiv (\Omega_1(\boldsymbol{\sigma}), \Omega_2(\boldsymbol{\sigma})) = (m(\boldsymbol{\sigma}), r(\boldsymbol{\sigma})) \ . \tag{13}$$

The resultant $P_t[\boldsymbol{\Omega}^s]$ does indeed satisfy a Liouville equation in the thermodynamic limit, yielding the deterministic flow equations

$$\frac{dm}{dt} = \int dz D_{m,r;t}(z)\tanh\beta(J_o m + Jz + \theta) - m \tag{14}$$

$$\frac{dr}{dt} = \int dz D_{m,r;t}(z) z \tanh \beta(J_o m + Jz + \theta) - 2r \tag{15}$$

where $D_{m,r;t}(z)$ is the sub-shell averaged distribution of the disorder contributions to the local fields

$$D_{m,r;t}(z) = \lim_{N\to\infty} \frac{\sum_{\boldsymbol{\sigma}} p_t(\boldsymbol{\sigma})\delta(m - m(\boldsymbol{\sigma}))\delta(r - r(\boldsymbol{\sigma}))N^{-1}\sum_i \delta(z - z_i(\boldsymbol{\sigma}))}{\sum_{\boldsymbol{\sigma}} p_t(\boldsymbol{\sigma})\delta(m - m(\boldsymbol{\sigma}))\delta(r - r(\boldsymbol{\sigma}))} \tag{16}$$

$$h_i(\boldsymbol{\sigma}) = J_o m(\boldsymbol{\sigma}) + J z_i(\boldsymbol{\sigma}) + \theta + 0(N^{-1}) \qquad z_i(\boldsymbol{\sigma}) = N^{-1/2}\sum_j z_{ij}\sigma_j \ . \tag{17}$$

As yet, because of the $p_t(\boldsymbol{\sigma})$ in (16), equations (14), (15) are not closed except in the disorder-free case $J = 0$. To close the equations we introduce two simple ansätze: (i) we assume that the evolution of the macrostate $(m, r)$ is self-averaging with respect to the specific microscopic realization of the disorder $\{z_{ij}\}$, (ii) as far as evaluating $D(z)$ is concerned we assume equipartitioning of the microstate probability $p_t(\boldsymbol{\sigma})$ within each $(m, r)$ shell. The first of these ansätze is well borne out by computer simulations of the microscopic dynamics and permits averaging over pattern choices. The second, which can only be judged *a posteriori*, eliminates memory effects beyond their reflection in $m, r$ and removes explicit time-dependence from $D$. Together they give

$$D_{m,r;t}(z) \to D_{m,r}(z) = \left\langle \frac{\sum_{\boldsymbol{\sigma}} \delta(m - m(\boldsymbol{\sigma}))\delta(r - r(\boldsymbol{\sigma}))N^{-1}\sum_i \delta(z - z_i(\boldsymbol{\sigma}))}{\sum_{\boldsymbol{\sigma}} \delta(m - m(\boldsymbol{\sigma}))\delta(r - r(\boldsymbol{\sigma}))} \right\rangle_{\{z_{ij}\}} \tag{18}$$

whose insertion into (14) and (15) yields the required closure.

The actual evaluation of $D_{m,r}(z)$ from (18) remains a non-trivial exercise, but one which is amenable to solution by replica theory as developed for the investigation of local field distributions in spin glasses[6]. After several manipulations $D$ can be expressed in the form

$$D_{m,r}(z) = \lim_{n\to 0} \int \prod_{i,j} \prod_{\alpha,\beta=1\ldots n} dx_i^\alpha dy_j^{\alpha\beta} \exp[-N\Phi(m, r, z; \{x_i^\alpha\}, \{y_j^{\alpha\beta}\})] \tag{19}$$

where the number of indices $i, j$ is finite and $\Phi$ is $O(N^0)$. Because of the exponential scaling as $N$, the integral can be evaluated by steepest descents.

The extremization of (19) is discussed in detail elsewhere[4] but we note that it involves a parameter $q^{\alpha\beta}$ which is a dynamical analogue of the usual (static) spin-glass order parameter and has a similar, but now dynamically constrained, interpretation in terms of overlaps; specifically, the disorder-averaged probability distribution for the mutual overlap between microscopic configurations confined to the same macroscopic

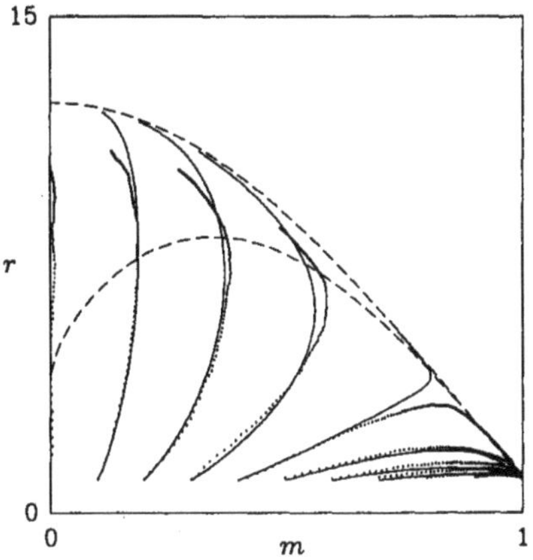


Fig. 1. Macroscopic flow trajectories for a Hopfield model with storage capacity $\alpha = 0.1$ and deterministic microscopic dynamics ($\beta = \infty$); dots indicate simulations ($N = 32000$), solid lines indicate analytic RS theory. The outer dashed line is the boundary predicted by RS theory; the inner dashed line indicates the onset of instability against RS-breaking fluctuations.

$(m, r)$ sub-shell is given by

$$P_{m,r}(q) \equiv$$

$$\left\langle \frac{\sum_{\boldsymbol{\sigma},\boldsymbol{\sigma}'} \delta(q - N^{-1}\sum_i \sigma_i\sigma_i')\delta(m-m(\boldsymbol{\sigma}))\delta(m-m(\boldsymbol{\sigma}'))\delta(r-r(\boldsymbol{\sigma}))\delta(r-r(\boldsymbol{\sigma}'))}{\sum_{\boldsymbol{\sigma},\boldsymbol{\sigma}'} \delta(m-m(\boldsymbol{\sigma}))\delta(m-m(\boldsymbol{\sigma}'))\delta(r-r(\boldsymbol{\sigma}))\delta(r-r(\boldsymbol{\sigma}'))} \right\rangle_{\{z_{ij}\}}$$

$$= \lim_{n\to 0} \frac{1}{n(n-1)} \sum_{\alpha\neq\beta} \delta(q - q^{\alpha\beta}) \ . \qquad (20)$$

The steady state condition $dm/dt = dr/dt = 0$ is satisfied by $m, r, \{q^{\alpha\beta}\}$ which obey the usual self-consistency equations as obtained from equilibrium analysis[1,7,8,9], including all replica-symmetry breaking aspects.

For general $m, r$ the explicit extremization and limiting procedure for (19) is greatly simplified within the replica-symmetric (RS) ansatz, which already yields a non-Gaussian form for $D_{m,r}(z)$ in qualitative accord with the results of simulation[3,4]. Substituting into (14) and (15) there result closed equations of the form

$$dm/dt = F(m,r) \qquad dr/dt = G(m,r) \qquad (21)$$

where $F, G$ are complicated functions taking the form of integrals over other functions involving parameters self-consistently determined from $m, r$ via non-linear equations. Figs 1 and 2 show examples of their predictions for the Hopfield model compared

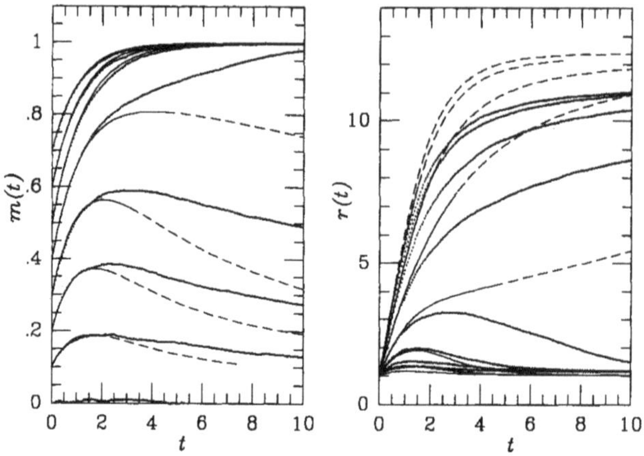


Fig. 2. Temporal dependence of the order parameters for a Hopfield model with storage $\alpha = 0.1$ and zero-temperature dynamics ($\beta = \infty$); dots indicate simulations ($N = 32000$), the other lines indicate RS theory shown with solid lines where stable, dashed lines where unstable. Time is measured in Monte Carlo steps per spin.

with microscopic simulations[10]. These figures show several points of note; first, they demonstrate the concept of basins of attraction - there is a critical locus of $(m(0), r(0))$ which separates flows which retrieve ($m(\infty) \sim \mathcal{O}(1)$) from those which do not ($m(\infty) \to 0$); second, the critical $m_c(0)$ depends on $r(0)$; third, the present version of the theory and ansätze yield qualitatively reasonable results for $r$ versus $m$ along flow lines away from the critical locus, but miss the slowing-down of $m(t), r(t)$ seen in simulations of non-retrieving flows. Fig. 1 also shows (dashed) two other special loci; the upper one is a theoretical limit of the boundary of the physical region within the RS ansatz and corresponds to the limit of maximum $q = 1$; the lower dashed locus is the upper boundary of the limit of stability of the RS ansatz for calculations of $D$, a dynamical analogue of the de Almeida-Thouless[11] (AT) line.

Although the crossing of the AT line could be expected to herald slowing-down due to replica-symmetry breaking, the figures show this occurs earlier, suggesting a different origin for the discrepancy between theory and simulation. Results of a similar quality are obtained for finite temperature and for the SK spin glass[4]. A study of a related RS toy model[12] further emphasises the qualitative usefulness but quantitative incompleteness of the simple theory.

## 5. The sophisticated version of the theory: order function dynamics

To improve on the theory requires broadening the range of order parameters. For a qualitative improvement we anticipate the need for a qualitative change in that range and so, instead of two order parameters, we consider a continuous order function.

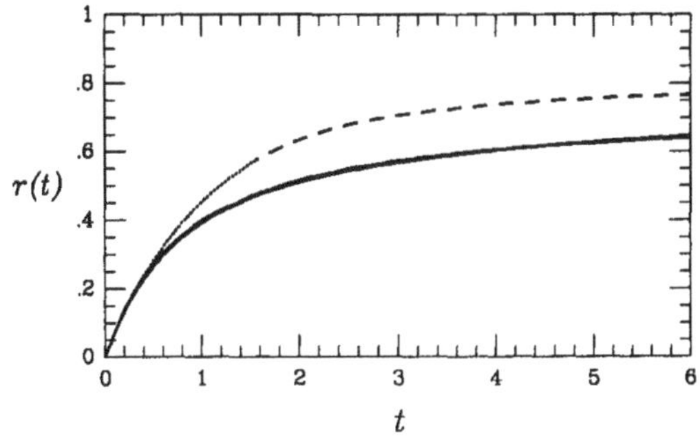


Fig. 3. Evolution of the binding energy of the Sherrington-Kirkpatrick spin glass ($J_o = 0$) from a random microscopic start. Comparison of simulations ($N = 8000$, solid line) and predictions of the simple two-parameter $(m, r)$ theory of section 4 (RS stable, dotted; RS unstable, dashed) and of the advanced order-function theory of section 5 (solid), for $\beta = \infty$. Note that the two solid lines are almost coincident.

Since the microscopic dynamics is formulated entirely in terms of the states of the spins $\sigma_i$ and the fields $h_i$ we choose for $\boldsymbol{\Omega}(\boldsymbol{\sigma})$ the joint distribution

$$\mathcal{D}[\zeta, h; \boldsymbol{\sigma}] = N^{-1} \sum_i \delta_{\zeta \sigma_i} \delta(h - h_i(\boldsymbol{\sigma})) \quad . \tag{22}$$

This choice automatically includes the positive features of the simple version, since $m, r$ follow straightforwardly from $\mathcal{D}$ and hence a formulation in terms of $\mathcal{D}$ will automatically be correct asymptotically for systems whose Hamiltonians can be expressed completely in terms of $m, r$. It has further advantages in being applicable also to systems without detailed balance (and therefore not expressible by Boltzmann statistical equilibrium) and of being extendable to analogue spins.

As discussed in §2 we consider the evolution of the probability distribution for the order function $P_t[\mathcal{D}]$. Provided we first discretize the $h$-distribution at a finite number of values, only passing later to the continuum limit, we do indeed find a Liouville form corresponding to deterministic evolution of $\mathcal{D}_t[\zeta, h]$ in the thermodynamic limit. The equation of motion for $\mathcal{D}_t[\zeta, h]$ involves a further noise distribution over a sub-shell with $\mathcal{D}[\zeta, h; \boldsymbol{\sigma}] = \mathcal{D}_t[\zeta, h]$, again weighted by $p_t(\boldsymbol{\sigma})$. Thus once more we make an equipartitioning ansatz to eliminate $p_t(\boldsymbol{\sigma})$, as well as self-averaging, but now the relevant sub-shell is much more restricted and contains much more memory information. This provides a closed equation for the evolution of the order function $\mathcal{D}_t[\zeta, h]$. To evaluate the 'noise-term' again requires replica analysis and even within replica-symmetric theory yields a complicated self-consistent set of equations to determine the flow. The results, however, are impressive, as fig. 3 illustrates for the flow of the binding energy of the SK model at $T = 0$[5,13]; while there is clear difference between the predictions of the simple theory and the slower evolution obtained in computer simulations, within the numerical accuracy shown $\mathcal{D}_t[\zeta, h]$ captures the slowing-down

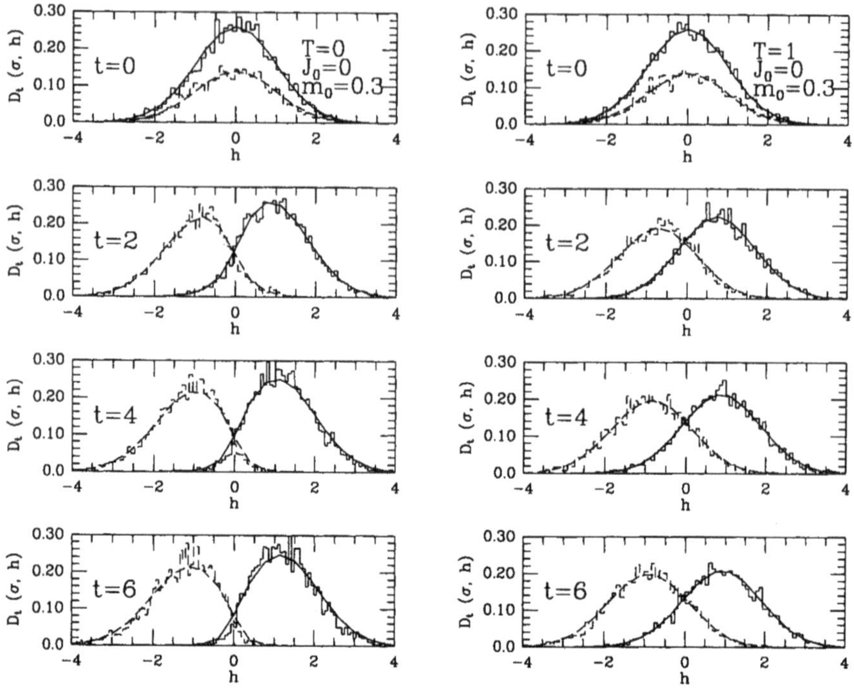


Fig. 4. Evolution of $\mathcal{D}_t(\sigma, h)$ for SK models with $J_o = 0$, $J = 1$, $T = 0, 1$. The lines are predictions of the sophisticated theory within RS ansatz, the histograms are obtained from microscopic simulations ($N = 8000$).

well, although clearly it cannot reproduce perfectly asymptotically the equilibrium ensemble result which is replica-symmetry broken. To be more quantitative we note that the true RSB equilibrium binding energy at $T = 0$ is 0.7633 while RS theory gives 0.798, a difference which is small compared with that between the simple theory and the sophisticated theory/simulation over the time range shown (except for the initial times; both theories are exact at $t = 0$). On the other hand, it should be noted that it is also possible that the dynamical equations yield an additional asymptotic steady state solution different from that of Gibbsian ensemble theory, but we have not yet investigated this question.

Fig. 4 shows an example of the evolution of $D_t[\zeta, h]$ itself [5,13] for $J_o = 0, T = 0, 1$. Note that for $T = 0$ $D_t[\xi, h]$ tends to zero for $(\zeta h) < 0$ as $t \to \infty$, where as for $T > 0$ it remains non-zero over both signs of $(\zeta h)$ for all times.

## 6. Conclusion

We have demonstrated that with an appropriate choice of order parameters and two simple ansätze one can derive closed macroscopic flow equations for range-free disordered and frustrated systems in good accord with the results of microscopic simulations. In its more sophisticated version, in terms of an order function describing the distribution of spins and fields, and including RSB effects the theory may even be exact. Discussion has been restricted to Ising spins and detailed balanced dynamics, but the sophisticated version of the theory, can be extended beyond these restrictions. In the case of neural networks both versions can also be extended to regions of phase space having finite overlaps with more than one pattern[13].

## 7. Acknowledgement

We would like to thank the Engineering and Physical Sciences Research Council of the United Kingdom for partial support under grant GRH 73028 (ACCC) and scholarship 923.0910 (SNL).

## 8. References

1. D. Sherrington and S. Kirkpatrick, *Phys. Rev. Lett.* **35** (1975) 1792
2. J.J. Hopfield, *Proc. Nat. Acad. Sci. USA* **79** (1982) 2554
3. A.C.C. Coolen and D. Sherrington, *Phys. Rev.* **E49** (1994) 1921
4. A.C.C. Coolen and D. Sherrington, *J. Phys* **A27** (1994) 7687
5. S.N. Laughton, A.C.C. Coolen and D. Sherrington, in preparation
6. M. Thomsen, M.F. Thorpe, T.C. Choy, D. Sherrington and H-J. Sommers, *Phys. Rev* **B33** (1986) 1931
7. D.J. Amit, H. Gutfreund and H. Sompolinsky, *Phys. Rev. Lett.* **55** (1985) 1530
8. G. Parisi, *J. Phys* **A13** (1980) 1101
9. M. Mézard, G. Parisi and M. Virasoro, *Spin Glass Theory and Beyond* (World Scientific, Singapore 1987).
10. D. Sherrington and A.C.C. Coolen, to be published in *Proceedings of 1994 Sitges Conference on '25 years of Non-Equilibrium Statistical Mechanics'* (Springer-Verlag, Berlin)
11. J.R.L. de Almeida and D.J. Thouless, *J. Phys* **A11** (1978) 983
12. A.C.C. Coolen and S. Franz, *J. Phys* **A27** (1994) 6947
13. S.N. Laughton, D. Phil Thesis, Oxford 1995

# Towards a Theory of Anomalous Scaling in Turbulence

Victor L'vov and Itamar Procaccia

*Department of Chemical Physics,*

*The Weizmann Institute of Science, Rehovot 76100, Israel*

In this short note we present a brief references-assisted overview of our recent progress in understanding the universal statistics of fully developed turbulence, with a stress on anomalous scaling.

Modern concepts about high Re number turbulence started to evolve with Richardson's insightful contributions [1] which contained the famous "poem" that paraphrased J. Swift: "*Big whirls have little whirls that feed on their velocity, and little whirls have lesser whirls and so on to viscosity -in the molecular sense*". In this way Richardson conveyed an image of the creation of turbulence by large scale forcing, setting up a cascade of energy transfers to smaller and smaller scales by the nonlinearities of fluid motion, until the energy dissipates at small scales by viscous mechanisms. This picture led in time to innumerable "cascade models" that tried to capture the statistical physics of turbulence by assuming some thing or other about the cascade process. Possibly the most ingenious attempt is due to Kolmogorov who in 1941 [2] proposed the idea of universality (turning the study of small scale turbulence from mechanics to fundamental physics) based on the notion of the inertial range. The idea is that for very large values of Re there is a wide separation between the scale of forcing and the typical scale at which dissipative processes become important. In this "inertial" range the only parameter is the mean energy flux per unit time per unit mass, denoted as $\bar{\epsilon}$, and the value of the kinematic viscosity is irrelevant. This assumption allows one to develop dimensional analysis that leads to detailed predictions about the statistical physics of turbulence. This is important; no one in their right mind is interested in the full solution of the turbulent velocity field at all points in space-time. The interest is in the statistical properties of the turbulent flow. For example, quantities that were considered by Kolmogorov and later attracted decades of experimental and theoretical attention (see for example [3–7]) are the structure functions of velocity differences, denoted as $S_n(R) = \langle |\mathbf{u}(\mathbf{r}+\mathbf{R},t) - \mathbf{u}(\mathbf{r},t)|^n\rangle$ where $\langle\ldots\rangle$ stands for a suitably defined ensemble average. If one considers $R$ to be in the inertial range, i.e $\eta \ll R \ll L$, the only combination of $\bar{\epsilon}$ and $R$ that gives the right dimension for $S_n$ is $(\bar{\epsilon}R)^{n/3}$. In particular for $n = 2$ this is the famous Kolmogorov "2/3" law which in Fourier representation is also known as the "-5/3" law. This ideology that one extracts universal properties by focusing on statistical quantities can be applied also to the correlations of gradients of the velocity field. An important example is the rate $\epsilon(\mathbf{r},t)$ at which energy is dissipated into heat due to viscous damping. This rate is roughly $\nu|\nabla\mathbf{u}(\mathbf{r},t)|^2$. The often-studied correlation function of $\hat{\epsilon}(\mathbf{r},t) = \epsilon(\mathbf{r},t) - \bar{\epsilon}$ is $K_{\epsilon\epsilon}(R) = \langle\hat{\epsilon}(\mathbf{r}+\mathbf{R},t)\hat{\epsilon}(\mathbf{r},t)\rangle$. Within the Kolmogorov theory one estimates $K_{\epsilon\epsilon}(R) \simeq \nu^2\bar{\epsilon}^{4/3}R^{-8/3}$.

Experimental measurements show that Kolmogorov was remarkably close to the truth. The major aspect of his predictions, i.e that the statistical quantities depend on the length scale $R$ as power laws is corroborated by experiments. On the other hand, the predicted exponents seem not to be exactly realized. For example, the correlation $K_{\epsilon\epsilon}(R)$ decays according to a power law, $K_{\epsilon\epsilon}(R) \sim R^{-\mu}$ for $\eta \ll R \ll L$, with $\mu$ having a numerical value of $0.2 - 0.3$ instead of $8/3$ [5]. The structure functions also behave as power laws, $S_n(R) \simeq R^{\zeta_n}$, but the numerical values of $\zeta_n$ deviate progressively from $n/3$ when $n$ increases [4,6]. Something fundamental seems to be missing.

Theoretical studies of the universal small scale structure of turbulence can be classified broadly into two main classes. Firstly there is a large body of phenomenological models that attempt to achieve agreement with experiments on the basis of some ad hoc assumption on the nature of the cascade or the statistics of the turbulent fields [7]. Since these models are not derived on the basis of the equations of fluid mechanics, one is always left with uncertainties about the validity or relevance of these models. The second class of approaches is based on the equations of fluid mechanics. Typically one acknowledges the fact that fluid mechanics is a (classical) field theory and resorts to field theoretic methods in order to compute statistical quantities. Even though there had been a continuous effort of almost

50 years in this direction, the analytic derivation of the scaling laws for $K_{\epsilon\epsilon}(R)$ and $S_n(R)$ from the Navier-Stokes equations and the calculation of the numerical value of the scaling exponents $\mu$ and $\zeta_n$ have been among the most elusive goals of theoretical research. Why did it turn out to be so difficult?

To understand the difficulties, we need to elaborate a little on the nature of the field theoretic approach [13]. Suppose that we want to calculate the average response of a turbulent fluid at some point $\mathbf{r}_0$ to forcing at point $\mathbf{r}_1$. The field theoretic approach allows us to consider this response as an infinite sum of all the following processes: firstly there is the direct response at point $\mathbf{r}_0$ due to the forcing at $\mathbf{r}_1$. This response is instantaneous if we assume that the fluid is incompressible (and therefore the speed of sound is infinite). Then there is the process of forcing at $\mathbf{r}_1$ with a response at an intermediate point $\mathbf{r}_2$, which then acts as a forcing for the response at $\mathbf{r}_0$. This intermediate process can take time, and we need to integrate over all the possible positions of point $\mathbf{r}_2$ and all times. This is the second-order term in perturbation theory. Then we can force at $\mathbf{r}_1$, the response at $\mathbf{r}_2$ acting as a forcing for $\mathbf{r}_3$ and the response at $\mathbf{r}_3$ forces a response at $\mathbf{r}_0$. We need to integrate over all possible intermediate positions $\mathbf{r}_2$ and $\mathbf{r}_3$ and all the intermediate times. This is the third-order term in perturbation theory. And so on. The actual response is the infinite sum of all these contributions. In applying this field theoretical method one encounters three main difficulties:

(A) The theory has no small parameter. The usual tack is to develop the theory perturbatively around the linear part of the equation of motion. The expansion parameter is then Re, and since we are interested in Re$\gg 1$, naive perturbation expansions are badly divergent. In other words the size of the various processes described above increases as $(\mathrm{Re})^n$ with the number $n$ of intermediate points in space-time.

(B) The theory suffers from two types of nonlinear interactions. Both are hidden in the nonlinear term $\mathbf{u}\cdot\nabla\mathbf{u}$ in the Navier-Stokes equations. The larger of the two is the sweeping of small "eddies" by all the larger ones. Physically this sweeping couples any given scale of motion to all the larger scales scales. Technically in perturbation theory these interactions cause infrared divergences of the integrals; the divergences become stronger with the order of of the contribution. This may lead to a belief that the theory of turbulence is not renormalizable. However, sweeping is just a kinematic effect that does not lead to energy redistribution between scales, and one may hope that if the effect of sweeping is taken care of in a consistent fashion a renormalizable theory might emerge. This redistribution of energy results from the second type of interaction, that stems from the shear and torsion effects that are sizeable only if they couple fluid motions of comparable scales. The second type of nonlinearity is smaller in size but crucial in consequence, and it may certainly lead to a scale-invariant theory.

(C) Nonlocality of interaction in **r** space. The effect of the pressure in the incompressible approximation is to instantaneously couple distant points.

Indeed, these difficulties seemed to complicate the application of field theoretic methods to such a degree that a consensus appeared to the effect that it is impossible to gain valuable insight into the universal properties of turbulence along these lines, even thought they proved so fruitful in other field theories. The present authors think differently, and in the rest of this paper we will explain why.

The first task of a successful theory of turbulence is to overcome the existence of the interwoven nonlinear effects with different scale dependence that rule out a scale invariant formulation even in the inertial range. This is not achieved by directly applying a formal field-theoretical tool to the Navier-Stokes equations. It does not matter whether one uses standard diagrammatics [8], path integral formulation, renormalization group [9] $\epsilon$-expansion, large $N$-limit [10] or one's formal method of choice. One needs to take care of the particular nature of hydrodynamic turbulence as embodied in difficulty (B) *first*, and *then* proceed using formal tools.

The removal of the effects of sweeping is based on Richardson's remark that universality in turbulence is expected for the statistics of velocity *differences across a length scale* $R$ rather than for the statistics of the Eulerian field itself. The Eulerian fields are dominated by the large scale winds that are not universal. This remark was developed by Kraichnan who attempted to cast the field theoretic approach in terms of Lagrangian paths that automatically remove the large scale contributions [11]. Kraichnan's approach was fundamentally correct, and gave rise to important and influential insights in the description of turbulence, but did not provide a convenient technical way to consider all the orders of perturbation theory. The theory was not "diagrammatic" in the sense that there were no transparent rules governing how to consider a arbitrarily high term in the perturbation theory. Only low order truncations were considered.

A way to overcome difficulty (B) was suggested by Belinicher and L'vov [12] who introduced a novel transformation that allowed on one hand the elimination of the sweeping that leads to infrared divergences, and on the other hand produced a diagrammatic theory for the universal statistics. The use of this transformation allowed the examination of the structure functions of velocity differences $S_n(R)$ to all orders in perturbation theory. Of course, difficulty (A) remains; the perturbation series still diverges rapidly for large values of Re, but now standard field theoretic methods can be used to resum infinite subsets of contributions (Dyson line resummation) to achieve a new perturbation theory in terms of renormalized quantities [14]. In particular the damping coefficient in the renormalized theory changes from the bare kinematic viscosity to the eddy viscosity. The result is that the effective expansion parameter is no longer Re but an expansion parameter of O(1). Of course, such a series may still diverge as a whole. Notwithstanding it is crucial to examine first the order-by-order properties of series of this type.

Such an examination leads to a major surprise: every term in this perturbation theory remains finite when the outer scale $L$ goes to $\infty$ and the inner scale $\eta$ goes to 0 [14]. The meaning of this is that the perturbative theory for $S_n$ does not indicate the existence of a renormalization length that can be used in order to describe anomalous scaling. In other areas of theoretical physics in which anomalous scaling has been found the perturbative series indicates this by exhibiting (resummable) logarithmic divergences that must be tamed by truncating the integrals at some renormalization length. Hydrodynamic turbulence is different. The nonlinear Belinicher-L'vov transformation changes the underlying linear theory such that the line resummation results in full renormalization of the perturbative scheme [12,14]. The physical meaning of this result is two-fold. Firstly it means that in terms of the reposnse at $\mathbf{r}_0$ to forcing at $\mathbf{r}_1$ as discussed above there exists a "ball of locality" of radius $R = |\mathbf{r}_1 - \mathbf{r}_0|$ such that if the intermediate points $\mathbf{r}_2 \ldots \mathbf{r}_n$ are taken far outside it, the contributions to the final result die out. In addition if the intermediate points come too em close to one other, the contributions of these processes remain finite. The main contribution comes when all the intermediate points are separated from one other by a distance of the order of $R$. This is the perturbative justification of the Richardson-Kolmogorov cascade picture in which widely separated scales do not interact.

But the main question still remains: how does a renormalization scale appear in the statistical theory of turbulence?

The first answer comes from considering the statistics of gradient fields rather than the velocity differences themselves [18,15]. As indicated above the discrepancy in the exponent $\mu$ between the experimental value and the naive expectation is large. This gives hope that the explanation must lie close to the surface. Indeed, considering the perturbative scheme for $K_{\epsilon\epsilon}$ leads immediately to the discovery of logarithmic ultraviolet divergences with some ultraviolet viscous scale $\tilde{\eta}$ acting as the renormalization scale. The summation of this infinite series results in a factor $(R/\tilde{\eta})^{2\Delta}$ with some anomalous exponent $\Delta$ which is, generally speaking, of the order of unity. This changes the exponent $\mu$ from 8/3 to $8/3 - 2\Delta$. The physical explanation of this renormalization [18,19] is the effect of the multi-step interaction of two $R$-separated small eddies of scale $\tilde{\eta}$ with a large eddy of scale $R$ via an infinite set of eddies of intermediate scales. At this point it is important to understand what is the numerical value of the anomalous exponent $\Delta$. It can be shown using elementary inequalities that $\Delta \leq 2 - \zeta_2$. The structure of the theory is however qualitatively different for $\Delta$ smaller than $2 - \zeta_2$ (referred to as the "subcritical scenario" [19]) and for $\Delta = 2 - \zeta_2$ (which is the "critical scenario" [14,15].

In the subcritical scenario the perturbation series for the structure functions can be shown to converge, with the sum of the series being of the order of each one of the terms.

In this scenario no new length scale enters in the limit Re $\to \infty$, and the K41 exponents $\zeta_n$ are expected to be exact. However, we showed in [14,15] that the subcritical scenario is *not* realized, and the theory of turbulence follows the critical scenario. The technical step leading to this is the derivation of an exact sum rule that forces $\Delta$ to attain precisely the critical value $2-\zeta_2$. Physically this sum rule is a consequence of the existence of a universal nonequilibrium stationary state that supports an energy flux from large to small scales. The criticality of $\Delta$ has far-reaching implications for the theory of the structure functions. Now the series for the structure functions $S_n(R)$ diverge when the outer scale $L$ approaches $\infty$ as powers of $L$, $(L/R)^{\delta_n}$. The anomalous exponents $\delta_n$ are the deviations of the exponents of $S_n(R)$ from their K41 value. This is a very delicate and important point. It is not simple to see why a series each of whose terms converges when $L \to \infty$ still diverges as a whole. One can understand this as in resumming the series of a desired function one finds an inhomogeneous equation whose inhomogeneous solutions are indeed $L$-independent. However, the equation possesses also homogeneous solutions which are inherently nonperturbative in nature and may have anomalous scaling. Homogeneous solutions must be matched with the boundary conditions, and this is the way that the outer scale appears in the theory. The next step in the theoretical development is to understand how to compute the anomalous exponents $\delta_n$.

The divergence of the perturbation theory for $S_n(R)$ with $L \to \infty$ forces us to seek a nonperturbative handle on the theory. One finds this in the idea that there exists always a global balance between energy input and dissipation, which may be turned into a non-perturbative constraint on each $n$-th order structure function [16]. Using the Navier-Stokes equations one derives the set of equations of motion

$$\frac{\partial S_n(R,t)}{\partial t} + D_n(R,t) = \nu J_n(R,t) , \tag{1}$$

where $D_n$ and $J_n$ stem from the nonlinear and the viscous terms in the Navier Stokes equations respectively. In the stationary state the time derivative vanishes and one has the balance equation $D_n(R) = \nu J_n(R)$. The evaluation of $D_n(R)$ requires dealing with the difficulty (C) of the nonlocality of the interaction, but it does not pose conceptual difficulties.

It was shown [16] that it is of the order of $dS_{n+1}/dR$. On the other hand, the evaluation of $J_n(R)$ raises a number of very interesting issues whose resolution lies at the heart of the universal scaling properties of turbulence. Presently not all of these issues have been resolved, and we briefly mention here some ground on which progress has been made.

The term $J_n(R)$ consists of a correlation of $\nabla^2 u$ with $n-2$ velocity differences across a scale $R$. The question is how to evaluate such a quantity in terms of the usual structure functions $S_n(R)$. Recall that a gradient of a field is the difference in the field values at two points divided by the separation when the latter goes to zero. In going to zero one necessarily crosses the dissipative scale. It becomes therefore crucial to formulate so-called "fusion rules" for turbulence, which govern the properties of $n$-point correlation functions when two or more coordinates "fuse" together. The answer is [20] that when $p$ coordinates are separated by a small distance $r$, and the remaining $n-p$ coordinates are separated by a large distance $R$, then the $r$ dependence is like that of $S_p(r)$, i.e. $r^{\zeta_p}$. This is true until $r$ crosses the dissipative scale. Assuming that below the dissipative scale derivatives exist and the fields are smooth, one can estimate gradients at the end of the smooth range by dividing differences across the dissipative scale over that scale. The question is, what is the dissipative scale? Is there just one dissipative scale, or is there a multiplicity of such scales, depending on the function one is studying? For example, when does the above $n$-point correlator become differentiable as a function of $r$? Is that typical scale the same as the one exhibited by $S_p(r)$ itself, or does it depend on $n$ and on the remaining distances $R$?

The answer is that there is a multiplicity of corssover scales. For the $n$-point correlator discussed above we denote the dissipative scale as $\eta(p, n, R)$, and it depends on each of its arguments [20,17]. In particular it depends on the inertial range variables $R$ and this dependence must be known when one attempts to determine the scaling exponents $\zeta_n$ of the structure functions. In brief, this line of thoughts leads to the scaling relation $\mu = 2 - \zeta_6$ (in close agreement with the experimental values) and the prediction that although the $\zeta_n$ are not K41, they are nevertheless linear in $n$ for large $n$.

In summary, it appears that there are four conceptual steps in the construction of a theory of the universal anomalous statistics of turbulence on the basis of the Navier-Stokes equations. First one needs to take care of the sweeping interactions that mask the scale invariant theory [12,14]. After doing so the perturbation expansion converges order by order, and K41 scaling of the velocity structure functions is found as a perturbative solution. Secondly one understands the appearance of the inner renormalization scale in the theory of the correlation functions of the gradient fields [18,15]. This step is similar to critical phenomena and it leads to a similarly rich theory of anomalous behaviour of the gradient fields. Only the tip of the iceberg was considered above. The third step is the understanding of the criticality of the theory, which results in the divergence of the diagrammatic series for the structure functions [16]. This sheds light on the appearance of the outer renormalization scale in the theory of turbulence. Last but not least is the formulation of the fusion rules and the multiplicity of the dissipative scales which should eventually result in a satisfactory description of all the scaling properties [17]. The road ahead is not fully charted, but it seems that some of the conceptual difficulties have been surmounted. We believe that the crucial building blocks of the theory are now available, and they begin to delineate the structure of the theory.

# REFERENCES


[1] L.F. Richardson, *Weather Prediction by Numerical Process*, (Cambridge, 1922).

[2] A.N. Kolmogorov, Dokl. Acad. Nauk SSSR, **30**, 9 (1941).

[3] A. S. Monin and A. M. Yaglom. *Statistical Fluid Mechanics: Mechanics of Turbulence*, volume II. (MIT Press, Cambridge, Mass., 1973).

[4] F. Anselmet, Y. Gagne, E.J. Hopfinger, and R.A. Antonia, J. Fluid Mech. **140**, 63 (1984).

[5] K.R. Sreenivasan and P. Kailasnath. Phys. Fluids A **5**, 512, (1993).

[6] R. Benzi, S. Ciliberto, R. Tripiccione, C. Baudet, F. Massaioli and S. Succi, Phys. Rev. E **48**, R29 (1993).

[7] Uriel Frisch. *Turbulence: The Legacy of A.N. Kolmogorov.* (Cambridge University Press, Cambridge, 1995).

[8] H. W. Wyld. Ann. Phys., **14**, 143 (1961).

[9] V. Yakhot and S.A. Orszag, Phys.Rev. Lett **57**,1722 (1986).

[10] C-Y. Mou and P. B. Weichman. Phys.Rev E **52**, 3738 (1995).

[11] R.H. Kraichnan, Phys. Fluids, **8**, 575 (1965).

[12] V. I. Belinicher and V. S. L'vov. Sov. Phys. JETP, **66**, 303 (1987).

[13] V.S. L'vov and I. Procaccia "Exact Resummation in the Theory of Hydrodynamic Turbulence": 0. Line Resummed Diagrammatic Perturbation Approach" in F. David P. Ginsparg, and J. Zinn-Justin, eds. Les Houches session LXII, 1994, Fluctuating Geometries in Statistical Mechanics and Field Theory (Elsevier, 1995).

[14] V.S. L'vov and I. Procaccia, "Exact Resummation in the Theory of Hydrodynamic Turbulence": I. The Ball of Locality and Normal Scaling",Phys. Rev. E, 52, 3840 (1995).

[15] V.S. L'vov and I. Procaccia, "Exact Resummation in the Theory of Hydrodynamic Turbulence": II. The Ladder to Anomalous Scaling",Phys. Rev. E, 52, 3858 (1995).

[16] V.S. L'vov and I. Procaccia, "Exact Resummation in the Theory of Hydrodynamic Turbulence": III.Scenarios for Multiscaling and Intermittency. Phys.Rev. E, in press.

[17] V.S. L'vov and I. Procaccia, "Exact Resummation in the Theory of Hydrodynamic Turbulence":IV. Fusion Rules and the Multiplicity of Viscous Scales", Phys.Rev E, to be published.

[18] V.S. L'vov and V. Lebedev, Europhys. Lett. **29**, 681 (1995).

[19] V.S. L'vov and I. Procaccia, Phys.Rev. Lett. **74**, 2690 (1995) .

[20] V.S. Lvov and I. Procaccia, Fusion Rules in Turbulent Systmes with Flux Equilibrium, Phys. Rev. Lett. Submitted.

# CHIRAL POTTS MODEL: EXACT RESULTS AND IMPLICATIONS

HELEN AU-YANG and JACQUES H.H. PERK
*Department of Physics, Oklahoma State University,*
*145 Physical Sciences, Stillwater, OK 74078-3072, USA*
E-mail: perk@okstate.edu

Since the original discovery of chiral Potts models, integrable by the star-triangle equations in terms of higher-genus curves, much progress has been achieved. The critical exponents and certain scaling functions are exactly known, together with exact results for free energies, order parameters, and interfacial tensions. There are also several ways to reformulate these models, for example as a composition of chiral clock models. Among such composite models, a new model with only two parameters is proposed for further numerical studies, which is not very different from the integrable chiral Potts model in certain regimes.

## 1 Introduction

### *1.1 Chiral Versus Next-Nearest-Neighbor Interactions*

The Chiral Clock Model was first introduced by Ostlund and others [1] as a model for commensurate–incommensurate phase transitions. Here the word chirality, or handedness is used after the Greek word $\chi\epsilon\acute{\iota}\rho$ meaning hand. As the left hand differs from the right hand, so Ostlund's model has manifestly broken parity built in, to which one may want to object as the microscopic interactions are parity invariant.

One may say that in view of this objection one should consider the other class of models for commensurate–incommensurate phase transitions in terms of further neighbor interactions[2] including the ANNNI model.[3] However, it has been noted [4] that the ANNNI model is equivalent to a four-state chiral model and, more generally, [5] that — upon using exact block-spin transformations to reduce the range of the interactions — usually the parity operation mixes the internal degrees of freedom and the equivalent nearest-neighbor interaction energy is not manifestly parity invariant. Such an equivalence also explains why the Pearce–Scott model [6] may be a good effective model for the ripple phase in biomembranes.

From the viewpoint of integrability and many other theoretical approaches, nearest-neighbor interactions are more local and preferred. Therefore, based on the above, from now on we will consider nearest-neighbor models on a square lattice with chiral interactions.

The chiral clock model is defined by the interaction energy

$$\mathcal{E} = - \sum_{\substack{\langle n,n' \rangle \\ \text{horizontal}}} J \cos\left[\frac{2\pi}{N}(n - n' + \Delta)\right] - \sum_{\substack{\langle n,n' \rangle \\ \text{vertical}}} \bar{J} \cos\left[\frac{2\pi}{N}(n - n' + \bar{\Delta})\right], \quad (1)$$

where $n, n' = 1, \cdots, N$ enumerate the states on pairs of neighboring sites on the lattice. Ostlund's original model[1] has $\bar{\Delta} = 0$.

If the number of states per site $N \geq 4$, using single cosines in (1) does not describe the generic case, and we should generalize this model to the Chiral Potts Model having general pair interactions depending on the differences of spin states modulo $N$.

### *1.2 Supersolvability Alias Superintegrability*

It is well-known that the two-dimensional Ising model and the corresponding transverse Ising chain are "supersolvable." For this $N = 2$ model one can calculate many properties explicitly. Its correlation functions satisfy Painlevé II, III, V, VI, $\cdots$, and various Toda lattice type equations.[7] Are there similar models for $N > 2$?

The first evidence for an affirmative answer to this question is the work of Howes, Kadanoff, and Den Nijs[8] for a special case of the $N = 3$ chiral clock quantum spin chain. They conjectured linear mass gaps and a simple Onsager-Yang type formula for the spontaneous magnetization. Von Gehlen and Rittenberg[9] discovered a generalization for $N \geq 2$ with the exact linear mass gap property for small chains using the Dolan-Grady criterium which is now understood to be equivalent to Onsager's 1944 loop algebra relations.[10] Their model is contained in the integrable chiral Potts model[11] as the so-called superintegrable[12] case, which is both Yang-Baxter (star-triangle) integrable[13] *and* Onsager algebra integrable.

### *1.3 Quantum Lax Pair Alias Star-Triangle Equation*

The discovery of the integrable chiral Potts models was started with the simple observation that the quantum Lax pair equations[14] are solved by inspection — zero conditions to be satisfied — for the $N = 3$ self-dual chiral clock model.[11] This was followed by the discovery of a genus 10 curve for the non-self-dual $N = 3$ case[11] and the conjecture of a product formula for the $N$-state self-dual case,[15] generalizing the non-chiral result of Fateev and Zamolodchikov.[16] The explicit general solution of the star-triangle equations was constructed in collaboration with Baxter.[13] A detailed account exists to which we refer the reader.[17]

## 2 Integrable Chiral Potts Model

The first ingredient in the construction [13] of the integrable chiral Potts model is the appearance of cyclic $q$-products, $q^N = 1$, [12]

$$f(n) = f(0)\prod_{j=1}^{n}\Big(\frac{\omega x_1 - x_2\omega^j}{x_4 - x_3\omega^j}\Big) \equiv f(n+N), \quad \omega \equiv e^{2\pi i/N}, \quad i \equiv \sqrt{-1}. \qquad (2)$$

This is only consistent if the "Fermat relation"

$$x_1^N + x_3^N = x_2^N + x_4^N. \qquad (3)$$

holds, whence the first appearance of higher-genus curves in the solution of the Yang-Baxter equations. From the linear recurrence relation [13] for the product weights (2) we obtain again such a product form after duality transformation, which here is given by the discrete Fourier transform

$$\hat{f}(m) \equiv \sum_{n=0}^{N-1}\omega^{mn} f(n) = \hat{f}(0)\prod_{j=1}^{m}\Big(\frac{x_4 - x_1\omega^j}{x_3 - x_2\omega^j}\Big). \qquad (4)$$

The other ingredient in the construction [13] is Onsager's elliptic function parametrization of the Boltzmann weights of the two-dimensional Ising model

$$\begin{aligned}\frac{W_{pq}(1)}{W_{pq}(0)} &= e^{-2J/k_BT} = \frac{\mathrm{cn}(q) - \mathrm{sn}(p)\mathrm{dn}(q)}{\mathrm{cn}(p) - \mathrm{dn}(p)\mathrm{sn}(q)} = \frac{\mu_p}{\mu_q}\frac{y_q + x_p}{y_p + x_q}, \\ \frac{\bar{W}_{pq}(1)}{\bar{W}_{pq}(0)} &= e^{-2\bar{J}/k_BT} = \frac{k'\big(\mathrm{sn}(p) - \mathrm{sn}(q)\big)}{\mathrm{dn}(p)\mathrm{cn}(q) + \mathrm{cn}(p)\mathrm{dn}(q)} = \mu_p\mu_q\frac{-x_p + x_q}{y_q + y_p}, \end{aligned} \qquad (5)$$

with sn, cn, and dn the usual elliptic functions of modulus $k$,

$$k' = \sqrt{1-k^2}, \quad 1/k' = \sinh(2J/k_BT)\sinh(2\bar{J}/k_BT), \qquad (6)$$

and $p$ and $q$ representing spectral variables [17] on the elliptic curve. We can also represent point $p$ (and similarly $q$) on this curve by

$$x_p = -\sqrt{k}\,\mathrm{sn}(p), \quad y_p = \sqrt{k}\,\mathrm{cn}(p)/\mathrm{dn}(p), \quad \mu_p = \sqrt{k'}/\mathrm{dn}(p). \qquad (7)$$

Even though (5) corresponds to products with only one factor, one can obtain the general form [13] of the Boltzmann weights

$$\begin{aligned} W_{pq}(n) &= W_{pq}(0)\prod_{j=1}^{n}\Big(\frac{\mu_p}{\mu_q}\Big)\Big(\frac{y_q - x_p\omega^j}{y_p - x_q\omega^j}\Big), \\ \bar{W}_{pq}(n) &= \bar{W}_{pq}(0)\prod_{j=1}^{n}(\mu_p\mu_q)\Big(\frac{\omega x_p - x_q\omega^j}{y_q - y_p\omega^j}\Big), \end{aligned} \qquad (8)$$

with $n$ the difference of two neighboring spins $n', n'' = 1, \cdots, N$. From the periodicity modulo $N$ of these weights we have a higher genus curve given by[a]

$$\mu_p^N = k'/(1 - kx_p^N) = (1 - ky_p^N)/k', \quad x_p^N + y_p^N = k(1 + x_p^N y_p^N), \tag{9}$$

determining point $p$. The variables $\mu_p$, $\mu_q$ in (8) are gauge variables and often can be eliminated, possibly at the expense of the periodicity property.

It can be proved[17] that the weights (8) for each triple of points $\{p, q, r\}$ on this curve satisfy the star-triangle equation

$$\sum_{d=1}^{N} \bar{W}_{qr}(b-d)\, W_{pr}(a-d)\, \bar{W}_{pq}(d-c) = R_{pqr}\, W_{pq}(a-b)\, \bar{W}_{pr}(b-c)\, W_{qr}(a-c), \tag{10}$$

where the constant

$$R_{pqr} = \frac{f_{pq} f_{qr}}{f_{pr}}, \quad f_{pq}{}^N \equiv \prod_{m=1}^{N} \sum_{k=1}^{N} \omega^{mk} \bar{W}_{pq}(k) \Big/ \prod_{m=1}^{N} W_{pq}(m), \tag{11}$$

plays an important role in the further development of the theory.

## 3 Selected Exact Results

For $N > 2$, there are two physical regimes within the integrable chiral Potts model. The first regime[13,17,18] has real positive Boltzmann weights. To be more specific, if

$$y_p = \omega^{1/2} e^{i\theta_p}, \quad 0 < \theta_p < \theta_q < \theta_r < \pi/N, \quad (0 < k, k' < 1), \tag{12}$$

for a triple $\{p, q, r\}$ of points on curve (9) with suitable choices of the branches of the $N$th roots, the six sets of weights in the star-triangle equation (10) are all positive. The 3-state antiferromagnetic Potts model on the square lattice at zero temperature is not of this form as its dual has unphysical weights; it is included, however, as an isolated point of a family with alternating spectral parameters $p$, $p'$, $q$, $q'$, which generically has some negative Boltzmann weights.

The second regime corresponds to hermitean quantum spin chains,[11,13,17] obtained by applying the limit "$q \to p$" to the logarithmic derivative of the transfer matrix with respect to the variable $q$ parametrizing the commuting family. The simplest hermitean Hamiltonian, generalizing the transverse-field

[a] In the original work,[13] homogeneous variables were used, rather than the more economic affine variables $x_p = a_p/d_p$, $y_p = b_p/c_p$, $\mu_p = d_p/c_p$.

Ising chain, is given by $k'$ real and ${y_p}^N = \pm {x_p^*}^N$, with + or − if $k$ is real or imaginary and the asterisk denoting complex conjugation. More complicated Hamiltonians require periodically varying spectral parameters. This second regime includes the superintegrable cases. We note that only for the zero-chirality case of Fateev and Zamolodchikov [16] do the two regimes coincide. So the two regimes describe in general different physics, with phase transitions driven by level crossings [19] in the quantum chain case that are forbidden in the first regime by the Perron-Frobenius theorem.

The simplest exact result for the integrable chiral Potts model is a general conjecture for the order parameters in the ordered state, [12]

$$\langle\sigma_0^n\rangle = (1-{k'}^2)^{\beta_n}, \quad \beta_n = \tfrac{n(N-n)}{2N^2}, \quad (1 \le n \le N-1, \quad \sigma_0^N = 1), \tag{13}$$

which still remains to be proved. There are also several exact results on free energies, [20] ground state energies, and excitation energies, [19] based on exact functional equations for transfer matrix eigenvalues. [21,22] From these results, we find for example the critical exponent $\alpha = 1 - 2/N$.

From Baxter's recent work [23] we can extract a very explicit formula [18] for the interfacial tensions in the fully symmetric, $W \equiv \bar{W}$, integrable chiral Potts model, i.e.

$$\frac{\epsilon_r}{k_{\mathrm{B}}T} = \frac{8}{\pi}\int_0^\eta dy \, \frac{\sin(\pi r/N)}{1+2y\cos(\pi r/N)+y^2} \, \mathrm{artanh}\sqrt{\frac{\eta^N - y^N}{1-\eta^N y^N}}, \tag{14}$$

where $\eta \equiv [(1-k')/(1+k')]^{1/N}$ is a temperature-like variable and $r$ is the difference of the spin variables across the interface, $(r = 1, \cdots, N-1)$.

In the low-temperature region, $k' \to 0$, we can expand (14) as [18]

$$\frac{\epsilon_r}{k_{\mathrm{B}}T} = -\frac{2r}{N}\log\frac{k'}{2} - \log\left(\frac{1}{\cos^2 r\bar{\lambda}} \prod_{j=1}^{[r/2]} \frac{\cos^4(r-2j+1)\bar{\lambda}}{\cos^4(r-2j)\bar{\lambda}}\right) + \mathrm{O}(k'), \tag{15}$$

where $\bar{\lambda} \equiv \pi/(2N)$ and the constant term comes from a dilogarithm integral. For the critical region, $\eta \approx (k/2)^{2/N} \sim (T_{\mathrm{c}} - T)^{1/N} \to 0$, (14) reduces to [18]

$$\begin{aligned} \frac{\epsilon_r}{k_{\mathrm{B}}T} &= \frac{8\sin(\pi r/N)B(1/N,1/2)}{\pi(N+2)}\eta^{1+N/2} \\ &- \frac{8\sin(2\pi r/N)B(2/N,1/2)}{\pi(N+4)}\eta^{2+N/2} + \mathrm{O}(\eta^{3+N/2}) \\ &\approx \eta^{N\mu}D_r(\eta) = \eta^{N\mu}D_r(\Delta/\eta^{N\phi}), \end{aligned} \tag{16}$$

with $B(x,y)$ the Beta function, $D_r$ a scaling function, and $\Delta \sim (T_c - T)^{1/2}$ the chiral field strength on the integrable line, giving the critical exponents

$$\mu = \tfrac{1}{2} + \tfrac{1}{N} = \nu, \quad \phi = \tfrac{1}{2} - \tfrac{1}{N}. \tag{17}$$

These results are for the case with real positive Boltzmann weights. For the spin chain case there is directional dependence and one seems to be led to $\nu_1 = 1$, related to the linear mass gaps, and $\nu_2 = 2/N$. [12,20] The results for the cases of positive Boltzmann weights and hermitean spin chains agree with analytic continuation of results for the superintegrable case (with non-positive Boltzmann weights) taking different boundary conditions, [20,23,18] and obey the hyperscaling relation $2 - \alpha = \nu_1 + \nu_2$. However, in the quantum chain case, this describes only a critical point for the analytic continuation of the ordered phase; the true ground state is lower due to level crossings. [19]

## 4 Generalizations and Reformulations

The chiral Potts model can be reformulated in many ways, as spin models with pair interactions or with three-spin interactions, as interaction-around-a-face (IRF) model, vertex model, or generalized Ashkin-Teller model. All this structure is closely related to fusion constructions for affine quantum groups at roots of unity to obtain functional relations for transfer matrices, [21,22] and we shall present a detailed account of this elsewhere.

There are several generalizations. The first one is the sl($n$) chiral Potts model, which has been reinterpreted as a three-dimensional model solving the tetrahedron equations. [24]

The large-$N$ limit of the chiral Potts model has been worked out, [25] which may be of interest in the context of conformal field theories based on $W_\infty$ algebras. Also, there are several chiral Potts models that can be viewed as multicomponent models. [26,27] We may have only seen the tip of an iceberg of new solvable models.

## 5 New Model for Further Study

The interaction energy of the symmetric integrable chiral Potts model with real positive Boltzmann weights is, within very small corrections, [18] equal to

$$-\frac{\mathcal{E}(n)}{k_{\mathrm{B}}T} = K \sum_{j=1}^{N-1} \frac{\sin(\pi/N)}{\sin(\pi j/N)} \cos\frac{2\pi}{N}\Big(nj + (N-2j)\Delta\Big), \tag{18}$$

with $K$ and $\Delta$ related variables. Relaxing this condition, (18) is giving us a new model with only two free parameters, [18] suitable for numerical studies.

## 6 Conclusions

To our knowledge, the chiral Potts model is the first of a relatively new class of exactly solvable models based on higher genus curves, which provide us with a natural generalization of Zamolodchikov's solution of the tetrahedron equations and new structures relating to cyclic representations of affine quantum groups and cyclic basic hypergeometric series.[28] Many physical quantities have been evaluated already and further progress is expected in the understanding of commensurate–incommensurate transitions.

## Acknowledgments

We thank Professor Hao Baolin and the STATPHYS 19 Organizing Committee for their kind hospitality. We also thank Professor R.J. Baxter for valuable correspondence. This work has been supported in part by NSF Grants No. PHY-93-07816 and PHY-95-07769.

## References

1. S. Ostlund, *Phys. Rev.* B **24**, 398 (1981); D.A. Huse, *Phys. Rev.* B **24**, 5180 (1981); J.M. Yeomans and M.E. Fisher, *J. Phys.* C **14**, L835 (1981).
2. A. Yoshimori, *J. Phys. Soc. Japan* **14**, 807 (1959); J. Villain, *J. Chem. Phys. Solids* **11**, 303 (1959); T.A. Kaplan, *Phys. Rev.* **116**, 888 (1959).
3. R.J. Elliott, *Phys. Rev.* **124**, 346 (1961); J. von Boehm and P. Bak, *Phys. Rev. Lett.* **42**, 122 (1979); W. Selke and M.E. Fisher, *Phys. Rev.* B **20**, 257 (1979); M.E. Fisher and W. Selke, *Phys. Rev. Lett.* **44**, 1502 (1980); W. Selke, *Phys. Repts.* **170**, 213 (1988).
4. M.N. Barber, *J. Phys.* A **15**, 915 (1982); D.A. Huse and M.E. Fisher, *J. Phys.* C **15**, L585 (1982).
5. H. Au-Yang and J.H.H. Perk, *Physica* A **177**, 139 (1991).
6. W.S. McCullough, J.H.H. Perk, and H.L. Scott, *J. Chem. Phys.* **93**, 6070 (1990).
7. H. Au-Yang and J.H.H. Perk, *Physica* A **144**, 44 (1987).
8. S. Howes, L.P. Kadanoff, and M. den Nijs, *Nucl. Phys.* B **215**[FS7], 169 (1983).
9. G. von Gehlen and V. Rittenberg, *Nucl. Phys.* B **257**[FS14], 351 (1985).
10. J.H.H. Perk, *Proc. Symp. Pure Math.* **49**, part 1, 341 (1989); B. Davies, *J. Math. Phys.* **32**, 2945 (1991).
11. H. Au-Yang, B.M. McCoy, J.H.H. Perk, S. Tang, and M.L. Yan, *Phys. Lett.* A **123**, 219 (1987).

12. G. Albertini, B.M. McCoy, J.H.H. Perk, and S. Tang, *Nucl. Phys.* B **314**, 741 (1989).
13. R.J. Baxter, J.H.H. Perk, and H. Au-Yang, *Phys. Lett.* A **128**, 138 (1988).
14. Yu.A. Bashilov and S.V. Pokrovsky, *Comm. Math. Phys.* **76**, 129 (1980); F.C. Alcaraz and A. Lima Santos, *Nucl. Phys.* B **275**[FS17], 436 (1986).
15. H. Au-Yang, B.M. McCoy, J.H.H. Perk, and S. Tang, in *Algebraic Analysis*, Vol. 1, eds. M. Kashiwara and T. Kawai, (Academic, San Diego, 1988), p. 29.
16. V.A. Fateev and A.B. Zamolodchikov, *Phys. Lett.* A **92**, 37 (1982).
17. H. Au-Yang and J.H.H. Perk, *Adv. Stud. Pure Math.* **19**, 57 (1989).
18. H. Au-Yang and J.H.H. Perk, *J. Stat. Phys.* **78**, 17 (1995).
19. G. Albertini, B.M. McCoy, and J.H.H. Perk, *Phys. Lett.* A **135**, 159 (1989), **139**, 204 (1989), *Adv. Stud. Pure Math.* **19**, 1 (1989); B.M. McCoy and S.-S. Roan, *Phys. Lett.* A **150**, 347 (1990).
20. R.J. Baxter, *J. Stat. Phys.* **52**, 639 (1988), **57**, 1 (1989), and in press.
21. V.V. Bazhanov and Yu.G. Stroganov, *J. Stat. Phys.* **59**, 799 (1990).
22. R.J. Baxter, V.V. Bazhanov, and J.H.H. Perk, *Intern. J. Mod. Phys.* B **4**, 803 (1990).
23. R.J. Baxter, *J. Stat. Phys.* **73**, 461 (1993), *J. Phys.* A **27**, 1837 (1994); M.J. O'Rourke and R.J. Baxter, *J. Stat. Phys.*, (in press).
24. V.V. Bazhanov and R.J. Baxter, *J. Stat. Phys.* **69**, 453 (1992), **71**, 839 (1993); R.M. Kashaev, V.V. Mangazeev, and Yu.G. Stroganov, *Intern. J. Mod. Phys.* A **8**, 587 (1993); S.M. Sergeev, V.V. Mangazeev, and Yu.G. Stroganov, *J. Stat. Phys.*, (in press); Z.-N. Hu and B.-Y. Hou, *J. Stat. Phys.*, (in press).
25. H. Au-Yang and J.H.H. Perk, *Intern. J. Mod. Phys. (Proc. Suppl.)* **3**A, 430 (1993).
26. H. Au-Yang and J.H.H. Perk, *Intern. J. Mod. Phys.* A **7**, Suppl. 1B, 1007, 1025 (1992).
27. T. Miwa, *Nucl. Phys.* B **270**[FS16], 50 (1986).
28. H. Au-Yang and J.H.H. Perk, *Intern. J. Mod. Phys., (Proc. of Nankai Satellite of STATPHYS 19)*, to be published.

# Numerical Simulations of Quantum Phase Transitions

A. P. Young
*Physics Department,*
*University of California, Santa Cruz, CA 95064*

**Abstract**

In this talk I will describe some examples of quantum phase transitions with disorder and explain in what ways these transitions differ from conventional phase transitions at finite temperature. It will be explained that the $\epsilon$-expansion does not work and so much of what we know has come from numerical simulations. A novel feature is the importance of "Griffiths singularities", which come from rare, strongly correlated, regions of the system. Some recent results on the quantum spin glass will be discussed in a little more detail.

## 1 Introduction

Quantum phase transitions occur at $T = 0$, driven by quantum, rather than thermal, fluctuations as some parameter in the system is varied. Until recently, such transitions have received less attention than the more conventional "thermal" phase transitions where one goes through the transition by varying the temperature. However, there has recently been an upsurge of interest in quantum phase transitions, partly because there are experiments, and partly because, *in the presence of quenched disorder*, such transitions exhibit new physics which is rather poorly understood.

Let us first discuss why one can generally set $\hbar = 0$ in treatments of critical phenomena. The thermal energy per degree of freedom is $k_B T$, where $k_B$ is Boltzmann's constant, while the zero-point quantum of energy is $\hbar\omega_c$, for a fluctuation of characteristic frequency $\omega_c$. As one approaches a phase transition, the correlation length, $\xi$, generally diverges with a power law, *i.e.*

$$\xi \sim \delta^{\nu} , \tag{1}$$

where $\delta$ is the deviation from criticality. As the critical point is approached, not only do the fluctuations extend over large spatial scales, they also become very slow, and there is a divergent relaxation time $\tau$. This effect is known as critical slowing down. It is conventional [1] to define a dynamical exponent, $z$, by

$$\tau \sim \xi^{z} . \tag{2}$$

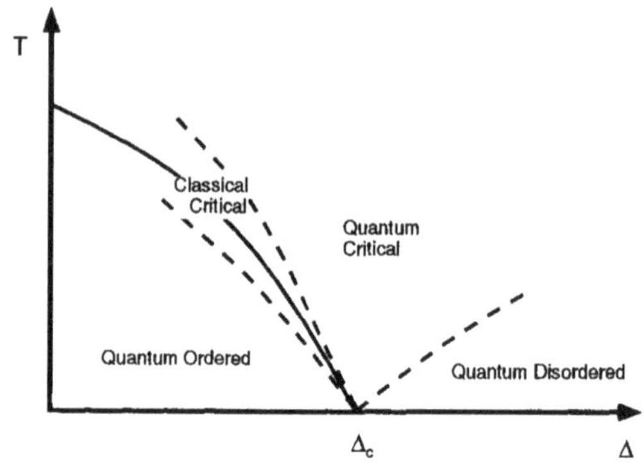


Figure 1: A sketch of the various regions which occur when a parameter $\Delta$ drives $T_c$ to zero. The phase boundary is indicated by the solid line. The dashed lines indicate regions where crossovers occur between different types of behavior. The region of classical critical behavior shrinks to zero as $T_c \to 0$. The quantum universality class is obtained as one goes through the quantum critical point at $T = 0, \Delta = \Delta_c$ from the quantum disordered to the quantum ordered regime.

The appropriate quantum of energy is then $\hbar\omega_c$ with $\omega_c \sim 1/\tau$. Hence, if the transition is at a finite temperature, $T_c$, quantum mechanics is irrelevant for critical phenomena, *i.e.* $\hbar\omega_c \ll k_B T$, provided

$$\delta \ll T_c^{1/z} , \tag{3}$$

which is *always* satisfied sufficiently close to $T_c$, *provided* $T_c$ *is finite*. In this regime one observes the critical behavior of the classical (thermal) transition.

Note, however, that the lower is $T_c$, the closer to criticality one has to be to observe classical critical behavior. If $T_c$ is forced to zero by changing some parameter in the Hamiltonian, and the transition at $T = 0$ is traversed by changing this parameter, then one is *never* in the classical regime. Instead, one observes critical behavior of a *quantum universality class* which is different, in general from the *classical universality class* of the finite $T_c$ transition. Fig. 1 sketches a typical phase diagram [2], in which $\Delta$ denotes the parameter which drives $T_c$ to zero. One sees that the regime of classical critical behavior shrinks to zero as $T_c \to 0$. The new quantum universality class is observed in the quantum disordered and quantum ordered regimes. In the intermediate regime (sometimes called the quantum critical regime) the only energy scale is $k_B T$ and the there are interesting predictions [3] for the dynamical scaling functions which turn out to depend only on the ratio $\hbar\omega/k_B T$. I shall not, however, discuss this regime here.

It is well known [4] that the quantum problem can be described in terms of an effective classical action in $d$ space plus one imaginary time dimensions. The size of the system in the imaginary time direction is $\hbar/k_B T$. In this picture it is easy to see why quantum mechanics is only important at $T = 0$. At any finite temperature we are dealing with a slab, infinite in the space directions but finite in the time direction. When the correlation "length" in the time direction reaches the thickness of the slab, the growth of correlations with time becomes limited by the slab thickness, and can be described by the techniques of finite size scaling [5]. The critical behavior is then just controlled by the diverging correlation length in the space directions, *i.e.* we have a $d$-dimensional classical problem. However, when $T$ is zero, the size of the system in the time direction is infinite, just as it is in the space directions, so we have to study a $d+1$ dimensional problem, which will generally have different critical behavior from that of the $d$-dimensional model. Thus the quantum to classical crossover is similar to a dimensional crossover. Note that for systems with quenched disorder, the disorder is independent of (imaginary) time and so is *perfectly correlated* in the time direction.

In simple cases, and with no disorder, one ends up with a classical model in one higher dimension [6], in which space and time are equivalent. Another way of saying this is that the dynamical exponent, $z$, is equal to one. However, in the cases of greatest interest, *i.e.* those with quenched disorder, space and time are *not* equivalent, so usually $z \neq 1$. This complicates the analysis of numerical data because scaling is *anisotropic*. Data analysis then has to be done using anisotropic finite-size scaling, see *e.g.* [7, 8].

In the rest of this talk I will give some examples of quantum phase transitions with disorder, discuss what new physics arises in these systems, and summarize why this is rather poorly understood. Because analytical work has proved hard, much of what we know has come from simulations. I will discuss briefly some of the recent numerical work on a particular problem, the quantum spin glass.

# 2 Examples

The most commonly studied quantum phase transition is the metal-insulator transition [9]. In a disordered system, some or all of the electron states are *localized*, *i.e.* they have a finite amplitude to be on a site in a particular region of the lattice, and negligible amplitude to be outside this region. If the Fermi energy lies in a region of localized states, the system is an insulator. (At finite temperature however, conduction can occur because *inelastic* scattering processes, *e.g.* due to phonons, can kick an electron from one localized state to another.) In three and more dimensions, and for weak disorder, only the states at the edge of the band are localized, and there is a "mobility edge" separating these states from extended states at the center. If the degree of disorder, or the density of electrons, is changed, the Fermi energy can go from being in a region of extended states to a region of localized states: this is the metal-insulator transition. At a critical value of disorder, *all* the states become localized. It turns out that in two dimensions, all states are localized for *any* non-zero disorder, *i.e.* the critical value of disorder is zero.

Denoting by $\delta$ the deviation from criticality then the $T = 0$ conductivity is found to vanish on the metallic side like

$$\sigma \sim \delta^{\mu} , \tag{4}$$

and, of course, to be identically zero on the insulating side. One goal of theory is to calculate the exponent $\mu$ and compare with experiment. Numerical work [10] finds $\mu \simeq 1.4$ and the experimental values [10] range from about this down to much lower values. This wide range reflects the difficulty of extrapolating the data to $T = 0$ and possibly also indicates a rather narrow critical region (the region in which asymptotic scaling behavior is seen).

Analytical work has been developed for this model, based on a $2+\epsilon$ expansion [11], but this does not seem to converge well in $d = 3$. Unfortunately, there is no expansion away from an upper critical dimension (which, in the case of classical critical phenomena, is better behaved), and it is not known whether an upper critical dimension (above which the exponents would stick at some "mean field" value) even exists.

One difficulty with the metal-insulator transition is that there is no obvious order parameter. This is not the case for my second example, the superconductor-insulator transition in disordered thin films. I have already mentioned that there is no metallic state at $T = 0$ in a two-dimensional system with disorder because all the electron states are localized. However, superconductivity (which involves a coherent propagation of a *pair* of electrons) is not ruled out, though it too will presumably be eliminated if the disorder is strong enough. In 2-$d$ one therefore has the possibility of a direct transition from a superconductor to an insulator without going through an intervening metallic phase, as one changes the degree of disorder. The order parameter is just the amplitude of the pair condensate, as usual in a superconducting transition, and so one can formulate standard methods, such as Ginzburg-Landau theory.

An interesting prediction [12] is that the system *is* metallic (*i.e.* has a finite conductivity) precisely *at* the critical point and that the value of the conductivity there, $\sigma_c$, is *universal*. To be more precise,

$$\sigma_c = C\sigma_Q , \tag{5}$$

where $C$ is a universal constant and

$$\sigma_Q = \frac{(e^{\star})^2}{h} \tag{6}$$

is the quantum of conductance, with $e^{\star} = 2e$, the charge of a Cooper pair. A goal of the theory is to calculate this universal conductance as well as standard universal quantities such as critical exponents. Experimental estimates [13] of $\sigma_c$ vary over a range of about a factor of three, which, if correct, would mean that $\sigma_c$ is not universal after all, but it is possible that many of the experiments did not probe the asymptotic critical region, often very small in superconductors.

Although this model can be reformulated in a Ginzburg-Landau picture, an $\epsilon$-expansion about an upper critical dimension does not work [14], since the renormalization group flows go to strong coupling. Since the analytic renormalization group approach doesn't work, much of what we know has been learned from Monte Carlo simulations, *e.g.* [7].

My third and final example of a quantum phase transition is a magnetic system. Suppose we have an Ising magnet, where the spins lie up or down. These are eigenstates of the Pauli spin matrix $\sigma^z$, so the Hamiltonian can be written as

$$\mathcal{H} = -\sum_{\langle i,j\rangle} J_{ij}\sigma_i^z\sigma_j^z \ . \tag{7}$$

We shall assume that the interactions $J_{ij}$ are random and *frustrated*, *i.e.* have random signs so that no spin configuration simultaneously minimizes all the terms in the Hamiltonian. Such a system is called a *spin glass* [15]. Typically one chooses an Edwards-Anderson model, considering only nearest neighbor interactions which are independent random variables chosen from a symmetric distribution, *i.e.*

$$[J_{ij}]_{\rm av} = 0 \ ; \quad [J_{ij}^2]_{\rm av} = J^2 \ . \tag{8}$$

So far, all operators commute so one could just replace the spin operators by $c$-numbers, $\pm 1$, which is the usual way to present the Ising model. Next, however, add a transverse field which couples to $\sigma^x$ so the Hamiltonian becomes

$$\mathcal{H} = -\sum_{\langle i,j\rangle} J_{i,j}\sigma_i^z\sigma_j^z - \Gamma\sum_i \sigma_i^x \ . \tag{9}$$

The problem is now quantum mechanical since $\sigma^x$ and $\sigma^z$ don't commute. If the system has a finite-temperature transition with $\Gamma = 0$, then turning on $\Gamma$ will decrease $T_c$ since the transverse field cants the spins away from their ordering direction (the $z$-direction). If $\Gamma$ is big enough, $T_c$ can be depressed to zero and a quantum critical point results.

An experimental system for which Eq. (9) should be a good model has been studied in Ref. [16]. One advantage of this system is that one can tune through the quantum transition by varying the transverse field, a continuous parameter, whereas in the other systems I have mentioned one generally has to vary the degree of disorder, which cannot be changed continuously and for each data point one has to grow a new sample.

For the quantum spin glass, as for the other examples mentioned above, we do not have an $\epsilon$-expansion away from an upper critical dimension [17], so the only estimates for critical exponents, at least for short range interactions, have come from Monte Carlo simulations [8, 18] on the model described by Eq. (9). These will be described in section 4.

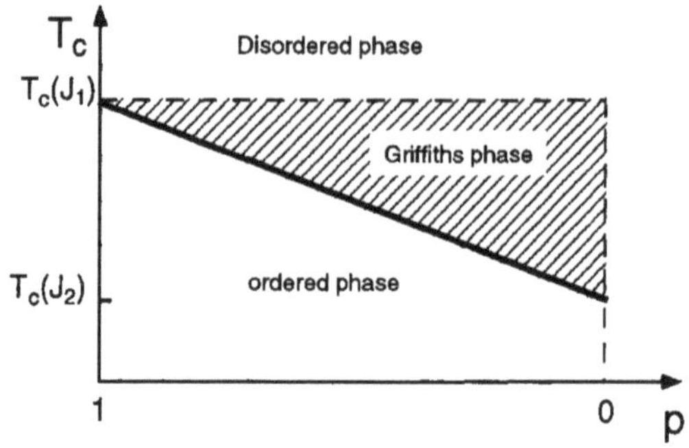


Figure 2: A sketch of the phase diagram of the model described by Eqs. (7) and (10), showing the Griffiths phase, where the free energy is a non-analytic function of magnetic field $h$. The phase boundary is indicated by the thick line. For this classical model the Griffiths singularity is an essential singularity, *i.e.* very weak. However, the corresponding singularity is much stronger for quantum transitions at $T = 0$.

# 3 Griffiths Singularities

In this section I will discuss an aspect of phase transitions in disordered systems which has no counterpart in pure systems. These effects, called Griffiths singularities [19], are rather weak at thermal phase transitions but can give surprising and spectacular effects in the quantum case.

To illustrate Griffiths singularities, consider first a simple *classical* spin model, as in Eq. (7), where the interaction can take one of two values, *i.e.*

$$J_{ij} = \begin{cases} J_1 & \text{prob. } p \\ J_2 & \text{prob. } 1-p \end{cases} \tag{10}$$

Assume that $J_1$ and $J_2$ are both positive, so there is no frustration, and $J_1 > J_2$. The resulting phase diagram is sketched in Fig. 2. The Griffiths phase is the region which would be below $T_c$ if all the interactions had their largest possible value ($J_1$ in this case) but is actually above the $T_c$ of the random system. Griffiths singularities arise in the Griffiths phase because there is a probability of having a large region with all $J_1$ bonds which would be locally ordered. Since an ordered system has a large response to a magnetic field, $h$, the free energy turns out not to be an analytic function of $h$ at $h = 0$ in the Griffiths phase [19]. The probability of having a large region with all $J_1$ bonds is exponentially small, and, as a result, the singularity is very weak at a thermal transition, just an essential singularity [20]: all derivatives of the free energy with respect to $h$ are finite.

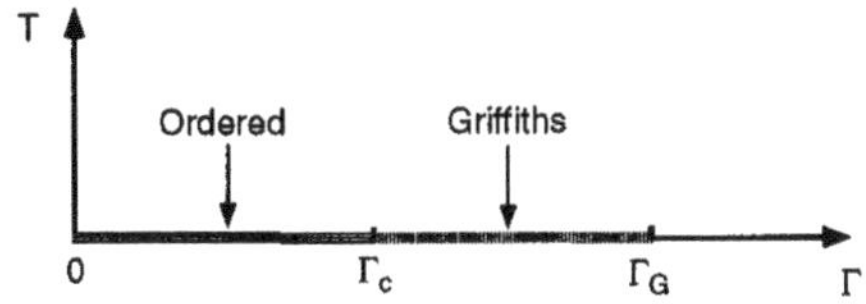


Figure 3: A sketch of the phase diagram of the one dimensional random transverse Ising model. The ordered phase only occurs at $T = 0$. In the Griffiths phase the free energy is a non-analytic function of magnetic field $h$, the singular part varying as in Eq. (11), where $\delta$ is a function of $\Gamma$ tending to unity as $\Gamma \to \Gamma_c^+$ and tending to infinity as $\Gamma \to \Gamma_G^-$.

Although the effects of Griffiths singularities on static properties are probably unobservably weak for thermal transitions, they have a much bigger effect on quantum transitions. For the one dimensional version of the quantum spin glass model of Eq. (9) these can be worked out in detail and are quite spectacular. This was first studied by McCoy and Wu [21] in the equivalent (1+1)-dimensional classical version, an Ising model with disorder which is perfectly correlated along one direction. Subsequently, new results and a great deal of insight were obtained by Fisher [22]. A one-dimensional system with short range interactions has no long range order except at $T = 0$, so the ordered region is entirely along the $\Gamma$ axis, as shown in Fig. 3. The system is ordered up to a critical value $\Gamma_c$. However, there is a finite region in the disordered phase, $\Gamma > \Gamma_c$ where the susceptibility is infinite. This is most unusual because the susceptibility normally diverges *at* the critical point. In fact the free energy is a non-analytic function of field, $h$, in the disordered region, up to $\Gamma = \Gamma_G$, the end of the Griffiths phase. $\Gamma_G$ is the value $\Gamma_c$ would have if all the interactions had their largest value. Unlike the essential singularity in the case of thermal transitions, one has a much stronger effect here:

$$f_{sing} \sim |h|^{\delta(\Gamma)} , \tag{11}$$

where $\delta$ is a function of $\Gamma$ tending to unity as $\Gamma \to \Gamma_c^+$ and tending to infinity as $\Gamma \to \Gamma_G^-$. The value of $\Gamma$ at which the susceptibility diverges is now seen to have no special significance, but is just the point where $\delta = 2$. Griffiths singularities also give striking effects in the dynamics. The dynamical exponent varies continuously in the Griffiths phase and *diverges* at the critical point [22]. This has recently been confirmed in numerical studies [23] using the representation of the problem in terms of non-interacting fermions [24].

These results occur because of rare regions which are more strongly correlated than average and are locally in the ordered phase. A rough argument, which applies for arbitrary $d$, goes as follows: The probability of finding a region of volume $L^d$ which is locally in the ordered phase is of order

$$P(L) \sim e^{-\mu L^d} , \tag{12}$$

where $\mu$ is some constant. Consider now the spins as a function of (imaginary) time in this region. To change the sign of the magnetization requires inserting a domain wall, whose energy is $\sigma L^d$, where $\sigma$ is the surface tension. Hence the relaxation time in this region is of order

$$\tau \sim e^{\sigma L^d} . \tag{13}$$

In other words, there is an exponentially small probability of having an exponentially large relaxation time. Consequently the distribution of relaxation times follows a power law,

$$P(\tau) \sim \frac{1}{\tau^{1+\frac{\mu}{\sigma}}} , \tag{14}$$

as does the local time dependent spin-spin correlation function,

$$\langle \sigma_i^z(t) \sigma_i^z(0) \rangle \sim \frac{1}{t^{\frac{\mu}{\sigma}}} , \tag{15}$$

where $t$ is imaginary time. The local susceptibility is obtained by integrating over $t$ and so diverges if $\mu/\sigma < 1$.

This argument, while crude, suggests that the form of the singularity in Eq. (11) should apply quite generally to the Griffiths phase at $T = 0$ in Ising systems, though the dependence of $\delta$ on $\Gamma$ presumably depends on the model. In particular, it is not clear whether the susceptibility will diverge in the Griffiths phase except in $d = 1$. It is in part to try to answer this question that numerical work on the quantum spin glass Ising Hamiltonian in Eq. (9) has been undertaken [8, 18]. This will be described in the next section.

# 4 Ising Spin Glass in a Transverse Field

Recently large scale simulations have been carried out on the quantum spin glass Hamiltonian in Eq. (9), both in two [8] and three [18] dimensions. As described in those references, the simulation involves an effective classical action in $d$ spin plus one imaginary time dimensions. The interactions in the space directions have a Gaussian distribution with zero mean and standard deviation of unity and are *independent* of imaginary time, while the interactions between the "time slices" are ferromagnetic and all equal to unity. Increasing the effective temperature in the classical action disorders the system and so corresponds to increasing the strength of the transverse fields.

The system in (9) is very anisotropic, so, according to a finite size scaling hypothesis [5] extended to anisotropic systems [25], thermodynamic quantities close to the critical point depend on *two* independent scaling variables, which we can take to be $\delta L^{1/\nu}$ and the aspect ratio $L_\tau / L^z$. The scaling analysis is straightforward only if it depends on a single parameter, so it is necessary to fix the aspect ratio. Since $z$ is unknown, one has to scan several different sample shapes to see which choice for $z$ scales best.

As in standard spin–glass theory [15], we define the overlap between the configurations of two replicas, 1 and 2, with the same disorder as

$$Q = \frac{1}{L^d L_\tau} \sum_{i,\tau} S_i^{(1)}(\tau) S_i^{(2)}(\tau) \quad , \tag{16}$$

and for each disorder realization we calculate the dimensionless combination of moments

$$g = 0.5\left[3 - \langle Q^4\rangle / \langle Q^2\rangle^2\right] \quad . \tag{17}$$

The disorder averaged quantity, $g_{\text{av}} = [g]_{\text{av}}$ obeys the finite size scaling form

$$g_{\text{av}}(K, L, L_\tau) = \tilde{g}_{\text{av}}(\ \delta L^{1/\nu},\ L_\tau/L^z\ ) \ , \tag{18}$$

and has the property [26] that it vanishes in the disordered phase for $L \to \infty$, and tends to a finite value in the ordered phase. Consequently, $\tilde{g}(x,y)$ vanishes at fixed $x$ both for $y \to 0$ (where the system is a classical two–dimensional spin glass at finite "temperature", which is disordered) as well as for $y \to \infty$ (where the system is effectively a long one–dimensional chain along the $\tau$ direction, which is also disordered). Hence, $\tilde{g}(x,y)$ must have a maximum at some value of $y$ for fixed $x$. We used this to estimate the "temperature" of the effective classical action which we find is given by $T_c^{cl} = 3.275 \pm 0.025$. Furthermore, at the critical point, the values of $L$ and $L_\tau$ for which $\tilde{g}$ is a maximum are related by $L_\tau \sim L^z$. By this method we determine the dynamical exponent and get $z = 1.50 \pm 0.05$.

One can also calculate the non-linear susceptibility, which is measured in experiments. Results for the dependence of $\chi_{nl}$ on $L$ at the critical point in Fig. 4 show that this divergenes as $L^{4.7}$, which, using finite-size scaling, translates into a divergence with $L_\tau$ ($\propto 1/T$) at fixed $L$ of $L_\tau^{4.7/z}$. Using $z \simeq 1.5$ this leads to the prediction that the non-linear susceptibility at the critical value of $\Gamma$ should diverge roughly like $1/T^3$. A similar strong divergence is also found in 3-d [18]. The reason for the discrepancy with experiment [16], where only a weak divergence was found, is unclear.

Recent studies have also been carried out [27] in the disordered phase of the 2-$d$ model to look at Griffiths singularities. The distributions of the local susceptibility and local non-linear susceptibility (i.e. the non-linear response of a spin on site i to a field on that site) show power law tails as expected from Eq. (14). The power decreases (showing that there are more long relaxation times) as one approaches the critical point. However, the effects of Griffiths singularities are less spectacular than in 1-$d$. For example, the dynamical exponent is finite at the critical point, as already noted above. Whereas in 1-$d$ Griffiths singularities seem to control the critical behavior, in 2-$d$ they seem to ride along on top of the critical behavior and simply contribute additional singularities in the scaling functions in the limit $L_\tau/L^z \to \infty$.

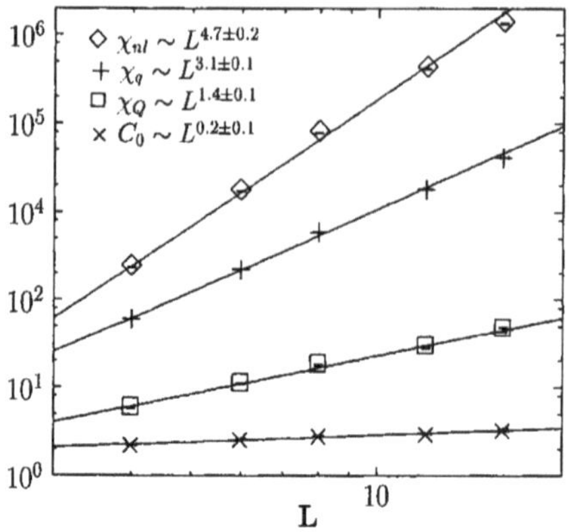


Figure 4: The diamonds show the non-linear susceptibility as a function of system size at the critical point, $T^{cl} = 3.30 \simeq T_c^{cl}$, on a double logarithmic plot. The slope is about 4.7, indicating a strong divergence. From ref. [8].

# 5 Conclusions

The physics of quantum phase transitions with disorder remains poorly understood compared with our rather complete understanding of thermal phase transitions. We have seen that non-perturbative effects (Griffiths singularities) have a significant effect, especially in one-dimension. The simulations seem of the critical behavior of the quantum spin glass seem to be under fairly good control, but indicate a large divergence in the non-linear susceptibility, in contrast to experiments. This discrepancy remains to be understood.

**Acknowledgements:** Much of this work has be done in collaboration with H. Rieger. I would also like to thank R. Bhatt and D. Huse for many extremely valuable suggestions and discussions. This work is supported by NSF DMR-94-11964.

# References


[1] P. C. Hohenberg and B. I. Halperin, Rev. Mod. Phys. **49**, 435 (1977).

[2] S. Chakravarty, B. I. Halperin and D. R. Nelson, Phys. Rev. Lett. **60**, 1057 (1988); Phys. Rev. B **39**, 2344 (1989).

[3] S. Sachdev and J. Ye, Phys. Rev. Lett. **69**, 2411 (1992).

[4] R. P. Feynman, *Statistical Mechanics*, (Addison-Wesley, 1972).

[5] V. Privman, in *Finite Size Scaling and Numerical Simulation of Statistical Systems*, ed. V. Privman (World Scientific, Singapore, 1990), p. 1.

[6] P. Pfeuty, Ann. Phys. (NY) **27**, 79 (1970).

[7] M. Wallin, E. S. Sørensen, S. M. Girvin and A. P. Young, Phys. Rev. B, **49**, 12115 (1994).

[8] H. Rieger and A. P. Young, Phys. Rev. Lett. **72**, 4141 (1994).

[9] P. A. Lee and T. V. Ramakrishnan, Rev. Mod. Phys. **57**, 287 (1985).

[10] B. Kramer and A. MacKinnon, Rep. Prog. Phys. **56**, 1469 (1993).

[11] E. Abrahams, P. W. Anderson, D. C. Licciardello, and T. V. Ramakrishnan, Phys. Rev. Lett. **42**, 673 (1979).

[12] M. P. A. Fisher, G. Grinstein, and S. M. Girvin, Phys. Rev. Lett. **64**, 587 (1990).

[13] Y. Liu, K. A. McGreer, B. Nease, D. B. Haviland, G. Martinez, J. W. Halley, and A. M. Goldman, Phys. Rev. Lett. **67**, 2068 (1991); Y. Liu and A. M. Goldman (submitted to Int. J. Mod. Phys. B); A. F. Hebard and M. A. Paalanen, Phys. Rev. Lett. **65**, 927 (1990).

[14] M. P. A. Fisher, P. B. Weichman, G. Grinstein, and D. S. Fisher, Phys. Rev. B **40**, 546 (1989).

[15] For a review on spin glasses see *e.g.* K. Binder and A. P. Young, Rev. Mod. Phys. **58**, 801 (1986).

[16] W. Wu, B. Ellmann, T. F. Rosenbaum, G. Aeppli and D. H. Reich, Phys. Rev. Lett. **67**, 2076 (1991); W. Wu, D. Bitko, T. F. Rosenbaum and G. Aeppli; Phys. Rev. Lett. **71**, 1919 (1993).

[17] N. Read, S. Sachdev and J. Ye (unpublished).

[18] M. Guo, R. N. Bhatt, and D. Huse, Phys. Rev. Lett. **72**, 4137 (1994).

[19] R. B. Griffiths, Phys.Rev.Lett., **23**, 17 (1969).

[20] A. B. Harris, Phys. Rev. B **12**, 203 (1975).

[21] B. M. McCoy and T. T. Wu, Phys. Rev. **176**, 631 (1968); **188**, 982 (1969); **188**, 1014 (1969).

[22] D. S. Fisher, Phys. Rev. Lett. **69**, 534 (1993); Phys. Rev. B, **51**, 6411 (1995).

[23] A. P. Young and H. Rieger (unpublished).

[24] E. Lieb, T. Schultz and D. Mattis, Ann. Phys. **16**, 407 (1961).

[25] K. Binder and J. S. Wang, J. Stat. Phys. **55**, 87 (1989).

[26] R. N. Bhatt and A. P. Young, Phys. Rev. B **37**, 5606 (1988).

[27] H. Rieger and A. P. Young (unpublished).

# Large Scale Simulations of Lattice Gas Models of Capillary Condensation and Wetting Transitions in Confined Fluids

D. P. LANDAU
*Center for Simulational Physics, The University of Georgia*
*Athens, GA 30602, USA*
E-mail: dlandau@uga.cc.uga.edu

and

K. BINDER
*Institut für Physik, Johannes Gutenberg-Universität Mainz, D-55099 Mainz, Germany*

ABSTRACT

In the presence of surface fields (walls), Ising magnets (lattice gases) serve as a laboratory for the study of wetting phenomena. We use Monte Carlo simulations to examine simple cubic lattices with two free $L \times L$ surfaces a distance $D$ apart. For large D we find a range of different wetting transitions, and in thin films we describe the capillary condensation and localization-delocalization transitions which occur for different combinations of surface fields. These results provide tests of the validity of theoretical predictions.

## 1. Introduction

When a saturated vapor in contact with a wall is raised above some characteristic "wetting temperature" a macroscopically thick liquid film may form at the wall. As shown in Fig. 1, a rich variety of phenomena may occur, including both first order and second order transitions, and a full understanding of the detailed nature of the behavior has been difficult to achieve. Since the pioneering work by Nakanishi and Fisher [1] drew the comparison between the problem of wetting and that of lattice gas (Ising) models subject to both bulk and surface fields, a variety of theoretical and simulational explorations of wetting phenomena have occured [2-14,16-18,22-27]. Wetting with short range forces poses exciting challenges to the theoretical understanding of interface unbinding phenomena in general, and Monte Carlo simulations can help to examine these issues. After first reviewing the nature of the wetting transitions which may occur in "thick films", we shall focus on the statistical mechanics of phase transitions in a thin film geometry, studying the interplay between confinement (i. e. a "finite size effect") and competing wall forces[2-5,11,12] (i.e. a "surface effect").

## 2. Model and simulation technique

We study the Ising hamiltonian on the simple cubic lattice in a $L \times L \times D$ geometry, applying periodic boundary conditions in $x$ and $y$ directions. At the two free $L \times L$

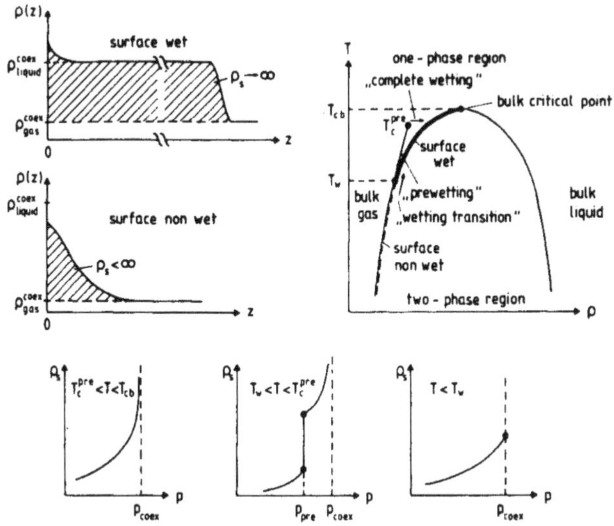


Fig. 1. (Upper left) Schematic profiles of the density as a function of the distance from the wall; (upper right)schematic phase diagram of a semi-infinite fluid in contact with a wall; (Bottom)schematic adsorption isotherms near different transitions.

surfaces at layer numbers $n = 1$ and $n = D$ fields $H_1$, $H_D$ are applied:

$$\mathcal{H} = -J \sum_{\langle i,j \rangle \epsilon \, bulk} S_i S_j - J_s \sum_{\langle i,j \rangle \epsilon \, surfaces} S_i S_j - H \sum_i S_i - H_{1(D)} \sum_{i \epsilon \, surfaces} S_i \tag{1}$$

where $S_i = \pm 1$. Typically $L$ varied from 32 to 128, but occasionally both smaller ($L = 16$) and larger ($L = 192, 256$) sizes were used. Thicknesses studied varied from $D = 6$ to $D = 160$.

In addition to standard bulk quantities we measured layer properties[6] including profiles of magnetization $m_n$, energy $U_n$, and the layer susceptibilities $\chi_n$ and $\chi_{nn}$

$$\chi_{nn} = L^2(\langle \sigma_n^2 \rangle - \langle \sigma_n \rangle^2)/k_B T \ , \ \chi_n = L^2 D(\langle \sigma_n M \rangle - \langle \sigma_n \rangle \langle M \rangle)/k_B T \ , \tag{2}$$

where $m_n = \langle \sigma_n \rangle$, $\sigma_n = \frac{1}{L^2} \sum_{i \epsilon \, layer \, n} S_i$ , $M = \frac{1}{L^2 D} \sum_i S_i$ .
Runs were typically $2 \times 10^5$ Monte Carlo steps/site, although near $T_c(D)$ considerably longer runs were also carried out, particularly for the larger choices of $L$. We also carried out several independent runs with different random numbers to obtain error estimates. In most cases, the error bars for the order parameter and energy are smaller than the size of the symbols; the errors in the fluctuation quantities, e.g. susceptibilities and specific heats, are often larger.

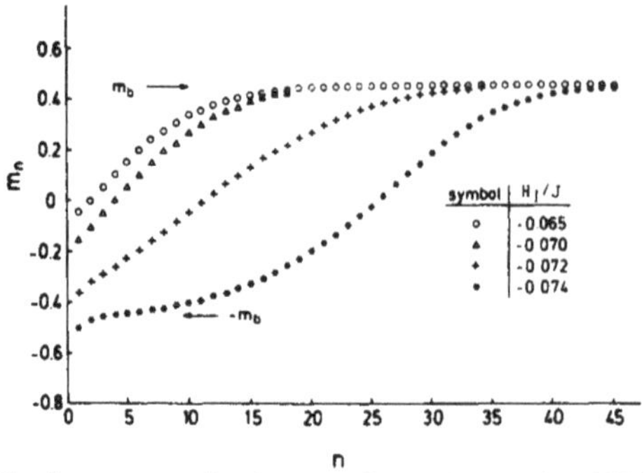


Fig. 2. Profiles of the layer magnetizations $m_n$ for a $128 \times 128 \times 160$ lattice with $J_s/J = 1.33$, $J/k_BT = 0.226$. The arrows show the values of the bulk magnetization.

This investigation encountered considerable technical difficulty and rather specialized Monte Carlo techniques were used but many thousands of hours of cpu time were required nonetheless. (Details can be found elsewhere[11-13,16,17].)

## 3. Wetting in Thick Films

For very thick films the behavior closely approximates that of a semi-infinite system with a single wall or surface. For magnetic systems, of course, it is most natural to consider films whose surfaces do not favor any sign of the order parameter, i.e. surface magnetic fields are zero. However, when we reinterpret an Ising magnet as a lattice gas model for a fluid (spin $S_i = \pm 1$ corresponds to site $i$ being occupied or empty) the situation is completely different: binding energies to the wall translate into surface magnetic fields for the corresponding Ising model [6]. Even if such binding energies are absent (considering e.g. a "free" surface against vacuum) the effect of "missing neighbors" also contributes to such a surface field in the Ising spin representation { see e.g. Refs. 6,7}.

In a semi-infinite fluid (or equivalently a semi-infinite Ising magnet with a surface field) wetting transitions occur where the surface is coated with an infinitely thick layer of the phase which it prefers, separated by an interface from the bulk phase. Examples of the profiles which appear as the wetting transition is approached are shown in Fig. 2. {In lattice gas language, $m_n$ is related to the layer density $\rho_n = (1 - m_n)/2$}. (A more complete description of our data as well as extensive references were given elsewhere[13,16,17].) We find that the transition from the nonwet state to the wet state may be either first order, critical, or tricritical. At sufficiently low temperature Monte Carlo simulations[17] reveal the existence of a sequence of layering transitions for which the formation of the thick layer proceeds by a series of jumps where individual layers become coated as a function of temperature or wall strength.

Renormalization group studies of critical wetting [22] predict that Mean Field theory

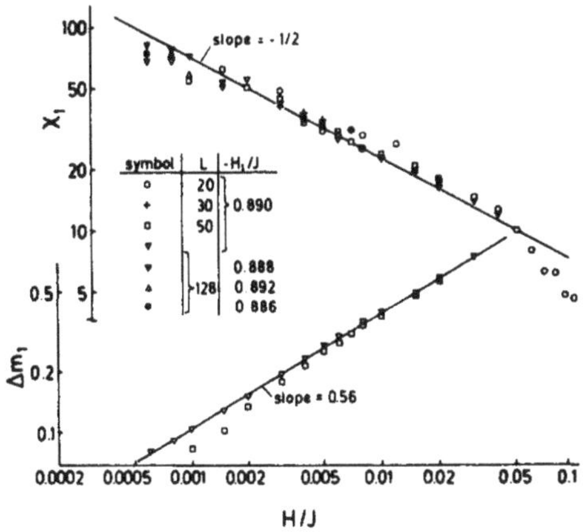


Fig. 3. Critical wetting behavior of the surface-layer susceptibility $\chi_1$ (upper part) and excess magnetization $\Delta m_1$ (lower part) vs bulk field at $H_1 = H_{1c}$, $J_s = J$, and $J/k_BT = 0.35$.

is valid above the marginal dimension $d^* = 3$, but in $d = 3$ critical wetting is non-universal and depends on the dimensionless parameter $\omega = k_BT/4\pi\sigma\xi_b{}^2$ where $\sigma$ is the interface tension of the free interface and $\xi_b$ is the bulk correlation length. Monte Carlo data[13,23], however, yield strong evidence for mean field behavior; in Fig. 3 we show both the surface layer susceptibility and the magnetization excess for $J/kT = 0.35$. The temperature is sufficiently low that $\xi_b$ is quite small but high enough that no layering takes place. Both quantities are described well by mean field exponents. The Monte Carlo data have helped inspire numerous attempts to explain the discrepancy[24-27]. A recent and compelling suggestion by Boulter and Parry[18] is that coupling between fluctuations in the order parameter near the surface and near the depinning interface could renormalize the singular behavior at complete wetting as the bulk field tends to zero. But it is also possible that the wetting transition is, in fact, weakly first order[27].

## 4. Thin Films with Equal Walls: Capillary Condensation

In a thin film geometry the walls can be coated only by layers of finite thickness: it is thus no surprise that in thin films confined between two symmetric walls, both second order and first order wetting transitions are rounded off [1]. Monte Carlo data have been used[7] to construct plots of the free energy as a function of the surface field, and the intersection points of free energy curves obtained when approaching the transition from opposite directions show that the surface fields shift the transition between the coexisting phases, which correspond to the coexisting phases in the bulk, away from the bulk location (i.e. of pressure in the fluid, or chemical potential in the

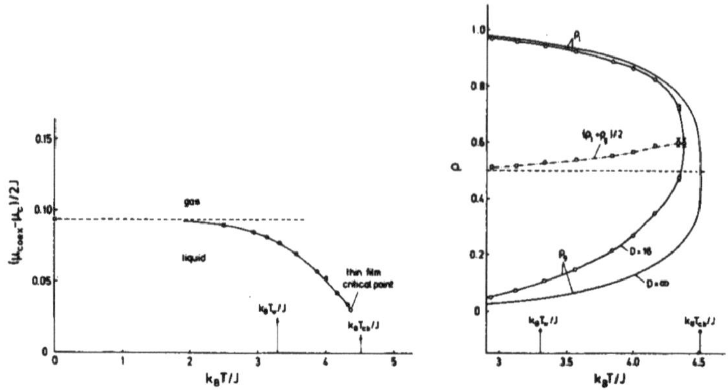

Fig. 4. Phase diagram of a capillary of thickness $D = 16$ and wall potential $H_1/J = -0.75$: (left)chemical potential vs temperature; The dashed line shows the chemical potential for gas-liquid condensation at $T = 0$; (right) Liquid-gas coexistence curve compared to the bulk coexistence curve ($D = \infty$). Arrows show the bulk wetting temperature $T_w$ and the bulk critical temperature $T_{cb}$ [14].

mixture, respectively). Thus, for a fluid which tends to wet the walls, the pressure $P$ at which condensation occurs in a thin film is reduced relative to its bulk coexistence value $P_{coex}$: "capillary condensation" [1,7,10]. In Fig. 4 we show the resultant phase diagram for a thin capillary. The liquid-gas coexistence curve for this situation is also shown in Fig. 4. The curve is now asymmetric about 0.5 and lies wholly within the curve for an infinitely thick film which is shown for comparison. In the magnetic terminology of the Ising model, capillary condensation means a change of sign of the total magnetization, and this first order transition no longer occurs at zero bulk field (as it would in an infinite system) but at a bulk field, inversely proportional to film thickness, which effectively "cancels" the effect of the surface fields [1,7].

## 5. Thin Films with Competing Walls: Interface Delocalization

Competing surface fields, i.e. one positive and one negative (in lattice gas terminology, one wall favors liquid while the other wall favors the gas) lead to rather unusual behavior. In Fig. 5 profiles of the layer magnetization $m_n$ are shown for a variety of temperatures. Even above the bulk transition temperature $T_{cb}$ there is a precursor of an interface, since the surface fields induce a local magnetization near the walls. But we expect that this surface-induced order decays exponentially fast with the distance from the walls as one moves into the interior of the film. Of course, this is true only if $\xi_b \ll D/2$, so that the two walls are essentially non-interacting, and the profiles of all quantities of interest are quite flat in the center of the film.

The behavior gradually changes, when $D \approx 2\xi_b$, i.e. near $T \approx T_{cb}$, where the two

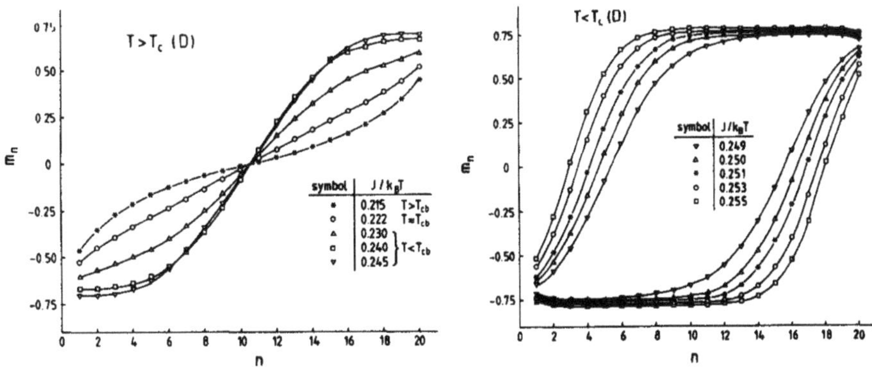


Fig. 5. Profiles of the layer magnetization $m_n$ plotted versus layer number $n$, for asymmetric walls, $H_1 = -H_D = 0.55$, for $D = 20$, $L = 128$.

interfaces start to "feel" each other, the profile of $m_n$ then deviates distinctly from an exponential decay. The energy profile no longer has any extended flat part, and the layer susceptibility profile $\chi_{nn}$ develops a pronounced peak in the center of the film. However, this change of behavior leading to the gradual build-up of an interface in the center of the film is non-singular; the transition that occurs at $T_{cb}$ in the bulk system ($D \to \infty$) is smeared out for all finite $D$.

However, at a much lower temperature $T_c(D)$ {which for the present choice of surface field is already rather close to the wetting transition temperature of a semi-infinite system, $J/k_BT_W = 0.250 \pm 0.005$ for $|H_1|/J = 0.55$ [13]} we see symmetry breaking in the film: the interface between the phases with negative and positive magnetization is bound either to the left or to the right wall (for $H_1 = -H_D$ the two states are degenerate). This interface localization - delocalization for $D \to \infty$ is predicted [2] to converge to the wetting transition temperature $T_W(H_1/J)$ and not to $T_{cb}$! Note, however, the transition at $T_c(D)$ is the only one which we observe, and the antisymmetry of the profiles $m_n$ in Fig. 5 for $T > T_c(D)$ with respect to the center of the film implies that the total magnetization of the film is zero for $T > T_c(D)$. In contrast, the two states for $T < T_c(D)$ have a spontaneous magnetization $M \neq 0$ of equal magnitude but opposite sign. Using the total magnetization as the order parameter, we find a transition from a disordered state to a twofold degenerate state. The transition is thus expected [2,4] to belong to the universality class of the $d = 2$ Ising model, for all values of $D < \infty$. The shape of the interface changes dramatically as we move through the transition; this is shown in Fig. 6.

As the film is made thinner the behavior of the profiles changes. For $D = 6$, even far above $T_{cb}$ the profiles of order parameter and energy are not flat. Clearly for smaller $D$ the bulk correlation length $\xi_b$ must be smaller for interactions between the two walls to be negligible. Similarly, while for $D \geq 12$ and $T < T_c(D)$ the order

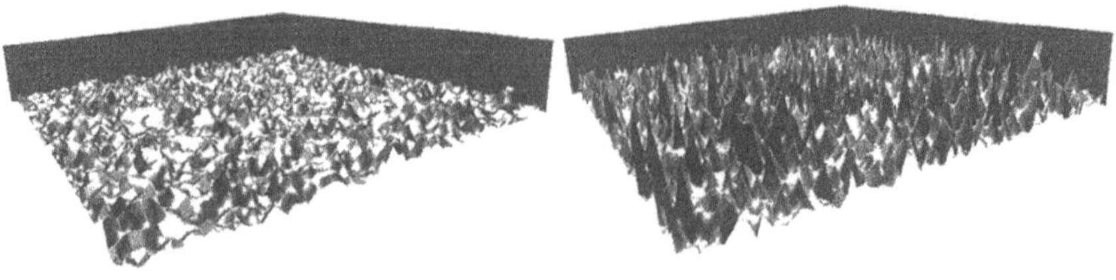

Fig. 6. "Snapshots" of the interface for a film with asymmetric walls, $H_1 = -H_D = 0.55$. $L = 256$ and $D = 20$: (left) $T < T_c(D)$: (right) $T > T_c(D)$.

parameter profile quickly develops a flat region where $m_n$ is quite close to the "bulk" value , no such behavior can be identified for $D = 6$. For such thin films, all layers "feel" the effects of the two competing walls rather strongly.

For $D = 20$ the interface produces peaks in the layer energy and the layer susceptibility $\chi_{nn}$. For $T < T_c(D)$ we expect the interface separating bulk $d = 3$ phases to be bound either to the right or to the left wall; Fig. 6 gives clear evidence for this description, and one can see how the interface increasingly "depins" from the surface, as $T$ approaches $T_c(D)$ from below.

Cumulant crossing can be used to locate the transition. For $D = 12$ and for $L \leq 64$ we find[12] a well-defined intersection point at $J/k_BT \approx 0.2497$, rather close to $T_w$ [13]. However, the data exhibit two disturbing features: (i) the ordinate $U^*$ of this cumulant crossing is rather close to $U^* \approx 0.45$ rather than to the value $U^*$ (2d Ising) $\approx 0.615$ characterizing the universality class of the $d = 2$ Ising model [19]. (ii) The data for $L = 128$ seem to yield a intersection point at a somewhat higher temperature and at a slightly larger value $U^* \approx 0.47$.

The behavior of other quantities is also somewhat unusual: the specific heat has only a very weak anomaly; the susceptibility $\chi$ for temperatures far above $T_c(D)$ is roughly compatible with $d = 2$ Ising behavior , but the finite size-rounding of this anomaly near $T_c(D)$ does not scale with $d = 2$ Ising exponents.

A possible solution to this puzzle lies in the growth of a very large length scale[4] $\xi_{||} \propto \exp(D/4\xi_b)$ in the phase with the delocalized interface ($T_c(D) < T < T_{cb}$): the true nature of the transition can probably be observed only when $L$ is clearly larger than $\xi_{||}$ . Of course, when we approach the Ising transition the correlation length $\xi_{||}$ diverges with exponent $\nu(d = 2) = 1$. For $L \lesssim \xi_{||}$ the interfacial fluctuations are not yet restricted significantly by the finite $D$, instead the wetting transition is rounded due to the finiteness of $L$ and the "effective" thickness is quasi-infinite. Since at $J/k_BT = 0.25$ the true correlation length is expected to be only of the order of one lattice spacing [20,21], for $D = 12$ the the regime $L \gg \xi_{||}$ is not reached for our smaller choices of $L$. Thinner films, e.g. $D = 8$ and $D = 6$, are also interesting in this regard. Fig. 7 shows that the temperatures of the cumulant crossings $T_{U_{cross}}$ and of

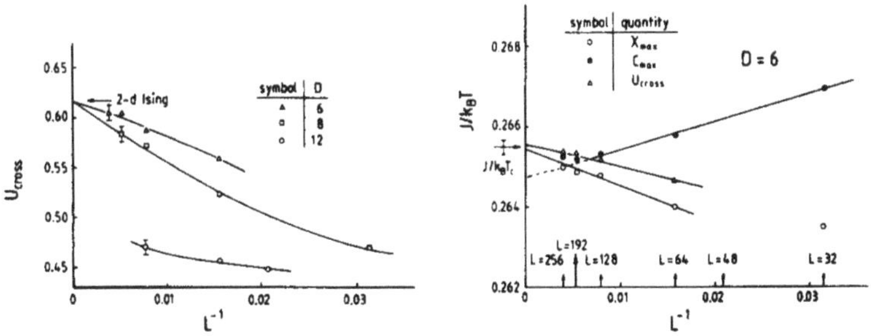


Fig. 7. Finite size effects in films with asymmetric walls: (left)Cumulant crossing values $U_{cross}$ vs $L^{-1}$ for $|H_1|/J = 0.55$. The arrow shows the $2D$ Ising value; (right)finite lattice estimates for the inverse critical temperature for $D = 6$. The arrow shows the final estimate for $T_c(D)$.

the maxima of specific heat ($T_{C_{max}}$) and susceptibility ($T_{\chi' max}$) are roughly compatible with a linear extrapolation in the variable $L^{-1}$, yielding $J/k_BT_c(D) = 0.2578 \pm 0.0002$.

The scatter in the extrapolated values comes from the inaccuracy in locating $T_{C_{max}}$, $T_{\chi' max}$ because of the statistical errors in the data. However, the cumulant intersections $U_{cross}(L)$, defined from $U(L) = U(L/2) \equiv U_{cross}(L)$, extrapolate quite smoothly towards the expected $d = 2$ Ising value[19] $U^* = 0.615$, see Fig. 7. The increase of the susceptibility maxima with the linear dimension $L$ is also compatible with the expected finite size scaling behavior $\{(\chi' J)_{max} \propto L^{\gamma/\nu}$ with $\gamma/\nu = 1.75$ for the two-dimensional Ising models$\}$.

For the thinnest film, ($D = 6$), the cumulant intersection plot still suffers from crossover effects, but the intersections for the larger sizes converge rather nicely to the Ising value ($U^* = 0.615$). Both susceptibility $\chi'$ and specific heat $C$ have rather well-developed peaks, and an extrapolation of their locations yields a value of $T_c(D)$ which is compatible with the cumulant extrapolation. Fig. 7 also includes an extrapolation of the values $U_{cross}$ observed at these cumulant crossings. Very large values of $L$ would be needed for $D \geq 12$, to see the asymptotic 2-dim Ising behavior. Finite size analyses of the susceptibility maximum and $\langle |M| \rangle$ yield the same conclusion..

In the regime between $T_c(D)$ and $T_{cb}$ for large $D$ the system is already ordered locally everywhere except in the immediate vicinity of the interface, which on average is located in the center of the film. If D were infinite we would have a freely floating interface which could be shifted as a whole without free energy cost. For finite $D \gg \xi_b$, this free energy cost is of order $\exp\{-D/(2\xi_b)\}$ in mean field theory[4]. These free energy costs can be understood from the fact that in mean field theory the interface is not rough and is described by the familiar tanh profile,

$$m(z) \approx -m_b \tanh(z/l) \ , \quad l = 2\xi_b \ , \tag{3}$$

where we take the zero of the $z$-coordinate in the center of the interface. The intrinsic width $l/2$ is simply identical to the true correlation range in the direction perpendicular to the interface. Of course, near the walls the tanh profile is somewhat distorted due to the response to the surface fields $\pm H_1$. The application of fields $H$ uniformly in the bulk to induce a magnetization (or a local field $H_n$ in the $n^{th}$ layer, respectively) implies a displacement of the interface, and therefore the response to these fields is also predicted to scale proportional to the inverse of the above energy cost [4], $\chi_n \propto \chi_{nn} \propto \exp\{D/(2\xi_b)\}$ , $D/\xi_b \gg 1$ . While the data show the predicted form, $l$ disagrees with the prediction of mean field theory, $2\xi_b$. Here $\xi_b$, of course, was not calculated from molecular field theory but rather was taken from the analysis of low temperature series expansions [20]. The difference between the actual value of $l^{-1}$ and $(2\xi_b)^{-1}$ decreases somewhat when the temperature is lowered. The interface in $d = 3$ dimensions is not strictly localized as assumed by mean field theory, but rather delocalized (rough) due to capillary wave excitations, which in our confined geometry exist only up to wavelengths, of order $\xi_{||}$.

An interesting tentative explanation for this difference between $l$ and $\xi_b$ has been proposed recently[18].

## 6. Conclusions

We have reviewed the rich variety of wetting phenomena which have been studied by extensive Monte Carlo simulations in systems with Ising-type symmetry confined in films of various thickness between walls of different types. There has been an interplay between theory and simulation which has succeeded in describing the different types of behavior which might result. For thick films Monte Carlo simulations yielded the unexpected result that the critical wetting seems to be mean field-like. In thin films with symmetric walls we find a shift in the location of the condensation transition. For the case of antisymmetric walls a single phase transition occurs at a temperature $T_c(D)$ close to the wetting transition temperature $T_W(H_1)$ from a phase with a fluctuating interface centered in the middle of the film to a state where the interface is bound to either one of the walls. In agreement with predictions of Parry and Evans [2,4], we find that this transition belongs to the universality class of the $d = 2$ Ising model, but the observability of this Ising critical behavior is limited to a tiny region very close to $T_c(D)$ if $D$ is large. For $T < T_C(D)$ the gradual unbinding of the interface from the wall at which it has been localized clearly can be interpreted as a wetting phenomenon. Lastly, we confirm the prediction of Parry and Evans [9] that the phase at $T_c(D) < T < T_{cb}$ is characterized by susceptibilities varying exponentially with film thicknesses ($ln\chi_n \propto D$, $ln\chi_{nn} \propto D$), but we fail to find the prefactor (obtained from a mean field theory) in these relations.

Acknowledgements:

We wish to thank S. Wansleben and A.M. Ferrenberg for their contributions to portions of this work and to S.Pal and J.-F. Nielsen for their efforts in visualization. This research was supported by NSF grant DMR-9405018, DFG grant No. SFB 262/D1, the Humboldt Foundation , and NATO grant No. CRG 921202.


1. H. Nakanishi and M.E. Fisher, J. Chem. Phys. **78**, 3279 (1983)
2. A.O. Parry and R. Evans, Phys. Rev. Lett. **64**, 439 (1990)
3. M.R. Swift, A.L. Owczarek, and J.O. Indekeu, Europhys. Lett. **14**, 475 (1991)
4. A.O. Parry and R. Evans, Physica **A181**, 250 (1992)
5. J. Rogiers and J.O. Indekeu, Europhys. Lett. **24**, 21 (1993)
6. K. Binder, in **Phase Transitions and Critical Phenomena, Vol. 8** (C. Domb and J.L. Lebowitz, eds.) Ch. 1 (Academic, New York 1983)
7. K. Binder and D.P. Landau, J. Chem. Phys. **96**, 1444 (1992)
8. For reviews of wetting phenomena, see S. Dietrich, in **Phase Transitions and Critical Phenomena, Vol. 12** (C. Domb and J.L. Lebowitz, eds.) Ch. 1 (Academic, New York 1988).
9. M.E. Fisher, J. Stat. Phys. **34**, 667 (1984); J. Chem. Soc. Faraday Trans. **282**, 1569 (1986)
10. R. Evans J. Phys.: Condens. Matter **2**, 8989 (1990), and references therein.
11. K. Binder, A.M. Ferrenberg and D.P. Landau, Ber. Bunsenges. Phys. Chemie **98**, 340 (1994).
12. K. Binder, D.P. Landau and A.M. Ferrenberg, Phys. Rev.Lett. **74**, 298 (1995); Phys. Rev. **E51**, 2823 (1995)
13. K. Binder, D.P. Landau, and. S. Wansleben, Phys. Rev. **B40**, 6971 (1989)
14. A.M. Ferrenberg and D.P. Landau, Phys. Rev. **B44**, 5081 (1991)
15. K.K. Mon, D.P. Landau, and D. Stauffer, Phys. Rev. **B42**, 545 (1990)
16. K. Binder and D.P. Landau, Phys. Rev. **B37**, 1745 (1988)
17. K. Binder and D.P. Landau, Phys. Rev. **B46**, 4844 (1992)
18. C. J. Boulter and A. O. Parry, Phys. Rev. Lett. **74**,3403 (1995)
19. D.P. Landau and D. Stauffer, J. Physique **50**, 509 (1989) and references therein
20. A.J. Liu and M.E. Fisher, Physica **156A**, 35 (1989)
21. M. Hasenbusch and K. Pinn, Physica **A192**, 342 (1993)
22. E. Brezin, B.I. Halperin, and S. Leibler, Phys. Rev. Lett. **50**, 1387 (1983); R. Lipowsky, D.M. Kroll, and R.K.P. Zia, Phys. Rev. **B27**, 4499 (1983); D.S. Fisher and D.A. Huse, Phys. Rev. **B32**, 247 (1985); R. Lipowsky and M.E. Fisher, Phys. Rev. **B36**, 2126 (1987)
23. K. Binder, D.P. Landau, and D.M. Kroll, Phys. Rev. Lett. **56**, 2276 (1986)
24. E. Brezin and T. Halpin-Healey, Phys. Rev. Lett. **58**, 1220 (1987); J. Phys. (Paris) **48**, 757 (1987)
25. A.O. Parry and R. Evans, Phys. Rev. **B39**, 12336 (1989)
26. M.E. Fisher and H. Wen, Phys. Rev. Lett. **68**, 3654 (1992); K. Binder, D.P. Landau, and D.M. Kroll, Phys. Rev. Lett. **68**, 3655 (1992)
27. A. J. Jin and M. E. Fisher, Phys. Rev. **B47**, 7365 (1993)

# PROBABILITY OF PHASE SEPARATION FOR THE BOSE GAS WITH DELTA INTERACTION


ALEXANDER R. ITS
*Department of Mathematical Sciences, Indiana University-Purdue University at Indianapolis (IUPUI)*
*Indianapolis, IN 46202-3216, U.S.A.*
E-mail: itsa@math.iupui.edu

and

VLADIMIR E. KOREPIN
*Institute for Theoretical Physics, State University of New York at Stony Brook*
*Stony Brook, NY 11794-3840, U.S.A.*
E-mail: korepin@max.physics.sunysb.edu

and

ANDREW K. WALDRON
*Institute for Theoretical Physics, State University of New York at Stony Brook*
*Stony Brook, NY 11794-3840, U.S.A.*
E-mail: wally@max.physics.sunysb.edu



ABSTRACT

We consider the quantum Non-linear Schrödinger equation $i\partial_t\Psi = -\partial_x^2\Psi + 2c\Psi^\dagger\Psi^2$ with positive coupling constant $c$ varying from zero to infinity. We study quantum correlation functions of this model using the determinant representation of these correlation functions. We consider the case of a finite density ground state and evaluate, using the Riemann Hilbert problem, the asymptotics of the probability that no particles are present in the space interval $[0, x]$ in the large $x$ limit. We call this the probability of phase separation or alternately the emptiness formation probability.


## 1. Introduction

The one-dimensional Bose Gas is described by quantum Bose fields $\Psi(x,t)$ with canonical equal time commutation relations $\left[\Psi(x), \Psi^\dagger(y)\right] = \delta(x-y)$, The field $\Psi$ annihilates the Fock vacuum $\Psi|0\rangle = 0$ and the Hamiltonian of the model is

$$H = \int dx \left\{\partial_x\Psi^\dagger\partial_x\Psi + c\Psi^\dagger\Psi^\dagger\Psi\Psi\right\}. \tag{1}$$

The equation of motion is the celebrated nonlinear Schrödinger equation $i\partial_t\Psi = -\partial_x^2\Psi + 2c\Psi^\dagger\Psi^2$. The Hamiltonian and the number operator $Q = \int dx\Psi^\dagger\Psi$ commute whereby the model for a given number of particles $N$ is equivalent to the quantum mechanical Hamiltonian with delta interactions

$$\mathcal{H}_N = -\sum_{j=1}^{N} \frac{\partial^2}{\partial x_j^2} + 2c \sum_{1\le j<k\le N} \delta(x_j - x_k). \tag{2}$$

The Bethe Ansatz for this model was constructed by Lieb and Liniger [1]. The ground state for the system in a box of length $L$ containing $N$ particles in the thermodynamic limit ($N \to \infty$, $L \to \infty$, $N/L = D =$ const.) is a Fermi "sphere" of radius $q$. Introducing the density of particles in momentum space $\rho(\lambda)$ so that $D = \frac{N}{L} = \int_{-q}^{q} d\lambda \rho(\lambda)$, the ground state at zero temperature is then described by the linear integral equation

$$\frac{1}{2\pi} = \rho(\lambda) - \frac{1}{2\pi}\int_{-q}^{q} d\mu K(\lambda,\mu)\rho(\mu) = \rho(\lambda) - \frac{1}{2\pi}\left(\hat{K}\rho\right)(\lambda), \tag{3}$$

where the kernel of the integral operator $\hat{K}$ is

$$K(\lambda,\mu) = \frac{2c}{c^2 + (\lambda-\mu)^2}. \tag{4}$$

Let us now lay out the structure of this paper. We begin in section 2 by defining the probability of phase separation correlation function $P(x)$ which we express in terms of the determinant of a particular Fredholm integral operator. In section 3 we use the identity $ic/(\lambda - \mu + ic) = c\int_0^\infty ds e^{is(\lambda-\mu+ic)}$ to write this integral operator as a special Fredholm integral operator related to a Riemann–Hilbert problem (RP). The introduction of the $s$–integration means that this RP is infinite-dimensional, or more precisely, integral operator-valued. In section 4 we present a new representation of $Gl(2, \mathrm{C})$ in terms of such integral operators which reduces the jump condition (see 24) of the operator RP to the jump condition (see 46) of a $2 \times 2$ matrix-valued RP. Finally in section 5 we solve the $2 \times 2$ RP for large $x$. Indeed, we find that this $2 \times 2$ RP is related to that encountered if one studies the case $c = \infty$ for which the $s$–integration is unnecessary. The results obtained in section 5 lead us to a conjecture (see 16) for the asymptotic behaviour of the phase separation probability correlation function. Presently we are unable to make this conjecture completely rigorous.

## 2. Determinant Representation for Probability of Phase Separation

Although the ground state is homogeneous and translationally invariant, due to quantum fluctuations there is a non-vanishing probability that no particles will be found in the space interval $[0, x]$ which we denote

$$P(x) = \text{Probability that no particles are found in the interval } [0, x]. \tag{5}$$

In the thermodynamic limit the probability of an empty interval may be expressed as the ratio of two Fredholm determinants [2]:

$$P(x) = \frac{(0|\det(I+\hat{V})|0)}{\det(I-\hat{K}/2\pi)}. \tag{6}$$

The integral operator $\hat{K}$ appearing in the denominator is given above in (4) and is $x$ independent. Therefore when calculating the large $x$ asymptotics we need consider only the numerator $(0|\det(I+\hat{V})|0)$. The definition of the integral operator $\hat{V}$ is more involved. $\hat{V}$ acts on the interval $C=[-q,q]$ in the same fashion as $\hat{K}$, namely

$$\left(\hat{V}f\right)(\lambda) = \int_{-q}^{q} d\mu V(\lambda,\mu)f(\mu). \tag{7}$$

However in addition to depending on the parameters $x$ and $c$ (the length of the empty interval and coupling respectively) $\hat{V}$ and in turn $\det(I+\hat{V})$ are functionals of a dual quantum field $\overset{\circ}{\varphi}(\lambda)$. Let us give the kernel of $\hat{V}$

$$\begin{aligned} V(\lambda,\mu) &= -\frac{c}{2\pi}\left[\frac{\exp\{ix\lambda/2+\overset{\circ}{\varphi}(\lambda)/2\}\exp\{-ix\mu/2-\overset{\circ}{\varphi}(\mu)/2\}}{(\lambda-\mu)(\lambda-\mu+ic)}\right. \\ &\quad \left. +\frac{\exp\{ix\mu/2+\overset{\circ}{\varphi}(\mu)/2\}\exp\{-ix\lambda/2-\overset{\circ}{\varphi}(\lambda)/2\}}{(\mu-\lambda)(\mu-\lambda+ic)}\right]. \end{aligned} \tag{8}$$

The dual field acts in a Hilbert space with "dual" Fock vacuum $|0)$ so that $(0|\det(I+\hat{V})|0)$ denotes the dual vacuum expectation value of the dual field valued determinant $\det(I+\hat{V})$ which may be computed from the definition of the dual field $\overset{\circ}{\varphi}(\lambda)$ given below:

$$\overset{\circ}{\varphi}(\lambda) = \overset{\circ}{p}(\lambda) + \overset{\circ}{q}(\lambda), \tag{9}$$

where $\forall\, \lambda,\mu \in C$

$$[\overset{\circ}{p}(\lambda), \overset{\circ}{q}(\mu)] = \log\left(\frac{c^2}{c^2+(\lambda-\mu)^2}\right), \tag{10}$$

$$[\overset{\circ}{p}(\lambda), \overset{\circ}{p}(\mu)] = 0 = [\overset{\circ}{q}(\lambda), \overset{\circ}{q}(\mu)]\ ;\ \overset{\circ}{p}(\lambda)|0) = 0 = (0|\,\overset{\circ}{q}(\lambda). \tag{11}$$

¿From the symmetry of $\log(c^2/(c^2+(\lambda-\mu)^2))$ in $\lambda$ and $\mu$ we have the crucial property

$$[\overset{\circ}{\varphi}(\lambda), \overset{\circ}{\varphi}(\mu)] = 0 \qquad \forall\, \lambda,\mu \in C. \tag{12}$$

This allows us to treat $\overset{\circ}{\varphi}(\lambda)$ simply as some function on $C$ in our calculation of $\det(I+\hat{V})$. Only at the end of this calculation do we take the dual vacuum expectation value.

Obtained using the Riemann Hilbert problem in the large $x$ limit, our conjecture for the determinant as a functional of $\overset{\circ}{\varphi}(\lambda)$ is[a]:

$$\det(I+\hat{V}) \overset{x\to\infty}{=} (\text{const}) \exp\left\{-(xq)^2/8 - \frac{x}{2\pi}\int_{-q}^{q} \frac{\overset{\circ}{\varphi}(\mu)\mu d\mu}{_+\sqrt{\mu^2-q^2}} + o(x)\right\}. \tag{13}$$

Let us now analyse the dual vacuum expectation value of (13). Using (9)-(11) and the identity $e^{A+B} = e^A e^B e^{-\frac{1}{2}[A,B]}$ valid $\forall$ $A$, $B$ which commute with $[A,B]$ we obtain

$$\begin{aligned}(0|\det(I+\hat{V})|0) &\sim \exp\left\{-\frac{(xq)^2}{8} - \frac{x^2}{8\pi^2}\int_{-q}^{q}\frac{\mu d\mu}{_+\sqrt{\mu^2-q^2}}\int_{-q}^{q}\frac{\nu d\nu}{_+\sqrt{\nu^2-q^2}}\log\frac{c^2+(\mu-\nu)^2}{c^2}\right\}\\ &\equiv \exp\left\{-\frac{(xq)^2}{8}[1+I(c^2/q^2)]\right\},\end{aligned} \tag{14}$$

where we denote the dimensionless integral

$$I(c^2/q^2) = \frac{2}{\pi^2}\int_{-1}^{1}\frac{ydy}{\sqrt{1-y^2}}\int_{-1}^{1}\frac{zdz}{\sqrt{1-z^2}}\log\left(\frac{c^2/q^2+(y+z)^2}{c^2/q^2+(y-z)^2}\right). \tag{15}$$

$I(c^2/q^2)$ has the following properties: $I(c^2/q^2) > 0$, $\frac{\partial I}{\partial(c^2/q^2)} < 0$, $I(c^2/q^2 \to \infty) = 0$ and $I(c^2/q^2 \to 0) = 1$. Therefore we find as our conjecture for the phase separation probability correlation function

$$P(x) \overset{x\to\infty}{=} (\text{const}) \exp\left\{-\frac{(xq)^2}{8}[1+I(c^2/q^2)] + o(x^2)\right\}. \tag{16}$$

The probablility of phase separation is Gaussian in the length $x$ and the rate of decay in $x$ decreases monotonically as the coupling $c$ varies from zero to infinity.

## 3. Integrable Linear Integral Operators and $s$–Integration

Integral operators $I+\hat{V}$ (acting on an interval $C$) where the kernel of $\hat{V}$ has the form

$$V(\lambda,\mu) = \frac{1}{\lambda-\mu}\sum_{j=1}^{N} e_j(\lambda)E_j(\mu), \tag{17}$$

with $0 = \sum_{j=1}^{N} e_j(\lambda)E_j(\lambda)$, belong to a special class which we will call "integrable" operators (the functions $e_i(\lambda)$ are $N$ linearly independent functions, continuous and integrable on the interval $C$, similarly for $E_j$). Indeed such integrable integral operators form a group (where multiplication is defined in the usual way for integral

---

[a] $_+\sqrt{\lambda^2-q^2}$ denotes the limit approaching the contour $C$ from above. Further, we remind the reader that the Hardy small "$o$" symbol $o(x)$ denotes any function of $x$ increasing slower than $x$ as $x \to \infty$.

operators) and are closely related to a certain $N \times N$ matrix Riemann–Hilbert problem [2].

Therefore to employ the RP for finite $c$ we must express the kernel (8) in the standard form above. To this end we rewrite the second factor in the denominators using an old identity made famous by Feynman $\frac{1}{\lambda-\mu+ic} = -i\int_0^\infty ds e^{is(\lambda-\mu+ic)}$. The kernel (8) becomes

$$V(\lambda,\mu) = \int_0^\infty \frac{e_+(\lambda|s)e_-(\mu|s) - e_-(\lambda|s)e_+(\mu|s)}{\lambda-\mu} ds \tag{18}$$

$$= \int_0^\infty \frac{e_1(\lambda|s)E_1(\mu|s) + e_2(\lambda|s)E_2(\mu|s)}{\lambda-\mu} ds \tag{19}$$

where $e_\pm(\lambda|s) = \sqrt{\frac{i}{2\pi}} \exp\{\pm(ix\lambda/2 + \overset{\circ}{\varphi}(\lambda)/2)\}\sqrt{c}\exp\{\pm is\lambda - cs/2\}$, (20)

and $(e_1(\lambda|s), e_2(\lambda|s)) = (e_+(\lambda|s), e_-(\lambda|s))$, (21)

$(E_1(\lambda|s), E_2(\lambda|s)) = (e_-(\lambda|s), -e_+(\lambda|s))$, (22)

which is in the form (17) with the sum $\sum_{j=1}^N$ generalized to an integral $\int_0^\infty ds \sum_{j=1}^2$. We must, therefore study a $2 \times 2$ operator valued RP which we now define.

The operator valued RP is to find the $2 \times 2$ matrix $\widehat{\chi}(\lambda)$ whose entries are integral operators acting on the interval $[0,\infty)$ and functions of the complex plane $\mathbf{C}$ with kernels

$$\chi(\lambda|s,t) = \begin{pmatrix} \chi_{11}(\lambda|s,t) & \chi_{12}(\lambda|s,t) \\ \chi_{21}(\lambda|s,t) & \chi_{22}(\lambda|s,t) \end{pmatrix}, \quad \lambda \in \mathbf{C};\, s,t \in [0,\infty), \tag{23}$$

such that:

1. $\chi(\lambda|s,t)$ is analytic $\forall\, \lambda \in \mathbf{C} \setminus C$.

2. $\chi(\lambda|s,t)$ is, however, not analytic across the conjugation contour $C$, rather the limits $\chi_+(\lambda|s,t)$ and $\chi_-(\lambda|s,t)$, approaching $C$ from above and below respectively[b], satisfy the jump condition

$$\chi_-^{ik}(\lambda|s,t) = \int_0^\infty \sum_{j=1}^2 \chi_+^{ij}(\lambda|s,r)G^{jk}(\lambda|r,t)dr, \quad \forall\, \lambda \in C,\ (i,k=1,2), \tag{24}$$

where the conjugation matrix $G_{ik}(\lambda|s,t)$ is given by

$$G_{ik}(\lambda|s,t) = \delta_{ik}\delta(s-t) - 2\pi i e_i(\lambda|s)E_k(\lambda|t), \quad (i,k=1,2). \tag{25}$$

[b] We consider $C = [-q,q]$ as an oriented contour so that $\int_C \equiv \int_{-q}^q$. In general for any oriented contour we will define "+" to be the limit from the left side of the direction of travel.

3. $\chi(\lambda|s,t)$ is canonically normalized:

$$\chi(\infty|s,t) = \begin{pmatrix} 1 & 0 \\ 0 & 1 \end{pmatrix} \delta(s-t). \tag{26}$$

In fact, if $\widehat{\chi}(\lambda)$ solves the RP defined above, then at $\lambda = \infty$ it has the whole Laurent series expansion $\chi(\lambda|s,t) = \begin{pmatrix} 1 & 0 \\ 0 & 1 \end{pmatrix} \delta(s-t) + \sum_{n=1}^{\infty} \frac{M_n(s,t)}{\lambda^n}$ where the integral operators $\widehat{M}_n$ are independent of $\lambda$.

In order to extract the determinant $\det(I+\hat{V})$ from the operator RP we consider the formal solution to the above RP

$$\chi_{ik}(\lambda|s,t) = \delta_{ik}\delta(s-t) + \int_C \frac{f_i(\mu|s)E_k(\mu|t)d\mu}{\mu-\lambda} \tag{27}$$

$$\text{where} \quad e_i(\lambda|s) = [(I+\hat{V})f_i](\lambda|s) = f_i(\lambda|s) + \int_C V(\lambda,\mu)f_i(\mu|s)d\mu. \tag{28}$$

For $\lambda$ large we have

$$\chi_{ik}(\lambda|s,t) \overset{\lambda\to\infty}{\sim} \delta_{ik}\delta(s-t) - \frac{1}{\lambda}\int_C f_i(\mu|s)E_k(\mu|t)d\mu + O(\lambda^{-2}) \tag{29}$$

$$= \delta_{ik}\delta(s-t) + \frac{M_1^{ik}(s,t)}{\lambda} + O(\lambda^{-2}). \tag{30}$$

Let us calculate [2] now the logarithmic derivative of $\det(I+\hat{V})$ w.r.t. $x$

$$\begin{aligned} \frac{\partial}{\partial x}\log\det(I+\hat{V}) &= \frac{i}{2}\int_C\int_0^{\infty}(f_1(\mu|s)e_2(\mu|s) + f_2(\mu|s)e_1(\mu|s))dsd\mu \\ &= -\frac{i}{2}\int_0^{\infty}(M_1^{11}(s,s) - M_1^{22}(s,s))ds = -i\mathrm{tr}\widehat{M}_1^{11}. \end{aligned} \tag{31}$$

Here "tr" denotes the trace operation in the space of integral operators on $[0,\infty)$ and we used formula (19). Therefore $\det(I+\hat{V})$ can be obtained from the large $\lambda$ asymptotics of the solution to the above RP.

At this point the prospect of solving an operator valued RP should seem daunting since even the solution for the matrix case is not in general known. However we now construct a representation of $Gl(2,\mathbf{C})$ which will allow us to consider a (known) $2\times 2$ RP.

## 4. $Gl(2,\mathbf{C})$ Represention by Integral Operators

Let us consider some $2\times 2$ matrix of integral operators $\widehat{\mathcal{O}}$ with kernels

$$\mathcal{O}(s,t) = \begin{pmatrix} \mathcal{O}_{11}(s,t) & \mathcal{O}_{12}(s,t) \\ \mathcal{O}_{21}(s,t) & \mathcal{O}_{22}(s,t) \end{pmatrix}, \quad s,t\in[0,\infty), \tag{32}$$

and multiplication defined by[c]

$$\mathcal{O}_{\rm III}(s,t) = \left(\widehat{\mathcal{O}}_{\rm I}\widehat{\mathcal{O}}_{\rm II}\right)(s,t) = \begin{pmatrix} \mathcal{O}^{\rm III}_{11}(s,t) & \mathcal{O}^{\rm III}_{12}(s,t) \\ \mathcal{O}^{\rm III}_{21}(s,t) & \mathcal{O}^{\rm III}_{22}(s,t) \end{pmatrix}, \quad s,t \in [0,\infty), \tag{33}$$

where

$$\begin{aligned} \mathcal{O}^{\rm III}_{11}(s,t) &= \int_0^\infty dr \left(\mathcal{O}^{\rm I}_{11}(s,r)\mathcal{O}^{\rm II}_{11}(r,t) + \mathcal{O}^{\rm I}_{12}(s,r)\mathcal{O}^{\rm II}_{21}(r,t)\right) \\ \mathcal{O}^{\rm III}_{12}(s,t) &= \int_0^\infty dr \left(\mathcal{O}^{\rm I}_{11}(s,r)\mathcal{O}^{\rm II}_{12}(r,t) + \mathcal{O}^{\rm I}_{12}(s,r)\mathcal{O}^{\rm II}_{22}(r,t)\right) \\ \mathcal{O}^{\rm III}_{21}(s,t) &= \int_0^\infty dr \left(\mathcal{O}^{\rm I}_{21}(s,r)\mathcal{O}^{\rm II}_{11}(r,t) + \mathcal{O}^{\rm I}_{22}(s,r)\mathcal{O}^{\rm II}_{21}(r,t)\right) \\ \mathcal{O}^{\rm III}_{22}(s,t) &= \int_0^\infty dr \left(\mathcal{O}^{\rm I}_{21}(s,r)\mathcal{O}^{\rm II}_{12}(r,t) + \mathcal{O}^{\rm I}_{22}(s,r)\mathcal{O}^{\rm II}_{22}(r,t)\right). \end{aligned} \tag{34}$$

Let us now construct a special class of such operators $\widehat{\mathcal{O}}$ which form a representation of $Gl(2,\mathbf{C})$ (the generalization to $Gl(N,\mathbf{C})$ is straightforward). Consider a pair of functions $\alpha(s)$ and $\beta(s)$ on $[0,\infty)$ which we write in Dirac notation as $\alpha(s) \equiv |1\rangle$ and $\beta(s) \equiv |2\rangle$. In this notation we may write left multiplication by $\widehat{\mathcal{O}}_{ik}$ as

$$\widehat{\mathcal{O}}_{ik}|1\rangle = \int_0^\infty dt \mathcal{O}_{ik}(s,t)\alpha(t). \tag{35}$$

Further suppose the functions $A(s) \equiv \langle 1|$ and $B(s) \equiv \langle 2|$ (right multiplication by $\widehat{\mathcal{O}}_{ik}$ is defined integrating over the first argument of $\mathcal{O}_{ik}(s,t)$) satisfy

$$\langle 1|1\rangle \equiv \int_0^\infty ds A(s)\alpha(s) = 1 = \langle 2|2\rangle \equiv \int_0^\infty ds B(s)\beta(s). \tag{36}$$

Observe now that one may define a representation $\widehat{\mathcal{A}}$ of $Gl(2,\mathbf{C})$ via

$$M \in Gl(2,\mathbf{C}) \longmapsto \widehat{\mathcal{A}}(M) = \begin{pmatrix} I - |1\rangle\langle 1| & 0 \\ 0 & I - |2\rangle\langle 2| \end{pmatrix} + \begin{pmatrix} M_{11}|1\rangle\langle 1| & M_{12}|1\rangle\langle 2| \\ M_{21}|2\rangle\langle 1| & M_{22}|2\rangle\langle 2| \end{pmatrix}, \tag{37}$$

($M_{11}$, $M_{12}$, $M_{21}$ and $M_{22}$ are complex numbers and $I$ is the identity operator in the space of integral oprators on $[0,\infty)$). Multiplication by the integral operators (projectors) $|1\rangle\langle 1|$, $|1\rangle\langle 2|$, $|2\rangle\langle 1|$ and $|2\rangle\langle 2|$ is given, for example, by

$$|1\rangle\langle 2|f(s) = \left(\int_0^\infty ds B(s)f(s)\right)|1\rangle. \tag{38}$$

[c] We assume that the integral operators $\widehat{\mathcal{O}}_{ik}$ act on the interval $[0,\infty)$ but this is of course not necessary.

In particular, we have $[I - |1\rangle\langle 1|]\,|1\rangle\langle 1| = 0$. Indeed for any $M,\ N \in Gl(2,\mathrm{C})$ the representation $\widehat{\mathcal{A}}$ has the following properties:

$$\widehat{\mathcal{A}}(MN) = \widehat{\mathcal{A}}(M)\widehat{\mathcal{A}}(N)\ ;\ \widehat{\mathcal{A}}(I) = I\ ;\ \widehat{\mathcal{A}}(M^{-1}) = \widehat{\mathcal{A}}^{-1}(M) \tag{39}$$

$$\det \widehat{\mathcal{A}}(M) = \det M = M_{11}M_{22} - M_{12}M_{21} \tag{40}$$

$$\mathrm{tr}\left(\widehat{\mathcal{A}}(M) - \begin{pmatrix} I - |1\rangle\langle 1| & 0 \\ 0 & I - |2\rangle\langle 2| \end{pmatrix}\right) = \mathrm{tr}M = M_{11} + M_{22} \tag{41}$$

Notice that (40) expresses an infinite dimensional determinant as the determinant of the $2\times 2$ matrix $M$. Let us conclude this section by rewriting the conjugation matrix of our RP using the representation $\widehat{\mathcal{A}}$. We begin by defining functions

$$\begin{cases} |1\rangle = \sqrt{c}e^{is\lambda - cs/2} = \alpha(s), & \langle 1| = \sqrt{c}e^{-is\lambda - cs/2} = A(s) \\ |2\rangle = \sqrt{c}e^{-is\lambda - cs/2} = \beta(s), & \langle 2| = \sqrt{c}e^{is\lambda - cs/2} = B(s) \end{cases} \tag{42}$$

The vectors $|1\rangle$ and $|2\rangle$ are normalized but not orthogonal, so that their inner products are $\langle 1|1\rangle = c\int_0^\infty e^{-cs}ds = 1 = \langle 2|2\rangle$ and $\langle 1|2\rangle = \langle 2|1\rangle^* = c\int_0^\infty e^{-2is\lambda - cs}ds = \frac{c}{c+2i\lambda}$. (for $\lambda$ real the bra and ket notation denote complex conjugation in the usual fashion although this is not necessary). Their outer products are of course integral operators in terms of which we will write the conjugation matrix. We now have

$$\begin{cases} e_1(\lambda|s) = \sqrt{\frac{i}{2\pi}}e^{(ix\lambda/2 + \mathring{\varphi}(\lambda)/2)}|1\rangle \equiv e_1^\circ(\lambda)|1\rangle, \\ E_1(\lambda|t) = \sqrt{\frac{i}{2\pi}}e^{(-ix\lambda/2 - \mathring{\varphi}(\lambda)/2)}\langle 1| \equiv E_1^\circ(\lambda)\langle 1|, \\ e_2(\lambda|s) = \sqrt{\frac{i}{2\pi}}e^{(-ix\lambda/2 - \mathring{\varphi}(\lambda)/2)}|2\rangle \equiv e_2^\circ(\lambda)|2\rangle, \\ E_2(\lambda|t) = -\sqrt{\frac{i}{2\pi}}e^{(ix\lambda/2 + \mathring{\varphi}(\lambda)/2)}\langle 2| \equiv E_2^\circ(\lambda)\langle 2|. \end{cases} \tag{43}$$

Hence, from (25) we have for the conjugation matrix (where from now on the $s$ and $t$ dependence will be taken as understood)

$$\widehat{G}(\lambda) = \begin{pmatrix} I - 2\pi i e_1^\circ(\lambda)E_1^\circ(\lambda)|1\rangle\langle 1| & -2\pi i e_1^\circ(\lambda)E_2^\circ(\lambda)|1\rangle\langle 2| \\ -2\pi i e_2^\circ(\lambda)E_1^\circ(\lambda)|2\rangle\langle 1| & I - 2\pi i e_2^\circ(\lambda)E_2^\circ(\lambda)|2\rangle\langle 2| \end{pmatrix} = \widehat{\mathcal{A}}(G^\circ(\lambda)). \tag{44}$$

where the $2\times 2$ matrix $G^\circ(\lambda)$ which here plays the rôle of $M$ is readily computed to be

$$G^\circ(\lambda) = \begin{pmatrix} 2 & -e^{ix\lambda + \mathring{\varphi}(\lambda)} \\ e^{-ix\lambda - \mathring{\varphi}(\lambda)} & 0 \end{pmatrix}. \tag{45}$$

Except for the additional exponents $\pm \overset{\circ}{\varphi}(\lambda)$, $G^\circ(\lambda)$ is the conjugation matrix one finds for the $2 \times 2$ RP encountered in the case $c = \infty$.

## 5. Asymptotic Analysis of the Riemann Hilbert Problem

Firstly let us sketch the derivation of the asymptotic solution to the $2 \times 2$ RP with conjugation matrix $G^\circ(\lambda)$ given above in (45) (we follow the scheme developed in [3] for the case $\overset{\circ}{\varphi}(\lambda) \equiv 0$). I.e. we are searching for the $2 \times 2$ matrix function $\chi^\circ(\lambda)$ normalized to $\begin{pmatrix} 1 & 0 \\ 0 & 1 \end{pmatrix}$ at infinity and analytic everywhere in the complex plane $\mathbf{C}$ save for a discontinuity across the contour $C = [-q, q]$ where $\chi^\circ(\lambda)$ satisfies

$$\chi^\circ_-(\lambda) = \chi^\circ_+(\lambda) G^\circ(\lambda), \quad \forall \lambda \in C. \tag{46}$$

The idea is to make a "gauge transformation" under which the above RP limits for large $x$ to a new RP with conjugation matrix $\tilde{G} = \begin{pmatrix} 0 & -1 \\ 1 & 0 \end{pmatrix}$. Define, therefore

$$\tilde{\chi}^\circ(\lambda) = \exp\{\psi_\infty \sigma_3\} \chi^\circ(\lambda) \exp\{[(ix/2)(\lambda - g(\lambda)) - \psi(\lambda)]\sigma_3\}, \tag{47}$$

($\sigma_3 = \mathrm{diag}(1, -1)$). The function $g(\lambda) = \sqrt{\lambda^2 - q^2}$ satisfies the scalar RP $g_-(\lambda) = -g_+(\lambda)$ across $C$ with normalization $g(\lambda) \overset{\lambda \to \infty}{\simeq} \lambda - q^2/2\lambda + O(\lambda^{-3})$ and

$$\psi(\lambda) = -\frac{\sqrt{\lambda^2 - q^2}}{2\pi i} \int_C \frac{\overset{\circ}{\varphi}(\mu) d\mu}{{}_+\sqrt{\mu^2 - q^2}(\mu - \lambda)}. \tag{48}$$

satisfies the affine scalar RP $\psi_+(\lambda) + \psi_-(\lambda) + \overset{\circ}{\varphi}(\lambda) = 0$ across the contour $C$ with normalization $\psi(\lambda) \overset{\lambda \to \infty}{\longrightarrow} \psi_\infty = \frac{1}{2\pi i} \int_C \frac{\overset{\circ}{\varphi}(\mu) d\mu}{{}_+\sqrt{\mu^2 - q^2}}$. Notice that $\psi(\lambda)$ is well defined on all of $\mathbf{C}$ even though it depends on the dual quantum field $\overset{\circ}{\varphi}(\lambda)$. One may verify that $\tilde{\chi}^\circ(\lambda)$ satisfies a canonically normalized RP across $C$ with conjugation matrix

$$\tilde{G}^\circ(\lambda) = \exp\{-[(ix/2)(\lambda - g_+(\lambda)) - \psi_+(\lambda)]\sigma_3\} G^\circ(\lambda) \exp\{[(ix/2)(\lambda - g_-(\lambda)) - \psi_-(\lambda)]\sigma_3\}$$

$$= \begin{pmatrix} 2e^{\Delta\psi(\lambda) + ixg_+(\lambda)} & -1 \\ 1 & 0 \end{pmatrix} \overset{x \to \infty}{\longrightarrow} \begin{pmatrix} 0 & -1 \\ 1 & 0 \end{pmatrix}, \tag{49}$$

where $\Delta\psi(\lambda) = \psi_+(\lambda) - \psi_-(\lambda)$. The canonically normalized RP for $\tilde{\chi}^\circ(\lambda)$ is easily solved in the large $x$ limit, we find

$$\tilde{\chi}^\circ(\lambda) \overset{x \to \infty}{\simeq} \begin{pmatrix} \frac{1}{2}(a(\lambda) + a^{-1}(\lambda)) & \frac{i}{2}(a(\lambda) - a^{-1}(\lambda)) \\ -\frac{i}{2}(a(\lambda) - a^{-1}(\lambda)) & \frac{1}{2}(a(\lambda) + a^{-1}(\lambda)) \end{pmatrix}, \tag{50}$$

where $a(\lambda) = (\frac{\lambda+q}{\lambda-q})^{1/4}$ satisfies the canonically normalized RP $a_-(\lambda) = ia_+(\lambda)$ ($\forall\lambda \in C$). Inverting the transformation (47) we obtain the large $x$ solution to the original RP,

$$\chi^\circ(\lambda) \overset{x\to\infty}{\sim} \begin{pmatrix} \frac{1}{2}(a(\lambda)+a^{-1}(\lambda))e^{\psi(\lambda)-\psi_\infty-ix(\lambda-g(\lambda))/2} & \frac{i}{2}(a(\lambda)-a^{-1}(\lambda))e^{-\psi(\lambda)-\psi_\infty+ix(\lambda-g(\lambda))/2} \\ -\frac{i}{2}(a(\lambda)-a^{-1}(\lambda))e^{\psi(\lambda)+\psi_\infty-ix(\lambda-g(\lambda))/2} & \frac{1}{2}(a(\lambda)+a^{-1}(\lambda))e^{-\psi(\lambda)+\psi_\infty+ix(\lambda-g(\lambda))/2} \end{pmatrix}$$

$$= I + \frac{1}{\lambda}\begin{pmatrix} \psi_1 - ixq^2/4 & \frac{iq}{2}e^{-2\psi_\infty} \\ -\frac{iq}{2}e^{2\psi_\infty} & -\psi_1 + ixq^2/4 \end{pmatrix} + O(\lambda^{-2}), \tag{51}$$

where $\psi_1 = \frac{1}{2\pi i}\int_C \frac{\overset{\circ}{\varphi}(\mu)\mu d\mu}{+\sqrt{\mu^2-q^2}}$.

We may now try to utilise our $Gl(2,\mathbf{C})$ representation to write the solution to the finite $c$ operator valued RP as

$$\widehat{\chi}(\lambda) = \widehat{\mathcal{A}}(\chi^\circ(\lambda)), \tag{52}$$

since by property (39)

$$\widehat{\chi}_-(\lambda) = \widehat{\mathcal{A}}(\chi^\circ_-(\lambda)) = \widehat{\mathcal{A}}(\chi^\circ_+(\lambda)G^\circ(\lambda)) = \widehat{\mathcal{A}}(\chi^\circ_+(\lambda))\widehat{\mathcal{A}}(G^\circ(\lambda)) = \widehat{\chi}_+(\lambda)\widehat{G}(\lambda). \tag{53}$$

The observant reader may notice that this solution does not obey the normalization at infinity since the projectors $|1\rangle\langle 1|, |1\rangle\langle 2|, |2\rangle\langle 1|$ and $|2\rangle\langle 2|$ are essentially singular at $\lambda = \infty$. Moreover, as bounded operator valued functions they are analytic in the strip $|\mathrm{Im}\lambda| < c/2$ only. Therefore strictly speaking equation (52) is not correct. The correct version is

$$\widehat{\chi}(\lambda) = \widehat{\Phi}(\lambda)\widehat{\mathcal{A}}(\chi^\circ(\lambda)), \tag{54}$$

where $\widehat{\Phi}(\lambda)$ is an operator valued function analytic in the strip $\mathcal{S} : |\mathrm{Im}\lambda| < c/2$. The last equation itself may be treated as a Riemann Hilbert Problem. Indeed, let $\Gamma$ be a simple contour belonging to the strip $\mathcal{S}$ encircling the interval $C$ in the counterclockwise direction. Introducing

$$\widehat{Y}(\lambda) = \begin{cases} \widehat{\chi}(\lambda) \text{ if } \lambda \text{ is outside } \Gamma \\ \widehat{\Phi}(\lambda) \text{ if } \lambda \text{ is inside } \Gamma, \end{cases} \tag{55}$$

we can rewrite our initial operator RP 1-3 as the following new RP[d]:

$1^\Gamma$. $\widehat{Y}(\lambda)$ is analytic $\forall\lambda \in \mathbf{C}\setminus\Gamma$.

$2^\Gamma$. $\widehat{Y}_-(\lambda) = \widehat{Y}_+(\lambda)\widehat{\mathcal{A}}(\chi^\circ(\lambda))$, $\lambda \in \Gamma$.

---

[d]To avoid confusion, in agreement with our conventions let us reiterate that $+$ is the limit from <u>inside</u> the anticlockwise contour $\Gamma$.

$3^\Gamma$. $\widehat{Y}(\infty) = I \begin{pmatrix} 1 & 0 \\ 0 & 1 \end{pmatrix}$.

From (31) and the equation

$$\det \widehat{\chi}_{11}(\lambda) = 1 + \mathrm{tr}\widehat{M}_1^{11}/\lambda + O(\lambda^{-2}), \tag{56}$$

we see that to calculate $\det(I+\hat{V})$ we need $\det \widehat{Y}^{11}(\lambda)$ for large $\lambda$. It follows from $2^\Gamma$ that

$$\det \widehat{Y}_+^{11}(\lambda) = \det \widehat{Y}_-^{11}(\lambda)\,(\overset{\circ}{\chi}_{22}(\lambda) - \overset{\circ}{\chi}_{21}(\lambda)c(\lambda))\,, \tag{57}$$

where "det" indicates the determinant of integral operators on the space $[0,\infty)$. and $c(\lambda) = \langle 1|[\widehat{Y}_-^{11}(\lambda)]^{-1}\widehat{Y}_-^{12}(\lambda)|2\rangle = \langle 1|[\widehat{\chi}^{11}(\lambda)]^{-1}\widehat{\chi}^{12}(\lambda)|2\rangle$. Let us now assume that

(a) $c(\lambda)$ is analytic outside $\Gamma$,

(b) $c(\lambda) = O(1/\lambda)$ as $\lambda \to \infty$.

Note that the product $[\widehat{Y}_-^{11}(\lambda)]^{-1}\widehat{Y}_-^{12}(\lambda)$ $(= [\widehat{\chi}_-^{11}(\lambda)]^{-1}\widehat{\chi}_-^{12}(\lambda))$ has these properties.

Statements (a) and (b) constitute our main hypothesis concerning the solution $\widehat{\chi}(\lambda)$ of the RP 1-3. By virtue of this hypothesis, equation (57) implies that

$$\det \widehat{Y}^{11}(\lambda) = (\overset{\circ}{\chi}_{22}(\lambda) - \overset{\circ}{\chi}_{21}(\lambda)c(\lambda))^{-1}, \tag{58}$$

for all $\lambda$ outside of the contour $\Gamma$. In addition property (b) leads to the equation

$$\lim_{\lambda\to\infty} \lambda(\det \widehat{Y}^{11}(\lambda) - 1) = -\lim_{\lambda\to\infty} \lambda(\overset{\circ}{\chi}_{22}(\lambda) - 1). \tag{59}$$

Taking (56) and (51) into account we obtain

$$\mathrm{tr}\widehat{M}_1^{11} \overset{x\to\infty}{=} \psi_1 - ixq^2/4 + o(1), \tag{60}$$

which yields our conjecture for the determinant

$$\det(I+\hat{V}) \overset{x\to\infty}{=} (\mathrm{const})\exp\left\{-(xq)^2/8 - \frac{x}{2\pi}\int_C \frac{\overset{\circ}{\varphi}(\mu)\mu d\mu}{+\sqrt{\mu^2-q^2}} + o(x)\right\}, \tag{61}$$

as quoted in section 2. Finally let us observe that for $c=\infty$, $[\overset{\circ}{p}(\lambda), \overset{\circ}{q}(\mu)] = 0$ so that for any functional $F[\overset{\circ}{\varphi}(\lambda)]$ the dual vacuum expectation value $(0|F[\overset{\circ}{\varphi}(\lambda)]|0) = F\big|_{\overset{\circ}{\varphi}=0}$. Hence $(0|\det(I+\hat{V})|0)\,|_{c=\infty} \sim e^{-(xq)^2/8}$. An old result first obtained in [4] and then extended to the complete asymptotic expansion in [5].

## 6. Acknowledgements

This work was supported by NSF grant PHY-9321165 and DMS-9501559.

## 7. References


1. E. H. Lieb and W. Liniger, *Phys. Rev.* **130** (1963) 1605. E. H. Lieb, *Phys. Rev.* **130** (1963) 1616.
2. V. E. Korepin, N. M. Bogoliubov and A.G. Izergin, *Quantum Inverse Scattering Method* (Cambridge University Press, 1993).
3. P. A. Deift, A. R. Its and X. Zhou (to be published), *A Riemann–Hilbert approach to Asymptotic Problems Arising in the Theory of Random Matrix Models, and in the Theory of Integrable Statistical Mechanics.*
4. J. des Cloizeaux and M. L. Mehta, *J. Math. Phys.* **14** (1973) 1648.
5. F. Dyson, *Commun. Math. Phys.* **47** (1976) 171.

# Generalized Ideal Gas and Strongly Correlated Systems

Yong-Shi Wu

Department of Physics, University of Utah
Salt Lake City, UT 84112, U.S.A.

ABSTRACT

Generalized ideal gas is introduced with a new combinatoric rule for state counting, generalizing Pauli's exclusion principle. Here I review progress made by me and collaborators on its statistical mechanics and realization in strongly correlated systems. Particularly, the low-temperature fixed point of Luttinger liquids is shown to be described by ideal excluson gas.

## 1. Introduction

Since early days of quantum mechanics, an outstanding problem has been to search for a generalization or interpolation between Bose-Einstein and Fermi-Dirac statistics which is *physically relevant*, i.e. realizable in systems existing in Nature. Such a search in particle physics in the fifties and sixties has not produced a real case for known fundamental particles. However, after continuing efforts in last two decades, now we feel sure that fractional statistics do occur to elementary excitations in lower dimensional condensed matter systems.

There have been two distinct ways to define fractional statistics: 1) by examining the change in phase of a wave function due to exchanging two identical particles, and 2) by counting the number of multi-particle quantum states, so-called statisitcal weight. Because of their conceptual difference, I will call the two definitions distinguishably as "exchange statistics" and "*exclusion statistics*", and the corresponding particles as "anyons" and "*exclusons*". Fractional exchange statistics or anyons have been first explored in the study of quasiparticles in two (space) dimensional systems, such as fractional quantum Hall fluids and anyon superconductivity. In this review I will concentrated on fractional exclusion statistics (FES) and associated generalized ideal gases (GIG), and their realization in strongly correlated systems. The applications include 1d quantum fluids described by integrable many-body models, and quasiparticle excitations in the fractional quantum Hall effect (FQHE). Moreover, we suggest that ideal excluson gases (IEG) may be used to characterize the low-energy (or low temperature) fixed points of some quantum liquids exhibiting non-Fermi behavior, such as 1d Luttinger liquids, just like Landau's Fermi liquids having ideal Fermi gas as their low-temperature fixed point.

## 2. Statistical Distribution in a Generalized Ideal Gas

We consider the ideal situation, what we call as a "generalized ideal gas" [1], in which the system satisfies the following two conditions: 1) The total energy (eigenvalue) is always of the form of a simple sum, $E = \sum_i N_i \varepsilon_i$, with $N_i$ being particle number, and $\varepsilon_i$ identified as the energy of a particle of species $i$; 2) The many-body state counting or the statistical weight $W$ is given by the combinatoric rule:

$$W = \prod_i \frac{[G_i + N_i - 1 - \sum_j \alpha_{ij}(N_j - \delta_{ij})]!}{(N_i)!\,[G_i - 1 - \sum_j \alpha_{ij}(N_j - \delta_{ij})]!} \,. \tag{1}$$

Here $G_i$ is the number of states when the system consists of only a single particle.

We note that eq. (1) is applicable to usual Bose or Fermi ideal gas with species index $i$ labeling different energy levels and with $\alpha_{ij} = \alpha\delta_{ij}$ and $\alpha = 0, 1$. By definition, the diagonal $\alpha_{ii}$ is the "self-exclusion" statistics of species $i$, while the non-diagonal $\alpha_{ij}$ (for $i \neq j$) is the mutual-exclusion statistics. The interpretation of the second condition is the following: By assumption, the statistical weight remains to be *a single combinatoric number* for each species, so one can count the states by thinking of the particles *effectively either as bosons or as fermions*, with the number of available single-particle states for species $i$ *linearly dependent on all particle numbers* $N_j$:

$$G^{(b)}_{eff,i} = G_i - \sum_j \alpha_{ij}(N_j - \delta_{ji}), \quad \text{or} \quad G^{(f)}_{eff,i} = G^{(b)}_{eff,i} + N_i - 1. \tag{2}$$

This is equivalent to Haldane's original definition [2], which uses the difference relation

$$\Delta G^{(b)}_{eff,i} = -\sum_j \alpha_{ij}\, \Delta N_j \,, \tag{3}$$

The term "generalized" is used here to emphasize that mutual exclusion is allowed between particles of the same kind but with different quantum numbers (say, energies or momenta), a situation common in 1d quantum fluids and FQHE (see below).

For our definition to apply, on one hand, the interaction energies between particles should be negligible. On the other hand, the particles have to be strongly correlated so that the generalized fractional exclusion principle, as formulated by eq. (1), applies. We expect that realization of the two conditions can be found perhaps only in *elementary excitations of some strongly correlated systems.*

Consider a grand canonical ensemble with the partition function

$$Z = \sum_{\{N_i\}} W(\{N_i\}) \, \exp\{\sum_i N_i(\mu_i - \varepsilon_i)/kT\} \,. \tag{4}$$

By usual procedure, I have derived [1] the statistical distribution for mean occupation numbers, defined by $n_i \equiv \bar{N}_i/G_i$, which are determined by

$$\sum_j (\delta_{ij} w_j + \beta_{ij}) n_j = 1 \,, \tag{5}$$

with $\beta_{ij} \equiv \alpha_{ij} G_j / G_i$, and $w_i$ being determined by the functional equations

$$(1 + w_i) \prod_j \Big(\frac{w_j}{1 + w_j}\Big)^{\alpha_{ji}} = e^{(\varepsilon_i - \mu_i)/kT}. \tag{6}$$

The corresponding thermodynamic potential $\Omega \equiv -kT \log Z$ is given by

$$\Omega \equiv -PV = -kT \sum_i G_i \log \frac{1 + n_i - \sum_j \beta_{ij} n_j}{1 - \sum_j \beta_{ij} n_j} . \tag{7}$$

Above results *apply to cases when some* $G_j$ *vanish*, with non-zero $N_j$ due to *negative* value of certain mutual statistics $\alpha_{jk}$: We rewrite $\sum_j (\delta_{ij} w_j + \alpha_{ij}) \bar{N}_j = G_i$ , and

$$\Omega \equiv -PV = -kT \sum_i G_i \log \frac{G_i + \bar{N}_i - \sum_j \alpha_{ij} \bar{N}_j}{G_i - \sum_j \alpha_{ij} \bar{N}_j} . \tag{8}$$

As an example, consider ideal gas (with no mutual statistics) consisting of only one kind of particles, with $i$ labeling energy levels and $\alpha_{ij} = \alpha \delta_{ij}$ and $\mu_i = \mu$. Then

$$n_i = \frac{1}{w(e^{(\varepsilon_i - \mu)/kT}) + \alpha} , \tag{9}$$

where the function $w(\zeta)$ satisfies the functional equation

$$w(\zeta)^\alpha [1 + w(\zeta)]^{1-\alpha} = \zeta \equiv e^{(\varepsilon - \mu)/kT} . \tag{10}$$

Let us consider several limiting situations: 1) Eq. (9) recovers the usual Bose and Fermi cases, respectively, with $\alpha = 0$ and $\alpha = 1$. 2) In the dilute limit, $\zeta$ is very large and $w(\zeta) \approx \zeta$, so we recover the Boltzmann distribution independently of $\alpha$: $n_i = e^{-(\varepsilon_i - \mu)/kT}$, at sufficiently low densities, as it should be. 4) For $\alpha > 0$, $\zeta$ is non-negative, so is $w$. Thus, $n_i \leq 1/\alpha$ , exhibiting effects of fractional exclusion. Especially, at $T = 0$, the occupation numbers for a continuous spectrum is step-like:

$$n_i = \begin{cases} 0, & \text{if} \quad \varepsilon_i > E_F, \\ 1/\alpha, & \text{if} \quad \varepsilon_i < E_F. \end{cases} \tag{11}$$

Note the fractional occupation number $1/\alpha$ below the Fermi surface. 5) There is an interesting duality [4] between statistics $\alpha$ and $1/\alpha$, i.e.

$$w_i(T; 1/\alpha) = 1/w_i(-T/\alpha; \alpha). \tag{12}$$

The sign change in $T$ means a sign change in $\epsilon_i - \mu$, i.e. exchanging particle and hole.

The case with quadratic dispersion, $\varepsilon_i = k_i^2/2m$, is of particular interests: $G_i = V k_i \Delta k_i / 2\pi$, say in two dimensions. Treating momentum as continuous, one gets the Fermi momentum at $T = 0$, which is $\alpha$ *times larger* than the usual one:

$$k_F^2 = (4\pi\alpha)(N/V). \tag{13}$$

In the dilute limit, up to first order in density we have the equation of state

$$PV = NkT[1 + (2\alpha - 1)N\lambda^2/4V] , \tag{14}$$

where $\lambda = (2\pi/mkT)^{1/2}$. So the "statistical interactions" are attractive or repulsive depending on whether $\alpha < 1/2$ or $\alpha > 1/2$.

## 3. Integrable Models for One Dimensional Quantum Fluids

For physical realization of generalized ideal gas, the first class of models we want to discuss is the *integrable* (continuum or lattice) interacting many-body models that describe quantum fluids in one dimension. Suppose that particles have *no internal degrees of freedom.* The $N$-body eigenstates on a circle are labeled by pseudo-momenta $\{k_r\}$ $(r = 1,, \cdots, N)$, which are solutions of the Bethe Ansatz Equations (BAE)

$$L\,k_r = \sum_{s\neq r} \theta(k_r - k_s) + 2\pi\, I_r\ , \tag{15}$$

with $L$ the length of the system, and $\theta(k)$ the two-body scattering phase shift. Here $\{I_r\}$ is a set of integers or half integers, depending on N being odd or even, which can be chosen to be all different. The energy of the eigenstates is indeed always of the form of a sum: $E = \sum_r \epsilon^0(k_r)$ .

For state counting, we follow Yang and Yang [3] to explore the BAE (15). Divide the $k$-space into intervals with equal size $\Delta k$, then treat the particles with pseudo-momentum in the $i$-th interval as belonging to the $i$-th species. Then introduce the density $\rho(k)$ of particles of momentum $k$ and the density of holes $\rho_h(k)$ by

$$\rho(k) = \frac{1}{L(k_{r+1} - k_r)}\,; \quad \rho_h(k) = \frac{I_{r+1} - I_r - 1}{L(k_{r+1} - k_r)}. \tag{16}$$

Then from (15) it follows that

$$\rho_h(k) + \rho(k) + \int_{-\infty}^{+\infty} \frac{dk'}{2\pi}\,\theta'(k - k')\rho(k') = \rho^0(k) \equiv \frac{1}{2\pi}. \tag{17}$$

Here the prime over $\theta$ stands for the derivative. The linear dependence of the hole density $\rho_h(k)$ on the particle density $\rho(k')$ implies the statistics matrix [4]

$$\alpha(k, k') = \delta(k - k') + \frac{1}{2\pi}\,\theta'(k - k')\ . \tag{18}$$

This shows that the dynamical interactions have been transmuted into statistical interactions. Hence these integrable models are indeed *generalized ideal gases with non-trivial exclusion statistics between identical or different pseudo-momenta.* Our formulas (7) and (6) for thermodynamics of GIG, when applied to these models, coincide with those of Yang-Yang's theormodynamics [3]. In particular,

$$w_i \longleftrightarrow \exp\,[\epsilon(k)/k_B T]; \quad G^{(b)}_{eff,i} \longleftrightarrow L\rho_h(k). \tag{19}$$

$\epsilon(k)$ is Yang-Yang's dressed energy, and $\rho_h(k)$ *represents density of available states.*

For the repulsive $\delta$-function boson gas with two-body potential $V(x) = 2c\delta(x)$ $(c \geq 0)$, we have $\theta(k) = -2\,tan^{-1}(k/c)$, and the statistical matrix

$$\alpha(k,k') = \delta(k-k') - \frac{1}{\pi}\frac{c}{c^2+(k-k')^2}\,. \tag{20}$$

When $c \to 0$ or $c \to \infty$, we recover respectively the ideal Bose or Fermi gas. But for intermediate values of $c$, the thermodynamics is that of a generalized Fermi ideal gas with a specific mutual statistics between different pseudo-momenta.

In the Calogero-Sutherland model, fermions interact through a long-range two-body potential $\lambda(\lambda-1)/sin^2 x$. Since $\theta(k) = \pi(\lambda-1)\,sgn(k)$ , the statistical matrix is $\alpha(k,k') = \lambda\,\delta(k-k')$, describing an IEG (with no mutual statisitcs between different pseudo-momemnta). Bosonization [5] shows that the low-temperature behavior of the system can be described by a $c=1$ conformal field theory with "compactified radius" equal to $1/\sqrt{\lambda}$; and the duality $\lambda \longleftrightarrow 1/\lambda$ (eq. (12)) has a counterpart in $c=1$ conformal field theories.

For models with internal degrees of freedom, such as spin, it has been shown that the Haldane-Shastry quantum spin chain (with $1/x^2$ exchange interactions) is exactly solvable, and has an infinite-dimensional symmetry, called *Yangian* [6]. Moreover, the elementary excitations, called spinons, are exclusons with statistics $\alpha = 1/2$, and the system can be characterized as ideal gas of these semionic spinons [7].

Another interesting model is the Hubbard chain. It has two classes of excitations (charge-spin separation). From the BAE one reads [8] off the statistics matrix

$$\alpha_{cc}(k,k') = \delta(k-k'), \quad \alpha_{ss}(\lambda,\lambda') = \delta(\lambda-\lambda') + \frac{1}{\pi}\frac{1}{1+(\lambda-\lambda')^2}, \tag{21}$$

$$\alpha_{cs}(k,\lambda) = -\frac{4}{\pi U}\frac{\cos k}{1+4\,(2\sin k/U-\lambda)^2}, \tag{22}$$

$$\alpha_{sc}(\lambda,k) = -\frac{1}{\pi}\frac{2}{1+4\,(\lambda-2\sin k/U)^2}\,. \tag{23}$$

where $k$ (or $\lambda$) is momentum (rapidity) of a charge (spin) excitation. Thus, the Hubbard chain describes *a GIG with mutual statistics between charge and spin excitations.* At low temperatures, the thermodynamic potential $\Omega$ is a sum of the charge and the spin parts: $\Omega = \Omega_c + \Omega_s$, indicating charge-spin separation [8,9]:

$$\frac{\Omega_c}{L} = -\frac{T}{2\pi}\int_{-\pi}^{\pi} dk \log\left[1+e^{-\epsilon_c(k)/T}\right]; \quad \frac{\Omega_s}{N} = -\frac{T}{2\pi}\int_{-\infty}^{\infty} d\chi \log\left[1+e^{-\epsilon_s(\chi)/T}\right]. \tag{24}$$

## 4. Self and Mutual Exclusion Statistics for FQHE Quasiparticles

Another physical example of exclusons is quasiparticles (including both quasiholes and quasielectrons) in incompressible FQHE fluids, which are known to be fractionally charged anyons. As we will see, they are also exclusons with both fractional self (diagonal) and mutual (nondiagonal) exclusion statistics. Actually the Haldane [10] and the Jain [11] hierarchy schemes lead to the same assignment for self-statistics, but to different assignments for mutual statistics between quasiholes and quasielectrons.

Haldane' scheme is based on the picture that the FQHE quasiparticles are vortex-like excitations, for which the number of available states is determined by $N_e$, the number of fluid particles (electrons):

$$G^{(b)}_{eff,-} = N_e, \qquad G^{(b)}_{eff,+} = N_e - 2(N_+ - 1). \tag{25}$$

Here the subscripts $-$ and $+$ refer to quasiholes and quasielectrons respectively. The second term in $G^{(b)}_{eff,+}$ is appropriate due to the *hard-core* nature of quasielectrons. Now let us eliminate $N_e$ from these equations in terms of the external flux, $N_\phi = mN_e + N_- - N_+$, for a FQH liquid with electron filling $\nu_0 = 1/m$ ($m$ odd):

$$\begin{aligned} G^{(b)}_{eff,-} &= \frac{1}{m}N_\phi - \frac{1}{m}N_- + \frac{1}{m}N_+ \,, \\ G^{(b)}_{eff,+} &= \frac{1}{m}N_\phi - \frac{1}{m}N_- - (2 - \frac{1}{m})N_+ \,. \end{aligned} \tag{26}$$

The first term on the right side gives $G_+ = G_- = (1/m)N_\phi$, implying fractional charge $1/m$ (in absolute value) for the quasiparticles. The coefficients in other terms give the exclusion statistics:

$$\begin{aligned} \alpha_{--} &= 1/m, & \alpha_{-+} &= -1/m, \\ \alpha_{+-} &= 1/m, & \alpha_{++} &= 2 - 1/m \,. \end{aligned} \tag{27}$$

Jain has proposed [11] a different picture for the FQHE quasiparticles: they are composite fermions, i.e. electrons each combined with $2p$ flux quanta, and fill the Landau levels in the residue magnetic field $B_{eff} = B - B_{1/2p}$, where $B_{1/2p}$ is the value at exact electron filling $1/2p$. When a certain number, say $n$, of Landau levels in $B_{eff}$ are completely filled,, the system is in an incompressible state, which is identified as an FQH hierarchical state with $\nu = n/(2pn+1)$. Consider, say, the case of $n = 1$ and write $2p = m - 1$. The Landau degeneracy for the composite fermion is given by the effective flux $N_{\phi,eff} = N_\phi - 2pN_e$, while the numbers of excitations are constrained by $N_{\phi,eff} = N_e + N_- - N_+$. The number of available single-particle states for unit-charged composite-fermion excitations is obviously

$$G_{eff,\mp} = N_{\phi,eff} - N_\mp \,. \tag{28}$$

Here we have assumed that the magnetic field is so strong that we can ignore the possibility for quasielectron to fill a Landau level higher than the lowest available one. Eliminating $N_{\phi,eff}$ in eq. (28) in term of the external $N_\phi$, we have the statistics:

$$\begin{aligned} \alpha_{--} &= 1/m, & \alpha_{-+} &= 1 - 1/m, \\ \alpha_{+-} &= -(1 - 1/m), & \alpha_{++} &= 2 - 1/m \,. \end{aligned} \tag{29}$$

Comparing this equation with eq. (27), we see that the Jain scheme leads to the same diagonal statistics for quasiholes and quasielectrons (with the latter being automatically *hard-core*) as Haldane's. Thus the two schemes give the same predictions for the occurrence of hierarchical states at $T = 0$, for which only diagonal statistics is relevant. Indeed, at $T = 0$ only one species of quasiparticles exist. Say, if $N_\phi < mN_e$, there are only quasielectrons, whose number is $N_+ = mN_e - N_\phi$. When their density is not too high, we ignore interactions between them and apply eq. (7), resulting in the equation of state

$$\frac{PV}{NkT} = \left(\frac{\rho_+}{\rho_0}\right)^{-1} \log \frac{1 + (1 - \alpha_{++})(\rho_+/\rho_0)}{1 - \alpha_{++}(\rho_+/\rho_0)} . \tag{30}$$

It implies that the pressure $P$ diverges at the critical density

$$\rho_{+,c} = (1/\alpha_{++})\rho_0, \tag{31}$$

leading to a new incompressible state. With statistics $\alpha_{++} = 2 - 1/m$, the critical density of quasielectrons is reached when electron filling is $\nu \equiv N_e/N_\phi = 2/(2m-1)$, corresponding to the well-known hierarchical states. So *the experimental existence of FQHE hierarchical states can be viewed as evidence for the exclusonic nature of the FQH quasiparticles.*

How to test the difference in mutual statistics for FQHE quasiparticles in the Haldane and the Jain schemes? For a situation with coexistence of both quasielectrons and quasiholes, we may examine thermal activation of quasiparticle pairs *at finite temperatures.* The densities $\rho_\pm$ of the excitations are given by

$$n_\pm = \frac{\rho_\pm}{\rho_0} = \frac{w_\mp + \alpha_{\mp\mp} - \alpha_{\pm\mp}}{(w_+ + \alpha_{++})(w_- + \alpha_{--}) - \alpha_{+-}\alpha_{-+}} , \tag{32}$$

where $\rho_0 \equiv G_\pm/V$; $w_\pm$ satisfy the functional equations

$$w_\pm{}^{\alpha_{\pm\pm}} (1 + w_\pm)^{1-\alpha_{\pm\pm}} \left(\frac{w_\mp}{1 + w_\mp}\right)^{\alpha_{\mp\pm}} = e^{(\varepsilon_\pm - \mu_\pm)/kT} . \tag{33}$$

Here $\varepsilon_\pm$ is the creation energy of a single excitation. Due to the charge conservation, only pairs of quasielectron and quasihole are thermally activated, leading to the condition $N_+ - N_- = mN_e - N_\phi$. Furthermore, the chemical potentials for the two species should satisfy $\mu_+ + \mu_- = 0$. Multiplying the two equations in eq. (33) and using the above constraints, one can derive a polynomial equation that uniquely determines the $T$-dependent densities of both excitations. Thermodynamic quantities, such as specific heat and magnetization per unit area, are calculable. The numerical results will be reported in details somewhere else [12]. The hope is that someday these theoretical predictions could be put to experimental tests, despite the tremendous difficulties in measuring thermodynamic quantities of a thin layer of electron gas.

## 5. Characterization of Luttinger Liquids by Ideal Excluson Gas

An important issue is the relevance of GIG to realistic gapless systems for which state counting is somewhat obscure due to strong correlations among particles. Recently we have suggested[5] that at least *for some strongly correlated systems or non-Fermi liquids, their low-energy or low-temperature fixed point may be described by a GIG*, similar to the way that of Landau-Fermi liquids by ideal Fermi gas. A well-established class of non-Fermi liquids is the 1d Luttinger liquids, proposed by Haldane[13]. We have shown that the low-$T$ critical properties of Luttinger liquids are exactly reproduced by those of 1d IEG, if one identifies Haldane's controlling parameter of the former with the statistics $\lambda$ of the latter. A main development here is that we have succeeded in bosonizing 1d excluson systems at low temperatures [5].

To sketch, consider an IEG of $N_0$ particles on a ring with size $L$, with total energy and momentum given by $E = \sum k_i^2$ and $P = \sum k_i$. The thermodynamic potential is

$$\Omega = -\frac{T}{2\pi} \int_{-\infty}^{\infty} dk \, \ln(1 + w(k,T)^{-1}), \tag{34}$$

with the function $w(k,T) \equiv \rho_h(k)/\rho(k) = 1/2\pi\rho(k) - \lambda$ satisfying

$$w(k,T)^{\lambda}[1 + w(k,T)]^{1-\lambda} = e^{(k^2-\mu)/T}. \tag{35}$$

In the ground state, there is a Fermi surface such that $\rho(k) = 1/2\pi\lambda$ for $|k| < k_F$ and $\rho(k) = 0$ for $|k| < k_F$, with Fermi momentum $k_F = \pi\lambda N_0/L$. There are three classes of excitations: density fluctuations due to particle-hole excitations, i.e. sound waves with velocity $v_s = v_F \equiv 2k_F$, charge excitations obtained by adding extra $M$ particles to the ground state, and persistent currents by Galileo boost. By calculating corresponding changes in ground state energy, the velocities of these three classes of elementary excitations can be determined to be

$$v_N = v_s\lambda, \quad v_J = v_s/\lambda, \quad v_s = \sqrt{v_N v_J}. \tag{36}$$

These coincide with the well-known relations[13] in Luttinger liquid theory, if we identify $\lambda$ with Haldane's controlling parameter $\exp(-2\varphi)$.

To calculate the exponents in IEG, we have developed a bosonization approach. Following Yang and Yang[3], we introduce the dressed energy $\epsilon(k,T)$ by writing

$$w(k,T) = e^{(\epsilon(k,T)-\mu)/T}. \tag{37}$$

In the low-T limit, according to eq. (35), $\epsilon(k,T) = \epsilon(k) + O(e^{-|\epsilon|/T})$, where

$$\epsilon(k) = \begin{cases} (k^2 - k_F^2)/\lambda + k_F^2, & |k| < k_F, \\ k^2, & |k| > k_F. \end{cases} \tag{38}$$

Thus, the low-$T$ grand partition function can be obtained from the effective Hamiltonian given by $H_{eff} = \sum_k \epsilon(k) \, c_k^\dagger c_k$, where $c_k^\dagger$ are fermionic creation operators. To bosonize this Hamiltonian, we construct the density fluctuation operator at $k \sim k_F$ as follows

$$\rho_q^{(+)} = \sum_{k>k_F} :c_{k+q}^\dagger c_k: + \sum_{k<k_F-\lambda q} :c_{k+\lambda q}^\dagger c_k: + \sum_{k_F-\lambda q<k<k_F} :c_{\frac{k-k_F}{\lambda}+k_F+q}^\dagger c_k: \tag{39}$$

for $q > 0$. A similar density operator $\rho_q^{(-)}$ can also be defined for $k \sim -k_F$. Within the Tomonaga approximation, in which commutators are taken to be their ground-state expectation value, we obtain the current algebra

$$[\rho_q^{(\pm)}, \rho_q^{(\pm)\dagger}] = \frac{\lambda L}{2\pi} q\, \delta_{q,q'}, \quad [H_\pm, \rho_q^{(\pm)}] = \pm v_F q \rho_q. \tag{40}$$

which describe 1d free phonons with the sound velocity $v_s = v_F \equiv 2k_F$. Introducing normalized boson annihilation operators $b_q = \sqrt{2\pi/\lambda q L}\, \rho_q^{(+)}$, $\tilde{b}_q = \sqrt{2\pi/\lambda q L}\, \rho_q^{(-)\dagger}$, the bosonized Hamiltonian satisfying (40) is given by

$$H_B = v_s\{\sum_{q>0} q(b_q^\dagger b_q + \tilde{b}_q^\dagger \tilde{b}_q) + \frac{1}{2}\frac{\pi}{L}[\lambda M^2 + \frac{1}{\lambda} J^2]\}. \tag{41}$$

where $M$ and $J$ are operators with integral eigenvalues, describing quantum numbers of charge and current excitations. This bosonized Hamiltonian coincides with Haldane's harmonic fluid description of Luttinger liquids[14], again with identification $\lambda = \exp(-2\varphi)$. So the critical exponents of IEG reproduce those of Luttinger liquids. In this sense, *IEG can be used to characterize the fixed points of Luttinger liquids.*

Effects of mutual statistics between different momenta and of Luttinger density-density interactions are also studied. In both cases, the low-$T$ behavior is controlled by an effective statistics $\lambda_{eff}$ for excitations near the Fermi points. In one dimension both the momentum independent part of interactions and change in chemical potential $\mu$ are *relevant* perturbations, leading to a continuous shift in the fixed-point line parametrized by $\lambda_{eff}$ and to "non-universal" exponents.

## 6. Discussions

The new features of exclusion statistics include: i) It does not require two spatial dimensionality, in contrast to anyons. ii) It allows fractional exclusion as a generalization of Pauli's exclusion principle. In particular, at $T = 0$, there is always a Fermi surface except for bosons. iii) Mutual exclusion is allowed either between non-identical particles, such as quasielectrons and quasiholes in the FQHE, or between identical particles at different energy levels, as in certain 1d integrable models. iv) The statistics parameters $\alpha_{ij}$ are not periodic as the exchange phase for anyons. These new features of FES are expected to lead to rich and novel physics for elementary excitations in strongly correlated systems.

We emphasize that with $\alpha = 1/n$, the statistical weight (1) is *not* the same as that for the following problem: Put $N$ balls into a fixed number, $G$, of boxes subject to the constraint that at most $n$ balls are allowed in one box.

There are a lot of interesting questions to be investigated, a few of them being

1. What is the algebraic, or group theoretical, characterization of the exclusion statistics? In 1d, somehow Yang-Baxter equations should be relevant. The most general characterization of exclusion statistics may go beyond Yangians, which is known to be responsible, e.g. , for semion statistics of spinons in the long-range quantum spin chain.

2. What is the field theoretical or operator formulation of the "free" exclusons?

3. Thermodynamic properties of FQH edge states should be studied by using statistical mechanics of IEG or GIG.

4. What would be the applications of FES and GIG in higher dimensions?

To conclude, we have formulated quantum statistical mechanics of generalized ideal gas of exclusons (particles or elementary excitations obeying fractional exclusion statistics including mutual statistics), by adopting a new combinatoric rule for many-body state counting. In one-dimensional quantum fluids, dynamical interactions are transmuted into exclusonic statistical interactions between elementary excitations. The Luttinger-liquid fixed points have been identified with ideal excluson gases. On the other hand, in the FQH liquids anyonic quasiparticles are also exclusons with fractional self and mutual statistics. Thus we have seen that generalized ideal gas has provided a unified conceptual framework for these well-understood strongly correlated systems. It would be interesting to look for realization in higher dimensional systems.

The author is deeply indebted to Prof. C.N. Yang for drawing his attention to quantum statistical mechanics in 1d systems. He acknowledges pleasant collaboration with D. Bernard, Y. Hatsugai, M. Kohmoto and Yue Yu, and helpful discussions with B. Sutherland, D. Mattis, and Fu-Chun Zhang. This work was supported in part by U.S. NSF grant PHY-9309458.

## References

1. Y.S. Wu, *Phys. Rev. Lett.* **73**, (1994) 922.
2. F.D.M. Haldane, *Phys. Rev. Lett.* **67**, (1991) 937.
3. C.N. Yang and C.P. Yang, *J. Math. Phys.* **10**, (1969) 1115.
4. D. Bernard and Y.-S. Wu, in "*Recent Developments on Quantum Integrable Models*", Proceedings of the Sixth Nankai Workshop (June 1994, Tianjin, China), ed. M.L. Ge and Y.S. Wu; World Scientific, Singapore, 1995.
5. Y.S. Wu and Yue Yu, *Phys. Rev. Lett.* **75**, (1995) 890.
6. F.D.M. Haldane et al., *Phys. Rev. Lett.* **69**, (1992) 2021.
7. F.D.M. Haldane, in *Proc. of XVI Taniguchi Symposium*, (Japan, 1993); ed. Kawakami and Okiji (Springer Verlag, 1994).
8. Y. Hatsugai, M. Kohmoto, T. Koma and Y.S. Wu, cond-mat/9504112.
9. Y.S. Wu and Yue Yu, in preparation.
10. F.D.M. Haldane, *Phys. Rev. Lett.* **51**, (1983) 605.
11. J.K. Jain, *Phys. Rev. Lett.* **63**, (1989) 199.
12. Y. Hatsugai, M. Kohmoto, Y.S. Wu and Y. Yu (in preparation).
13. F.D.M. Haldane, *J. Phys. C* **14**, (1981) 2585.
14. F.D.M. Haldane, *Phys. Rev. Lett.* **25**, (1982) 1840.

# STATISTICAL PHYSICS AND PROTEIN FOLDING

PETER G. WOLYNES
*School of Chemical Sciences, University of Illinois*
*Urbana, Illinois 61801, USA*
*E-mail: wolynes@aries.scs.uiuc.edu*

Statistical physicists have contributed to the study of biomolecules for many years. The discovery in the early 1950s that DNA was a nearly one-dimensional double helix with a simple repetitive average structure in which the complexity or "disorder" of the sequences was compensated via base pairing was quite remarkable. Schröedinger, in his essay "What is Life?" suggested that genetic information must be contained in a long-lived structure, an "aperiodic crystal."[1] The DNA structure seemed to realize this dream and statistical mechanicians turned to the question of the phase transitions of nucleic acid polymers that allowed this aperiodic crystal to form.[2]. The study of the so-called helix-coil transitions also contributed to the understanding of basic phase transition theory itself. Those biomolecules were a realization of the one-dimensional systems which are so different from three-dimensional systems that had been studied previously. The crucial role of dimensionality highlighted early in the study of these phase transitions became ever more apparent throughout statistical mechanics. In recent years, the mutually beneficial interaction between statistical physics and molecular biology has been renewed in the study of the protein folding problem. How the information in the amino acid sequence of a protein allows the folding of that protein into a very well organized three dimensional structure, capable of functioning, raises entirely new problems for statistical physicists. These new problems are connected with the complexity and individuality of protein sequences and structures. Unlike DNA with its double helix motif with compensated disorder, there are many three-dimensional structures of proteins. Somehow the details of sequence are not averaged over entirely but rather encode the choice of structure. Protein folding, therefore, presents us with the general conceptual problem of the statistical mechanics of discrimination, a delicate information processing task. The individual nature of each sequence needs to be understood, but we wish to have a general theory. Therefore, theory must distinguish between the general behavior of folding processes and the specific phenomena that must occur for a given molecule. A second set of issues of importance in folding arise from the significant interplay between questions of kinetics and thermodynamics. While achieving a stable structure is important, getting to it could be quite problematic. How can the discrimination process be carried out so efficiently when there are a cosmological number of possibilities for protein conformation? This is an issue known as Levinthal's paradox.[3]

These conceptual issues about the physics of protein folding have been refined considerably and partially resolved by using advances in the statistical mechanics of disordered systems, such as spin glasses.[4-7] The apparent disorder in the sequence of biomolecules resembles the quenched disorder of spins in a dilute magnetic alloy superficially. The issues of discrimination require us, however, to ask the question whether proteins really are completely random or not. While the answer is probably no, the study of completely random heteropolymers provides an important backdrop to thinking about the folding problem for natural proteins. Similarly, the study of the nonequilibrium aspects of folding has benefitted from the use of the statistical description of energy landscapes that is inspired by equilibrium spin glass theory.[8] Many of the subtler features of spin glass theory, such as the nature of replica symmetry breaking are of central importance in the molecular biology context. Indeed, these issues are probably more testable experimentally for biopolymers than they are for the magnetic alloys because of the reproducibility of the "disorder" in proteins made possible through genetic engineering.

The fundamentals of the statistical energy landscape view of protein folding have been reviewed extensively recently.[5] There has been an exciting interplay between the analytical theory and detailed computer simulation of simplified models that treat proteins as necklaces of beads residing often on lattices. The energy landscapes of the simple models can be characterized with just a few parameters. A completely random heteropolymer is described with only two parameters – an entropy measure that quantifies the size of the configurational search and an energetic parameter measuring the fluctuations in energy from state to state. A completely random heteropolymer is "frustrated" like a spin glass. The conflicting interactions between different segments of the heteropolymer give rise to many local minima and at a low enough temperature, lead to a phase transition into a glasslike state with quite a few long-lived configurations. The non-selfaveraging properties of the glassy state, well known in formal spin glass theory, are directly related to the questions of how individual peculiarities in sequence are reflected in the thermodynamic and kinetic behavior of a macromolecule. The glasslike nature of the state, however, suggests that the dynamics of accessing the low energy states becomes a problem. Many of the issues in glass theory connected with the barriers between pure states resurface then in our thinking about the kinetics of protein folding. In the thermodynamic limit these barriers would be infinite but are finite and can be slowly overcome for a finite protein molecule which is in sense mesoscopic. The energy landscape theory suggests that real proteins fold much faster than would completely random heteropolymers. Protein-like heteropolymers must have somewhat rare and untypical sequences in this respect. The landscape theory suggests sequences are chosen in such a way that they are less frustrated than expected.[9-11] This "principle of minimal frustration" gives rise to an energy landscape in which a single valley dominates. The landscape can be thought of as a many-dimensional funnel.[12] Kinetically the protein is guided through a myriad of pathways to the native structure while being impeded along the way by trapping in minima. These trapping minima are statistically not too different from those expected in the random heteropolymer and so can be described still by the energetic ruggedness. The funnel topography indicates that a protein-like heteropolymer needs an additional parameter in describing its energy landscape. This parameter is the stability gap between the groundstate structure at the bottom of the funnel and the dissimilar collapsed structures near the top..

The energy landscape theory with its few parameters can function for protein folding physics much as the Landau theory of Fermi liquids does for the physics of metals. Experimental information can be used to quantify the entropy, energetic ruggedness and stability gap and then a correspondence can be set up between the real system and the idealized model. Recently, such a correspondence was developed on a phenomenological basis between the simple lattice models and real α-helical proteins. Using a theory of helix formation in collapsed polymers,[13] Onuchic, et al., were able to relate the measured amount of helix in collapsed states to the conformational entropy associated with chain topology.[14] The smaller helical proteins with 60 residues then have entropy parameters similar to those of small lattice models containing only 27 beads.

Dynamical measurements on collapsed denatured proteins give an estimate of the energetic ruggedness. Overall folding thermodynamics then determines the stability gap. These estimates together suggest that small helical proteins have landscape parameters similar to the 27 bead lattice models with a three letter code of amino acids. The theory then allows the establishment of several landmarks in the folding process. Most important of these is that there is a thermodynamic bottleneck for folding with only a very modest barrier, of order $3k_BT$. In this bottleneck region, only three-fifths of the contacts in the protein are made in a flickering fashion.

Recently this prediction of the theory was dramatically confirmed by NMR measurements of Huang and Oas on the submillisecond folding of $\lambda$ repressor.[15] Their results showed that the bottleneck for folding contains roughly half the structure of the final protein.

Like low temperature physics, the abinitio determination of the landscape parameters raises a whole set of different issues that require powerful computational techniques. It may be quite sometime before this can be done reliably in all cases. Nevertheless, some confirmation of the landscape ideas has come from a first principle molecular dynamics simulation of Boczko and Brooks on a three-helix bundle protein.[16] Their study shows that the overall barrier for folding of staphylococcal protein A agrees rather closely with the results obtained by the correspondence analysis of Onuchic, et al. They were also able to statistically analyze the similarity between different structures in a manner much like that done in spin glass theory and were able to establish that folding occurs by descending through a funnel-like series of conformational states rather than through the sequential pathway that many experimentalists had envisioned.

While the energy landscape perspective seems to be a good overall view of the folding process, it is clear that there are many more things for statistical physicists to do in this field. The mesoscopic nature of a protein molecule puts a great deal of emphasis on finite size corrections to the usual phase transition theory. Many of these contributions to the thermodynamics, which are generally non-selfaveraging, can be important in determining whether a given sequence is foldable. Understanding these effects can help quantify the chances of making novel foldable molecules in the laboratory. Many of these questions can now be addressed by chemists who are beginning to be able to make large numbers of well defined sequences for molecules through a technology called "combinatorial synthesis." Similarly the exact scaling of trapping activation barriers with system size and temperature is an open issue. Another set of challenges for statistical physicists arise because the partially folded protein is in a low entropy density state. As in the analogous situation of low temperature physics, the low entropy density allows very subtle forces to be able to cause subtle forms of partial ordering. Thus theory leads us now to believe that the partially folded protein is liquid crystalline,[13] but other types of order, such as microphase separation, also likely play a role. The role of substructures in folding[17] as well as of local signals like those so prominent in the helix-coil problem,[18] is also of great interest.

The statistical physics view of the protein folding problem has already lead to new approaches for practical problems in protein molecular biology, such as structure prediction.[19] The partial successes have made experimentalists willing to think about the problems posed from the statistical physics perspective. Thus we can soon expect the combined theoretical and experimental study of protein folding to be an important area for addressing issues of statistical physics that may be of even greater general interest to those interested in the physics of complexity.

## References

1. E. Schrödinger, "*What is Life?*," (Cambridge University Press, 1944).
2. B. H. Zimm and J. K. Bragg, *J. Chem. Phys.* **31** (1959) 52; S. A. Rice and P. Doty, *J. Am. Chem. Soc.* **79** (1957) 3937.
3. C. Levinthal in *Proceedings of a Meeting held at Allerton House, Monticello, Illinois*, P. DeBrunner, J. Tsibiris and E. Munch, editors (University of Illinois Press, Urbana, Illinois, 1969) 22-24.
4. P. G. Wolynes in "*Spin Glasses and Biology*," Dan Stein, editor (World Scientific Press, 1990)
5. J. Bryngelson, J. Onuchic, N. Socci and P. G. Wolynes, *Proteins: Structure, Function and Genetics* **21** (1995) 167.
6. P. G. Wolynes, J. Onuchic and D. Thirumalai, *Science* **267** (1995) 1619.
7. K. A. Dill, et al., *Protein Science* **4**, (1995) 561; T. Garel, H. Orland and D. Thirumalai in "New Developments in Theoretical Studies of Protein," *Advanced Studies in Physical Chemistry*, Vol 7, Ron Elber, editor (World Scientific, Singapore, 1995).
8. M. Mezard, G. Parisi and M. Virasoro, "*Spin Glass Theory and Beyond*," (World Scientific, 1987).
9. J. D. Bryngelson and P. G. Wolynes, *Proc. Natl. Acad. Sci.* **85**, (1987) 7524.
10. J. D. Bryngelson and P. G. Wolynes, *J. Phys. Chem.* **93**, (1989) 6902.
11. E. Shakhnovich, *Phys. Rev. Lett.* **72**, (1994) 3907.
12. P. E. Leopold, M. Montal and J. N. Onuchic, *Proc. Natl. Acad. Sci.* **89**, (1992) 872.
13. Z. Luthey-Schulten, B. E. Ramirez and P. G. Wolynes, *J. Phys. Chem.* **99** (1995) 2177.
14. J. N. Onuchic, P. G. Wolynes, Z. Luthey-Schulten and N. Socci, *Proc. Natl. Acad. Sci.* **92** (1995) 3626.
15. G. S. Huang and T. G. Oas, *Proc. Natl. Acad. Sci.* **92** (1995) 6878.
16. E. M. Boczko and C. L. Brooks, III, *Science* **269** (1995) 393.
17. A. Panchenko, Z. Luthey-Schulten and P. G. Wolynes, *Proc. Natl. Acad. Sci.*, to appear.
18. J. Saven and P. G. Wolynes, *J. Mol. Biol.* preprint
19. R. G. Goldstein, Z. Luthey-Schulten and P. G. Wolynes in "New Developments in Theoretical Studies of Protein," *Advance Series in Physical Chemistry* **Vol 7**, Ron Elber, editor (World Scientific, Singapore (1995).

# SOME STATISTICAL MECHANICS ASPECTS OF MITOSIS

STANISLAS LEIBLER
*Department of Molecular Biology, Lewis Thomas Laboratory, 355*
*Department of Physics, Jadwin Hall, 360*
*Princeton University, Princeton, NJ 08544, USA*

I will present a short summary of recent efforts to understand the collective processes underlying the separation of chromosomes in eukaryotic cells. Mitosis can be viewed as a prototype of a regulated assembly phenomenon which is: (i) stochastic; (ii) self-correcting and (iii) synchronized. I will illustrate these aspects of mitosis by discussing recent theoretical and experimental results obtained in vitro and in vivo. Special emphasis will be put on the quantitative study of the behavior of molecular components, such as microtubules and molecular motors. I will also sketch possible future directions in the study of complex assembly phenomena.

# TOPICS IN GLASS TRANSITIONS AND SUPERCOOLED LIQUIDS

**Kyozi Kawasaki**

Department of Natural Science and Mathematics
Chubu University, Kasugai, Aichi 487, Japan
E-mail: kawasaki@isc.chubu.ac.jp

**Abstract**

Two topics related to glass transitions are discussed. The first topics is concerned with the recently developed mode coupling theory. After brief introduction we reformulate and extend the theory with the aid of dynamical extension of the density functional theory of inhomogeneous liquid. In this connection we briefly describe a general formulation of the irreducible memory function where the mode coupling approximation for slow dynamics is sensibly applied. The second topics is about the excess scattering and the associated ultraslow mode of diffusive type. We propose a new interpretation of the latter in terms of phase dynamics applied to irregular spatial heterogeneities.

# 1 Introduction

Despite the long history of research related to structural glass transitions[1], our understanding still remains rather elusive since much of the theoretical considerations is based on artificial models, and computer simulations for more realistic models are limited in their scope. Therefore, it is natural that the serious first principle attempts incorporating the mode coupling idea are drawing a great deal of attention[2]. It is true that at the present stage these mode coupling theories (MCT) make use of uncontrolled approximations and simplifying assumptions which severely limit validity of MCT to the region well before glass transitions. Nevertheless, there is at least one example, namely, the dense colloidal suspension where the quantitative predictions of MCT are successfully tested by experiments[3]. Thus it is fair to say that MCT provides a valid starting point for the lofty goal of first principle understanding of structural glass transitions.

In this article we first describe our new derivation of the MCT equation via the dynamical density functional theory (DDFT) [4]. An advantage of this approach is that DDFT offers a new direction going beyond the current MCT. Next we present our recent tentative interpretation[5] of the so-called ultraslow mode with the relaxation time much longer than that of the $\alpha$-relaxation[6]. This mode is associated with the excess light scattering observed in many so-called fragile glass-formers whose origin is not understood.

## 2 Mode coupling theories for glass transitions

There have been at least two distinct approaches to MCT for glass transitions. One is due to Götze and the coworkers[7], and may be called a single-step theory. They deal directly with the density auto-correlation function (DAF) defined microscopically. With the use of the Zwanzig-Mori projector technique and a factorization approximation they arrive at a self- consistent equation for DAF. This nonlinear MCT equation contains as input a set of control parameters through the static scattering structure factor of supercooled liquid. As one moves along a path in the space of the control parameters, one encounters a point where the density correlation function changes its character. Namely, on one side of this bifurcation point DAF vanishes in the long time limit whereas this is not so on the other side. In earlier days this ergodic-nonergodic transition was identified as the glass transition, but later it was regarded as indicating a smooth cross-over from one behavior to another where truly non-ergodic behavior disappears when the theory is refined by including thermally activated processes which were neglected in the above mentioned MCT equation. Note however that thermally activated processes are greatly reduced for colloidal suspensions. In this case the bifurcation point of MCT is expected to coincide with the glass transition point[3]. This should be one of the reasons for good agreement of MCT predictions and experiments.

The second approach due to Mazenko and the coworkers[8] may be called the two-step theory and start from the nonlinear stochastic equation for a set of field variables of the type (1) and (2) below. This equation called by Mazenko *et al* the fluctuating nonlinear hydrodynamics can in principle be "derived" microscopically by an appropriate projector technique. Mazenko and the coworkers applied the field theoretic technique to deal with nonlinearities and obtained the self-consistent equation for DAF which is similar to the original equation of Götze *et al* except that effects of thermally activated processes are included.

# 3 Dynamical density functional theory (DDFT)

The self-consistent MCT briefly described in the preceding section contains drastic approximations such as factorization. Also thermally activated processes enter the theory in a rather awkward way. Thus there is a need of theory which avoids such drastic approximations and in which thermally activated processes enter in a more natural way. We propose a dynamical density functional theory as a candidate of such a theory. Our approach differs from the related attempt by Kirkpatrick and Wolynes[9]. Since mathematical details of this theory were published elsewhere[4], we will be rather brief. We first consider the two field variables: $\rho(\mathbf{r})$, the local mass density, and $\mathbf{g}(\mathbf{r})$, the local momentum density. Our starting point is the generalized Fokker-Planck equation for the probability density functional $P(\{\rho\},\{\mathbf{g}\};t)$ as follows,

$$\frac{\partial}{\partial t}P(\{\rho\}\{\mathbf{g}\};t) = \Omega(\{\rho\},\{\mathbf{g}\})P(\{\rho\},\{\mathbf{g}\};t) \tag{1}$$

where $\Omega$ is the time-displacement operator given by

$$\Omega(\{\rho\},\{\mathbf{g}\}) \equiv \int d\mathbf{r}\left\{\frac{\delta}{\delta\rho(\mathbf{r})}\nabla\cdot\mathbf{g}(\mathbf{r}) + \frac{\delta}{\delta\mathbf{g}(\mathbf{r})}\cdot\left[\rho(\mathbf{r})\cdot\nabla\frac{\delta H\{\rho\}}{\delta\rho(\mathbf{r})} + \nabla\cdot\frac{\mathbf{g}(\mathbf{r})\mathbf{g}(\mathbf{r})}{\rho(\mathbf{r})}\right] + \cdots\right\} \tag{2}$$

where $H\{\rho\}$ is the free energy density functional and the ellipsis stands for the Fokker-Planck operator associated with microscopic momentum relaxation and contains microscopic viscosities. We shall not need its explicit expression except to say that it is chosen in such a way that (1) admits as its stationary solution the following equilibrium distribution;

$$P_e(\{\rho\}\{\mathbf{g}\}) = Z^{-1} exp\Big\{ -\beta\Big[H\{\rho\} + \int \frac{\mathbf{g}(\mathbf{r})^2}{2\rho(\mathbf{r})} d\mathbf{r}\Big]\Big\}, \tag{3}$$

$Z$ and $\beta$ being the normalization constant and the inverse of the Boltzmann constant times the absolute temperature, respectively. The various terms in the curly bracket of (2) can be interpreted as follows. The first term comes from the continuity equation for $\rho(\mathbf{r})$. The second term contains the reversible part of the equation of motion for $\mathbf{g}(\mathbf{r})$ where the first and second terms in the square bracket represent negative of the local body force due to the inhomogeneous mass density and the continuity equation for the momentum density, respectively. The equivalent equation was also given in Ref. [8].

As one approaches the glassy state from the supercooled liquid side, a part of the mass density fluctuation begins to slow down whereas the momentum density still fluctuates rapidly. An intuitive argument supporting this fact was given by Cohen and de Schepper[10]. If this is the case, the momentum density can be adiabatically eliminated from (1). Formally this can be done by a projector technique where an arbitrary momentum density distribution is projected onto a local Maxwell distribution. Skipping algebraical details[4], we give in the following the final result in which the single relaxation time approximation for the momentum distribution is used. Introducing the mass density distribution functional $P(\{\rho\},t)$ defined by

$$P(\{\rho\},t) = \int d\{\mathbf{g}\} P(\{\rho\},\{\mathbf{g}\},t) \tag{4}$$

where the integral is over the function space of $\mathbf{g}(\mathbf{r})$, we have[4,11]

$$\frac{\partial}{\partial t} P(\{\rho\},t) = -\tau \int d\mathbf{r} \frac{\delta}{\delta\rho(\mathbf{r})} \nabla \cdot \rho(\mathbf{r}) \nabla \Big[\beta^{-1} \frac{\delta}{\delta\rho(\mathbf{r})} + \frac{\delta H\{\rho\}}{\delta\rho(\mathbf{r})}\Big] P(\{\rho\},t) \tag{5}$$

where $\tau$ is the relaxation time of momentum density fluctuation. This equation looks similar to more familiar time-dependent Ginsburg-Landau equation for a conserved order parameter except for the extra factor of $\rho(\mathbf{r})$ between the two $\nabla$. However, we should note that here we are dealing with small-scale fluctuations of $\rho(\mathbf{r})$. A consequence of this is that the form of $H\{\rho\}$ is quite different from the more usual Ginsburg-Laudau square-gradient type. An example is the following Ramakrishnan-Yusouff form[12]:

$$\beta H\{\rho\} = \int d\mathbf{r}\rho(\mathbf{r})\left[ln\frac{\rho(\mathbf{r})}{\rho_0} - 1\right] - \frac{1}{2}\int\int d\mathbf{r}d\mathbf{r}'C(|\mathbf{r}-\mathbf{r}'|)[\rho(\mathbf{r})-\rho_0][\rho(\mathbf{r}')-\rho_0] \quad (6)$$

where $\rho_0$ is the density of the reference liquid and $C(\mathbf{r})$ is the direct correlation function of the liquid theory. The density functional theories of freezing and glass transition seek local minima of $H\{\rho\}$. Eq. (5) is thus a dynamical extension of these density functional theories. In passing we point out that the form of (5) is uniquely determined by the form of the reversible body force and the equilibrium distribution density functional.

Before deriving the self-consistent MCT equation from (5) we point out that (5) is itself a good candidate for direct computer simulation to study approach of supercooled liquids to glass transition where we are concerned with the long time behavior of the small scale density fluctuations. Thermally activated processes enter (5) in a very natural way, and the rapidly varying momentum variable is eliminated. The latter feature is of great advantage as compared with the direct simulation of (1) and (2) [13,14].

# 4 Self-consistent mode coupling approximation

The mode coupling approximation was originally developed when nonlinearities entering the starting equation is reversible[15]. However, here we want to apply it to (5) which is purely dissipative. This difference is important enough that naive applications of mode coupling approximation (MCA) can lead to various difficulties as pointed out by Jäckle and the coworkers[16]. In order to clarify this problem in a general way we recently introduced an irreducible memory function in addition to the usual memory function for which MCA is often applied[17]. We consider a general dissipative stochastic system with detailed balance. Let C(z) be the Laplace transform of the autocorrelation function of order parameter C(t). The usual memory function M(t) or its Laplace transform M(z) is related to C(z) through

$$C(z) = [z - E - M(z)]^{-1} \tag{7}$$

where $E$ is the first moment given by

$$E = \lim_{t\to 0+} \frac{d}{dt} C(t). \tag{8}$$

The irreducible memory function $M^{ir}$ is related to $M$ through

$$M(z) = [1 - E^{-1} M^{ir}(z)]^{-1} M^{ir}(z). \tag{9}$$

There are explicit formally exact expressions for $M$ and $M^{ir}$ which will not be reproduced here[17].

The value of this formal result will become apparent near freezing where C(t) tends to remain non-vanishing at the infinity or C(z) develops a singularity $z^{-1}$ at $z = 0$. This implies vanishing of $M(z) + E$ at $z = 0$. This then requires delicate cancelations of various terms in $M(z)$ which result from some approximate calculation of $M(z)$. On the other hand, from (9), $M^{ir}(z)$ only needs to diverge as $z$ vanishes. Hence even a drastic approximation for $M^{ir}$ does not destroy delicate cancelation of $M(z) + E$, which is not the case if $M(z)$ itself is approximated. If only the order parameter entering $C(t)$ is relevant, MCA for $M^{ir}(t)$ takes the form

$$M^{ir}(t) \cong M^{ir}_{MCA}(t) = F(C(t)). \tag{10}$$

The function $F(x)$ is typically a polynomial in $x$ with positive coefficients. Substitution of (10) into (7) yields

$$C(z)/[1 - zC(z)] = E^{-2} M^{ir}_{MCA}(z) - E^{-1}. \tag{11}$$

The last term on the rhs of (11) can be neglected near freezing where $M^{ir}_{MCA}(z)$ tends to diverge as $z$ vanishes. This is the self-consistent MCT equation for $C$ derived for the case of supercooled liquids by Leutheusser[18], Götze and others[7]. Extension of this result to the cases with many relevant variables is straightforward.

# 5 Ultraslow modes and phase dynamics

One of the puzzling problems in glass-forming systems, in particular, in the so-called fragile glass-forming systems is the appearance of excess scattering at long wave lengths up to a few thousand Angstroms and the very slow diffusive decay of fluctuations associated with this excess scattering[6]. Such excess scattering is not always observed, namely, a sample of the same system can be prepared which does not show excess scattering under the same condition[19]. This sorts of excess scattering have been known in other systems than glass-formers. Perhaps the earliest observation on excess scattering was reported for a polymer solution in semi-dilute regimes[20]. Subsequently, this is also found in concentrated polyelectrolytes solutions with low salt or salt free solvents[21]. There is as yet no generally accepted theoretical understanding for any of these observations although it is believed that some sorts of spatial heterogeneities are responsible for these observations, which are certainly not artifacts due to dusts, etc. These observations suggest existence of delicate structure for the complex systems mentioned above, which are yet to be discovered. We cannot go into further details about the nature of these structures in view of the limited information currently available. Instead here we focus on one aspect of the problem, namely, on the relationship between random spatial heterogeneity and the diffusive relaxation.

Recently, dynamics of non-equilibrium patterns is receiving a great deal of attention, which, however, is mostly concerned with regular spatial structures with periodicity[22]. In particular, the idea of phase dynamics to describe relaxation of slow modulation of spatial patterns was successfully applied to the roll pattern of Rayleigh-Benard convection[23], and was extended in various different ways. Here we apply this idea to irregular stationary structures, which are likely to occur in glass-forming systems.

We start from the assumption that an undeformed stationary structure is described by some scalar order parameter $S_0(\mathbf{r})$ which is a local minimum of a free energy functional $H\{S\}$ satisfying

$$\delta H\{S\}/\delta S(\mathbf{r}) = 0. \tag{12}$$

The order parameter field is assumed to obey the purely dissipative equation of motion like

$$\frac{\partial}{\partial t} S(\mathbf{r},t) = -\Omega(\mathbf{r},\{S(t)\}) \frac{\delta}{\delta S(\mathbf{r},t)} H\{S(t)\} \tag{13}$$

where $\Omega(\mathbf{r},\{S\})$ is a self-adjoint kinetic operator that can functionally depend on $\{S\}$ and the translational invariance of this equation is assumed. Clearly $S_0(\mathbf{r})$ is a stationary solution of (13). The assumed translational invariance of (13) then implies that, with $\mathbf{u}$ an arbitrary vector, $S_0(\mathbf{r}-\mathbf{u})$ is also a stationary solution. Now imagine that $\mathbf{u}$ is not a constant vector but is replaced by a slowly varying function of $\mathbf{r}$, $\mathbf{u}(\mathbf{r})$. The thermodynamic driving force

$$X(\mathbf{r},\{S\}) \equiv -\delta H\{S\}/\delta S(\mathbf{r}) \tag{14}$$

is no longer zero but is still very small. We assume that $\mathbf{u}(\mathbf{r})$ itself is small and linearize (13) in $\mathbf{u}(\mathbf{r})$ to obtain

$$\nabla S_0(\mathbf{r}) \cdot \frac{\partial}{\partial t}\mathbf{u}(\mathbf{r},t) = -\int d\mathbf{r}'\Omega(\mathbf{r},\{S_0\})\left[\frac{\delta^2 H\{S\}}{\delta S(\mathbf{r})\delta S(\mathbf{r}')}\right][\nabla' S_0(\mathbf{r}')]\cdot \mathbf{u}(\mathbf{r}',t) \tag{15}$$

where now $\mathbf{u}$ should depend on the time $t$ as well, and the suffix zero indicates that the quantity should be evaluated with $S(\mathbf{r}) = S_0(\mathbf{r})$. In fact, $\mathbf{u}(\mathbf{r})$ itself need not be small but we must require its gradients to be small. Linearization (15) is justified if $\mathbf{u}(\mathbf{r}) = 0$ for a particular $\mathbf{r}$. Otherwise an equation similar to (15) holds if $S_0(\mathbf{r})$ is replaced by $S_0(\mathbf{r}-\mathbf{u}(\mathbf{r}))$ on the lhs and $S_0(\mathbf{r}')$ by $S(\mathbf{r}'-\mathbf{u}(\mathbf{r}))$ everywhere on the rhs after performing all the operations where $\mathbf{u}(\mathbf{r})$ can be treated as though a constant. The final result is the same and hence we simply use the linearized equation (15). Here we should note that $S(\mathbf{r})$ is not a conserved quantity. Otherwise we must require $\nabla \cdot \mathbf{u}(\mathbf{r}) = 0$ which is not.

From now on we proceed somewhat differently from[5], which is more in line with our previous works where we derived interfacial equations from field equations[25] and has the advantage of separating the static force and the frictional force acting upon the deformation field $\mathbf{u}(\mathbf{r})$. For this purpose we first operate the inverse of $\Omega$ on both sides of(15) assuming its existence:

$$\Omega^{-1}(\mathbf{r}, \{S_0\})[\nabla S_0(\mathbf{r})] \cdot \frac{\partial}{\partial t}\mathbf{u}(\mathbf{r}, t) = -\int d\mathbf{r}' \left[\frac{\delta^2 H\{S\}}{\delta S(\mathbf{r})\delta S(\mathbf{r}')}\right]_0 [\nabla' S_0(\mathbf{r}')] \cdot \mathbf{u}(\mathbf{r}', t). \quad (16)$$

Next we multiply by $\nabla S_0(\mathbf{r})$ on both sides of (16) and integrate to yield

$$\int\int d\mathbf{r}d\mathbf{r}'\mathbf{W}(\mathbf{r}, \mathbf{r}') \cdot \frac{\partial}{\partial t}\mathbf{u}(\mathbf{r}', t) = -\int\int d\mathbf{r}d\mathbf{r}'\mathbf{E}(\mathbf{r}, \mathbf{r}') \cdot \mathbf{u}(\mathbf{r}', t) \quad (17)$$

where $\mathbf{W}$ and $\mathbf{E}$ are the symmetric tensor functions given by

$$\mathbf{W}(\mathbf{r}, \mathbf{r}') \equiv [\nabla S_0(\mathbf{r}')][\Omega^{-1}(\mathbf{r}, \{S_0\})\delta(\mathbf{r} - \mathbf{r}')][\nabla' S_0(\mathbf{r}')] \quad (18)$$

and

$$\mathbf{E}(\mathbf{r}, \mathbf{r}') \equiv [\nabla S_0(\mathbf{r})]\left[\frac{\delta^2 H}{\delta S(\mathbf{r})\delta S(\mathbf{r}')}\right]_0 [\nabla' S_0(\mathbf{r}')]. \quad (19)$$

The symmetry of (18) is a consequence of self-adjointness of $\Omega$.

Assuming that $\mathbf{u}(\mathbf{r}, t)$ changes slowly in space compared with the scale on which $S_0(\mathbf{r})$, which is assumed to be random, varies, we can course-grain both sides over the length scale which is in between these two scales. Some mathematical details of this procedures are presented in Appendixes of Ref. [5], which can be basically applied here and we will be very brief. We then observe that the rhs of (17) after coarse-graining is the static force acting on the field $\mathbf{u}(\mathbf{r})$ where $t$ is omitted for brevity,

$$-\delta\Delta H\{\mathbf{u}\}/\delta\mathbf{u}(\mathbf{r}) \quad (20)$$

where

$$\Delta H\{\mathbf{u}\} = \frac{1}{2}\int\int d\mathbf{r}d\mathbf{r}'\mathbf{u}(\mathbf{r})\cdot\mathbf{E_c}(\mathbf{r}-\mathbf{r}')\cdot\mathbf{u}(\mathbf{r}') \tag{21}$$

in the harmonic approximation, $\mathbf{E_c}$ being the course-grained $\mathbf{E}$. If the pattern $S_0(\mathbf{r})$ is statistically isotropic and $\mathbf{E_c}(\mathbf{r})$ is short-ranged, (21) is further reduced to

$$\Delta H\{\mathbf{u}\} = \frac{1}{2}\int d\mathbf{r}\Big\{\epsilon_1[\nabla_\alpha u_\beta(\mathbf{r})][\nabla_\alpha u_\beta(\mathbf{r})] + \epsilon_2[\nabla\cdot\mathbf{u}(\mathbf{r})]^2\Big\} \tag{22}$$

where $\epsilon_1$ and $\epsilon_2$ are the two elastic moduli of pattern deformation and are expressible in terms of $\mathbf{E_c}$. The lhs of (17) after coarse-graining is the friction force acting on $\mathbf{u}(\mathbf{r})$ and is written, denoting the partial time derivative by the suffix $t$

$$\delta R\{\mathbf{u}_t\}/\delta\mathbf{u}_t(\mathbf{r}) \tag{23}$$

where $R\{\mathbf{u}_t\}$ is the Rayleigh dissipation functional associated with the rate of change of the field $\mathbf{u}$ and is given by

$$R\{\mathbf{u}_t\} = \frac{1}{2}\int\int d\mathbf{r}d\mathbf{r}'\mathbf{u}_t(\mathbf{r})\mathbf{W}_c(\mathbf{r}-\mathbf{r}')\cdot\mathbf{u}_t(\mathbf{r}'), \tag{24}$$

$\mathbf{W_c}$ being the coarse-grained $\mathbf{W}$. For statistically isotropic patterns $S_0(\mathbf{r})$ and short-ranged $\mathbf{W_c}(\mathbf{r})$, we get

$$R\{\mathbf{u}_t\} = \frac{1}{2}\int d\mathbf{r}w\ \mathbf{u}_t(\mathbf{r})^2 \tag{25}$$

where $w$ is the friction constant expressible as the integral of $3^{-1}Tr\mathbf{W}_c(\mathbf{r})$. The equation of motion for $\mathbf{u}(\mathbf{r},t)$ which is the phase dynamic equation, is obtained by balancing static and friction forces:

$$\frac{\delta}{\delta\mathbf{u}_t(\mathbf{r})}R\{\mathbf{u}_t\} = -\frac{\delta}{\delta\mathbf{u}(\mathbf{r})}\Delta H\{\mathbf{u}\}, \tag{26}$$

or

$$\int d\mathbf{r}'\mathbf{W_c}(\mathbf{r}-\mathbf{r}')\cdot\mathbf{u}_t(\mathbf{r}') = -\int d\mathbf{r}'\mathbf{E_c}(\mathbf{r}-\mathbf{r}')\cdot\mathbf{u}(\mathbf{r}'). \tag{27}$$

If the local approximations (22) and(25) are valid, we find

$$w\frac{\partial}{\partial t}\mathbf{u}(\mathbf{r}) = \epsilon_1\nabla^2\mathbf{u}(\mathbf{r}) + \epsilon_2\nabla\nabla\cdot\mathbf{u}(\mathbf{r}). \tag{28}$$

This can be decomposed into two separate diffusion equations for the longitudinal and transverse components $\mathbf{u}_{||}(\mathbf{r})$ and $\mathbf{u}_{\perp}(\mathbf{r})$, respectively:

$$\frac{\partial}{\partial t}\mathbf{u}_{||}(\mathbf{r}) = D_{||}\nabla^2\mathbf{u}_{||}(\mathbf{r}), \tag{29}$$

$$\frac{\partial}{\partial t}\mathbf{u}_{\perp}(\mathbf{r}) = D_{\perp}\nabla^2\mathbf{u}_{\perp}(\mathbf{r}) \tag{30}$$

where

$$D_{||} \equiv (\epsilon_1 + \epsilon_2)/w, \tag{31}$$

$$D_{\perp} \equiv \epsilon_1/w. \tag{32}$$

An advantage of this derivation is the direct proportionality of the phase diffusion constants $D_{||}$ and $D_{\perp}$ to the rigidity moduli of the pattern deformation. Hence we can understand the experimental observation that $D_{||}$ decreases as the temperature is lowered accompanying softening of the patterns as revealed through the in course of excess scattering intensity[6,19], provided that $w$ is not sensitive to the temperature variation.

Unfortunately we are unable to say much at this time about the origin and nature of the assumed spatial heterogeneities although there are many indications in the literature that some structures are likely as the glass transition is approached. For instance Jónsson and Andersen have demonstrated by molecular dynamic calculation of the dense Lennard-Jones system the emergence of icosahedral structure which percolates throughout the system[26].

# 6 Discussion

Despite all the successes achieved by MCT in accounting some aspects of structural glass transition the theory is still at a rather primitive stage. This is apparent if one compares this with the vastly sophisticated developments of various lattice models of glass transitions such as spin glass models[27]. This means that substantially new developments of structural glass transitions can be expected. One possible direction would be to pursue further the analogy of structural glass and Potts glass as was already started by Kirkpatrick, Wolynes and others[9,28].

The discovery of the excess scattering and the accompanying ultraslow modes may lead us , at least in some cases, to hitherto unsuspected soft structures in complex systems as we have conjectured elsewhere[5].

# References


[1] J. Jäckle, Rep. Prog. Phys. 49(1986)171.

[2] S. Yip, ed. Relaxation Kinetics in Supercooled Liquids–Mode Coupling Theory and its Experimental Tests, Theme issue of Transp. Theor. and Stat. Phys. (1995)

[3] W. van Megen and S. M. Underwood, in Ref. 2.

[4] K. Kawasaki, Physica A 208(1994)35.

[5] K. Kawasaki, Physica A (1995) in press.

[6] E. W. Fischer, Physica A 201(1993)183.

[7] W. Götze, in Liquids, Freezing and the Glass Transition, eds. J. P. Hansen, D. Levesque and J. Zinn-Justin (North-Holland, Amsterdan, 1991).

[8] B. Kim and G. Mazenko, Advances in Phys. 78(1990)129.

[9] T. R. Kirkpatrick and P. G. Wolynes, Phys. Rev. A35(1987)3072.

[10] E. G. D. Cohen and I. M. de Schepper, in Recent Progress in Many-Body Theories, Vol. 3, eds. T. L. Ainsworth *et al* (Plenum, New York, 1992).

[11] T. Munakata, J. Phys. Soc. Jpn. 43(1977)1723
B. Bagchi, Physica A 145(1987)273.

[12] T. V. Ramakrishnan and M. Yusouff. Phys. Rev. B 19(1979)2775.

[13] L. M. Lust, O. T. Valls and C. Dasgupta, Phys. Rev. E 48(1993)1787.

[14] C. Dasgupta and O. T. Valls, Phys. Rev. E 50(1994)3916.

[15] K. Kawasaki, Phys. Rev. 150(1966)291; Ann. Phys. (N. Y.) 61(1970)1.

[16] J. Jäckle, J. Non-Crystalline Solids 172-174 (1994)104.

[17] K. Kawasaki, Physica A 215(1995)61; Also see B. Cichocki and W. Hess, Physica A 141(1987)475.

[18] E. Leutheusser, Phys. Rev. A 29(1984)2765.

[19] T. Kanaya, A. Patkowski, E. W. Fischer, J. Seils, H. Gläser and K. Kaji, Acta, Polymer. 45(1994)137 and preprint.

[20] H. Benoit and C. Picot, Pure Appl. Chem. 12(1966)545.

[21] K. S. Schmitz, Biopolymers, 33(1993)953 and the references quoted therein.

[22] S. Kai, ed. Pattern Formation in Complex Dissipative Systems (World Scientific, Singapore 1992).

[23] Y. Pomeau and P. Manneville, J. de Physique Lett. 40(1979)L-609.

[24] H. R. Brand, in Propagation in Systems far from Equilibrium, eds. J. E. Wesfreid, H. R. Brand, P. Manneville, G. Abinet and N. Boccara, eds. (Springer, Heidelberg 1988).

[25] K. Kawasaki and T. Ohta, Physica A 118(1983)175.

[26] H. Jonsson and H. C. Andersen, Phys. Rev. Lett. 60(1988)2295.

[27] K. Binder and A. P. Young, Rev. Mod. Phys. 58(1986)801.

[28] P. G. Wolynes, in Proceedings of the International Symposium on Frontiers in Science, AIP Conference Proceedings Number 180, eds. S. S. Chan and P. G. Debrunner (American Institute of Physics, New York 1988).

# LIST OF CONTRIBUTORS

Aizenman M.
Au-Yang H.
Banavar J. R.
Barma M.
Binder K.
Botelho E.
Chan C. K.
Chandler D.
Chen N.-X.
Cieplak M.
Colaiori F.
Compagner A.
Coolen A. C. C.
Couder Y.
David F.
de Almeida R. M. C.
Derrida B.
Dhar D.
Dotsenko Vik.
Douady S.
Edwards S. F.
Feldman D. E.
Flammini A.
Giacometti A.
Hess O.
Hess S.
Huse D. A.
Iglesias J. R.
Indekeu J. O.
Its A. R.
Izrailev F. M.
Kaneko K.
Kantor Y.
Kardar M.
Kawasaki K.
Kim D.
Korepin V.
Landau D. P.
Laughton S. N.
Leibler S.
Ludwig A. W. W.
L'vov V. S.
Maritan A.
Maruyama K.
Miyazima S.
Mézard M.
Nechaev S.
Nelson D. R.
Noh J. D.
Okazaki A.
Okumura K.
Pearce P. A.
Perk J. H. H.
Procaccia I.
Rinaldo A.
Rodriguez-Iturbe I.
Sachdev S.
Schick M.
Shastry B. S.
Sheng P.
Sherrington D.
Spohn H.
Su Z.-B.
Tél T.
van Leeuwen J. M. J.
Velarde M. G.
Virasoro M.
Waldron A. K.
Wolynes P.
Wu F. Y.
Wu Y. S.
Yang C. N.
Young A. P.

# LIST OF ORAL AND POSTER PRESENTATIONS

**Aarão Reis F.D.A.** Finite-size scaling for random walks on fractals; see Chalub F.A.C.C.

**Abraham D.B., O'Connor D., Parry A.O. and Upton P.J.** Correlation functions on cylinders

**Achiwa N.** See Miura Y.

**Aharony A., Birgeneau R.J., Garland C.W., Kim Y.-J., Lebedev V.V., Netz R.R. and Young M.J.** Critical behavior of the structure factor for higher harmonics in density wave systems; see Stein J. and Hovi J.P.

**Akheyan A. Z.** (a) Critical properties of the B.E.G. model (b) Global Bethe lattice consideration

**Ala-Nissila T.** Theoretical studies of dynamics of spreading of tiny liquid droplets on solid surfaces; see Provatas N.

**Ala-Nissila T., Bukharev I. and Kosterlitz J.M.** Morphology of steps on vicinal surface

**Alabiso C., Besagni N., Casartelli M. and Marenzoni P.** Stochastic transition and geometrical features in Hamiltonian systems

**Alam M. A.** See Shah I.

**Alarcón-Waess** Time-dependent electrolyte friction on a charged non-spherical tracer within Fick approximation

**Alava M.J. and Niskanen K.J.** Structure of 3D Planar random fiber networks

**Albuquerque D.F., Sousa J.R. and Fittipaldi I.P.** D-vector model in the diluted site-bond-correlated version

**Alcaraz F.C. and Malvezzi A.L.** The XXZ chain in homogeneous and staggered magnetic fields

**Almeida R.L.** See Coutinho S.

**Alpatov P.** See Reichl L.E.

**Alstrφm P., Andersen J.S., Goldburg W.I., Levinsen M.T. and Schröder E.** Relative diffusion in capillary waves

**Alstrφm P., Berntsen N. and Vallespin A.** Conditioning and control in adaptive performance networks; see Sibani P.

**Altarelli M., Núñez-Regueiro M.D. and Papoular M.** Theoretical interpretation of the coexisting two scales in X-ray and neutron critical scattering

**Amaral L.A.N., Barabási A.-L., Makse H.A. and Stanley H.E.** Scaling properties of driven interfaces in disordered media

**Anawar G.** See Shah I.

**Andersen J.V.** See Leung K.T.

**Andrade Jr. J.S., Ito N. and Shibusa Y.** Percolation transition in conducting polymer networks

**Andrade Jr. J.S., Wainer I., Filho J.M. and Moreira J.E.** Self-organized criticality in the El Niño southern oscillation; see Lima F.W.

**Andresen B. and Gordon J.M.** Constant thermodynamic speed for thermodynamic processes and simulated annealing

**Andersen J.S.** See Alstr$\phi$m P.

**Andrey L.** Single neuron chaos is natural - What about consequences?

**Antoniou I. and Kuperin Yu.A.** An operator approach to few-body classical systems

**Antoranz J.C.** See Garcia-Ybarra P.L., Costa O.S. and Martin J.S.

**Aoki K.** Chaotic scattering and magneto-resistance in lateral surface super lattices

**Aoyagi Y.** See Iitaka T.

**Argyrakis P.** See Gallos L.

**Arimitsu T.** A canonical theory for dissipative quantum fields – Non-equilibrium thermo field dynamics –

**Arimitsu T. and Yamazzaki H.** A topological characterization of strange attractors for dissipative systems

**Artuso R., Casati G. and Guarneri I.** Billiards correlation functions

**Asakawa H. and Suzuki M.** Super-effective-field theory of 3-dimensional electron systems

**Asselmeyer T.** See Ebeling W.

**Attard P.J.** The electric double layer and the corresponding Poisson-Boltzmann description

**Baake M. and Grimm U.** Ferromagnetic Ising models on quasiperiodic labyrinths

**Barabási A.-L.** See Amaral L.A.N.

**Bariev R.Z., Klümper A., Schadschneider A. and Zittartz J.** New integrable fermion models in one dimension

**Barthélémy M.** Study of electro-magnetic properties of heterogeneous media through the replica method

**Bashmakov I.A.** See Butylina I.B.

**Basu C. and Gu B.Y.** A numerical study of the geometric scattering effect on magneto conductance in multiply connected mesoscopic rings

**Batchelor M.T. and Yung C.M.** Exact results for self-avoiding walks with an adsorbing boundary

**Batchelor M.T., Fridkin V. and Yung C.M.** Critical properties of the integrable spin-3 $G_2$ chain; see Yung C.M.

**Bausch R., Schmitz R. and Turski L.A.** Classical and quantum transport in the presence of topological defects
**Bellissent-Funel M.C.** See Zhao H. (FR)
**Ben-Aryeh Y.** See Jann A.
**Bennema P.** See Wang M.
**Berche B., Debierre J.-M., Eckle H.P. and Turban L.** Critical behavior in confined geometries: the percolation problem in two dimensions
**Berkolaiko G.** See Rabinovich S.
**Berntsen N.** See Alstr$\phi$m P.
**Berggren K.F. and Ji Z.L.** From strong to weak quantum chaos in billiards with soft walls
**Bertini L. and Cancrini N.** Rigorous results on the UV stability for Kardar-Parisi-Zhang equation
**Bertini L. and Giacomin G.** Rigorous microscopic derivation of Kardar-Parisi-Zhang equation
**Besagni N.** See Alabiso C.
**Betts D.D., Masui S., Vats. N. and Stewart G.E.** The improved finite lattice method for estimating properties of quantum systems on the square lattices at T=0
**Beysens D.** See Zhao H. (FR)
**Bhagavatula R., Chen K. and Jayaprakash C.** Earthquakes in a double-couple model of seismic zone with embedded pre-existing faults
**Bhatt R.N. and Guo M.Y.** Energy levels as a probe of integer quantum Hall phase transitions; see Guo M.Y.
**Bhattacharjee J.K.** See Hao H.
**Bird J.P.** See Iitaka T.
**Birgeneau R.J.** See Aharony A.
**Birman J.L.** See Bulatov A.
**Bishop R.F., Hale R.G. and Xian Y.** A microscopic many-body approach to quantum phase transition in spin-lattice systems; see Zeng C.
**Blacher S.** See Brouers F.
**Blossey R. and Indekeu J.O.** Interface potential for type-I superconductors
**Blöte H.W.J. and Heringa J.R.** Universality in3-d Ising models; see Luijten E., Shchur L.N. and Nightingale M.P.
**Bocquet J.L.** See Gallis C.
**Bollé D., Jongen G. and Shim G.M.** Optimal nonlinear training in the multi-class proximity problem
**Bonnier B., Leroyer Y. and Pommiers E.** Random sequential adsorption on a semi-continuous substrate

**Borštnik B.** Fractal character of sticky disk clusters vanishes as Baxter's limit is approached

**Böttger H. and Damker Th.** Heat transport and phonon statistics in vibrational hopping systems

**Böttger H., Bryksin V.V., Schulz F. and Bleibaum O.** Quantum-interference magnetotransport in Andersen insulators

**Böttger H. and Poetzsch R.H.** Molecular dynamics study of heat conduction: interplay of disorder and anharmonicity

**Böttger W. and Kastrup H.A.** Carathéodory's canonical framework applied to conformal field theories

**Böttger W., Wissowski H. and Kastrup H.A.** Covariant Hamiltonian formulations for integrable systems in 2-d field theories

**Boubcheur H., Loison D. and Diep H.T.** Phase diagram of the XY stacked triangular antiferromagnet

**Bray A.** See O'Connor D.

**Bréchet Y.** See Vitting Andersen J.

**Breymann W., Jung C. and Vollmer J.** Symbolic dynamics for 2D mappings describing chaotic scattering and spatial disorder

**Breymann W., Tèl T. and Vollmer J.** Diffusion and transport in aspatially - extended chaotic scattering system

**Brouers F., Blacher S. and Sarychev A.K.** Giant electric field fluctuations in semicontinuous films

**Brouers F. and Sarychev A.K.** Tunneling scaling in percolating composite materials

**Bru A.** See Rodriguez M.A.

**Bryksin V.V.** See Böttger H.

**Bualek S.** See Phaovibul O.

**Budde C.E.** See Cáceres M.O.

**Buendia G.M. and Novotny M.A.** Numerical study of a mixed Ising Ferromagnetic system

**Bukharev I.** See Ala-Nissila T.

**Bulatov A., Schmeltzer D. and Birman J.L.** An effective field approach to the Gaussian random matrix model

**Buldyrev S.V., Havlin S., Kertész J., Sadr R., Shehter A. and Stanley H.E.** Surface roughening with quenched disorder in high dimensions: Exact results for the Cayley tree; see Rabinovich S., Stanley H.E., Groβe I. and Viswanathan G.M.

**Bunde A., Dräger J. and Russ S.** Localization in random self-similar structures: beakdown of the self-averaging hypothesis and universal behavior; see Havlin S.

**Bünner M.J.** See Parisi J.

**Burioni R., Cassi D. and Regina S.** Recursion-renormalization technique for random walks on $n$-similar fractals

**Bussemaker H.J.** See Ernst M.H.

**Butylina I.B., Bashmakov I.A., Govor L.V., Grigorieva I.M., Kaputskii F.N., Ksenevich V.K. and Samuilov V.A.** Low dimensional permanent pattern formation based on Bernard-Marangoni effect due to chemical variations

**Cáceres M.O., Budde C.E. and Ré M.A.** Absorption probability in presence of a dynamic trap

**Cáceres M.O., Budde C.E. and Sibona G.J.** Stochastic escape process from potential forms

**Cáceres M.O. and Chattah A.K.** On the quantum dissipation: generalizing the van Kampen approach

**Cadot O.** See Douady S.

**Cai W.L.** See Ho F.C.

**Cai Z.P. and Stéphan G.** Master equations for polarization dynamics in a $Nd^{3+}$-doped fiber laser

**Cancrini N.** See Bertini L.

**Cannas S.A.** See De Magalhães A.C.N.

**Cao L.Y., Hong Y.G. and Fang H.P.** Predicting chaotic time series with wavelet networks

**Cao Z.J. and Guo Y.** The pressure-temperature relation for high-pressure gases with compressibility $e^{-1} < \zeta < e$ at constant volume

**Cartes J.** See Contreras S.

**Casartelli M., Granero M.I., Melloni L. and Porati A.** Non binary alphabetical strings and time series of rational partitions; see Alabiso C.

**Casati G.,** See Artuso R.

**Cassi D.** A further extension of the Mermin-Wagner theorem on graphs

**Cassi D. and Regina S.** Random walks on bundled structures; see Burioni R.

**Castillo J.L.** See Garcia-Ybarra P.L.

**Chakraborty B.** See Gu L.

**Chalub F.A.C.C., Aarão Reis F.D.A. and Riera R.** Series expansions for SAW in regular fractals

**Chan C.K.** See Liang N.Y.

**Chan W.L., Wang X.R. and Xie X.C.** Double-chain localization in the presence of random flux; see Wang X.R.

**Chandler D.** See Lum K.

**Chao C., Samuilov V.A. and Veselova T.** Nonlinear current dynamics in semi-insulating semiconductors: from macroscopic phenomena to microscopic level in complex systems; see Markevich M.I. and Samuilov V.A.
**Chaté H., Manneville P. and Marcq Ph.** Collective behavior, long range order and phase transitions in extensively chaotic dynamical systems
**Chattah A.K.** See Cáceres M.O.
**Chen C.** See and Wang Bao-H.
**Chen F., Ying H. and Li W.** The molecular orientation disorder and electronic localization of the doped $C_{60}$ solid; see Ying H.P.
**Chen F.X.** See Xu S.C.
**Chen G.Z.** See Liu Z.H.
**Chen H.J.** See Tseng H.C.
**Chen H.P.** See Ho F.C.
**Chen H.Y.** See Lin S.C.
**Chen J.** MD simulation of phase transitions in molecular clusters
**Chen J.Y. and Huang Z.Q.** An analytical formula for blood viscosity as a function of rouleaux size of erythrocytes and shear rate
**Chen K.** See Bhagavatula R.
**Chen L.X.** See Yan Z.J.
**Chen L.W. and Huang M.C.** The electronic properties and structures of $M@C_{60}(M = Li, He, Y)$
**Chen L.Y.** Nonequilibrum statistical mechanics in mesoscopic systems
**Chen L.Y., Goldenfeld N. and Oono Y.** Renormalization group theory for global asymptotic analysis
**Chen L.Y. and Weissman M.B.** Simulations of noise statistics and aging of a hierarchical spin-glass model
**Chen S.B.** See Zhou B.Z.
**Chen S.C.** See Zhou H.G.
**Chen S.G.** See Liu J., Wang W.J. and Yang Z.A.
**Chen T.L. and Wang Q.** The critical temperature of quantum spin model determined with variational-cumulant expansion; see Wang Q. and Li H.
**Chen X.S. and Dohm V.** Field theory of finite-size effects in $O(n)$ symmetric systems and its application to the Ising, XY, and Heisenberg models; see Koch W.
**Chen Y.** See Song Y.X.
**Chen Y.Z.** Calculation of chemical bond disruption in DNA, drug-DNA system by a microscopic statistical approach
**Chen Z.Y.** See Cui S.M. and Wang Z.C.

**Chen Z.X., Guo D.H., Liu R.T. and Wu B.X.** Chaotic-classified attractors of neural network with overstorage; see Wang Y.J., Liu Z.H. and Shuai J.W.

**Chen Z.H. and Takayasu H.** Numerical evidence of a percolation-like phase transition in oxygen-limited combustion; Lei H.B.

**Cheung H.F.** See Yip M.K.

**Chiriginskaia Y., Schertzer D. and Lovejoy S.** Dynamical space-time cascades and the Lie structure of turbulence

**Choi D.** See Fuchikami N.

**Choi H.J.** See Liang N.Y.

**Choi M.Y. and Lee H.Y.** Traffic flow and 1/f fluctuation; see Jeon G.S. and Yi J.

**Chowdhury D.** Immune network: an example of complex adaptive systems

**Choy T.C.** See Shannon S.R.

**Christensen K.** See Frette V.

**Clausen S., Pieranski P., Helgesen G. and Skjeltorp A.T.** Classifying periodic modes of motion of magnetic holes in a rotating magnetic field using notations of braid theory

**Cohen E.G.D.** See Rondoni L.

**Cohen M.H. and Natoli V.** Resolution of the N-differentiability problem in chemical reactivity theory for extended systems

**Compagner A.** Operational conditions for reliable random number generation

**Contreras S., Vogel E.E. and Cartes J.** Domains in frustrated Ising lattices with competing interactions

**Costa O.S., Vega Y.M., González L.J.J. and Antoranz J.C.** Criticality in fracture

**Costa U.M.S.** See Sarmento E.F.

**Couder Y.** See Douady S.

**Coutinho S., Nogueira E.Jr., Almeida R.L., Curado E.M.F. and Nobre F.D.** University and multifractality of short range Ising spinglass

**Coutinho S. and Silva L.** Chaotic magnetic phases on the Blume-Emery-Griffiths model with competing interactions; see Silva L.

**Coutinho-Filho M.D.** See Raposo E.P.

**Craubner H.** Statistical thermodynamics of polymer solutions

**Cressoni J.C.** Single gap densities of states determination within the recursion method; see Santos R.J.V. and Lyra M.L.

**Cruz H.R.** See Sarmento E.F.

**Cuccoli A., Macchi A., Pedrolli G., Tognetti V. and Vaia R.** The extrapolation to infinite Trotter number in path-integral Monte Carlo simulations of solid state systems

**Cuccoli A., Tognetti V., Verrucchi P. and Vaia R.** Monte Carlo simulations of the 2-dimensional classical XXZ ferromagnetic model

**Cui S.M. and Chen Z.Y.** Conformation of randomly branched polymers

**Cui Z.Y.** See Zhang K.X.

**Curado E.M.F.** See Coutinho S. and Silva L.

**Czerner P., Li Z.B., Ritschel U. and Zheng B.** Monte Carlo simulation of universal short-time behavior in critical dynamics

**Daboul J.** Application of neural networks versus simulated annealing to spin glass

**Daems D. and Nicolis G.** Cluster expansion of the Perron-Frobenius operator of coupled map lattices

**Dagotto E.** See Hu J.

**Dai J.H., Pu F.C., Wang Y.P. and Zhao H.W.** The Bethe Ansatz solutions for three-wave interaction models

**Damker Th.** See Böttger H.

**De Arcangelis L.** See Ghidaglia C.

**De Magalhães A.C.N. and Cannas S.A.** One-dimensional Potts model with long-range interactions: a renormalization group approach; see Mariz A.M.

**De Miranda L.C.B. and Riera R.** Storage capacity of an hierarchical neural network

**Debierre J.-M.** See Berche B.

**Dekeyser R.** See Drzewiński A.

**Dekker H., Leeuw G. and Maassen van den Brink A.** Boundary layer turbulence as a kangaroo process; see Maassen van den Brink A.

**Del Río J.A.** See Vázquez F.

**Deng C.L.** See Hu Q.F.

**Deng S.H.** Phase structure in a strong interacting matter and mechanism of QGP formation

**Deng Z.J., Liu C.X., Li J. and Liang Y.P.** The evolution of small partial aggregations

**Derrida B., Hakim V. and Pasquier V.** Phase ordering and first-passage exponents

**Desbois J., Furtlehner C. and Ouvry S.** Random magnetic impurities and the $\delta$ impurity problem

**Dezaneti L.M. and Oliveira F.A.** Coalescence of trajectories in complex systems

**Di Castro C.** Dimensional crossover from Fermi to Luttinger liquid

**Diehl H.W. and Shpot M.** Massive field-theory approach to surface critical behavior in three-dimensional systems

**Diep H.T.** See Quartu R. and Boubcheur H.

**Dietrich S.** See Groh B.

**Ding E.J.** See He D.R., Yang H.L. and Zhang S.D.

**Ding G.H. and Xu B.W.** The bosonized Hubbard model; see Xu B.W.

**Ding J. and Huang L.J.** Ergodic theorems for piecewise monotonic transformations

**Ding J. and Zhou A.H.** A finite element method for computing multidimensional absolutely continuous invariant measures

**Dittmar S. and Tang L.H.** Simulation studies of 2D vortex glass

**Dohm V.** See Chen X.S. and Koch W.

**Dokholyan N.** Power law distribution of random repeats in DNA sequences

**Dolan B.P.** Chaotic behavior of renormalisation flow in a complex magnetic field

**Domany E. and Mirny L.** Protein fold recognition and dynamics in the space of contact maps; see Schütz G.M.

**Dornic I.** See Sornette D.

**Douady S., Cadot O. and Couder Y.** Experiment study of strong vorticity filaments in turbulence

**Dougherty J.P.** Foundations of non-equilibrium statistical mechanics

**Dräger J.** See Bunde A.

**Drzewiński A. and Dekeyser R.** Renormalization of the anisotropic linear XY model

**Du C.G.** See Wang Z.C.

**Du C.Y., Zheng X., Zhou J.P. and Qi A.** Dynamical behaviors of a two-dimensional map describing industry-environment systems

**Du T. and Luo W.L.** Periodic breathing oscillations and instabilities in ferrofluids; see Luo W.L.

**Dufty J.W.** See Park Y.J.

**Dykman M.I., Mannella R., Luchinsky D.G., McClintock P.V.E., Short H.E., Stein N.D. and Stocks N.G.** The effect of fluctuations on two-photon absorption by a nonlinear oscillator

**Ebeling W., Neiman A. and Pöschel T.** Entropy and correlation functions of information-carrying sequences; see Militzer B.

**Ebeling W. and Asselmeyer T.** Mixing of thermodynamical and biological strategies in optimization

**Eckle H.P. and Truong T.T.** Corner transfer matrices of asymmetric vertex models; see Berche B. and Römer R.A.

**Einchcomb S.J.B. and McKane A.J.** Optimum paths for bistable systems subject to noise

**Eisenriegler E. and Ritschel U.** Long-range Casimir forces between spherical particles in a fluid and conformal invariance

**Entin-Wohlman O.** See Stein J.

**Eriksson A.B.** See Fodedby H.C.

**Ernst M.H. and Bussemaker H.J.** Statistical mechanics of dissipative systems

**Eto M., Katagi M. and Kawamura K.** Boundary effect on the KTB transition in an array of small tunnel junctions

**Evans J.W.** See Meng B.Q.

**Fan K.J.** See Zheng Y.

**Fang H.P.** Universal bifurcation property of the Hénon map in 2-dimensional parameter space; see Cao L.Y.

**Fang H.P. and Hao B.L.** Symbolic dynamics of the Lorenz equations

**Fang J.L.** See Zhou B.Z.

**Fang Z., Liu Z.L. and Yao K.L.** (a) The CDW transition in quasi-one-dimensional organic polymer ferromagnets (b) Study on the high-spin ground state in alternant $\pi$-conjugated organic polymer

**Feder J.** See Frette V. and Sun T.

**Feng D.** See Peng R.W.

**Feng S.** See Lebedev N.

**Ferraz A.** See Xue X.

**Ferrell R.A.** See Hao H.

**Feudel U.** See Kurths J.

**Filho J.M.** See Andrade Jr. J.S.

**Findenegg G.H.** See Schoen M.

**Fittipaldi I.P.** See Sousa J.R., Sarmento E.F. and Albuquerque D.F.

**Fleming R.J.** See Shannon S.R.

**Flesia S.** Rigorous and numerical results on oriented self-avoiding walks

**Fleurov V.** See Levanda M.

**Fodedby H.C., Eriksson A.B. and Mikheev L.V.** Continuum limit, Galilean invariance and solitons in the quantum equivalent of the noisy Burgers equation

**Fogedby H.C.** Lévy flights in quenched random force fields

**Foong S.K. and Kanno S.** Proof of Page's conjecture on the average entropy of a subsystem

**Freire F., O'Connor D. and Stephens C.R.** The specific heat of a ferromagnetic film

**Frette V., Christensen K., Malthe-S.A., Feder J., Jøssang T. and Meakin P.** Dynamics of a one-dimensional rice pile

**Fridkin V.** See Batchelor M.T.

**Friedmann W.** See Parisi J.

**Fritzsche S.** See Haberlandt R.

**Fröjdh P. and Johannesson H.** Kondo effect in a Luttinger liquid: exact results from conformal field theory

**Fu R.L.** See Rao X.S.

**Fu Z.W.** See Huang Y.

**Fuchikami N., Choi D. and Ishioka S.** Effective dimension of chaotic attractor in a spatially extended system

**Fujii K. and Kondo H.** Statistical theory of orientational phase transition

**Fujisaka H.** See Tutu H.

**Fujisawa S.** See Okumura K.

**Fukui Y. and Horiguchi T.** Finite temperature properties of 1-dimensional chiral XY-model

**Fukukawa D.** See Horiguchi T.

**Fulde P.** See Li Y.M.

**Furtlehner C.** See Desbois J.

**Galam S.** See Wang X.R.

**Gallos L., Argyrakis P. and Kehr K.W.** Search for multifractal properties of the distributions of the number of sites visited by random walks in disordered lattices

**Gallis C., Limoge Y. and Bocquet J.L.** Mass transport in disordered media

**Gammaitoni L., Marchesoni F., Saetta E.M. and Santucci S.** Stochastic resonance as a synchronization phenomenon

**Garcia V.H., Scherer C. and Mors P.M.** Growth kinetics of phase in solid-state diffusion couples

**Garcia-Ybarra P.L., Antoranz J.C., Sankovitch V. and Castillo J.L.** Experimental evidence of self-excited relaxation oscillations leading to homoclinic behavior in spreading flames

**Garland C.W.** See Aharony A.

**Garrido P.** See Gu L.

**Gaub M.** See Haberlandt R.

**Gaveau B., Moreau M. and Toth J.** Finite size scaling in stochastic kinetics: discrete master equation and continuous approximations

**Gervois A. and Mehta M.L.** Some trigonometric identities encountered by McCoy and Orrick

**Kastrup H.A.** See Böttger W.
**Katagi M.** See Eto M.
**Katori M.** See Inui N.
**Katsura S.** Spin-glass algebraic equation in the limit N tends to infinity
**Kawakatsu T.** Phase separation dynamics in the presence of block copolymers
**Kawakubo T., Kobayashi T. and Sakamoto S.** Randomness of the motion of granules in the cell of chara; see Kobayashi T.
**Kawamura K.** See Eto M.
**Kawasaki K., Maya N., Kouzuki A. and Nakamura K.** Spin dynamics of spin-Peierls transition in 1-d S=1/2 Heisenberg antiferromagnets; see Nakamura K.
**Kawasaki T. and Miyashita S.** Symmetry and dimension are relevant to metastability of random systems
**Kawashima N. and Gubernatis J.E.** Generalization of the Fortuin-Kasteleyn representation to quantum spin systems
**Kawazoe Y.** See Hu X., Ohno K. and Tao R.B.
**Ke S.H., Wang R.Z. and Huang M.C.** Theoretical study of energy discontinuities at strained alloy heterointerfaces
**Kehr K.W.** See Gallos L. and Murthy K.P.N.
**Kertész J.** See Buldyrev S.V.
**Kikuchi M.** See Tadaki S.I.
**Kim C.S.** See Park Y.J.
**Kim J.J.** See Mori S.
**Kim M.W.** See Lin M.Y.
**Kim S.** See Jeon G.S. and Reichl L.E.
**Kim Y.-J.** See Aharony A.
**Kimura M.** See Ohno K.
**Kinase W., Ishii A. and Maeda M.** Two-composite model for the anomaly of thermal expansion in water near $4^oC$
**Kinase W., Tsuchiya K., Tashiro M., Yamashita K. and Kuwata S.** Mobility of the p-type semiconductor in the process of tunneling-hopping
**King C.** See Wu F.Y.
**Kitatani H. and Ito N.** Dimensionality crossover in phase transitions Pure dimer ground state in a frustrated quantum Heisenberg spin system
**Kittel A.** See Parisi J.
**Klotz T.** See Kobe S.
**Klümper A.** Analytic calculation of correlation lengths of integrable quantum chains at finite temperature; see Bariev R.Z.

**Kobayashi M.** See Goda M.
**Kobayashi T. and Kawakubo T.** 1/f fluctuations due to the mode coupling in nonlinear systems; see Kawakubo T.
**Kobe S. and Klotz T.** Misfit of ordered and disordered frustrated systems
**Koch W., Dohm V., Stauffer D. and Chen X.S.** Field theory and Monte Carlo simulations of dynamic finite-size effects in Ising-like systems
**Koga S.J.** Scaling laws of free path lengths in Sinai's billiards
**Kohyama T.** Fractal growth of 1-dimensional matter
**Kolb M.** The NO-CO catalytic reaction: modeling oscillations and patterns
**Kolesik M. and Suzuki M.** On the low-temperature ordering in the 3D three-state antiferromagnetic Potts model
**Komiya M.** See Maruyama K.
**Kondo H.** See Fujii K.
**Konno N.** Rounds on order parameters of particle systems by Harris lemma
**Koo H.M.** Viscosity coefficients of perfectly oriented ellipsoids implied by the Kirkwood-Smolouchowski (KS) equation
**Korutcheva E.** See Nadal J.P.
**Kosterlitz J.M.** See Ala-Nissila T.
**Kotiya B.A. and Los V.F.** On the theory of transport phenomena in the polaron's systems
**Kouzuki A.** See Kawasaki K.
**Krech M. and Landau D.P.** Monte Carlo simulation of the Casimir effect in critical films
**Kröger M. and Hess S.** Nonequilibrium molecular dynamics computer simulations on polymers
**Ksenevich V.K.** See Butylina I.B.
**Kubo H., Goda M. and Lin Z.** Anomalous statistical properties of wave function in a Fibonacci lattice
**Kunold A.** See Pereyra P.
**Kuperin Yu.A.** See Antoniou I.
**Kuramoto Y.** See Han S.K.
**Kurata S.** See Nakamura K.
**Kurrer C.** See Han S.K.
**Kurths J., Pikovsky A.S., Zaks M.A. and Feudel U.** Singular continuous spectra in dissipative dynamics
**Kuwata S.** See Kinase W.
**Lagüe P., Miao L., Zhang Z.P., Tobochnik J. and Zuckermann M.** A two-dimensional muti-state model for hydrogen bonding and hydration in lipid bilayers
**Lai P.Y.** Stretch transition in a polymer brush

**Lakshmanan M.** Controlling and synchronization of chaos
**Landau D.P.** See Krech M.
**Landsberg P.T.** The classification of thermodynamics types by entropy properties
**Lang C.Y.** See Yu Z.G.
**Lawson N.S.** See Hendry P.C.
**Lebedev N., Maass P. and Feng S.** Diffusion and superdiffusion of a particle in a random potential with finite correlation time
**Lebedev V.V.** See Aharony A.
**Lebowitz J.L.** See Gu L.
**Lee F.T. and Huang M.C.** Critical properties on the Migdal-Kadanoff hierarchies
**Lee H.Y.** See Choi M.Y.
**Lee R.A.M.** See Hendry P.C.
**Leeuw G.** See Dekker H.
**Lei H.B., Xiao F.F., Chen Z.H. and Zhou B.Z.** Microstructure and effect of impurity (O) on the Er luminescence in the (Er-O) co-doped GaAs[1]; see Xiao F.F.
**Lemaitre A.** Hyperent approximation for collective behavior of large dynamic systems
**Leroyer Y.** See Bonnier B.
**Leung K.** Dynamical simulation of transport in one-dimensional quantum wires
**Leung K.T., Müller J. and Andersen J.V.** The two-dimensional nonlinear Burridge-Knopoff model of earthquakes
**Levanda M. and Fleurov V.** Quantum kinetic Boltzmann equations
**Levinsen M.T.** Experimental study of collective phenomena in pools of coupled oscillators; see Alstrøm P.
**Li B.W. and Robnik M.** Separating the regular and irregular energy levels and their statistics in Hamiltonian system with mixed classical dynamics
**Li B.Z.** See Zhang D.G. and Zhao P.B.
**Li H. and Chen T.L.** Analytical study of antiferromagnetic Heisenberg model
**Li H.C.** The statistical property of heart phythm under different condition
**Li J.** See Deng Z.J.
**Li R., Inaoka H. and Takayasu H.** Water erosion model with fractal geological features
**Li R., Qin G.R., Zhu H.J., Wen X.D.** Experimental control of chaos with two independent periodic signals

**Li S.H. and Yang Y.T.** Entropy and temperature of nonequilibrium radiation

**Li W.** See Chen F.

**Li Y.M. and Fulde P.** Effect of conduction electron correlations for the Andersen impurity model

**Li Z.B.** See Czerner P.

**Li Z.Y., Wang J.J. and Liu X.Y.** Effective dielectric function and high order nonlinear susceptibilities of granular composite medium; see Weng X.M.

**Lian S.Y., Zhou B.Z. and Xu J.** Physical origin for DX centers in III-V alloy semiconductors

**Liang N.Y., Choi H.J. and Chan C.K.** Bursting of soap films

**Liang Y.P.** See Deng Z.J.

**Liang Z.C. and Helfrich W.** Bilayer membrane shape by computer simulation

**Liao H.Z.** A statistical study for the effect of geomagnetic storm on VLF emission at cleft latitudes

**Lim S.C.** One-sided fractional Brownian motion

**Lima F.W., Andrade Jr. J.S. and Moreira J.E.** Controlling chaos in a neural network

**Lima J.P. and Goncalves L.L.** Static and dynamic properties of the XY-model on the one-dimensional superlattice

**Limoge Y.** See Gallis C.

**Lin A.Q.** See Hu Q.F.

**Lin G.X.** See Yan Z.J.

**Lin M.Y., Sinha S.K., Kim M.W., Peiffer D.G., Hanley H.J.M. and Straty G.C.** Worm-like micells - their rich structural information studies by scattering techniques

**Lin R.S.** See Lin S.C.

**Lin S.C., Lin R.S. and Chen H.Y.** Optimization problems and information based complexity

**Lin X.S.** See Hu Q.F.

**Lin Z.** See Kubo H. and Wan C.C.

**Lipowski A.** See Honda Y. and Horiguchi T.

**Lise S.** See Gonnella G.

**Lisyansky A.A. and Livdan D.** Diffusion coefficient of classical waves in absorbing random medium with internal resonances

**Lisyansky A.A. and Nicolaides D.** Phase transitions in disordered systems, exactly solvable model

**Liu C.X.** See Deng Z.J.

**Liu Ji-X. and Zhao X.G.** Inverse problems in statistical physics of a degenerate electron gas

**Liu Jun-X., Zheng W.M. and Hao B.L.** Neimark bifurcation and symbolic analysis of the forced Brusselator

**Liu J. and Chen S.G.** Stability of the invariant tori and curves in a three-dimensional measure-preserving mapping

**Liu K.F.** Stochastic estimation with $Z_2$ noise - application to lattice QCD

**Liu R.T.** See Chen Z.X. and Shuai J.W.

**Liu S.S., Zhao H. (CN), Yang J.Y. and Xiang J.** Laboratory simulation of Rossby solitary vortices and its relation to the LBE computation

**Liu W.M.** Bosonization in Kondo lattice model

**Liu W.M. and Pu F.C.** The finite-size spectrum of the multichannel Kondo model

**Liu X.Y.** See Li Z.Y.

**Liu Z.** See Wang N.Q.

**Liu Z.H. and Chen G.Z.** Targeting and controlling instabilities of cirle map; see Wang Y.J.

**Liu Z.H., Wang Y.J., Chen Z.X. and Qiu W.B.** Study on electroluminescent excited mechanism in ZnS:$Er^{+3}$ thin film

**Liu Z.L.** See Fang Z.

**Liu Z.R.** See Xu J.H.

**Livdan D.** See Lisyansky A.A.

**Loison D.** See Boubcheur H.

**Lopez J.M., Rodriguez M.A. and Pesquera L.** Self-averaging properties in the transport of particles through random media

**Los V.F.** See Kotiya B.A.

**Losson J.** Spectral properties of the transfer operators for deterministic and stochastic coupled map lattices

**Lovejoy S., Schertzer D., Naud C. and Watson B.** Radiative and diffusive transport in multifractals; see Chiriginskaia Y., Marsan D. and Schertzer D.

**Luchinsky D.G., McClintock P.V.E. and Soskin S.M.** Fluctuations in periodically-driven oscillators whose dependence of eigenfrequency on energy is nonmonotonic; see Dykman M.I.

**Lu J.P.** Orientational disorder in fullerenes and fullerides: effects on structural, electronic and transport properties

**Lu Q.K.** Relativistic BBGKY theory and collision integral of Boltzmann equation in a magnetized relativistic plasma

**Lu Q.S.** Spatio-temporal patterns in perturbed Cahn-Hilliard-Schlögl reaction diffusion systems

**Lu Z.Y., Su Z.B. and Yu L.** The doping induced quantum bound states within the Haldane gap

**Luijten E. and Blöte H.W.J.** (a) Ising models with long-range interactions in 1,2 and 3 dimensions (b) Novel Monte Carlo method for spin models with long-range interactions

**Lum K., Luzar A. and Chandler D.** Pathway to boundary-induced phase transition of a confined field

**Luo H.G.** See Yang K.Q.

**Luo J.L.** Stochastic thermodynamics of nonequilibrium electrode-reaction systems; see Song Y.X. and Tu J.

**Luo W.L., Zhang J.L. and Du T.** Structures in a phase separated ferrofluid; see Du T.

**Luzar A.** See Lum K.

**Lyra M.L. and Cressoni J.C.** Extended states in one-dimensional thermally correlated disordered harmonic chains

**Ma B.K.** See Huang Z.B.

**Maass P.** See Lebedev N.

**Maassen van den Brink A. and Dekker H.** Local temperature measurement and the Kapitza boundary resistance; see Dekker H.

**Macchi A.** See Cuccoli A.

**Madden P.A.** See Itoh S.

**Maeda M.** See Kinase W.

**Maekawa S.** See Nagamine T.

**Makse H.A., Havlin S. and Stanley H.E.** Modeling morphology of cities and towns; see Amaral L.A.N.

**Malthe-S.A.** See Frette V.

**Malvezzi A.L.** See Alcaraz F.C.

**Mann G.A.** See Nepomnyashchy Y.A.

**Mannella R.** See Dykman M.I.

**Manneville P.** See Chaté H.

**Mantegna R.N.** See Stanley H.E.

**Marcq Ph.** See Chaté H.

**Mao J.M.** Bifurcations in quantum systems

**Marchesoni F.** See Gammaitoni L.

**Marenzoni P.** See Alabiso C.

**Maritan A.** See Gonnella G.

**Mariz A.M. and De Magalhães A.C.N.** Percolation and n-component cubic model

**Piskounov F.A.** See Markevich M.I.
**Podol'tsev A.S.** See Markevich M.I.
**Poetzsch R.H.** See Böttger H.
**Pollak M.** See Möbius A.
**Pommiers E.** See Bonnier B.
**Poon K.L. and Young K.** Thermodynamics of a one-dimensional system of 'charged' bosons
**Porati A.** See Casartelli M.
**Pöschel T.** See Ebeling W.
**Provatas N. and Ala-Nissila T.** Dynamics of flame fronts in random and fibrous media
**Prügel-Bennett A.** See Herrmann M.
**Pu F.C.** See Dai J.H., Liu W.M. and Zhao P.B.
**Pu T.H.** See Visscher P.B.
**Pyragas K.** See Parisi J.
**Qi A.** See Du C.Y.
**Qian S.M.** See Hong S.M.
**Qin G.R.** See Li R.
**Qiu W.B.** See Liu Z.H. and Wang Y.J.
**Qiu X.M. and Wang A.K.** The dynamo-driven magnetic felicity transport
**Qu S.X.** See He D.R.
**Qu Z.L.** See Xie F.G. and Yang J.Z.
**Quartu R. and Diep H.T.** Magnetic properties of a ferromagnet: application to superlattices
**Rabinovich S., Berkolaiko G., Buldyrev S., Shehter A. and Havlin S.** Exact analytical solution of the "logistic map"
**Ramirez-Santiago G.** See González A.E.
**Ranganathan S. and Pathak K.N.** Binary collision contribution to current correlation functions of dense fluids
**Rao X.S., Zhang G.P., Fu R.L. and Sun X.** Relaxation processes in nondegenerate conjugated polymer
**Raposo E.P. and Coutinho-Filho M.D.** (a) Field theory of polymeric Hubbard chains (b) Field-induced domains in disordered antiferromagnets(AF): $Fe_xZn_{1-x}F_2$
**Rashid R.I.M.A.** The vacancy formation energies in hot rare gas solids within the statistical theory of dense liquids
**Raval A.** See Hu B.L.
**Ré M.A.** See Cáceres M.O.
**Regina S.** See Cassi D. and Burioni R.

**Reichl L.E., Alpatov P. and Kim S.** The effect of symmetry breaking on random walks and Brownian motion

**Reimann P.** Thermally driven escape with fluctuating potentials: a new type of resonant activation

**Ren Y.** See Xu J.H.

**Revuelta J., Valle A., Pesquera L. and Rodriguez M.A.** Pulse statistics in randomly modulated systems: Application to gas and semiconductor lasers

**Revzen M.** See Nepomnyashchy Y.A.

**Rieger H.** Glassy dynamics and slow domain growth

**Riera R.** See Chalub F.A.C.C. and De Miranda L.C.B.

**Ringsdorf H.** See Wang M.

**Ritschel U.** See Czerner P. and Eisenriegler E.

**Robnik M.** See Li B.W.

**Rodgers G.J. and Hassan M.K.** Fragmentation of particles with more than one degree of freedom

**Rodgers G.J. and Singh P.** Generalized random sequential absorption in one dimension

**Rodriguez M.A. and Bru A.** Diffusion, driving and capture in correlated disordered systems; see Revuelta J., Lopez J.M.

**Rojdestvenski I.V.** See Sarmento E.F.

**Römer R.A., Eckle H.P. and Sutherland B.** Gaps in the Heisenberg-Ising model

**Rondoni L., Morriss G.P. and Cohen E.G.D.** A dynamical partition function for the Lorentz gas; see Morriss G.P.

**Röpke G.** Bound states and superfluidity in strongly coupled fermion systems; see Morozov V.G.

**Rössler O.E.** See Parisi J.

**Russ S.** See Bunde A. and Sapoval B.

**Sadr R.** See Buldyrev S.V.

**Saetta E.M.** See Gammaitoni L.

**Saika Y.** (a) Competing bond alternations in spin-1/2 quantum $XY$ - chain (b) Order-from-disorder in the anisotropic $J_1 - J_2$ model

**Saika Y. and Nakamura T.** Study of the general Majumdar-Ghosh model

**Saika Y. and Okamoto K.** Dimer correlations in spin-1/2 alternating $XY$ chain

**Saito Y.** See Tanabe N.

**Sakamoto S.** See Kawakubo T.

**Sakashita H. and Horimoto E.** Melting of native structure in RNase A

**Sakashita H. and Miura Y.** Thermal unfolding as structural phase transition; see Miura Y., Horimoto E.

**Saleur H.** See Sornette D.

**Salinas D.** See Perez P.

**Sammis C.G.** See Sornette D.

**Samuilov V.A. and Chao C.** Nonlinear chaotic charge transport and low frequency current oscillations in semi-insulating semiconductors; see Chao C. and Butylina I.B.

**Sankovitch V.** See Garcia-Ybarra P.L.

**Santos R.J.V. and Cressoni J.C.** Decoration transformation and effective BEG model

**Santucci S.** See Gammaitoni L.

**Sapoval B.** A simple renormalization study of the Laplace transfer across fractal and irregular surfaces, electrodes, membranes and catalysts

**Sapoval B. and Russ S.** Super-localized character of the anomalous viscous damping of the vibrations of 2d fractal percolation clusters

**Sarmento E.F., Costa U.M.S. and Rojdestvenski I.V.** The damage spreading in an Ising BEG ultra thin film

**Sarmento E.F., Cruz H.R. and Tucker J.W.** Critical lines of ferromagnetic Ising films in a random surface field

**Sarmento E.F., Fittipaldi I.P. and Kaneyoshi T.** Phase diagram and paramagnetic susceptibility of a diluted spin-1 layered magnet in a random field

**Sarychev A.K.** See Brouers F.

**Sauermann G.** See Shimizu T.

**Schadschneider A.** See Bariev R.Z.

**Scherer C.** See Garcia V.H.

**Schertzer D., Lovejoy S. and Schmitt F.** Strong vs. weak multifractal universality, S.O.C. and multifractal phase transitions; see Chiriginskaia Y., Lovejoy S., Marsan D. and Schmitt F.

**Schmeltzer D.** See Bulatov A.

**Schmidt M.R.** See Sibani P.

**Schmitt F. and Schertzer D.** Vectorial multifractal turbulence: theory and data analysis; see Schertzer D.

**Schmitz R.** See Bausch R.

**Schoen M., Thommes M. and Findenegg G.H.** The microscopic structure of a pure near-critical fluid confined to a mesoscopic slit-pore

**Schröder E.** See Alstrøm P.

**Schulz F.** See Böttger H.

**Schütz G.M. and Domany E.** Phase transitions and competing length scales in the asymmetric exclusion process with open boundary conditions; see Stinchcombe R.B.

**Schwarz K.** See Sun T.

**Scotti A.** See Mezzadri F.

**Sear R. and Jackson G.** Theory for the phase behavior of a mixture of a rod-like colloid and a rod-like polymer

**Shah H.A.** See Shah I.

Shah I., Anawar G., Shah H.A. and Alam M. A. Chaotic evolution of a parametric instability in a piezoelectric semiconductor plasma

**Shang C.H. and Wang Y.J.** Study of macroscopic dense branching morphology in BiAIMn/SiO alloy films

**Shannon S.R., Choy T.C. and Fleming R.J.** Corrections to scaling in 2D polymer statistics

**Shapir Y.** See Wang X.R.

**Shchur L.N. and Blöte H.W.J.** Finite-size scaling of systematic deviations in cluster Monte-Carlo with shift-register random number generators

**Shehter A.** See Buldyrev S.V. and Rabinovich S.

**Shen Y.W. and Huang M.C.** Statistical supercell method for calculating electronic structure and ground-state properties of the mixed crystal

**Shi Y.** Constraints on pairings in the Hubbard model

**Shibusa Y.** See Andrade Jr. J.S.

**Shida K.** See Ohno K.

**Shim G.M.** See Bollé D.

**Shimizu T. and Sauermann G.** Unique Floquet decomposition theory for nonlinear response

**Shimizu Y. and Ono I.** Simulations on critical state of realistic sandpile model

**Shinjo K.** Intensity of motion in phase space

**Shiokawa K. and Hu B.L.** Decoherence, delocalization and irreversibility in quantum chaotic systems

**Shirakura T. and Matsubara F.** Replica exchange MC simulation of the two-dimensional $\pm J$ Ising spin glass; see Morishita K.

**Short H.E.** See Dykman M.I.

**Shpot M.** See Diehl H.W.

**Shrock R.E.** Complex-temperature singularities of Ising models

**Shuai J.W., Chen Z.X., Liu R.T. and Wu B.** The Q-state 3-dimension rotation angle neural network model

**Sibani P., Schmidt M.R. and Alstr$\phi$m P.** Fitness optimization and decay of extinction rate through biological evolution

**Sibona G.J.** See Cáceres M.O.
**Silva L., Curado E.M.F., Morgado W.A.M. and Coutinho S.G.** Criticality and multifractality of Potts ferromagnetic model in fractal lattices; see Coutinho S.
**Simons M.** See Stanley H.E.
**Singh P.** See Rodgers G.J.
**Sinha S.K.** See Lin M.Y.
**Sinpatanaphan B.** See Phaovibul O.
**Siurakshina L.** See Yarunin V.S.
**Skjeltorp A.T.** See Clausen S.
**Soga G., Guo H. and Zuckermann M.** Numerical simulation of polymer brushes in good and poor solvents
**Song Y.X., Chen Y., Luo J.L. and Xian D.C.** An analytical method for evaluating critical point of lattice model
**Sousa J.R. and Fittipaldi I.P.** Magnetic phase diagram of a randomly decorated Heisenberg model: application to $La_{2-x}Ba_xCuO_4$; see Albuquerque D.F.
**Sornette A.** See Sornette D.
**Sornette D., Johansen A., Dornic I., Sornette A. and Vanneste C.** Mapping self-organized criticality onto criticality
**Sornette D., Saleur H., Johansen A. and Sammis C.G.** Complex fractal dimensions in random systems: relevance for rupture and earthquake prediction
**Soskin S.M.** See Luchinsky D.G.
**Stanley H.E., Buldyrev S.V., Goldberger A.L., Havlin S., Mantegna R.N., Peng C.-K. and Simons M.** Fluctuation phenomena and evolution in noncoding DNA; see Buldyrev S.V., Große I., Viswanathan G.M., Peng C.-K., Makse H.A., Amaral L.A.N., Gyure M.F.
**Staples I.** See Zeng C.
**Stauffer D.** See Koch W.
**Stein J., Entin-Wohlman O. and Aharony A.** Spin-orbit coupling, Coulomb exchange interactions, and weak ferromagnetism of copper-oxide planes
**Stein N.D.** See Dykman M.I.
**Stewart G.E.** See Betts D.D.
**Stéphan G.** See Cai Z.P.
**Stephens C.R.** See Freire F., O'Connor D.
**Stopa M.** See Iitaka T.
**Straty G.C.** See Lin M.Y.

**Strilka R.** See Gyure M.F.

**Stinchcombe R.B. and Schütz G.M.** Operator algebra for stochastic dynamics and quantum spin chains

**Stocks N.G.** See Dykman M.I.

**Su W.H.** See Wu X.H.

**Su Z.B.** See Lu Z.Y.

**Sugano T.** See Iitaka T.

**Sun C., Wang M., Peng R.W. and Ming N.B.** Growth and melt of fractal-like domains in lipid monolayers; see Wang M.

**Sun R.G. and Zhang J.** Interaction between inorganic chemical matters and biomembrane in liquid crystal state; see Zhang J.

**Sun T., Meakin P., Jøssang T., Feder J. and Schwarz K.** Computer simulations of meandering rivers

**Sun X.** See Rao X.S., Xu X.T. and Yu Z.G.

**Sung W. and Park P.J.** Membrane stability and dynamics of pore growth; see Park P.J.

**Sutherland B.** See Römer R.A.

**Suzuki M.** Quantum effective-field theory of electron and spin systems; see Todo S., Asakawa H., Kolesik M., Nishiyama Y. and Totsuka K.

**Suzuki N. and Matsubara F.** Phase transition of the S=1/2 quantum anisotropic Heisenberg antiferromagnet on the triangular lattice

**Tadaki S.I. and Kikuchi M.** Jam phases in a two-dimensional cellular automaton model of traffic flow

**Taji Y.** Extended Boltzmann equation for irreversible momentum transport

**Takahashi M.** Susceptibility anomalies of low-dimensional antiferromagnets; see Ohno K.

**Takayasu H.** See Chen Z.H., Inaoka H. and Li R.

**Takemura S.** See Nishiuma S.

**Tanabe N. and Saito Y.** Polymer crystallization model with the connectivity of monomers along a chain

**Tang C.** Self-organized criticality in nonconserved systems

**Tang L.H.** See Dittmar S.

**Tanimoto N.** See Nakamura K.

**Tanimura Y.** Time dependent correlation functions of perturbed quantum brownian osilator

**Tao R.B.** Extended states and localization in generalized Thue-Morse chain

**Tao R.B., Hu X. and Kawazoe Y.** Quantum fluctuation and instability of spin wave in magnetic multilayer systems; see Zhou L. and Wan C.C.

**Tashiro M.** See Kinase W.

**Tejero C.F.** Freezing of simple fluids with short-ranged attractions: A density functional approach

**Tèl T.** See Breymann W.

**Terai A.** Numerical analysis of Haldane gaps for S=1,2 and 3 Heisenberg spin chains using twisted boundary conditions

**Thill M.J. and Hilhorst H.J.** The phase space of low-dimensional critical spin glass

**Thommes M.** See Schoen M.

**Tian G.S.** (a) The $\eta$-pairing ground states in the negative-U Hubbard model are supersolid (b) The magnetic long-range orders in the strongly correlated electron systems: some new rigorous results; see Yang S.P.

**Tildesley D.J.** See Peters G.H.

**Ting C.S.** See Xu J.H.

**Tobochnik J.** See Lagüe P.

**Todo S. and Suzuki M.** Effective exclusive volume for hard-core systems in the liquid region

**Tognetti V.** See Cuccoli A.

**Toth J.** See Gaveau B.

**Tong P., Shen Y. and Xia K.Q.** Measured local-velocity fluctuations in turbulent convection; see Xia K.Q.

**Tong Q.Y.** See Xu J.H.

**Totsuka K. and Suzuki M.** Spin-1/2 Heisenberg spin ladder with bond alternation

**Toxvaerd S.** See Peters G.H.

**Tran V.Q. and Nguyen M.A.** The T-dependent law of chemical reaction rate in the point of view of the nonequilibrium statistical operator theory

**Tretyakov A.Yu. and Inui N.** Some aspects of the critical behavior of the two-neighbor system

**Truong T.T.** See Eckle H.P.

**Tseng H.C. and Chen H.J.** Some characteristics of deterministic diffusion

**Tsuchiya K.** See Kinase W.

**Tu J. and Luo J.L.** Stochastic thermodynamics of finite chemical reaction systems under the influence of external noise

**Tu Y.H.** Long range order in a 2D dynamical XY model: How birds fly together

**Tucker J.W.** See Sarmento E.F.

**Turban L.** See Berche B.

**Turski L.A.** See Bausch R.

**Tutu H. and Fujisaka H.** Ordering dynamics in nearly 1D anisotropic XY spin system

**Uezu T., Kabashima Y., Nokura K. and Nakamura N.** Replica calculation of learning curve in binary choice problems in the framework of the replica symmetry breaking

**Upton P.J.** See Abraham D.B. and O'Connor D.

**Uzawa T.** See Inui N.

**Uzunov D.I.** A new theory of critical phenomena

**Vaia R.** See Cuccoli A.

**Valera M. and Mata G.J.** The surface transition in antiferromagnetic chromium

**Valle A.** See Revuelta J.

**Vallespin A.** See Alstrøm P.

**Vanneste C.** See Sornette D.

**Vats. N.** See Betts D.D.

**Vázquez F. and Del Río J.A.** Poisson structure for hyperbolic transport

**Vega Y.M.** See Costa O.S.

**Verrucchi P.** See Cuccoli A.

**Veselova T.** See Chao C.

**Vespignani A.** See Pietronero L.

**Vieira A.P. and Goncalves L.L.** One-dimensional lattice gas with random interactions

**Visscher P.B. and Pu T.H.** Efficient simulation of complex fluids via hierarchical potentials

**Viswanathan G.M., Buldyrev S. and Stanley H.E.** Physics of a complex system: studies of scale-invariance in Avian behavior

**Vitting Andersen J. and Bréchet Y.** New macroscopic Hamiltonian for wetting of a solid by a liquid in the immersion geometry; see Leung K.T.

**Vogel E.E.** See Contreras S.

**Vollmer J.** See Breymann W.

**Wainer I.** See Andrade Jr. J.S.

**Wan C.C., Lin Z.F. and Tao R.B.** A lattice gas simulation of the two immiscible fluids

**Wang A.K.** See Qiu X.M.

**Wang Bao-H. and Chen C.** The application of complex neural network to protein structure prediction

**Wang Bin-H.** The universal plateau structure of Lyapunov exponent for period doubling cascade attractors; see He D.R.

**Wang C.T. and Yu L.** Momentum transfer between $^3$He quasiparticles and the solid surface via vortices

**Wang G.R.** See Yang Z.A. and Wang W.J.

**Wang J.** See Gu Y., Wang Y.J., Zhao H. (CN)

**Wang Jian** The relation between the exponents of the random elastic network and the random resistor network

**Wang Jin, Panchenko, Nienhaus U. and Wolynes P.G.** Reaction dynamics and single molecules in fluctuating environments; see Wang Y.J.

**Wang J.J.** See Li Z.Y.

**Wang J.S.** Anisotropic finite-size scaling analysis of a two-dimensional driven diffusive system

**Wang M., Ming N.B., Bennema P. and Ringsdorf H.** Morphology and stability of faceted crystals in lipid monolayers

**Wang M., Peng R.W. and Ming N.B.** The influence of electroconvection on pattern formations in electrochemical deposition; see Peng R.W., Sun C.

**Wang M., Sun C., Peng R.W. and Ming N.B.** Dynamic characterization of fractal-like solid domains in lipid monolayers: an insight into the domain growth mechanism

**Wang N.Q. and Liu Z.** Particle orbits in model current sheets

**Wang Q. and Chen T.L.** Study of the quantum spin Heisenberg model with variational-cumulant expansion to the fourth order; see Chen T.L.

**Wang R.** Dynamics of bilayer membranes

**Wang R.Z.** See Ke S.H.

**Wang W.J., Wang G.R. and Chen S.G.** Controlling the optical field chaos in storage ring free-electron lasers

**Wang X.R., Chan W.L., Shapir Y. and Galam S.** Long-range-order in 2D diluted random field Ising model; see Chan W.L.

**Wang Y.H.** See Zhao H. (CN)

**Wang Y.J., Liu Z.H. Chen Z.X. and Qiu W.B.** Study of the relaxation luminescent peaks in zinc sulfide film devices; see Liu Z.H. and Shang C.H.

**Wang Y.J., Wang J., Zhu N.J. and Guo H.** Spectra of quasi-bound states in open quantum-dots

**Wang Y.P.** See Dai J.H.

**Wang Y.Q.** See Weng X.M.

**Wang Z.C., Xing H.Z., Du C.G. and Chen Z.Y.** Quantum theory of intensity fluctuations in the nondegenerate two-photon optical bistability

**Watson B.** See Lovejoy S.

**Wei Q.H., Zhou C.H. and Ming N.B.** The percolation transition in two -dimensional aggregation induced by an alternating electrical field

**Weinberg W.H.** See Meng B.Q.

**Weissman M.B.** See Chen L.Y.

**Wen X.D.** See Li R.

**Weng X.M., Li Z.Y. and Wang Y.Q.** Study of phase transition of longitudinal and transverse field Ising systems

**Westervelt R.M.** See Hu J.M.

**Wissowski H.** See Böttger W.

**Wong K.Y.M.** Microscopic equations and stability conditions in optimal neural networks

**Wu B.** See Shuai J.W.

**Wu B.X.** See Chen Z.X.

**Wu C.Q.** Quantum lattice fluctuations in MX complexes

**Wu F.Y., Pant P. and King C.** The chiral Potts model and its associated link invariant

**Wu H.J.** See Zhou H.G.

**Wu S.** See Yang Z.R.

**Wu S.Z.** See Wu X.H.

**Wu X.H., Zhang C.X., Wu S.Z. and Su W.H.** The effective dielectric function of the composite with random microgeometry

**Wu Z.B.** Symbolic dynamics analysis of chaotic time series with driven frequency

**Wu Z.G.** See Xu S.C.

**Xia K.Q., Zhang Y.B. and Tong P.** A phase-boundary study of droplet-droplet interaction in microemulsion-polymer mixtures; see Tong P.

**Xian D.C.** See Song Y.X.

**Xian Y.** See Bishop R.F.

**Xiang J.** See Liu S.S.

**Xiao F.F. and Lei H.B. and Zhou B.Z.** Energy transport and relaxation in the luminescence process of $Er^{+}$-implanted $InP^{1}$; see Lei H.B.

**Xie F.G., Hu G. and Qu Z.L.** On-off intermittency in a coupled-map lattice system

**Xie X.C.** See Chan W.L.

**Xie Y.Z.** On Euler-Lagrange shape equations for axisymmetric vesicles

**Xing H.Z.** See Wang Z.C.

**Xu B.W. and Ding G.H.** The lattice Coulomb gas; see Ding G.H.

**Xu G.O. and Gong J.B.** Dynamical aspects of the random-matrix theory

**Xu G.O., Gong J.B. and Jie Q.L.** How can quantum chaos arise dynamically?

**Xu G.O. and Jie Q.L.** Structural instability of quantum chaotic states

**Xu H.** Process of electronic localization in a disordered system

**Xu J.** See Lian S.Y.

**Xu J.H., Liu Z.R., Liu R. and Tong Q.Y.** Information transmission of human brain cortex

**Xu J.H., Ren Y. and Ting C.S.** Ginzburg-Landau equations and their applications in d-wave superconductor

**Xu L.B. and Gu W.Z.** Investigation of critical free-energy-function

**Xu S.C.** I-ching aided classification scheme of binary numbers in...

**Xu S.C., Lin S., Chen F.X. and Wu Z.G.** Logistic model in laser dynamics

**Xu S.S.** See Yang S.P.

**Xu X.T., Yang Y.L. and Sun X.** The scaling theory of a single polymer chain in strongly confined geometries

**Xu Y.** See Pakhomov A.B.

**Xue X. and Ferraz A.** Supersymmetric electron-hole excitations in semiconductors at finite temperature

**Yamada H. and Ikeda K.S.** Self-induced irreversible transport of energy in an electron-phonon system with a few degrees of freedom; see Goda M.

**Yamamoto S.** Critical behavior in quantum spin chains with alternating interaction

**Yamashita K.** See Kinase W.

**Yamazaki Y.** SUSY stochastic dynamics of neural systems

**Yamazzaki H.** See Arimitsu T.

**Yan J.** See Zhao H. (CN)

**Yan M.** See Hu B.L.

**Yan X.** See Pakhomov A.B.

**Yan Z.J., Chen L.X. and Lin G.X.** (a) The characteristics of entropy of complex magnetic materials and its application in the magnetic refrigeration technology (b) Theory of irreversible Carnot cycle and improved Rutgers' relation (c) Fluctuations and fundamental equation of thermodynamics

**Yang C.Z.** See Zhang G.M.

**Yang H.L. and Ding E.J.** On-off intermittency for synchronistic- unsynchronistic phase transition

**Yang J.Y.** See Liu S.S.

**Yang J.Z., Qu Z.L. and Hu G.** Suppressing nonautonomous chaos: the phase effect

**Yang K.Q. and Luo H.G.** The classical solutions of a class of nonlinear equations

**Yang S.P., Tian G. and Xu S.S.** Using nonlinear delayed feedback to control chaos

**Yang Y.L.** See Xu X.T.

**Yang Y.T. and Zheng Y.** Geometric aspects of integrability in Hilbert space of quantum systems; see Li S.H. and Zheng Y.

**Yang Z.R., Wu S. and Guo W.A.** Phase transitions of Ising and classical O(2) models on nice tree

**Yang Z.A., Chen S.G. and Wang G.R.** Jacobian analysis of reconstructing topology

**Yao K.L.** See Fang Z.

**Yao W.G.** Controlling chaos by non-continuous delayed feedback method

**Yao Y.C.** The solid-liquid phase transitions in two-dimensional Collin's system

**Yarunin V.S. and Siurakshina L.** Microscopic theory of non-ideal Bose-gas excitations

**Yi J. and Choi M.Y.** Interaction effects on persistent currents in magnetic fields

**Ying H.P., Ji D.R. and Chen F.** Non-local Monte Carlo updating schemes for the quantum spin systems; see Chen F.

**Yip M.K., Zheng J.R. and Cheung H.F.** Persistent of one-dimensional perfect rings under the canonical ensemble

**Yoshida Z.** See Ito N.

**Yoshino H.** Aging effect of directed polymer in random media

**Young K.** See Poon K.L.

**Young M.J.** See Aharony A.

**Yu K.W.** Percolation effects in nonlinear inhomogeneous media: crossover from linear to nonlinear behaviors

**Yu L.** See Lu Z.Y. and Wang C.T.

**Yu Y.Z.** $\Phi_{1,2}$ perturbed superconformal field theory and supersymmetric ZMS model

**Yu Z.G., Lang C.Y. and Sun X.** Effect of lattice quantum fluctuation in MX chains

**Yuan D.Q.** The statistical distribution law and fluctuation-dissipation relationship of particle motion in living cell

**Yuan X.F. and Hu B.B.** On the study of KAM tori in a class of generalized FK model

**Yung C.M. and Batchelor M.T.** Integrable loop models with open boundaries from reflection matrices; see Batchelor M.T.

**Zakrzewski W. J. and Zapotocky M.** Kinetics of phase ordering with topological textures

**Zaks M.A.** See Kurths J.

**Zapotocky M.** See Zakrzewski W. J.

**Zeng C. and Henley C.** Interface representations of classical spin models with critical ground states

**Zeng C., Staples I. and Bishop R.F.** Quantum phase transition in square- and triangular-lattice spin-1/2 antiferromagnets

**Zeng X.C.** Integral equation and Monte Carlo simulation studies of local clustering in infinitely dilute supercritical solutions

**Zhang B.G.** The entropy of contracted gas group

**Zhang B.T. and Pan H.Y.** Lattice model for the $SO_2 - O_2 - H_2O$ surface reaction: a partial simulation study; see Pan H.Y.

**Zhang C.X.** See Wu X.H.

**Zhang D.G. and Li B.Z.** The quantum Ising chain with multiple critical-off-critical interfaces and conformal invariance

**Zhang G.M. and Yang C.Z.** Particle-size effect on the percolation of dispersed ionic conductors: a challenge to percolation theory

**Zhang G.P.** See Rao X.S.

**Zhang J. and Sun R.G.** Comparing with the influence of organic chemical matters on the PE and PC structure of liquid crystal by SAXS; see Sun R.G.

**Zhang J.L.** See Luo W.L.

**Zhang K.X., Zhang P., Zhao Q.X. and Cui Z.Y.** Statistical physics approach of cell membrane potential and anomalous membrane potential of cancer cell

**Zhang P.** See Zhang K.X.

**Zhang S.D. and Ding E.J.** Numerical simulations on burst process of fiber bundles

**Zhang Y.B.** See Xia K.Q.

**Zhang Z.P.** See Lagüe P.

**Zhang Z.Y.** Atomic-scale manipulation of 2D island shapes at surfaces

**Zhao B.** See Pakhomov A.B.

**Zhao D.L.** See Huang Y.

**Zhao H. (CN) and Wang Y.H., Yan J. and Wang J.** A general method of controlling chaos; see Liu S.S.

**Zhao H. (FR) and Bellissent-Funel M.C.** Neutron-diffraction studies of the inter-molecular structures of supercritical water

**Zhao H. (FR) and Beysens D.** Probing the universal critical-adsorption profile by neutron reflectometry
**Zhao H.W.** See Dai J.H.
**Zhao P.B., Li B.Z. and Pu F.C.** On the reorientation of surface magnetic direction of semi-infinite ferromagnets
**Zhao Q.X.** See Zhang K.X.
**Zhao X.G.** See Liu Ji-X.
**Zheng B.** See Czerner P.
**Zheng J.R.** See Yip M.K.
**Zheng W.M.** See Liu Jun-X.
**Zheng W.H., Hamer C.J. and Oitmaa J.** Comparison of expansion methods for the Heisenberg antiferromagnet
**Zheng X.** See Du C.Y.
**Zheng Y., Fan K.J. and Yang Y.T.** Order inverse after avoided crossing from iteration
**Zheng Z.G. and Hu G.** Ergodicity in hard sphere systems and Boltzmann entropy
**Zhong J.X.** Temporal autocorrelation and quantum decay in quasiperiodic systems
**Zhou A.H.** See Ding J.
**Zhou B.Z., Fang J.L. and Chen S.B.** Investigation of the phase transition of GaAs(Er) in $Er^+$-implanted and annealed processes using PAL and PL[1]; see Lei H.B., Lian S.Y. and Xiao F.F.
**Zhou C.H.** See Wei Q.H.
**Zhou H.G., Chen S.C. and Wu H.J.** Optical bistability, multiple bistability and chaos of liquid crystal in phase transitions
**Zhou J.P.** See Du C.Y.
**Zhou L. and Tao R.B.** Magnetizations of easy-plane spin-one ferromagnet
**Zhou X.P. and Zhu S.Q.** Intensity and phase fluctuations in a laser field with coupling of noise term
**Zhu H.J.** See Li R.
**Zhu N.J.** See Wang Y.J.
**Zhu S.Q.** See Zhou X.P.
**Zittartz J.** See Bariev R.Z.
**Zuckermann M.** See Soga G. and Lagüe P.

# LIST OF PARTICIPANTS

## ARGENTINA

Caceres, Manuel O.
Inst. Balseiro
Centro Atomico Bariloche
8400 S. C. de Bariloche

Zheng, Weihong
School of Physics
Univ. of New South Wales
Sydney, NSW 2052

## ARMENIA

Akheyan, Aram Z.
Theoretical Division
Yerevan Physics Inst.
Yerevan 375036

## AUSTRALIA

Attard, Phillip John
Department of Physics
Australian Nat'l University
Canberra ACT 0200

Batchelor, Murray
Department of Mathematics
Australian Nat'l University
Canberra ACT 0200

Choy, Tuck Chuen
Department of Physics
Monash University
Clayton Victoria 3168

Morriss, Gary Phillip
School of Physics
University of New South Wales
Sydney 2052

Pearce, Paul A.
Mathematics Department
Melbourne University
Victoria 3052

Rondoni, Lamberto
School of Physics
University of New South Wales
Sydney 2052

Yung, Chong Ming
Dept. of Mathematics
Australian Nat'l Univ.
Canberra ACT 0200

## BANGLADESH

Islam, A.K.M. Azharul
Dept. of Physics
Rajshahi Univ.
Rajshahi–6205

Islam, Shamsunnahar
Dept. of Physics
Rajshahi Univ.
Rajshahi–6205

Rashid, R.I.M.A.
Physics Dept.
Univ. of Dhaka
Dhaka–1000

## BELARUS

Markevich, Marija
Inst. of Electronics
Belarus Academy of Sciences
Minsk 220090

Samouilov, Vladimir
Physics Dept.
State Univ. of Belarus
Minsk 220080

## BELGIUM

Antoniou, Ioannis
Int'l Solvay Inst. for Phys. & Chem.
1050 Bruxelles

Bi, Qiao
Int'l Solvay Inst. for Phys. & Chem.
1050 Bruxelles

Blossey, Ralf Klaus Rudi
Lab. voor VSM
Katholieke Universiteit Leuven
Leuven

Bollé, Desire
Inst. for Theoretical Physics
K.U. Leuven
B-3001 Leuven

Brouers, Francois
Inst. de Physique
Universite de Liege
Liege 4000

Daems, David
Ctr. for Nonlinear Phenomena & Complex Sys.
Univ. Libre de Bruxelles
1050 Bruxelles

Dekeyser, Rafael
Dept. of Physics
Katholieke Universiteit Leuven
B-3001 Heverlee

Indekeu, Joseph O.
Lab. voor VSM
Katholieke Universiteit Leuven
B-3001 Leuven

Losson, Jerome
Service de Chimie Physique
Univ. Libre de Bruxelles
Bruxelles B-1050

Reimann, Peter
Dept. WNI
Limburgs Universitair Centrum
3590 Diepenbeek

**BRAZIL**

Alcaraz, Francisco Castilho
Dpto. de Fisica
Univ. Federal de Sao Carlos
CEP 13560-905 Sao Carlos-SP

Chalub, Fabio Augusto
Dept. de Fisica
Pontificia Univ. Catolica
Copacabana-Riode Janeiro 22081-000

Coutinho, Sergio Galvao
Depto. de Fisica
Univ. Federal de Pernambuco
CEP 50670-901-Recife

Coutinho-Filho, Mauricio D.
Depto. de Fisica
Univ. Federal de Pernambuco
50670-901 Recife-PE

Cressoni, Jose Carlos
Depto. de Fisica
Univ. Federal de Alagoas
57072-790 Maceio-Al

de Almeida, Rita Maria Cunha
Inst. de Fisica
Univ. Federal do Rio Grande do Sul
91501-970 Porto Alegre RS

de Magalhães, Aglae Cristina N.
Ctr. Braisileiro de Pesquisas Fisica
22290-180, Rio de Janeiro

de Miranda, Lourengo C.B.
Dept. de Fisica
Pontificia Univ. Catolica
Rio de Janeiro 22421 020

Ferreira Chaves, Carlos Mauricio
Physics Dept.
P.U.C.
22452-Rio de Janeiro

Fittipaldi, Ivon P.
Dept. of Physics
Federal Univ. of Pernambuco
Recife-Pe

Freire, Rosane Riera
Depto. Fisica
Pontificia Univ. Catolica
22453-900 Rio de Janeiro

Goncalves, Lindberg Lima
Dept. de Fisica
Univ. Federal Do Ceara
60451-930 Fortaleza, Ceara

Mendonca Fleury Curado, Evaldo
Centro Brasileiro de Pesquisas Fis.
Rio de Janeiro
70.919-970, Brasilia, DF

Neves, Armando G.M.
Departamento de Matematica
Univ. Federal de Minas Gerais
30.161-970-Belo Horizonte(MG)

Oliveira, Fernando A.
Cetro Intern'l de Fis. da Materia
Condensada Univ. de Brasilia
70919–970

Reis, Fabio
Depto. de Fisica
Univ. Federal Fluminense
CEP 24210–340, RJ

Sarmento, Enaldo F.
Departamento de Fisica
Univ. Federal de Alagoas
57072 Maceido - AL

Scherer, Claudio
Inst. de Fisica
Univ. Federal do R.G. Sul
91501–970, Porto Alegre

Xue, Xun
Intern'l Ctr. of Condens. Matter Phys.
Univ. de Brasilia
70910–900 Brasilia–DF

## BULGARIA

Uzunov, Dimo
Inst. of Solid State Phys.
Bulgarian Academy of Sciences
1784 Sofia

## CANADA

Betts, Donald D.
Dept. of Phys.
Dalhousie Univ.
Halifax, B3H 3J5

Chen, Z.Y. Jeff
Dept. of Physics
Univ. of Waterloo
Ontario N2L 3G1

Guo, Hong
Dept. of Physics
McGill Univ.
Montreal PQ H3A 2T8

Ranganathan, Subramanian
Dept. of Mathematics
Royal Military College
Ontario K7K 5L0

Zuckermann, Martin J.
Physics Dept.
McGill Univ.
Montreal, Quebec H3A 2T8

## CHILE

Perez, Patricio
Departamento de Fisica
Univ. de Santiago de Chile
Correo 2, Santiago

Vogel, Eugenio
Dpto. de Fisica
Univ. de La Frontera
Casilla 54–D Temuco

## CHINA

Cao, Liang-Yue
Inst. of Theor. Physics
The Chinese Academy of Sciences
Beijing 100080

Cao, Zhi-Jue
Dept. of Physics
Hunan Educational Inst.
Changsha 410012

Chen, Chao
Dept. of Physics
Xiamen Univ.
Xiamen 361005

Chen, Chen
Inst. of Biophysics
The Chinese Academy of Sciences
Beijing 100101

Chen, Chuan-hong
Dept. of Physics
Xiamen Univ.
Xiamen 361005

Chen, Fen-Xian
Dept. of Physics
Fujian Normal Univ.
Fuzhou 350007

Chen, Feng
Dept. of Physics
Zhejiang Univ.
Hangzhou 310027

Chen, Guang-Zhi
President Office
Guangxi Univ.
Nanning 530004

Chen, Guo-yi
Inst. of Theoretical Physics
The Chinese Academy of Sciences Beijing
100080

Chen, Hui-Ping
Ctr. for Fundamental Physics
Univ. of Science & Tech. of China
Hefei 230026

Chen, Jia-Er
President Office
Peking Univ.
Beijing 100871

Chen, Jian
Dept. of Physics
Tsinghua Univ.
Beijing 100084

Chen, Jin-Yu
Inst. of Low-Energy Nuclear Physics
Beijing Normal Unvi.
Beijing 100875

Chen, Jin-Can
Dept. of Physics
Xiamen Univ.
Xiamen 361005

Chen, Li-Xuan
Dept. of Physics
Xiamen Univ.
Xiamen 361005

Chen, Nan-Xian
Inst. of Applied Physics
Beijing Univ. of Sci. & Tech.
Beijing 100083

Chen, Ren
Dept. of Physics
Shanghai Normal Univ.
Shanghai 200234

Chen, Si-Yu
Nat'l Natural Sci. Foundation of China
Beijing 100083

Chen, Tian-Lun
Dept. of Physics
Nankai Univ.
Tianjin 300071

Chen, Zhen-Xiang
Dept. of Physics
Xiamen Univ.
Xiamen 361005

Chen, Zhi-Jian
Inst. of Theoretical Physics
The Chinese Academy of Sciences
Beijing 100080

Dai, Jian-Hui
Zhejiang Inst. of Modern Physics
Zhejiang University
Hangzhou 310027

Deng, Sheng-Hua
Dept. for Basic Courses
Jianghan Petroleum Inst.
Jianghan 434102

Deng, Zhao-Jing
Dept. of Physics
Southwest China Normal Univ.
Chongqing 630715

Ding, Guo-Hui
Dept. of Physics
Shanghai Jiao Tong Univ.
Shanghai 200030

Dong, Ji-Yang
Dept. of Physics
Fujian Normal Univ.
Fuzhou 350007

Dong, Yu-Fei
Dept. of Mechanics
Peking Univ.
Beijing 100871

Dou, Jian-Ya
Inst. of Theoretical Physics
The Chinese Academy of Sciences
Beijing 100080

Du, Chan-Ying
Dept. of Physics
Beijing Normal Univ.
Beijing 100875

Fang, Hai-Ping
Dept. of Physics
Fudan Univ.
Shanghai 200433

Fang, Jiang-Ling
Dept. of Physics
Xiamen Univ.
Xiamen 361005

Fang, Zhong
Nat'l Lab. of Laser Tech.
Huazhong Univ. of Sci. & Tech.
Wuhan 430074

Fen, Xiao
Dept. of Physics
Xiamen Univ.
Xiamen 361005

Gong, Chang-De
Dept. of Physics
Nanjing Univ.
Nanjing 210008

Gong, Jiang-Bin
Dept. of Physics
Nanjing Univ.
Nanjing 210093

Gu, Yan
Ctr. for Fundamental Physics
Univ. of Sci. & Tech. of China
Hefei 230026

Guan, Shan
Division of Theoretical Mechanics
Beijing Univ. of Aeron. & Astron.
Beijing 100083

Hao, Bai-Lin
Inst. of Theoretical Physics
The Chinese Acedemy of Sciences
Beijing 100080

Heng, Yue-Kun
Inst. of Low Energy Nuclear Phys.
Beijing Normal Univ.
Beijing 100875

Hong, Ming-Yuan
Nat'l Natural Sci. Founda. of China
Beijing 100083

Hong, Shi-Ming
Inst. of Applied Physics
Sichuan Union Univ.
Chengdu 610065

Hong, Shi-Zhong
Seismological Bureau of Chengdu
Chengdu 610015

Hong, Shui-li
Dept. of Physics
Fuzhou Univ.
Fuzhou 350025

Hu, Bei-Lai
Dept. of Physics
Nankai Univ.
Tianjin 300071

Hu, Gang
Dept. of Physics
Beijing Normal Univ.
Beijing 100875

Hu, Qi-Fu
Dept. of Physics
Xiamen Univ.
Xiamen 361005

Hu, Xi-Wei
Dept. of Modern Physics
Univ. of Science & Tech. of China
Hefei 230026

Hu, Zhan-Ning
Inst. of Theoretical Physics
The Chinese Academy of Sciences
Beijing 100080

Huang, Mei-Chun
Dept. of Physics
Xiamen Univ.
Xiamen 361005

Huang, Yun
Dept. of Physics
Peking Univ.
Beijing 100871

Huang, Zhi-Feng
Dept. of Physics
Tsinghua Univ.
Beijing 100084

Huang, Zhong-Bing
Dept. of Physics
Beijing Normal Univ.
Beijing 100875

Huang, Zu-Qia
Inst. of Low Energy Nuclear Phys.
Beijing Normal Univ.
Beijing 100875

Ji, Yong-Hua
Inst. of Theoretical Physics
The Chinese Academy of Sciences
Beijing 100080

Jie, Cheng-Lin
Dept. of Physics
Nanjing Univ.
Nanjing 210093

Ke, San-Huang
Dept. of Physics
Xiamen Univ.
Xiamen 361005

Kong, Liang
Dept. of Physics
Univ. of Science & Tech. of China
Hefei 230026

Lei, Hong-Bing
Dept. of Physics
Xiamen Univ.
Xiamen 361005

Li, Hong
Dept. of Physics
Univ. of Nankai
Tianjin 300071

Li, Hong-Cai
Dept. of Physic
Fujian Normal Univ.
Fuzhou 350007

Li, Shu-Hua
Dept. of Physics
Fujian Normal Univ.
Fuzhou 350007

Li, Zhen-Ya
Dept. of Physics
Suzhou Univ.
Suzhou 215006

Lian, Shi-Yang
Dept. of Physics
Xiamen Univ.
Xiamen 361005

Liang, Zhong-Cheng
Dept. of Physics
Nanjing Normal Univ.
Nanjing 210024

Liao, Huai-Zhe
Ctr. for Space Sci. & Applied Res.
The Chinese Academy of Sciences
Beijing 100080

Lin, Guo-Xing
Dept. of Physics
Xiamen Univ.
Xiamen 361005

Liu, Ji-Xing
Inst. of Theoretical Physics
The Chinese Academy of Sciences
Bijing 100080

Liu, Jie
Ctr. for Nonlinear Studies
Inst. of Appl. Phys. &
Computat'l Math.
Beijing 100088

Liu, Jun-Xian
Inst. of Theor. Physics
The Chinese Academy of Sciences
Beijing 100080

Liu, Rang-Su
Dept. of Physics
Hunan Univ.
Changsha 410012

Liu, Rui-Tang
Dept. of Physics
Xiamen Univ.
Xiamen 361005

Liu, Run-Qiu
Inst. of Applied Mathematics
The Chinese Academy of Sciences
Beijing 100080

Liu, Shu-Sheng
Dept. of Physics
Tianjin Univ.
Tianjin 300072

Liu, Wu-Ming
Inst. of Physics
The Chinese Academy of Sciences
Beijing 100080

Liu, Yi-Chang
Inst. of Theoretical Physics
The Chinese Academy of Sciences
Beijing 100080

Liu, Zhao-Hong
Dept. of Physics
Xiamen Univ.
Xiamen 361005

Liu, Zong-Xiu
Dept. of Physics
Baoji College of Art and Science
Baoji 721007

Lu, Qi-Shao
Dept. of Appl. Math. & Phys.
Beijing Univ. of Aeron. & Astron.
Beijing 100083

Lu, Quan-Kang
Dept. of Physics
Fudan Univ.
Shanghai 200433

Lu, Zhi-Cheng
Zhejing Inst. of Modern Physics
Zhejing Univ.
Hangzhou 310027

Luo, Hong-Gang
Dept. of Physics
Lanzhou Univ.
Lanzhou 730000

Luo, Jiu-Li
Dept. of Chemistry
Sichuan Union Univ.
Chengdu 610064

Luo, Liao-Fu
Dept. of Physis
Inner Mongolia Univ.
Huhehaote 010021

Ou Yang, Bi-Yao
Beijing Inst. of Appl. Phys. &
Computat'l Math.
Beijing 100088

Ou Yang, Zhong-Can
Inst. of Theoretical Physics
The Chinese Academy of Sciences
Beijing 100080

Pan, Hui-Yun
Dept. of Chemistry
Zhengzhou Univ.
Zhengzhou 450052

Peng, Shou-Li
Ctr. for Nonlinear Complex Systems
Yunnan Univ.
Kunming 650091

Qin, Guang-Rong
Dept. of Physics
Beijing Normal Univ.
Beijing 100875

Qiu, Zhao-Ming
The General Office
The Chinese Academy of Sciences
Beijing 100864

Qu, Shi-Xian
Dept. for Fundamentals
Xi'an Petroleum Inst.
Xi'an 710061

Rao, Xue-Song
Dept. of Physics
Fudan Univ.
Shanghai 200433

Shang, Chang-He
Inst. of Physics
The Chinese Academy of Sciences
Beijing 100080

Shen, Yao-Wen
Dept. of Physics
Xiamen Univ.
Xiamen 361005

Shi, Xiao-Lei
Inst. of Theoretical Physics
The Chinese Academy of Sciences
Beijing 100080

Shi, Yu
Dept. of Physics
Fudan Univ.
Shanghai 200433

Shuai, Jian-Wei
Dept. of Physics
Xiamen Univ.
Xiamen 361005

Song, Yan
Inst. of Semiconductor Physics
The Chinese Academy of Sciences
Beijing 100083

Su, Zhao-Bin
Inst. of Theoretical Physics
The Chinese Academy of Sciences
Beijing 100080

Sun, Cheng
Dept. of Physics
Nanjing Univ.
Nanjing 210093

Sun, Hong-Yan
Inst. of Physics
The Chinese Academy of Sciences
Beijing 100080

Sun, Ren-Guang
Ctr. for Experiments
Shanxi Normal Univ.
Xi'an 710062

Tan, Yu
Beijing Inst. of Appl. Phys. &
Computat'l Math.
Beijing 100088

Tao, Rui-Bao
Dept. of Physics
Fudan Univ.
Shanghai 200433

Tian, Guang-Shan
Dept. of Physics
Peking Univ.
Beijing 100871

Wang, Ai-Ke
Southwestern Inst. of Physics
Chengdu 610041

Wang, Bao-Han
Inst. of Biophysics
The Chinese Academy of Sciences
Beijing 100101

Wang, Bing-Hong
Dept. of Modern Physics
Univ. of Science & Tech. of China
Hefei 230026

Wang, Cheng-Tai
Dept. of Applied Physics
Tsinghua Univ.
Beijing 100084

Wang, Chui-Lin
China Ctr. of Adv. Sci. & Tech.
Beijing 100080

Wang, Guang-Rui
Inst. of Appl. Phys. &
Computat'l Math.
Beijing 100088

Wang, Hao
Inst. of Theoretical Physics
The Chinese Academy of Sciences
Beijing 100080

Wang, Jiao
Ctr. for Fundamental Physics
Univ. of Sience & Tech. of China
Hefei 230026

Wang, Mu
Nat'l Lab. of Solid State Microstructures
Nanjing Univ.
Nanjing 210093

Wang, Nai-Quan
Ctr. for Space Science & Applied Res.
The Chinese Academy of Sciences
Beijing 100080

Wang, Qing
Dept. of Physics
Univ. of Nankai 16–101
Tianjing 300071

Wang, Rong
Inst. of Biophysics
The Chinese Academy of Sciences
Beijing 100101

Wang, Shu-Min
Inst. of Theoretical Physics
Chinese Academy of Sciences
Beijing 100080

Wang, Ying-Hai
Dept. of Theor. Physics
Univ. of Lanzhou
Lanzhou 730000

Wang, Yu-Jiang
Dept. of Physics
Xiamen Univ.
Xiamen 361005

Wang, Zhi-Cheng
Dept. of Physics
Lanzhou Univ.
Lanzhou 730001

Wei, Qi-He
Lab. of Molec. & Biomolec. Electronics
Southeast Univ.
Nanjing 210096

Weng, Xiao-Ming
Dept. of Physics
Suzhou Univ.
Suzhou 215006

Wu, Chang-Qin
Dept. of Physics
Fudan Univ.
Shanghai 200433

Wu, Jian
Dept. of Physics
Tsinghua Univ.
Beijing 100084

Wu, Xu-Hong
Dept. of Physics
Jilin Univ.
Changchun 130023

Wu, Zuo-Bing
Inst. of Theor. Physics
The Chinese Academy of Sciences
Beijing 100080

Xiao, Fang-Fang
Dept. of Physics
Xiamen Univ.
Xiamen 361005

Xiao, Fen
Dept. of Physics
Xiamen Univ.
Xiamen 361005

Xie, Fa-Gen
Inst. of Theoretical Physics
The Chinese Academy of Sciences
Beijing 100080

Xie, Hui-Min
Dept. of Mathematics
Suzhou Univ.
Suzhou 215006

Xie, Yu-Zhang
Dept. of Modern Applied Physics
Tsinghua Univ.
Beijing 100084

Xu, Gong-Ou
Dept. of Physics
Nanjing Univ.
Nanjing 210008

Xu, Hui
Dept. of Physics
Central South Univ. of Technology
Changsha 410083

Xu, Lun-Biao
Dept. of Phyiscs
Zhejiang Univ.
Hangzhou 310027

Xu, Qiao-Zhen
Dept. of Physics
Xiamen Univ.
Xiamen 361005

Xu, Shen-Chu
Dept. of Physics
Fujian Normal Univ.
Fuzhou 350007

Xu, Shu-Shan
Dept. of Physics
Hebei Normal College
Shijiazhuang 050091

Xu, Xue-Tong
Dept. of Physics
Fudan Univ.
Shanghai 200433

Yan, Su
Inst. of Theoretical Physics
The Chinese Academy of Sciences
Beijing 100080

Yan, Zi-Jun
Dept. of Physics
Xiamen Univ.
Xiamen 361005

Yang, Hong-Liu
Inst. of Low Energy Nuclear Physics
Beijing Normal Univ.
Beijing 100875

Yang, Qing-Xian
Dept. of Fundamental Sciences
Nan Ping Forrest College
Nan Ping 353001

Yang, Ya-Tian
Dept. of Physics
Fujian Normal Univ.
Fuzhou 350007

Yang, Zhan-Ru
Dept. of Physics
Beijing Normal Univ.
Beijing 100875

Yang, Zhi-An
The Graduate School
Chinese Academy of Engineering Physics
Beijing 100088

Yang, Zun-Zhong
Dept. of Physics
Beijing Normal Univ.
Beijing 100875

Yao, Wei-Guang
China Inst. of Atomic Energy
Beijing 102413

Yao, Yu-Cui
Teaching & Res. Section
Hefei 230031

Ying, He-Ping
Zhejiang Inst. of Modern Physics
Zhejiang Univ.
Hangzhou 310027

Yu, Lu
Condensed Matter
ICTP, P. O. Box 586
34100 Trieste

Yu, Yang-Zheng
Dept. of Physics
Xiamen Univ.
Xiamen 361005

Yu, Zhi-Gang
Dept. of Physics
Fudan Univ.
Shanghai 200433

Yuan, Du-Qi
Dept. of Physics
Baoji College of Arts & Scinece
Baoji 721007

Yuan, Xiao-Feng
Ctr. for Mathematical Sciences
CICA, The Chinese Academy of Sciences
Chengdu 610041

Zhang, Bang-Gu
Science Press
The Chinese Academy of Sciences
Beijing 100717

Zhang, Bo-Tao
Physics Group
Information & Engineering Inst.
Zhengzhou 450002

Zhang, De-Gang
Inst. of Solid State Physics
Sichuan Normal Univ.
Chengdu 610068

Zhang, Guo-Min
Dept. of Physics
Suzhou Univ.
Suzhou 215006

Zhang, Guo-Ping
T.D. Lee Lab. & Physics Dept.
Fudan Univ.
Shanghai 200433

Zhang, Kai-Xi
Dept. of Physics
Hebei Univ.
Baoding 071002

Zhang, Shu-Dong
Inst. of Low Energy Nuclear Physics
Beijing Normal Univ.
Beijing 100875

Zhang, Yong-Li
Inst. of Theoretical Physics
The Chinese Academy of Sciences
Beijing 100080

Zhang, Zhi-Yi
Dept. of Physics
Fujian Normal Univ.
Fuzhou 350007

Zhao, Hong
Dept. of Physics
Lanzhou Univ.
Lanzhou 730000

Zhao, Kai-Hua
Dept. of Physics
Peking Univ.
Beijing 100871

Zhao, Ping-Bo
Inst. of Physics
The Chinese Academy of Sciences
Beijing 100080

Zhao, Ri-An
Inst. of Physics
The Chinese Academy of Sciences
Beijing 100080

Zheng, De-Juan
Inst. of Physics
The Chinese Academy of Science
Beijing 100080

Zheng, Wei-Mou
Inst. of Theoretical Physics
The Chinese Acedemy of Sciences
Beijing 100080

Zheng, Zhi-Gang
Dept. of Physics
Beijing Normal Univ.
Beijing 100875

Zhou, Bi-Zhong
Dept. of Physics
Xiamen Univ.
Xiamen 361005

Zhou, Bing
Inst. of Theoretical Physics
The Chinese Academy of Sciences
Beijing 100080

Zhou, Fu-Shan
Inst. of Theoretical Physics
The Chinese Academy of Sciences
Beijing 100080

Zhou, Hai-Guang
Dept. of Physics
Xiamen Univ.
Xiamen 361005

Zhou, Lei
Dept. of Physics
Fudan Univ.
Shanghai 200433

Zhou, Liang-Jian
Inst. of Solid Physics
The Chinese Academy of Sciences
Hefei 230031

Zhu, Jian-Yang
Dept. of Physics
Jiangxi Teachers College
Nanchang 344000

Zhu, Shi-Qun
Dept. of Physics
Suzhou Universitry
Suzhou 215006

Zhu, Yu-Qing
Inst. of Theoretical Physics
The Chinese Academy of Sciences
Beijing 100080

Zhu, Zi-Zhong
Dept. of Physics
Xiamen Univ.
Xiamen 361005

**CZECH**

Andrey, Ladislav
Neural Networks Section
Inst. of Computer Science
182 07 Prague 8

**DENMARK**

Alstrøm, Preben
Ctr. for Chaos & Turbulence Studies
Niels Bohr Inst.
DK–2100 Copenhagen Ø

Andresen, Bjarne
Orsted Lab., Niels Bohr Inst.
Univ. of Copenhagen
DK-2100 Copenhagen Ø

Eriksson, Anders
Nordita
DK-2100 Copenhagen Ø

Fogedby, Hans C.
Inst. of Physics & Astronomy
Univ. of Aarhus
DK-8000 Aarhus C

Hertz, J.A.
Nordisk Inst. for Teoretisk Fysik
Nordita
DK-2100 Copenhagen

Levinsen, Mogens Tweddell
Niels Bohr Inst.
Univ. of Copenhagen
DK-2100 Copenhagen

Mooij, Germonda
FKI, Building 206
Danske Tekniske Universitet
Lyngby

Peters, Guenther H.
Chemistry Lab. III
H.C. Oester Inst.
DK-2100 Copenghagen Ø

Sibani, Paolo
Fysik Inst.
Odense Universitet
DK-5230 Odense M

Vallespin Gomez, Alejanoro
Ctr. for Chaos & Turbulence Studies
The Niels Bohr Inst.
Copenhagen Ø

**FINLAND**

Ala-Nissila, Tapio
TFT Res. Inst. for Theor. Phys.
Univ. of Helsinki
Helsinki

Alava, Mikko
Lab. of Physics
Helsinki Univ. of Technology
Fin-02150 Espoo

Provatas, Nikolaos
Inst. of Theoretical Physics
Univ. of Helsinki
Helsinki 00014

**FRANCE**

Barthélémy, Marc
Ctr. D'etudes de Limeil-Valenton
Commissariat a l'Energie Atomique
94195 Villeneuve St-Georges

Bertini-Malgarini, Lorenzo
Centre de Physique Theorique
Ecole Polytechnique
91128 Palaiseau Cedex

Brezin, E.
Dept. de Physique
Ecole Normale Superieure
75231 Paris Cedex 05

Cai, Z.P.
Lab. d'Optronique
Associee du CNRS(EP001)
Lannion

Cancrini, Nicoletta
Centre de Physique Theorique
Ecole Polytechnique
91128 Palaiseau cedex

Chaté, Hugues
Service de Phys. de l'Etat Condense
CEA-Centre d'Etudes de Saclay
91191 Gif-sur-Yvette

Chigirinskaya, Yulia
LMD
Universite P & M Curie
75252 Paris cedex 05

Couder, Yves
Lab. de Physique Statistique
Ecole Normale Superieure
75231 Paris Cedex

David, Francois
Service de Physique Theorique
Centre d'Etude de Saclay
F-91191 Gif-sur-Yvette Cedex

Derrida, Bernard
Lab. de Physique Statistique
Ecole Normale Superieure
75231 Paris Cedex 05

Diep, Hung The
Groupe de Physique Statistique
Universite de Cergy-Pontoise
95806 Cergy-Pontoise Cedex

Douady, Stephane
Lab. de Physique Statistique
Ecole Normale Superieure
75231 Paris Cedex 05

Duplantier, B.
Service de Physique Theorique
CEA Saclay
Saclay 91191, Gif-sur-Yvette Cedex

Eckle, Hans-Peter
Departement de Physique
Universite Francois Rabelais de Tours
F-37200 Tours

Hakim, Vincent
Lab. de Physique Statistique
CNRS
75231 Paris

Kolb, Max
Chimie Theorique
Ecole Normale Superieure de Lyon
69364 Lyon Cedex 07

Lemaitre, Anael
LADHYX
Ecole Polytechnique
91128 Palaiseau

Leroyer, Yves
Ctr. de Physique et de Modelisation
Universite de Bordeaux-I
33175 Gradignan Cedex

Limoge, Yves
Section de Recherches de Metallurgie Phys.
Centre d'Etudes de Saclay
91191 Gif Sur Yvette Cedex

Mehta, Madan Lal
Service de Physique Theoretique
C. E. de Saclay
91191 Gif-sur-Yvette Cedex

Mezard, Marc
Lab. de Physique Statistique
Ecole Normale Superieure
F-75231 Paris Cedex 05

Moreau, Michel
Physique Theorique des Liquides
Universite Pierre et Marie Curie
75252 Paris Cedex 05

Nadal, Jean-Pierre
Lab. de Physique Statistique
Ecole Normale Superieure
75231 Paris Cedex 05

Nunez-Regueiro, Dolores
European Synchrotron Radiation Facility
Grenoble 38043 Cedex

Ouvry, Stephane
Division de Physique Theorique
Universite Paris-Sud
F-91406 Orsay Cedex

Perez, Asher
Lab. de Physique
Ecole Normale Superieure de Lyon
69364 Lyon, Cedex 07

Sapoval, Bernard
Lab. de Phys. de la Matiere Condensee
Ecole Polytechnique
91128 Palaiseau

Schertzer, Daniel
Lab. de Meteorologie Dynamique
Universite Pierre et Marie Curie
F-75252 Paris Cedex 05

Schmitt, Francois
Lab. de Meteorologie Dynamique
CNRS
75252 Paris cedex 05

Sornette, Didier
Lab. de Phys. de la Matiere Condensee
Univ. de Nice-Sophia Antipolos
Nice 06108

Stephan, Guy
Ecole Nat'le Superieure des Sci.
Appliquees et de Technologie
(E.N.S.S.A.T.)
22305 Lannion Cedex

Thill, Markus Johannes
Lab. de Phys. Theor. et Hautes Energies
Universite de Paris Sud
F-91405 Orsay Cedex

Vitting Andersen, Jorgen
Lab. de Phys. de la Matiere Condensee
Universite de Nice-Sophia Antipolis
06108 Nice Cedex 2

Zhao, Hong
CEA, Saclay
Gif-sur-Yuette, Saclay 91191

**GEORGIA**

Kotia, Bejan Arkadij
Solid State Physics Dept. N5
Georgian Technical Univ.
380075 Tbilisi

**GERMANY**

Baake, Michael
Inst. für Theoretische Physik
Universität Tübingen
D-72076 Tübingen

Bausch, Richard
Inst. für Theoretische Physik IV
Heinrich-Heine-Universität Düsseldorf
D-40225 Düsseldorf

Böttger, Harald Oskar
Inst. für Theor. Physik
Otto-von-Guericke-Univ. Magdeburg
D-39106 Magdeburg

Böttger, Wulf
Inst. für Theoretische Physik
RWTH Aachen
Aachen D-52056

Bunde, Armin
Inst. für Theor. Physik
Universität Gießeu
D-35392 Gießeu

Chen, Xiao-Song
Inst. für Theoretische Physik
RWTH Aachen
52056 Aachen

Craubner, Hans
Inst. für Physikalische Chemie
Universität Düsseldorf
D-40225 Düsseldorf

Diehl, Hans Werner
Fachbereich Physik
Universität. GH Essen
D-45117 Essen

Ebeling, Werner
Inst. of Phys.
Humboldt Universität
D-10099 Berlin

Groh, Benito
Fachbereich Physik
Universität Wuppertal
Wuppertal D-42097

Große, Ivo
Inst. of Physics
Humboldt Univ.
D-10115 Berlin

Haberlandt, Reinhold
Faculty for Phys. & Geosciences
Univ. Leipzig
04109 Leipzig

Hess, Ortwin
Fachbereich Physik
Philipps-Universitaet Marburg
D-35032 Marburg

Janke, Wolfhard
Inst. für Physik WA 331
Johannes Gutenberg-Universität
D-55099 Mainz

Kehr, Klaus W.
Inst. für Festkörperforschung
Theorie II
Forschungszentrum Jülich Gmbh
D-52425 Jülich

Klümper, Andreas
Inst. für. Theoretische Physik
Universität zu Köln
D-50937 Köln

Kobe, Sigismund
Inst. für Theoretische Physik
Technische Universität Dresden
D-01062 Dresden

Kroger, Martin
Inst. für Theoretische Physik
Technische Universität Berlin
D-10623 Berlin

Kurths, Jurgen
Max-Planck-Arbeitsgruppe
Universität Potsdam
14415 Potsdam

Li, Yanmin
Max-Planck Inst. für Phys.
Komplexer Systeme
D-01187 Dresden

Militzer, Burkhard
Inst. of Physics
Humboldt Univ. Berlin
D-10437 Berlin

Möbius, Arnulf
Inst. für Festkörper-und
Werkstofforschung
D-01171 Dresden

Parisi, Jurgen
Physical Inst.
Univ. of Bayreuth
D-95440 Bayreuth

Rieger, Heiko
Inst. für Theoretische Physik
Universitut zu Köln
50926 Köln

Ritschel, Uwe
Fachbereich Physik
Universität Essen
D-45117 Essen

Röpke, Gerd
Fachbereich Physik
Univeresität Rostock
D-18055 Rostock

Schoen, Martin
Inst. für Physik
TU Berlin
10623 Berlin

Spohn, Herbert
Theoretische Physik
Universität München
D-80333 München

Sukumar, Nagamani
Inst. für Theoretische Chemie
Universität Bonn
D-53115 Bonn

Tang, Lei-Han
Inst. f. Theor. Physik
Universität zu Köln
D-50937 Köln

Zittartz, Johannes
Inst. für Theor. Physik
Universität zu Köln
D-50937 Köln

**HONG KONG**

Chen, Wu Po
22 ND FL. Alliance Bldg.
Treasure Maritime Ltd.
Hong Kong

Hu, Bei-Lok
Dept. of Physics
Hon Kong Univ. of Sci. & Tech.
Hong Kong

Kwan, Chung Keu
Dept. of Electric Engineering
Univ. of Hong Kong
Hong Kong

Mao, Jian-Min
Dept. of Mathematics
Hong Kong Univ. of Sci. & Tech.
Hong Kong

Ng, Sze Kui
Dept. of Mathematics
Hong Kong Baptist Univ.
Hong Kong

Pakhomov, Alexandre B.
Dept. of Physics
Hong Kong Univ. of Sci. & Tech.
Clear Water Bay, Kowloon

Poon, Kai Lok
Physics Dept.
The Chinese Univ. of Hong Kong
Hong Kong

Sheng, P.
Dept. of Physics
Hong Kong Univ. of Sci. & Tech.
Kowloon, Hong Kong

Shiokawa, Kazutomu
Dept. of Physics
Hong Kong Univ. of Sci. & Tech.
Hong Kong

Szeto, Kwok-Yip
Dept. of Physics
Hong Kong Univ. of Sci. & Tech.
Hong Kong

Wang, Jian
Dept. of Physics
Univ. of Hong Kong
Hong Kong

Wang, Xiang Rong
Dept. of Physics
Hong Kong Univ. of Sci. & Tech.
Hong Kong

Wong, Kwok Yee Michael
Dept. of Physics
Hong Kong Univ. of Sci. & Tech.
Hong Kong

Xia, Keqing
Dept. of Physics
Chinese Univ. of Hong Kong
Hong Kong

Yip, Man-Kit
Dept. of Phys. & Material Sci.
City Univ. of Hong Kong
Hong Kong

Young, Kenneth
Dept. of Physics
Chinese Univ. of Hong Kong
Hong Kong

Yu, Kin-Wah
Dept. of Physics
Chinese Univ. of Hong Kong
Hong Kong

**HUNGARY**

Iglói, Ferenc
Res. Inst. for Solid State Phys.
H-1525 Budapest

Solyom, Jeno
Res. Inst. for Solid State Phys.
H-1525 Budapest

Tel, Tamas
Inst. for Theoretical Physics
Eötvös Univ.
H-1088 Budapest

**INDIA**

Barma, Mustansir A.
Tata Inst. of Fundamental Res.
Bombay 400005

Chowdhury, Debashish
Dept. of Phys.
Indian Inst. of Technology
Kanpur-208016

Lakshmanan, Muthusamy
Ctr. for Nonlinear Dynamics
Bharathidasan Univ.
Tiruchirapalli 620 024, Tamilnadu

Ota, Snehadri Bihari
Inst. of Physics
Bhubaneswa-751005

Romer, Rudolf A.
J. Nehru Ctr. for Adv. Sci. Res.
India Inst. of Science
Banglore 560012

Shastry, B.S.
Dept. of Physics
Indian Inst. of Science
560 012 Bangalore

**IRAN**

Shahnas, Mohammad Hossein
Physics Dept.
Tarbiat Modarres Univ.
Tehran

**IRELAND**

Dolan, Brian P.
St. Patrick's College
Dublin 8

O'Connor, Denis Joseph
Dublin Inst. for Adv. Studies
Dublin 4

## ISRAEL

Aharony, A.
School of Physics & Astronomy
Tel-Aviv Univ.
69978 Tel Aviv

Ben-Aryeh, Yacob
Dept. of Physics
Technion-Israel Inst. of Tech.
Haifa 32000

Brender, Chava
Dept. of Physics
Bar-Ilan Univ.
Ramat-Gan 52900

Daboul, Jamil
Physics Dept.
Ben Gurion Univ.
Beer Sheva

Domany, Eytan
Dept. of Electronics
The Weizmann Inst. of Sci.
Rehovot 76100

Havlin, Shlomo
Dept. of Physics
Bar-Ilan Univ.
52900 Ramat-Gan

Hovi, Juha-Pekka
Sackler Faculty of Exact Sci.
Tel Aviv Univ.
Tel Aviv 69978

Kantor, Yacov
School of Phys. & Astronomy
Tel Aviv Univ.
Tel Aviv 69978

Levanda, Matty
School of Physics
Tel Aviv Univ.
69978 Tel Aviv

Nepomnyashchy, Alexander
Dept. of Math.
Technion-Israel Inst. of Tech.
Technion, 32000 Haifa

Nepomnyashchy, Yuri
Dept. of Physics
Technion-Israel Inst. of Tech.
32000 Haifa

Procaccia, Itamar
Dept. of Chemical Physics
Weizmann Inst. of Science
76100 Rehovot

Rabinovich, Savely
Dept. of Physics
Bar–Ilan Univ.
52900 Ramat-Gan

Stein, Jurgen
School of Physics & Astronomy
Tel Aviv Univ.
Tel Aviv 69978

## ITALY

Alabiso, Carlo
Dept. of Phys.
Universita di Parma
Parma 43100

Artuso, Roberto
Inst. di Scienze Matematiche
Fisiche e Chimiche
I–22100 Como

Basu, Chaitali
Condensed Matter
ICTP
34100 Trieste

Casartelli, Mario
Dipartimento di Fisica
Universita di Parma
43100 Parma

Casati, Giulio
Dept. of Phys.
Milano Univ.
22100 Como

Cassi, Davide
Dipt. Di Fisica
Universita Di Parma
43100 Parma

de Arcangelis, Lucilla
Physics Dept.
Universita dell Aquila
Napoli 80122

Di Castro, Carlo
Dipartimento di Fisica
Universita di Roma "La Sapienza"
I–00185 Roma

Gammaitoni, Luca
Dipt. di Fisica
Inst. Nazionale di Fisica Nucleare
I–06100 Perugia

Gonnella, Giuseppe
Dipartimento di Fisica
Universita di Bari
I–70126 Bari

Malescio, Gianpietro
Dipartimento di Fisica
Universita degli Studi Messina
98166 S.Agata(me)

Maritan, Amos
Dipartimento di Fisica
Univ. di Padova
35131 Padova

Regina, Sofia
Dipartimento di Fisica
Universita di Parma
43100 Parma

Scotti, Antonio
Dipartimento di Fisica
Universita di Parma
43100 Parma

Vaia, Ruggero
Inst. di Electronica Quantistica
C. N. R.
I–50127 Firenze

Virasoro, M.
ICTP
P. O. Box 586
34100 Trieste

Xue, She-Sheng
INFN, Section of Milan
Milan 20133

**JAPAN**

Andrade Jr., Jose Soares
Dept. of Production Tech.
Showa Denko K.K.
Tokyo 105

Aoki, Kazunori
Faculty of Engineering
Kobe Univ.
Kobe 657

Araki, Fujihiro
Research Inst. for Math. Sci.
Kyoto Univ.
Kyoto 606

Arimitsu, Toshihico
Inst. of Physics
Univ. of Tsukuba
Ibaraki 305

Asakawa, Hitoshi
Dept. of Physics
Univ. of Tokyo
Tokyo 113

Chen, Zhi Hua
Graduate School of Information Sci.
Tohoku Univ.
Sendai 980-77

Eto, Mikio
Dept. of Phys.
Keio Univ.
Yokohama 223

Foong, See Kit
College of Education
Univ. of the Ryukyus
Okinawa

Fuchikami-Ishioka, Nobuko
Dept. of Physics
Tokyo Metropolitan Univ.
Tokyo 192–03

Fujii, Katsuhiko
Dept. of Appl. Phys.
Okayama Univ. of Science
Okayama 700

Fukui, Yoshihiko
Graduate School of Information Sci.
Tohoku Univ.
Sendai 980–77

Goda, Masaki
Dept. of Material Sci. & Tech.
Niigata Univ.
Niigata 950–21

Harada, Isao
Dept. of Phys. Faculty of Sci.
Okayama Univ.
Okayama 700

Hasegawa, Yutaka
Dept. of Engineering Physics
Chubu Univ.
Kasugai, Aichi 487

Hattori, Masumi
Dept. of Physics
Nagoya Inst. of Technology
Nagoya 466

Honda, Yasushi
Dept. of Computer & Mathematical Sci.
Tohoku Univ.
Sendai 980–77

Horibe, Minoru
Dept. of Physics
Fukui Univ.
Fukui City 910

Horiguchi, Tsuyoshi
Dept. of Computer & Math. Sci.
GSIS, Tohoku Univ.
Sendai 980–77

Horimoto, Eiji
Interdiscipl. Graduate School
of Eng. Sci.
Kyushu Univ.
Kasuga 816

Iitaka, Toshiaki
Nanoelectronics Material Lab.
RIKEN
Saitama 351–01

Inaoka, Hajime
Graduate School of Information Sci.
Tohoku Univ.
Sedai 980–77

Inui, Norio
Graduate School of Information Sci.
Tohoku Univ.
Sendai 980–77

Ishii, Akira
Dept. Physics
Waseda Univ.
Tokyo 169

Ishii, Yasushi
Dept. of Material Science
Himeji Inst. of Technology
Hyogo 678–12

Ishioka, Shunya
Dept.of Information Science
Kanagawa Univ. at Hiratsuka
Kanagawa 259–12

Ito, Nobuyasu
Dept. of Applied Physics
Univ. of Tokyo
Tokyo 113

Iwayama, Takahiro
Dept. of Earth & Planetary Sci.
Kyushu Univ.
Fukuoka 812

Kaneko, Kunihiko
Dept. of Pure & Applied Sciences
Univ. of Tokyo
Tokyo 153

Kaneyoshi, Takahito
Dept. of Natural Sci. Informatics
Nagoya Univ.
464–01 Nagoya

Katsura, Shigetoshi
Tohoku College of Engn. & Info. Sci.
Sendai 981

Kawakatsu, Toshihiro
Dept. of Physics
Tokyo Metropolitan Univ.
Hachioji-City, Tokyo 192–03

Kawakubo, Tatsuyuki
Faculty of Engineering
Toin Univ. of Yokohama
Yokohama 225

Kawasaki, Kazuko
Dept. of Physics
Nara Women's Univ.
Nara 630

Kawasaki, Mitsuhiro
Dept. of Physics
Kyushu Univ. 33
Fukuoka 812

Kawasaki, Tatuo
Fundamental Sciences, FIHS
Kyoto Univ.
Kyoto 606

Kawashima, Naoki
Dept. of Physics
Univ. of Tokyo
Tokyo

Kazumasa, Shinjo
ATR Optical & Radio Comm. Res. Lab.
Kyoto 619-02

Kinase, Wataru
Dept. of Physics
Waseda Univ.
Tokyo 169

Kitatani, Hidetsugu
Dept. of Electrical Engineering
Nagaoka Univ. of Technology
940-21 Nagaoka

Kobayashi, Takashi
Faculty of Engn.
Toin Univ. of Yokohama
Yokohama 225

Koga, Shinji
Dept. of Physics
Osaka Kyoiku Univ.
Osaka 582

Kohyama, Tamotsu
Faculty of Education
Shiga Univ.
Otsu 520

Komiya, Minoru
Dept. of Engineering Physics
Chubu Univ.
Kasugai, Aichi 487

Kondo, Hironobu
Dept. of Applied Physics
Okayama Univ. of Science
Okayama 700

Konno, Norio
Dept. of Mathematical Sci.
Muroran Inst. of Tech.
Hokkaido 050

Kubo, Kenn
Inst. of Physics
Univ. of Tsukuba
Ibaraki 305

Kurrer, C.
Dept. of Physics
Kyoto Univ.
Kyoto 606

Kuwata, Seiichi
Faculty of Information Science
Hiroshima City Univ.
Hiroshima 731-31

Li, Rong
Graduate School of Information Sci.
Tohoku Univ.
Sendai 980-77

Maruyama, Kanegasu
Dept. of Engineering Physics
Chubu Univ.
Kasugai, Aichi 487

Matsubara, Fumitaka
Dept. of Applied Physics
Tohoku Univ.
Sendai 980-77

Matsui, Jun
Dept. of Physics
Kyushu Univ.
Fukuoka 812-81

Miura, Yoshinori
Ctr. for Adv. Instrumental Analysis
Kyushu Univ.
Kasuga 816

Miyashita, Seiji
GSHES
Kyoto Univ.
Kyoto, 606

Miyazima, Sasuke
Dept. of Engineering Physics
Chubu Univ.
Kasugai, Aichi 487

Momoi, Tsutomu
Dept. of Physics
Univ. of Tokyo
Tokyo 113

Mori, Shigeyoshi
Dept. of Physics
Okayama Univ.
Okayama 700

Morioka, Nozomi
Dept. Electric Engineering
Kokushikan Univ.
Tokyo 154

Nagamine, Tomomasa
Dept. of Engineering Physics
Chubu Univ.
Kasugai, Aichi 487

Naito, Hiroyoshi
Dept. of Physics & Electronics
Univ. of Osaka Prefecture
Osaka 593

Nakamura, Ki-Ichi
Dept. of Physics
Nihon Univ.
Tokyo 156

Nishimori, Hidetoshi
Dept. of Physics
Tokyo Inst. of Technology
Tokyo 152

Nishimura, Hisashi
Dept. of Electrical Engn.
Kyushu Tokai Univ.
Kumamoto 862

Nishimura, S.
Shiga Bunka College
Yokaichi, Shiga 527

Nishiuma, Saburou
Shiga Bunka College
Yokaichi City Shiga Prefecture 527

Nishiyama, Yoshihiro
Dept. of Physics
Univ. of Tokyo
Tokyo 113

Nokura, Kazuo
Shonan Inst. of Technology
Fujisawa, Kanagawa-Ken 251

Nonomura, Yoshihiko
Dept. of Physics
Tokyo Inst. of Technology
Tokyo 152

Ohno, Kaoru
Inst. for Materials Res.
Tohoku Univ.
Aoba-ku, Sendai 980–77

Okamoto, Hisao
Dept. of Information Science
Kochi Univ.
Kochi 780

Okamoto, Kiyomi
Dept. of Physics
Tokyo Inst. of Tachnology
Meguro-ku, Tokyo 152

Okazaki, Akihiko
Dept. of Engineering Physics
Chubu Univ.
Kasugai 487

Ono, Ikuo
Dept. of Physics
Tokyo Inst. of Technology
Tokyo 152

Osako, Kazuhiro
Dept. of Engineering Physics
Chubu Univ.
Kasugai, Aichi 487

Otsuka, Hiromi
Dept. Physics, Faculty of Science
Tokyo Metropolitan Univ.
Tokyo 192–03

Ozeki, Tomoko
Inst. for Phys. & Chem. Res.
RIKEN
Saitama 351–01

Ozeki, Yukiyasu
Dept. of Physics
Tokyo Inst. of Technology
152 Tokyo

Saika, Yohei
Dept. Electric Engineering
Wakayama Nat'l College of Tech.
Wakayama 644

Saikawa, Kazihiko
Dept. of Physics
Shinshu Univ.
Matsumoto 390

Sakashita, Hirofumi
Ctr. of Adv. Instrumental Analysis
Kyushu Univ.
Kasuga 816

Shigemoto, Kazuyasu
Dept. of Physics
Tezukayama Univ.
Nara 631

Shimizu, Toshihiro
Graduate School of Engineering
Kokushikan Univ.
Tokyo 154

Shinjo, Kazumasa
ATR Optical & Radio Comm. Res. Lab.
Kyoto 619–02

Shirakura, Takayuki
Faculty of Humanity & Social Sci.
Iwate Univrsity
Morioka 020

Suzuki, Masuo
Dept. of Physics
Univ. of Tokyo
113 Tokyo

Suzuki, Nobuo
Dept. of Applied Physics
Tohoku Univ.
Sendai 980–77

Tadaki, Shin-ichi
Dept. of Information Science
Saga Univ.
Saga 840

Taji, Yukichi
Adv. Science Res. Ctr.
Japan Atomic Energy Res. Inst.
Ibaraki-ken 319–11

Takahashi, Minoru
Inst. for Solid State Phys.
Univ. of Tokyo
Tokyo 106

Takemura, Shun-ichi
Shiga Bunka College
Yokaichi City Shiga Prefecture 527

Tanabe, Naoyuki
Dept. of Electronics & Informatics
Shohoku College
Kanagawa 243

Tanimura, Yoshitaka
Inst. for Molecular Science
Okazaki, Aichi 444

Terai, Akira
Dept. of Physics
Univ. of Tokyo
Bunkyo-ku, Tokyo 113

Todo, Shinji
Dept. of Physics
Univ. of Tokyo
Bunkyo-ku Tokyo 113

Totsuka, Keisuke
Dept. of Physics
Univ. of Tokyo
Tokyo 113

Tretiakov, Alexei
Graduate School of Information Sci.
Tohoku Univ.
Sendai 980

Tsuchiya, Kenichi
Dept. Physics
Waseda Univ.
Tokyo 169

Tutu, Hiroki
Dept. Physics
Kyushu Univ. 33
Fukuoka 812

Uezu, Tatsuya
Dept. Of Physics
Nara Women's Univ.
630 Nara

Yamada, Hiroaki
Faculty of Engineering
Niigata Univ.
Niigata 950–21

Yamamoto, Shoji
Dept. of Physics
Osaka Univ.
Osaka 560

Yamashita, Koichi
Dept. of Physics
Waseda Univ.
Yokohama

Yamazaki, Yoshitake
Computer Sci. & System Engn.
Kyushu Inst. of Technology
Iizuka, Fukuoka-ken 820

Yoshino, Hajime
Inst. of Physics
Univ. of Tsukuba
Tsukuba 305

**KOREA**

Choi, Moo Young
Dept. of Physics
Seoul Nat'l Univ.
Seoul 151-742

Jeon, Gunsang
Dept. of Physics
Seoul Univ.
Seoul 151-742

Kim, Chang Sub
Dept. of Physics
Chonnam Nat'l Univ.
Kwangju 500-757

Koo, Hyearn-Maw
Dept. of Physics
Chungnam Nat'l Univ.
Daejeon 305-764

Park, Heungwon
Dept. of Physics
Inha Univ.
Inchon 402-751

Park, Hyunggyu
Dept. of Physics
Inha Univ.
Inchon 402-751

Park, Pyeong Jun
Dept. of Physics
Pohang Univ. of Sci. & Tech.
Pohang

Yi, Juyeon
Dept. of Physics
Seoul Nat'l Univ.
Seoul 151-742

**MALAYSIA**

Lim, Swee Cheng
Dept. of Physics
Univ. of Kebangsaan
43600 Bangi

**MEXICO**

Alarcón-Waess, Olegario
Dept. de Fisica y Matematicas
Univ. de las Americas
Cholula 72820, Puebla

Barrio, Rafael Angel
Inst. de Fisica
Univ. Nacional Autonoma de Mexico
01000 Mexico D.F.

del Rio, Jesus Antonio
Lab. de Energia Solar IIM.
Univ. Nacional Autonoma de Mexico
Temixco, Morelos 62580

Gonzalez Flores, Agustin Eduardo
Inst. de Fisica
Univ. Nac. Autonoma de Mexico (UNAM)
01000 Mexico D.F.

Pereyra, Pedro
Dpto. de Ciencias Basicas
Univ. Autonoma Metropolitana-Azcapotzalco
C.P. 02200, Mexico D.F.

Vázquez, Federico
Facultad de Ciencias
Univ. Autonoma del Estado de Morelos
Chamilpa Cuernavaca, Mor. 62210

**NEPAL**

Niraula, Om Prakash
Physics Dept.
Amrit Campus
Kathmandu

**NETHERLANDS**

Blöte, Hendrik W.J.
Faculty of Applied Physics
Delft Univ.
2600 GA Deflt

Compagner, Aaldert
Lab. of Applied Physics
Univ. of Technology
2600 GA Delft

Dekker, Hans
Inst. for Theor. Phys. & TNO Phys. Lab.
Univ. of Amsterdam
2509 JG Den Hague

Ernst, Matthieu H.J.J.
Inst. for Theoretical Physics
Univ. of Utrecht
NL-3508 TA-Utrecht

Grimm, Uwe Gebhard
Inst. voor Theor. Fysica
Universiteit van Amsterdam
1018 XE Amsterdam

Luijten, Erik
Faculty of Applied Physics
Delft Univ. of Technology
2628 CJ Delft

Maassen van den Brink, Alexander
Inst. voor theoretische Fysica
Univ. of Amsterdam
1018 XE Amsterdam

van Leeuwen, J.M.J.
Inst.-Lorentz for Theore. Phys.
Univ. of Leiden
2311 SB Leiden

Vespignani, Alessandro
Lorentz Inst.
Leiden Univ.
2311 SB Leiden

**NORWAY**

Clausen, Sigmund
Inst. for Energy Technology
N-2007, Kjeller

Hemmer, Per Christian
Dept. of Theoretical Physics
Univ. of Trondheim
NTH-7034 Trondheim

Hilfer, Rudolf
Inst. of Physics
Univ. of Oslo
0316 Oslo

Joessang, Torstein
Dept. of Physics
Univ. of Oslo
Blindern N-0316 Oslo

Malthe-Sorenssen, Anders
Dept. of Physics
Univ. of Oslo
0316 Oslo

Nøst, Bjarne
Dept. of Physics
Univ. of Oslo
0316 Oslo

Olaussen, Kare
Gruppe for Teoretisk Fysikk NTH
Universitetet i Trondheim
N-7034 Trondheim

Sun, Tao
Dept. of Physics
Univ. of Oslo
Blindern, 0316 Oslo

**PAKISTAN**

Alam, M. Anis
Physics Dept.
Univ. of Punjab
Lahore-54590

Shah, Hassan Amir
Ctr. for Solid State Physics
Univ. of Punjab
Lahore

**RUSSIA**

Chtchour, Lev
Landau Inst. for Theor. Physics
Russian Academy of Sciences
Moscow 117940

Dotsenko, Viktor
Landau Inst. for Theor. Phys.
Russian Academy of Sciences
Moscow 117940

Izrailev, Felix M.
Budker Inst. of Nuclear Physics
630090 Novosibirsk

Lebedev, Nikolai
Radioengn., Electronics & Automat.
Moscow Inst. of REA
117454 Moscow

Nechaev, Sergei
L.D. Landau Inst. for Theor. Phys.
Moskow 117940

Yarunin, V.S.
Bogolubov Lab. of Theor. Phys.
Joint Inst. for Nuclear Res.
911621 Dubna

**SINGAPORE**

Chen, Kan
Physics Dept.
Nat'l Univ. of Singapore
Singapore 0511

Ho, Hwei Moon
World Scientific Publishing Co.
Tai Seng Industrial Estate
Singapore 1953

Phua, K.K.
1022 Hougang Avenue 1 05–3520
Tai Seng Industrial Estate
Singapore 1953

Wang, Jian-Sheng
Computational Science
Nat'l Univ. of Singapore
Singapore 0511

Wong, Ji-kan
World Scientific Publishing Co.
Tai Seng Industrial Estate
Singapore 1953

Wong, Lock Yee
World Scientific Publishing Co.
Tai Seng Industrial Estate
Singapore 1953

**SLOVAK**

Kolesik, Miroslav
Inst. of Physics
Slovak Academy of Sciences
Bratislava

**SLOVENIA**

Borštnik, Branimir
Nat'l Inst. of Chemistry
61115 Ljubljana

Li, Baowen
Ctr. for Appl. Math. & Theor. Phys.
Univ. of Maribor
SLO–62000 Maribor

**SPAIN**

Antoranz, J. Carlos
Depto.de Fisica Fundamental, L.C.D.I.
U.N.E.D.
Madrid 28080

Parga, Nestor
Depto de Fisica
Universidad Autonooma de Madrid
28049 Madrid

Pesquera Gonzalez, Luis
Dept. de Fisica Moderna
Univ. de Cantabria
E–39005 Santander

Rodriguez, Miguel Angel
Inst. de fisica de Cantabria
Universidad de Cantabria
Santander. E–39005

Tejero, Carlos F.
Facultad de Ciencias Fisicas
Universidad Complutense de Madrid
28040 Madrid

Velarde, Manuel G.
Inst. Pluridisciplinar
Universidad Complutense de Madrid
28040 Madrid

**SRI LANKA**

Jayananda, Malagalage
Dept. of Physics
Univ. of Sri Jayewardenepura
Gangadawila, Nugegoda

## SWEDEN

Berggren, Karl-Fredrik
Dept. of Phys. & Measurement Tech.
Linköping Univ.
S-581 83 Linköping

Johannesson, Henrik
Inst. of Theoretical Physics
Chalmers & Göteborg Univ.
S-41296 Göteborg

## SWITZERLAND

Breymann, A.W. Wolfgang
Inst. für Physik
Universität Basel
8056 Basel

Gruber, Christian
Inst. de Physique Theorique
Ecole Polytechnique Federale
de Lausanne
Lausanne, CH-1015

Hansmann, Ulrich H.E.
IPS, RZ F9
ETH-Zentrum
CH-8092 Zürich

## TAILAND

Phaovibul, Orapin
Dept. of Chemistry
Mahidol Univ.
Bangkok 10400

Poorakkiat, Chaisingh
Physica Dept.
Chulalongkorn Univ.
Bangkok 10330

## TAIWAN, CHINA

Chan, Chi-Keung
Inst. of Physics
Academia Sinica
Taipei 11529

Chen, Hsin-Yen
Computing Ctr.
Academia Sinica
Nangkang, Taipei 11529

Hsu, Hsiao-Ping
Dept. of Physics
Chung-Yuan Christian Univ.
Chung-Li

Huang, Ming Chang
Dept. of Physics
Chung Yuan Christian Univ.
Chung-Li

Lai, Pik-Yin
Inst. of Physics
Nat'l Central Univ.
Chung-li 32054

Lee, Feng-Tien
Dept. of Physics
Chung-Yuan Christian Univ.
Chung-Li

Leung, Kwan-Tai
Inst. of Physics
Academia Sinica
Taipei 11529

Liang, N.Y.
Dept. of Physics
Nat'l Taiwan Univ.
Taipei 11529

Lin, Simon C.
Inst. of Phys. & Computing Ctr.
Academia Sinica
Taipei 11529

Shieh, Jeng-Shian
Dept. of Physics
Chung-Yuan Christian Univ.
Chung-Li

Tseng, Hsen-Che
Dept.of Phys.
Nat'l Chung-Hsing Univ.
Taichung

## TURKEY

Silier, Ayse Erzan
Dept. of Physics
Istanbul Technical Univ.
Istanbul 80626

**UK**

Dougherty, John Peter
Dept. of Appl. Math. & Theor. Phys.
Univ. of Cambridge
Cambridge CB3 9EW

Edwards, Sam
Cavendish Lab.
Univ. of Cambridge
Cambridge CB3 0HE

Flesia, Silvia
Theoretical Physics
Univ. of Oxford
Oxford OX1 3NP

Haslam, Andrew John
Dept. of Chemistry
Univ. of Sheffield
Sheffield S3 7HF

Hassan, Kamrul
Dept. of Physics
Brunel Univ.
Middlesex UB8 3PH

Henkel, Malte
Dept. of Physics
Oxford Univ.
Oxford OX1 3NP

Itoh, Sumiko
Cavendish Lab.
Cambridge CB3 OHE

Landsberg, Peter T.
Faculty of Mathematical Studies
Univ. of Southampton
Southhampton SO9 5NH

McClintock, Peter
School of Physics & Chemistry
Lancaster Univ.
Lancaster LA1 4YB

McGrother, Simon
Dept. of Chemistry
Univ. of Sheffield
Sheffield S3 7HF

McKane, Alan John
Dept. of Theor. Physics
Univ. of Manchester
Manchester M13 9PL

Rodgers, Geoffrey J.
Physics Dept.
Brunel Univ.
Middlesex UB8 3PH

Schütz, Gunter M.
Dept. of Physics
Univ. of Oxford
Oxford OX1 3NP

Sear, Richard
Dept. of Chemistry
Univ. of Sheffield
Sheffield S3 7HF

Sherrington, David
Dept. of Physics
Univ. of Oxford
Oxford OX1 3NP

Stinchcombe, Robin Bernard
Dept. of Physics
Univ. of Oxford
Oxford OX1 3NP

Tucker, John Wesley
Physics Dept.
Univ. of Sheffield
Sheffield S3 7RH

Xian, Yang
Dept. of Math.
UMIST
Manchester M60 1QD

Zakrzewski, Wojciech J.
Dept. of Mathematical Science
Univ. of Durham
Durham DH1 3HE

Zeng, Chen
Dept. of Mathematics
Univ. of Manchester
Manchester M60 1QD

**USA**

Aizenman, Michael
Physics Dept., P.O. Box 708
Princeton Univ.
Princeton, NJ 08544

Amaral, Luis A. Nunes
Ctr. for Polymer Studies
Boston Univ.
Boston, MA 02215

Bhatt, Ravin
Dept. of Electrical Engineering
Princeton Univ.
Princeton, NJ 08544–5263

Birman, Joseph L.
Dept. of Physics
The City College
New York 10031

Bukharev, Igor
Dept. of Physics
Brown Univ.
Providence, RI 02912

Chandler, David
Dept. of Chemistry
Univ. of California at Berkeley
Berkeley, CA 94720

Chen, Liao-Yuan
Division of Earth & Physical Sci.
Univ. of Texas at San Antonio
San Antonio, TX 78249–0663

Chen, Lin-Yuan
Inst. for Theoretical Physics
Univ. of California
Santa Barbara, CA 93106

Chen, Yu-Zong
Dept. of Physics
Purdue Univ.
West Layayette IN 47907

Cohen, Morrel H.
Corporate Res., LG344
Exxon Res. & Eng. Company
Annandale New Jersey 08801–3059

Ding, Jiu
Dept. of Math.
Univ. of Southern Mississippi
Hattiesburg, MS 39406–5045

Dokholyan, Nikolay
Ctr. for Polymer Studies
Boston Univ.
Boston, MA 02215

Glotzer, Sharon C.
Material Sci. & Engineering Lab.
Nat'l Inst. of Standards & Tech.
Gaithersburg, MD 20899

Grebogi, Celso
Inst. for Plasma Res.
Univ. of Maryland
College Park, MD 20742

Gu, Lei
Physics Dept.
Brandeis Univ.
Waltham, MA 02254

Guo, Muyu
Dept. of Electrical Engn.
Princeton Univ.
Princeton, NJ 08544

Hao, Hong
Dept. of Physics
Univ. of Maryland
College Park, MD 20742

Harrington, Stephen T.
Dept. of Physics
Boston Univ.
Boston, MA 02215

Hatano, Naomichi
Dept. of Physics
Harvard Univ.
Cambridge, MA 021138

Hohenberg, Pierre.C.
Deaprtment of Physics
Yale Univ.
New Haven, CT 06520–8120

Hu, Bambi
Dept. of Physics
Univ. of Houston
Houston, TX 77204–5506

Hu, Jun
Nat'l High Magnetic Field Lab.
Tallahassee, FL 32306

Hu, Junmin
Gordon Mckay Lab.
Harvard Univ.
Cambridge, MA 02138

Hu, Xiao
Division 814.05
Nat'l Inst. of Standards & Tech.
Boulder, Colorado 80303

Huang, Liang Jiao
Dept. of Math.
Rockford College
Rockford IL 61111

Huse, David A.
ID–237
AT & T Bell Labs
Murray Hill, NJ 07974

Kim, Doochul
Dept. of Phys.
Univ. of Washington
Seattle, WA 98195

Korepin, Vladimir
Inst. for Theoretical Physics
SUNY at Stony Brook
Stony Brook, NY 11794–3840

Krech, Michael
Ctr. for Simulational Phys.
Univ. of Georgia
Athens GA 30602

Landau, David P.
Ctr. for Simulational Physics
Univ. of Georgia
Athens GA 30602

Lebowitz, Joel L.
Ctr. for Mathematical Res.
Rutgers Univ.
New Brunswick, NJ 08903

Leibler, Stanislas
Dept. of Phys. & Molec. Biology
Princeton Univ.
Princeton, NJ 08544

Leung, Kevin
Dept. of Chemistry
Univ. of Southern California
Los Angeles, CA 90089–0482

Lin, Min Yue
Nat'l Inst. of Standards & Tech.
Gaithersburg, MD 20899

Lin, Yukweng Michael
Ctr. for Appl. Stochastics Res.
Florida Atlantic Univ.
Boca Raton, FL 33431–0991

Lisyansky, Alexander
Dept. of Physics
Queens College–CUNY
Flushing, NY 11367

Liu, Keh-Fei
Dept. of Physics & Astronomy
Univ. of Kentucky
Lexington, KY 40506

Lu, Jian Ping
Dept. of Physics & Astronomy
Univ. of North Carolina
Chapel Hill, NC 27599

Ludwig, Andreas W.
Dept. of Physics
UC Santa Barbara
Santa Barbara, CA 93106–9530

Lum, Ka
Dept. of Chemistry
UC Berkeley
Berkeley, CA 94720

Luo, Weili
Dept. of Physics
Univ. of Central Florida
Orlando, FL 32816

Luzar, Alenka
Dept. of Chemistry
Univ. of California at Berkeley
Berkelay, CA 94720–1460

Makse, Hernan A.
Physics Dept.
Boston Univ.
Boston, MA 02215

Meng, Baoqi
Dept. Chem. Engineering
UC Santa Barbara
Santa Barbara, CA93106–5080

Nelson, David Robert
Lyman Lab. of Physics
Harvard Univ.
Cambridge, MA 02138

Ou Yang, Yee Sun
Dept. of Physics
Oklahoma State Univ.
Stillwater, OK 74078–0444

Pant, Prateek
Dept. of Physics
Northeastern Univ.
Boston, MA 02115

Perk, Jacques H.H.
Dept. of Physics
Oklahoma State Univ.
Stillwater, OK 74078-0444

Petrosky, Tomio
Ctr. for Studies in Stat. Mech.
& Compl. Sys.
Univ. of Texas at Austin
Austin TX 78712

Reichl, Linda Elizabeth
Dept. of Physics
Univ. of Texas at Austin
TX 78712

Sachdev, Subir
Dept. of Physics
Yale Univ. P.O. Box 208120
New Haven CT 06520-8120

Sadr, Reza
Physics Dept.
Boston Univ.
Boston, MA 02215

Schick, Michael
Dept. of Physics FM-15
Univ. of Washington
Seattle, WA 98195

Shrock, Robert E.
Inst. for Theor. Physics
SUNY at Stony Brook
Stony Brook, NY 11794-3840

Stanley, Eugene
Ctr. for Polymer Studies
Boston Univ.
Boston MA 02215

Tang, Chao
NEC Res. Inst.
Princeton NJ 08540

Tong, Penger
Dept. of Physics
Oklahoma State Univ.
Stillwater, OK 74078

Tu, Yuhai
T.J. Watson Reserach Ctr.
IBM
Yorktown Heights, NY 10598

Visscher, Pieter B.
Dept. of Physics
Univ. of Alabama
Tuscaloosa, AL 35487-0324

Wang, Jin
School of Chemical Sciences
Univ. of Illinois at Urbana
Urbana IL 61801

Wolynes, Peter Guy
Dept. of Chemistry
Univ. of Illinois, Urbana-Champaign
Urbana, IL 61801

Wu, Dane W.
Dept.of Mathematics
Tacoma
Washington 98447-0003

Wu, F.Y.
Dept. of Physics
Northeastern Univ.
Boston MA 02115

Wu, Yong-Shi
Dept. of Physics
Univ. of Utah
Salt Lake City, Utah 84112

Xu, Ji-Hai
Texas Ctr. for Superconductivity
Univ. of Houston
Houston, TX 77204-5932

Yang, C.N.
Inst. for Theoretical Physics
SUNY at Stony Brook
NY 11794-3840

Young, Allan Peter
Dept. of Physics
Univ. of California
Santa Cruz, CA 95064

Zeng, Xiao Cheng
Chemistry Dept.
Univ. of Nebraska
Lincoln, NE 68588

## VENEZUELA

Buendia, Gloria
Dept. de Fisica
Universidad Simon Bolivar
Caracas 1080

Mata, Gustavo J.
Dept. de Fisica
Universidad Simon Bolivar
Caracas 1080 A

## VIETNAM

Nguyen, Thi Thanh Huong
Faculty of Physics
Hanoi Pedagogical Univ.
Hanoi

Nguyen Manh, An
Dept. of Theoretical Physics
Hanoi Nat'l Pedagogic Univ.
Hanoi

# SUBJECT INDEX

www.ingramcontent.com/pod-product-compliance
Lightning Source LLC
LaVergne TN
LVHW070907300926
849162LV00010B/1840

* 9 7 8 9 8 1 0 2 2 3 1 4 4 *